AF358791

Biometric Systems

Biometric Systems

Editors

Loris Nanni
Sheryl Brahnam

MDPI • Basel • Beijing • Wuhan • Barcelona • Belgrade • Manchester • Tokyo • Cluj • Tianjin

Editors
Loris Nanni
University of Padua
Italy

Sheryl Brahnam
Missouri State University
USA

Editorial Office
MDPI
St. Alban-Anlage 66
4052 Basel, Switzerland

This is a reprint of articles from the Special Issue published online in the open access journal *Sensors* (ISSN 1424-8220) (available at: https://www.mdpi.com/journal/sensors/special_issues/ Biometric-Systems).

For citation purposes, cite each article independently as indicated on the article page online and as indicated below:

LastName, A.A.; LastName, B.B.; LastName, C.C. Article Title. *Journal Name* **Year**, *Volume Number*, Page Range.

ISBN 978-3-0365-1128-3 (Hbk)
ISBN 978-3-0365-1129-0 (PDF)

Contents

About the Editors

Loris Nanni is an associate professor in the dipartimento di ingegneria dell'informazione of the University of Padova. His research activity focuses on pattern recognition in machine learning, mainly focused on deep learning. He has also done extensive work in biometric sytems. He has an H-index of 50.

Sheryl Brahnam is a professor in the Information Technology & Cybersecurity Department, Missouri State University. Her interests are focused on machine learning and cultural and ethical aspects of computing. She has an H-index of 33.

Preface to "Biometric Systems"

Biometric recognition/verification continues to be one of the most widely studied pattern recognition problems. The field is being driven in large part by the recent rise in sophisticated nation-state hacking and the increasing need of advancing technological systems, such as the Internet and cellular phones, to secure personal identification. Rapidly fading are the days of password protection. Biometrics along with authenticator apps and multifactor authentication are being pushed to safeguard systems in the new threat environment by many industrial leaders, such as Microsoft.

Biometric recognition is defined by several critical issues involved in the problem, such as quality checking of sensor inputs, biodata security, aliveness detection, and multimodal authentication. Regardless of the biometric chosen, all recognition systems must also isolate and extract a set of features in the biometric image or pattern that offers the greatest amount of information, whether these features are engineered or automatically determined by the classifier system.

This book's chapters present the very best work in biometric recognition and verification. Topics run the gamut of research in this field: security issues, signature verification (online and on mobile touch-screens), fingerprint identification, wrist vascular biometrics, ear detection, face detection and identification (including a new survey of face recognition systems), person re-identification, electrocardiogram (ECT) recognition, and several multi-modal systems, including coverage of the latest equipment for multi-modal contactless verification, the fusion of fingerprint identification with either finger-vein and finger knuckle-print identification or ECT signals or hand vein recognition. A chapter is also dedicated to image quality as it affects performance in biometric systems based on images. Many chapters propose novel methods for biometric recognition and verification, including the introduction of new descriptors and feature sets, the addition of depth information and augmentation, and the application in this field of the latest deep learner architectures. In several chapters, algorithms are exhaustively compared and verified across many data sets.

Due to the accelerating progress in biometrics research, this book's publication is not only timely but also much needed. This volume contains seventeen peer-reviewed chapters reporting state of the art in biometrics research, covering essential topics in the contemporary scene. This book will be a valuable resource for graduate students, engineers, and researchers interested in understanding and investigating this important field of study.

Loris Nanni, Sheryl Brahnam
Editors

Article

Multi-Biometric System Based on Cutting-Edge Equipment for Experimental Contactless Verification

Lukas Kolda [1], Ondrej Krejcar [1,*], Ali Selamat [1,2,3,4], Kamil Kuca [1,2] and Oluwaseun Fadeyi [1,5]

[1] Faculty of Informatics and Management, University of Hradec Kralove, Hradec Kralove 50003, Czech Republic
[2] Malaysia Japan International Institute of Technology (MJIIT), Universiti Teknologi Malaysia Kuala Lumpur, Jalan Sultan Yahya Petra, Kuala Lumpur 54100, Malaysia
[3] Media and Games Center of Excellence (MagicX), Universiti Teknologi Malaysia, Skudai 81310, Malaysia
[4] School of Computing, Faculty of Engineering, Universiti Teknologi Malaysia (UTM), Skudai 81310, Malaysia
[5] Department of Geology, Faculty of Space and Environmental Science, University of Trier, 54296 Trier, Germany
* Correspondence: ondrej.krejcar@uhk.cz

Received: 28 April 2019; Accepted: 20 August 2019; Published: 26 August 2019

Abstract: Biometric verification methods have gained significant popularity in recent times, which has brought about their extensive usage. In light of theoretical evidence surrounding the development of biometric verification, we proposed an experimental multi-biometric system for laboratory testing. First, the proposed system was designed such that it was able to identify and verify a user through the hand contour, and blood flow (blood stream) at the upper part of the hand. Next, we detailed the hard and software solutions for the system. A total of 40 subjects agreed to be a part of data generation team, which produced 280 hand images. The core of this paper lies in evaluating individual metrics, which are functions of frequency comparison of the double type faults with the EER (Equal Error Rate) values. The lowest value was measured for the case of the modified Hausdorff distance metric—Maximally Helicity Violating (MHV). Furthermore, for the verified biometric characteristics (Hamming distance and MHV), appropriate and suitable metrics have been proposed and experimented to optimize system precision. Thus, the EER value for the designed multi-biometric system in the context of this work was found to be 5%, which proves that metrics consolidation increases the precision of the multi-biometric system. Algorithms used for the proposed multi-biometric device shows that the individual metrics exhibit significant accuracy but perform better on consolidation, with a few shortcomings.

Keywords: biometry; identification; bloodstream; image recognition; multi-biometrics

1. Introduction

Due to the rapid development in information technology, it has become possible to utilize biometrics for identifying and verifying persons [1–3]. Personal verification involves associating an identity with a specific individual. Verification or authentication of an identity is related to the authorization or refusal of an individual's personal identification, which is verified and confirmed side-by-side with the identity provided. This procedure is crucial for identifying a search query of a person.

In recent times, biometric verification and identification systems have gained popularity, which brought about their extensive usage [4]. Most significantly, it is common to see laptops with fingerprint readers, as well as a Windows 10 "hello" function, which both support biometric identification and verification [3,5–7]. The latter feature is available to users who sign up for biometry usage. Biometry saves the user the stress of regular logins into gadgets and wares (e.g., keys or cards) with privacy and identity theft challenges [8]. For instance, lost or forgotten login details can be accessed by a third party and be used illegally. Biometry eliminates such problems, and greatly reduces the risk of copying or falsifying them. Nevertheless, biometry is not a complete and perfect solution for user verification and identification. Gathering of an individual's biometric data to create the person's biometric template is a complex process, which can sometimes yield an indefinite outcome [9]. However, the probability of success with biometric systems may vary up to 99% in the best systems. As such, biometric systems are greatly beneficial in curbing security challenges such as privacy invasion and identity theft problems (the most common being attacks carried out using dummies or models of the given body part).

Hand recognition remains the earliest form of available biometric characteristics used to identify and differentiate humans, as well as verify their identities [10]. Using the hand, it is possible to recognize, for example, hand geometry, fingerprints, palm lines, lines on finger joints, and image of the bloodstream. [11]. As soon as a person attains adulthood, features of the individual's hands remain the same for the rest of his or her life. Hence, these characteristics can be used to identify and/or verify the person [12]. Moreover, these characteristics can be scanned (excluding the fingerprint) using a camera with a basic resolution (640 x 480 pixels), and saved in biometric memory, unlike the scanning of the eye retina [13] (which may not necessarily be saved) for future references. From the foregone, we can say that it is safe and inexpensive to embark on a project of creating small devices for the scanning biometric characteristics of the hand.

A number of biometric systems currently exist that work based on the principle of the shape and surface elevation of the hand (jointly referred to as "hand contour" in the context of this work), and serves the purpose of a person's identity features [14–20]. The design and implementation of such systems offer several benefits, especially given that hardware and software requirements are very easy to come by. However, biometric characteristics of a hand contour is often not sufficient to distinguish individuals. Consequently, identification security is compromised, which leads to a high false-match rate (FMR). As a result, optimized biometric systems are continuously designed based on the biometric characteristics that are hidden, and cannot be replicated. For example, ultramodern finger print scanning results cannot be replicated, but the cost associated with its design and implementation is very high. A second possibility for improving identification and verification by biometric systems is the use of two or more biometric characteristics within a system. For instance, joint scanning of hand and blood stream offer better identification results, which, in recent times, have become a lot cheaper, and have improved their security [21]. This is enabled by the fact that false match rates (FMR) of the individual biometric characteristics are multiplied, which, in turn, decreases FMR. Hence, the whole system becomes more secure.

Related Literature

As development in science and technology progresses, smarter ways of operating household and office equipment are becoming more popular. For instance, Reference [20] described a gesture-based technology that makes use of some specific grammatical terms in its operation. This is possible due to these technologies and continuous issues of fake identity behind finger print detection [22] and walking patterns (also known as "gait") [23].

According to Reference [24], traditional singular biometry identifiers are flawed due to varying interclass sensitivity, data that bears a lot of noise, or very high error margins. For instance, in today's world of mobile phones, Reference [25] tried to analyze using a mobile phone, hand segmentation, and fingertip readings of about a hundred subjects. It was observed that the sensitivity of the technique yielded around 52%, a figure which may have been more precise if the method was bi-biometric-based or multi-biometric based. As a result of the sensitivity shortcoming common to uni-biometric systems, Reference [24] proposed a multi-model technique by adopting fused faces and fingerprints, i.e., making use of a collaboration-based classifier to identify the faces of a different person. A selective neural network was jointly used with "Viola–Jones method-based PCA" by Reference [26] for a gridding system. Here, over 100 faces from a face-based database were distinguishable, which proves the success of the method. In comparison to the work of References [25,27], the merged vascular recognition with hand geometry in a multi-modal design derives an equal error rate of 6%. While singular biometric techniques are currently less preferred, the unique and greater sensitivity of multi-biometric technologies seem to be revolutionizing the field of biometry. In a fascinating technological twist, biometry has stepped forward beyond mere gesture identification to commence blood-based verification using physiological features, as reported by Reference [28]. Table 1 summarizes a few multi-modal biometric studies (within the last 10 years) with their corresponding error rates. From the table, HG denotes hand geometry, FV denotes finger vein (vascular structure), PP is the palm print, MFV is the multi-finger vein, FK is the finger knuckle, KS is the knuckle shape, and FG is the finger geometry.

Table 1. Some multi-modal biometric studies from 2009 until today.

Study Reference	Year	Combined Biometric Methods	Sample Size	Equal Error Rate (%)
Current work	2019	HG and FV	40	5
[27]	2013	FV and HG	100	0.06
[29]	2019	PP and FG	237	58
[30]	2014	HG and FV	204	0.02
[31]	2015	MFV	106	0.08
[32]	2017	FK and FV	100	0.35
[33]	2010	FV and FG	102	0.075
[34]	2009	FV and KS	100	1.14

Except otherwise stated, the following performance ratings (abbreviations) have been used throughout this paper and in similar biometric literature.

- EER (Equal Error Rate)
- FAR (False Acceptance Rate)
- FRR (False Rejection Rate)
- FIR (Far infrared)
- NIR (Near Infrared)
- ROI (Point or region of interest on the image)

The rest of this paper is organized as follows. Section 2 further describes the problems upon which the study lies, and for which more multi-modal techniques are required. Section 3 describes in detail the newly proposed multi-biometric method, while experimental testing of the system and discussion of the generated results are reflected in Section 4. The final section discusses the limitations of the new method and suggests ways to improve the limitation in future research studies.

2. Problem Definition

Users' hand-geometry-based identification systems can (on the basis of the image scanning evaluation method adopted) be categorized into three distinct types.

1. Contact with pins: This method makes use of "scanning hand fixation images," i.e., a set of pins that define the position of individual fingers. The hand is laid on a flat surface, which creates a contrast to the surroundings. The method makes evaluation rather simple, even though it is less comfortable for the user, with some unwanted deformations. In the past, "contact-with-pins" was mainly used in research identification systems. Today, it finds a commercial application. In a study by Reference [35], the authors used five pins for the fixation of the position of the hand before scanning from the top and the sides. For the purpose of the evaluation, 16 geometric characteristics were utilized, which yielded an EER of 6%.

2. Contact without pins: In this case, the hand is freely laid down onto a contrasting surface or scanner. The absence of pins implies a free movement of the hand, which eventually settles in its

natural shape. This way, unwanted deformations are eliminated [36]. In their work, [36] the researchers adopted a tabular scanner without pins fixed on it.

3. They evaluated the size and geometry of the finger tips, reaching an FRR value of 11.1% and a low FAR value of 2.2%. These values differ significantly from those obtained when pins are fixed (FRR value of 4% and FAR value of 4.9%). As such, "Contact without pins" is considered the most suitable for security applications due to the importance attached to FAR within domains of the method.

4. Contactless scanning: This method does not require pins or surfaces where the hand will be laid and is, by far, the most user-friendly technique of all. To create contactless scanning points, a standard 2D camera or 3D digitizer is used for scanning. Reference [37] carried out hand scanning in front of a camera in an open surrounding. Evaluation was deliberately fixed at the hand's center of gravity. As such, it was possible to create homocentric circles that intersect with the fingers. The author measured the fingers using these circles, with the results formed by measuring the size of the fingers in 124 points. FRR and FAR were recorded as 54.3% and 8.6%, respectively, with hand movement and its inclination to the camera surface causing some problems. Oftentimes, the movement from and to the camera leads to significant distortions in the size of the image generated. Instability of the surroundings is another problem associated with contactless scanning, which influences its measurement.

A novel way to identify or verify a person's identity is by detecting the distribution of veins around the wrist or on the palm surface. The main advantage of this approach is the difficulty in replicating or fabricating the human vein, since the vessels are hidden in the body. References [38] described two possible scanning methods in this regard, which include far infrared (FIR) and near infrared (NIR) [39,40]. FIR is a technology that scans thermal radiation with a wavelength of 15 to 1000 μm of the object (an individual's wrist or palm surface) under examination. A number of external and internal conditions influence FIR. For instance, temperature and/or moisture of the external environment as well as existing health conditions of the scanned individual can influence its sensitivity. Consequently, the scanned image may be unreliable. NIR on the other uses infrared radiation of the wavelength range of 0.76–1.4 μm of the scanned tissue(s). The technology has a penetration range of approximately 3 mm. Deoxidized hemoglobin found in the veins has the tendency to absorb maximum radiation with a wavelength range between 7.6×10^{-4} mm [41]. IR (Infra Red) radiation absorption increases with large veins as compared to tissues adjacent to it [42,43]. As such, the contrast between the object under investigation (large veins) and surrounding objects (tissue) is easily accomplished. Two essential conditions for effective near infrared biometric technology are appropriate camera with IR filter and suitable lighting of the area that uses the IR radiation source. Furthermore, a primary benefit of NIR technology is that the external and internal conditions do not affect the image scanning process. The process is also not affected by skin deviations and tones of the image.

In the light of the foregone analysis on biometric verification and identification, the current study aims at designing an experimental multi-biometric verification system, which is based on two biometric characteristics: hand geometry and blood flow from the upper part of the hand. Since it has been established in literature that a combination of two biometric features provide better verification and identification results. The proposed system has been selected on the basis that it is impossible to integrate the appraisal on coarse data level (sensor-level fusion) as well as on features absent on a collected biometric characteristics level (Feature-Level Fusion) [6,44]. Additionally, due to the vast misunderstanding between identification, verification, and authentication biometric devices [45], it is important to clarify that the proposed system is a verification system, which checks a user's biometric identity against those of a number of persons within a database (Figure 1) [45–47].

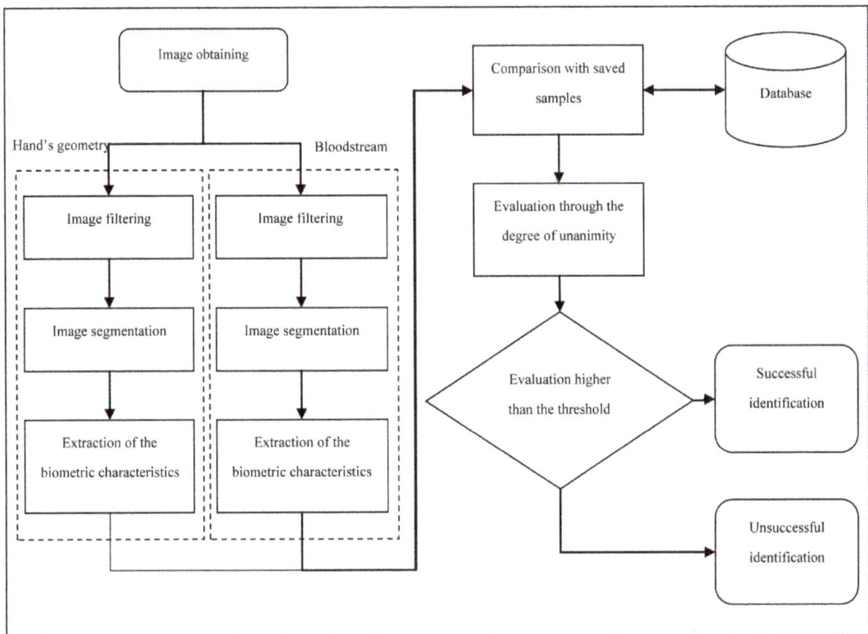

Figure 1. Schematic representation of the proposed identification and verification system.

3. System Description

The proposed user identification and verification system is comprised of a software that causes it to function according to the steps illustrated in Figure 1.

In this section, the extraction of hand and bloodstream data is discussed. The entire system commences with obtaining the image of the hand, and then that of the bloodstream (vascular structure) [48,49]. Next, background noise is removed from the images to make it fit for further processing. This process is followed by the first screening process (segmentation) for which every aspect of the scanned object is carefully verified by the system, which tries to match the image with existing or similar saved features (images) within the database. Extracted data from the image moves into a central database where the verification is completed. Next, the score-level fusion is obtained for the image (degree of unanimity) and, depending on the threshold, the user is either successfully verified or not. However, this is often done after the data has been normalized as a function of the number of tests carried out. The rest of this section is described in deeper terms and the different blocks are represented in Figure 1. Furthermore, some useful metrics such as EER is often calculated in order to have a full grasp of the efficiency of the system in comparison to other existing systems.

3.1. Hand Geometry

3.1.1. Image Pre-Processing

In order to reduce the noise in the image results generated by the system, a basic averaging method (a filtering technique) was used [50]. Although the method does not completely eliminate blurring especially at the edges, it works well for most parts of the image surface. Continuous application of the filter blurs out the entire image. However, once the filtering is executed to a level where it produces a preferred image (an image that is not multi-colored), the filtering process is stopped. The color of the image represents the mean values of the image, which is derived from using a convolution method of calculation. Consequently, the size of the convolution core acts as the variable (it has been identified as the parameter used in the method and the blurring effect is directly proportional to the growing size of the core) and the values are equivalent to 1.

The equation for calculating a 3×3 sized core is shown below.

$$f(i, j) = \frac{1}{9} \sum_{k=-1}^{1} \sum_{l=-1}^{1} g(i+k, j+l) \tag{1}$$

3.1.2. Image Segmentation

The image that is produced by the proposed system is segmented into components with similar characteristics. Segmentation aims at distinguishing scanned objects from one another, as well as from other objects within the surrounding [51]. A variety of methods are available for segmentation. For this study, the thresholding methods has been adopted due to its simplicity in its calculation [52]. Thresholding produces a complete segmentation of objects based on the transformation of the input image f on output binary image f' according to the relationship shown below.

$$f'(i, j) = \begin{array}{l} 1 \; for \; f(i, j) \geq T \\ 0 \; for \; f(i, j) < T \end{array} \tag{2}$$

where T is in an advance defined constant (threshold) and $f'(i,j) = 1$ for the image parts of the examined object. Thresholding also helps in testing the elements of an image in a progressive manner, by assigning values to elements of the image in accordance with identified requirements. Nevertheless, this segmentation method is flawed on the ground that selection of an accurate T (threshold value) may be difficult. To automatically set the brightness of the image generated by the system, a "global or local characteristics of the image" can be used. The global image characteristics method uses the information from all pixels in the image in order to determine the threshold value. The threshold is subsequently adjusted in accordance with the histogram generated or the mean value of the intensity of each image point obtained within the image [53]. On the other hand, the local image characteristics emphasize the usage of different thresholds for each element of the image, so that the threshold value is then calculated from the surrounding. The pixel value is calculated by using the following formula.

$$f'(i, j) = \begin{array}{l} 1 \; for \; f(i, j) \geq (\mu_{ij} - T_g) \\ 0 \; for \; f(i, j) < (\mu_{ij} - T_g) \end{array} \tag{3}$$

where μ_{ij} denotes an average value of all points from the surrounding. The surrounding size selection is done in accordance with the selected target of the segmentation, and it is mainly dependent on the size and shape of the examined object. T_g is a constant, whose value is dependent on the threshold value, and, therefore, it adjusts accordingly. This constant is often a positive number, but can also become negative depending on the prevailing situation. Segmentations are often selected with respect to the required results. As such, they are empirically identified for a given task.

The software system of the proposed identification and verification was used to experiment both image brightening techniques thresholding [52,54]. Solutions produced using the local image characteristics was not selected due to non-homogeneity of the examined object (hand). This non-homogeneity is caused, for example, by the wrinkles on the skin, different color of the skin, etc. Unfortunately, it was not possible to set up the thresholding parameters in the same way as the size of the surrounding, and the T_g constant. If it were possible, the resultant image would have been subjected to further processing (for determining the contour of the hand). As a result, the global threshold setting method has been adopted, with the average value of complete image intensity reduced by the constant. This method provides the best image that can be subjected to further processing.

3.1.3. Definition of Biometric Characteristics

In order to propose the use of biometric characteristics from this experimental investigation, the studies by References [27,36] were reviewed. The measured characteristics identified for this study, which is utilized in the design of the biometric device, are discussed and illustrated in Figure 2.

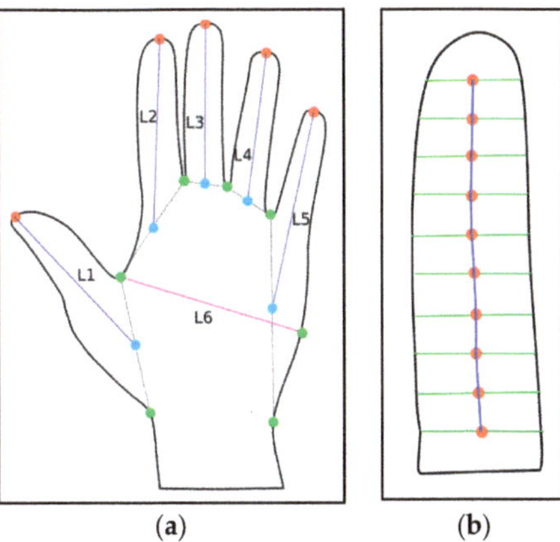

(a)   (b)

Figure 2. Depiction of measured biometric characteristics [9], where at (**a**) graphics shows the measurable characteristics of the palm of the hand and (**b**) described the base as the point of intersection of green lines.

Figure 2a shows the measurable characteristics of the palm of the hand. The individual length of the fingers would be measured from the red point (tip of the finger) to the blue point (the central part of the base of the hand). The base is described as the point of intersection of green lines, known as the neighboring values, as illustrated in Figure 2b. Furthermore, the distance between the fingertips are measured (excluding the distance between the thumb and the finger next to it). This is due to the huge variation in the size of the distance between both fingers. Experimenting this on the proposed scanner shows that the difference in the distance from single user images were found to have a greater value than the differences among other users. The distances between points "values" are also measured, but without the thumbs. The width of the palm, L6, is the last measured feature. For a single finger (Figure 2b), there are 10 points distributed along the finger length. The central red point position is determined from the angle located between the sections (represented by blue vectors).

3.1.4. Extraction of Biometric Characteristics

The result of segmentation comes in the form of a binary image (with a background value equal to 0 and the objects value is 255). An algorithm is used for searching the described characteristics of the examined object. Extraction of characteristics is essential to conserve data volume, and obtain higher processing speed. The extraction algorithm, in itself, is responsible for locating the hand contour, which is equivalent to the tested shape of the hand. Finding the contour is quite simple since the wanted line of the related point is visible from the binary image. Working of the biometric characteristics' extraction algorithm is a function of the contour found. The extraction algorithm causes the localization of the convex contour case, which is responsible for detecting the points on the finger tips. The convex contour case is a polygon, which contains the outermost points of the contour. Hence, the result is formed by a point sequence that creates the case. Consequently, the algorithm searches for the points in the values between the fingers.

The contour of the hand is again used for locating defects. Defects represent any point that is located on the contour, which is present on the part demarcated by the surrounding points. At least one defect is prevalent in between any pair of points. This follows the pattern of a convex case.

Hence, the identified defects are filtered to a minimal depth (distance from the case), which eliminates a number of small defects. The only defects left are often the ones corresponding to the values between the fingers. In the next step, measurement of the biometric characteristics takes place, as displayed in Figure 3. The characteristics for each finger, and the overall characteristics of the whole hand are measured.

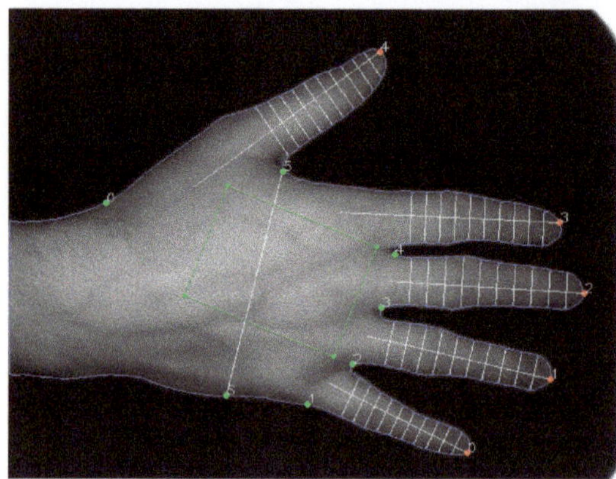

Figure 3. Depiction of all measured characteristics of an object (the human hand).

3.2. Bloodstream

3.2.1. Image Pre-Processing

In the identification and verification of blood flow at the upper part of the hand, the first stage is to locate the point of interest (ROI). It is identified as the part of image that has the object of interest. In this context, the object of interest is the bloodstream at the upper part of the hand (this is based on a pre-selected biometric characteristic). To define the location and the value of ROI, the position and the shape of the hand needs to be identified. Extracting blood stream biometric characteristics implies gathering information on the biometric characteristics of the hand. Figure 4 shows how ROI is derived (green rectangle). The ROI area starts from the joints level, and is associated with the vector that exists from finger valley point numbers 2 and 4, and up to the middle of the hand. This helps to differentiate (through ROI width) each user and/or specific individual image. The height of ROI is determined and expressed as a multiple value (precisely in multiples of 1.4) of the width.

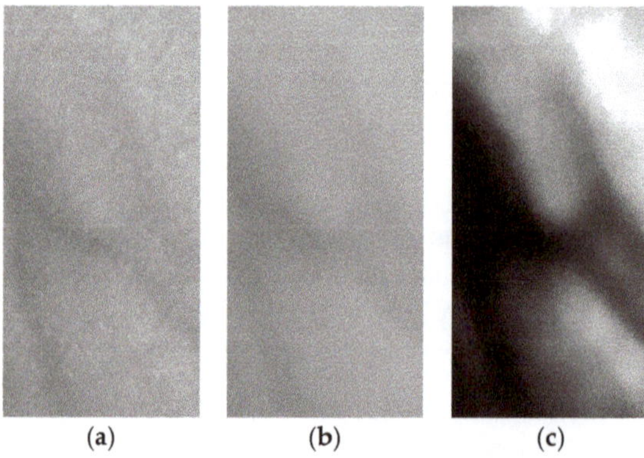

Figure 4. Process of bloodstream filtration: (**a**) before filtration (left), (**b**) after median filtration (middle), and (**c**) after equalization (right).

During the second phase, the ROI is used to develop a copy of the new image, which is subjected to rotation. The goal of image rotation is to ensure that the orientation of the bloodstream image is vertical and corresponds with the longer side. Figure 4 shows the result after copying ROI (Region of Interest) and rotating the image.

Rotation and equalization are followed by image filtering. In the context of the proposed system, the image is filtered using a median filter with a relatively large surrounding of the points (11 pixels). As such, significant image smoothening is achieved, so that skin wrinkles and body hair distortions are removed. However, the shape of the veins is retained on the image (Figure 4b).

To further treat the images, equalization of the histogram was carried out. This process adjusts the brightness of individual pixels relative to the histogram. For the images with a similar intensity of pixels, as in this case, the contrast becomes better. Using this procedure, the difference between the background and the object of interest (bloodstream) increases (Figure 4c).

3.2.2. Image Segmentation

The adaptive thresholding method was used for segmentation of the bloodstream [13]. The obtained image after thresholding (Figure 5a) corresponded to the bloodstream of large veins. Unfortunately, the image bears plenty of noise, with several unfinished lines, and calls for filtering.

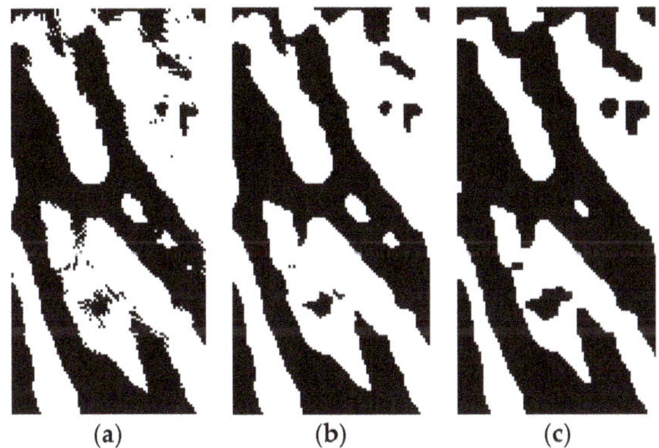

(a)   (b)   (c)

Figure 5. Process of bloodstream filtration: (**a**) after thresholding (left), (**b**) after median filtration (middle), and (**c**) after dilatation and erosion (right).

The median filter in combination with a morphologic filter was used. The median filter of 3 pixels mainly removes the small fragments in the image (Figure 5b). Furthermore, dilatation and erosion processes are carried out, so that some sections of the individual parts of veins connect. These were previously not connected due to noise and faults in the image (Figure 5c). The next step in the image processing is the extraction and evaluation of biometric characteristics of the bloodstream results. This is derived by comparing the obtained image and the image of the fingerprint during dactyloscopy, as discussed by Reference [20].

After the smoothening and thinning of the image, the skeletonization of the image is automatically achieved (the skeleton is extracted) [2]. Thinning is a morphologic operation that is responsible for the deletion of selected pixels from the binary images. The process shares similarity to opening and erosion. During thinning, pixels at the edges get deducted from the objects, but not in such a way that disturb the object results. Iterative thinning can be achieved using an algorithm. In 1986, Alberto Martin and colleagues [55] analyzed a few different thinning algorithms. The results of their study indicated that the best outcomes from the point of reliability and effectiveness are achieved by algorithms that are based on the method of samples and the method "sign and delete." One of the representative algorithms of this group that was chosen for this study is the thinning algorithm by Zhang-Suen [56].This thinning method is simple, and gives room for evaluation even when in low-quality contour objects. Figure 6 shows the process of thinning iterations.

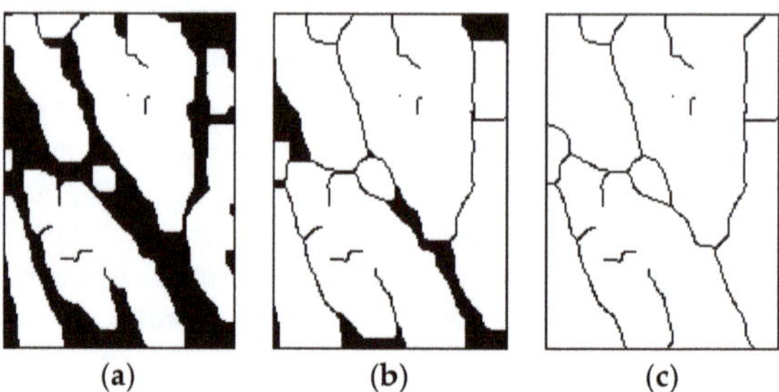

Figure 6. Process of thinning: (**a**) 2nd. Iteration (left). (**b**) 6th. Iteration (middle). (**c**) Result (right).

As noticeable on the last image (Figure 6c), due to the faults in the image, parts of the skeleton as well as the points or unfinished lines are still visible. These are veins that, in one place, run deeper into the tissue of the hand. Therefore, they become invisible for the camera. It is often necessary to eliminate such artifacts before the extraction.

For the elimination vein line artefacts, a self-created image filter was used. This filter is based on the number of the pixel connections. This is the combination of the number of individual connections of the examined pixels and the chosen object's pixels. The number of such connections may have values between 0 and 4. Figure 7 shows examples of the surrounding pixels. The filter's algorithm runs through all image pixels and searches for pixels that are at the end of the line (stand-alone pixels). The number of connections is usually 0 or 1 (Figure 8).

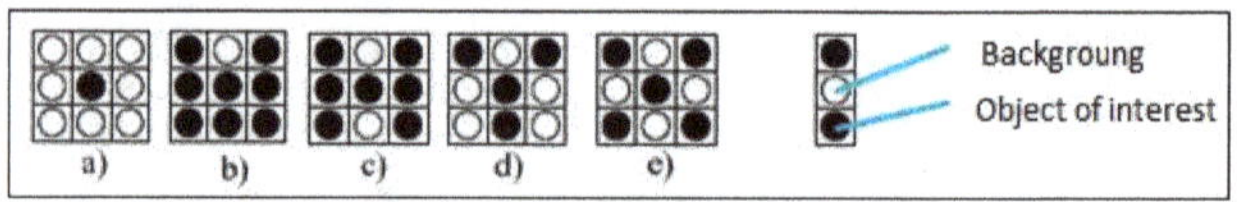

Figure 7. Examples of the number of pixel connections.

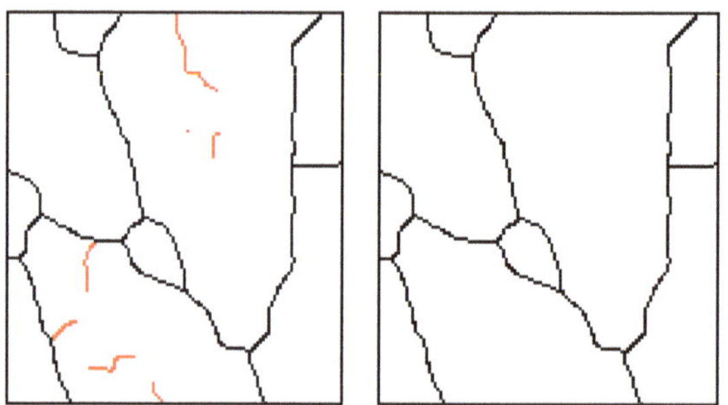

Figure 8. Skeleton before and after filtration.

3.2.3. Definition of Biometric Characteristics

The definition of the biometric characteristics for the verification of bloodstream results originate from the work [20]. For the definition of biometric characteristics in the context of this paper, the similarities of the bloodstream and fingerprints were used. In order to verify a person according to the fingerprints, the comparison of the critical point position (*minutiae*) is used instead of carrying out a comparison of the whole image (on the basis of the sample). For fingerprints' evaluation, the minutiae is used at the beginning and at the end of a dermal papillae, bifurcation (decoupling), hook, and eye.

Within this experiment, two types of minutiae were defined, which includes vein branching and vein ending at the top portion of the image. Quantity of located minutiae is expected to be different across all test subjects. However, for one user, the quantity of minutiae located during each scan should be the same. Moreover, the positions of these minutiae in the case of one user should be expected to maintain the same position, no matter the number of scanning. This is because all subsequent images are compared to the image on which the extraction of bloodstream biometric characteristics was done. Therefore, the experimental software works with the minutiae's coordinates, which are closely related to the image of the bloodstream.

3.2.4. Extraction of Biometric Characteristics

In order to extract minutiae from the images, we adopt the principle of the number of pixel connections. The algorithm scans the bloodstream skeletal image to go through each of the pixels, which is the property of the object of interest (ROI). The number of pixel connections, which is measured as the pixel of the object of interest, is scanned. Pixels with higher connections values (greater than 3) are responsible for creating a branching point, which is, otherwise, referred to as minutia. Pixels whose connections are equal to 1 and are found at the edge of the image are also marked as minutia and become the end point of the vein.

3.3. Evaluation Using a Degree of Unanimity

3.3.1. Calculation of Fractional Metrics Evaluation

In the proposed experimental project, a number of methods were used for calculating score-level fusion [57] (Figure 9).

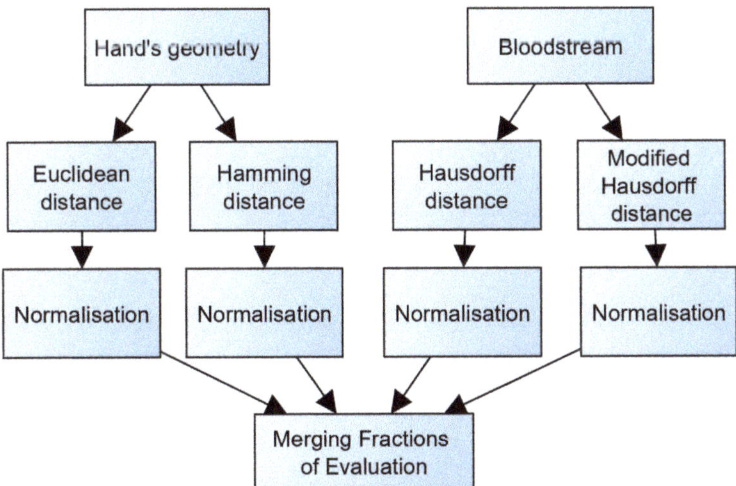

Figure 9. Calculation of fractional metrics evaluation.

- Euclidean distance, which is the distance measured between two points, is located in the 'N' dimension. The Euclidean distance is computed using the formula below.

$$d = \sqrt{\sum_{i=1}^{N} (x_i - t_i)^2} \qquad (4)$$

where N is the number of dimensions, which is (in the case of a template and testing data) the number of measured biometric characteristics, x_i is the i-th element of the tested data, and t_i is the i-th element of the template.

The number of template dimensions is equal to one testing data. The resulting value is the addition of all differences between the template and testing data.

- Hamming distance: This is another way of computing the score-level fusion. Hamming distance originates from the theory of information. In comparing two chains with the same length, the Hamming distance shows the lowest number of differentiation positions. In other words, it presents the number of substitutions that need to be established in order to change one chain into the second one. In their study, the researchers [6] generalized Hamming distance into a form suitable for evaluation of biometric data similarities. The authors suggested the use of a comparison on the basis of the number of non-unanimous biometric characteristics. The result is formed by a metric that does not measure deviation as in the case of Euclidean distance. Rather, it indicates the number of individual biometric characteristics for which there are higher deviations (during the comparison of the testing data and the template) than the root mean square (RMS) error of biometric characteristics. The RMS error is defined for each feature during template generation. Such an error is selected due to the presumption that the characteristics of one user during multiple photographing would never be completely identical. The presumed allocation of values for the given characteristic corresponds to the normal allocation. Hamming distance is calculated according to the following formula.

$$d(x_i, \overline{x}_i) = \#\{i \in \{1..N\} / |x_i - \overline{x}_i| > \sigma_i\} \tag{5}$$

where x_i is a biometric characteristic of the testing data with a serial number i, $\overline{x}_i$ is the average of the biometric characteristic (from the template) with the serial number i, N is the overall number of biometric characteristics for the given template, and σ_i is the RMS error (from the template) with the serial number i

- Hausdorff distance [58] is another method useful for a score-level fusion calculation. It determines the distance between two set of points found in the metric space. In simple terms, the two data sets that are in closest proximity to one another such that points of the second dataset can be found near the surroundings of the first dataset. Hausdorff distance - Helicity Violating (HV) is considered to be the longest distance among the existing distances between the set of points. It is created by joining one point of the first point set to another point on the second set and vice versa. If there are similarities between two sets of points, then HV has a lower value. Since the biometric characteristic of the bloodstream is composed of various sets of points, such a position on the image is essential. In this experimental set-up, HV is able to calculate score-level fusion, which naturally compares the similarity in shapes. HV is, however, flawed on the sensitivity to remote values. Oriented HV, which is marked $\overline{H}$ between the sets of points A and B, corresponds to the maximum distance from all pairs $x \in A$ and $y \in B$. The oriented HV is expressed by the equation below.

$$\overline{H}(A, B) = \max_{x \in A}\left\{\min_{y \in B}\{\|x, y\|\}\right\} \tag{6}$$

where $\|, \|$ is a random evaluation function, which is mainly a Euclidean distance.

The oriented HV is asymmetric. Therefore, $\overline{H}(A, B) \neq \overline{H}(B, A)$ applies. It also does not provide the distance between the sets A and B, but only provides the longest distance from the point $x \in A$ to the closest point $y \in B$. On the other hand, the non-oriented HV, which is marked H, is the maximum from $\overline{H}$ in both directions, and indicates the difference of the two sets of points. The formula for the calculating non-oriented HV is shown below.

$$H(A, B) = \max\left\{\overline{H}(A, B), \overline{H}(B, A)\right\} \tag{7}$$

- Modified Hausdorff distance: As earlier mentioned, HV is very sensitive to distant values. This implies that even a few points from the testing set of points that are outside of the template points would cause a large increase in the value of HV. This is regardless of whether or not the sets are very similar to each other. In order to find the solution to this weakness, researchers [58] looked at many different modifications of HV. Results from their analysis showed that, while using modified HV (further MHV), the problem of distant values is suppressed. In contrast to the previous formula, the non-oriented MHV can then be defined as:

$$H(A,B) = \frac{1}{N_A} \sum_{x \in A} \min_{y \in B}\{\|x, y\|\} \tag{8}$$

where N_A is the number of elements in the set A and $\|.\|$ is a random evaluation function, using mostly Euclidean distance.

3.3.2. Normalization of a Fractional Metric

Before merging the results of the individual metrics, it is necessary that the results undergo some form of normalization. Individual metrics provide results in different "dimensions." Normalization within this study is carried out using a 'min-max' method within the experimental software. It is calculated according to the formula below.

$$no = \frac{o - \min_{i=1}^{N} o_t^i}{\max_{i=1}^{N} o_t^i - \min_{i=1}^{N} o_t^i} \tag{9}$$

where o is a coarse evaluation, N is the number of elements in the set of testing data, and o_t^i is an element of the testing data.

3.3.3. Merging Fractions of Evaluation

The merging procedure in multiple biometric systems (blending of scans and results of different types of biometric characteristics) is carried out in different levels of processing. In the multi-biometric scanner proposed within this study, merging is done by recording fractional results from individual metrics. In this method, individual conformity assessments are combined after normalization. This method is most commonly used [59–61] while it provides clear and simple results processing.

In order to calculate the overall evaluation, it is first necessary to normalize the individual outputs from different metrics. Normalization ensures that all intermediate results have the same weight regardless of the method used.

The metrics merging itself is done using an arithmetic average. Furthermore, implementation is made possible. At the same time, the process provides the best results [62]. During the final verification phase, the template is tested with the best score-level feature. If the score level fulfills the requirement of the threshold (set to 50% in this case), then the process is tagged successful. If not, then it is tagged unsuccessful.

3.4. Image Scanning

3.4.1. Proposal of the Scanning Device

Image scanning is the first step toward the experimental implementation of the multi-biometric scanner. Effective image scanning positively influences the results to a great extent, especially during the image evaluation. To arrive at an appropriate image that will be subjected to further processing, configurations such as background lighting, direct lighting, and side lighting can be used (Figures 10–12). The image selection process is based on the task requirements. For instance, background lighting is considered to be an ideal option to measure the shape of the object. This is because it highlights the contour of the object (hand). In the context of the current work, direct lighting configuration was adopted.

For effective image processing, suppressing background noise (influence of the surrounding) is a very important requirement. Background noises negatively affect image processing and further assessment. Nevertheless, the use of additional lighting can resolve this problem. Additional lighting creates an improved scene and looks like an industrial light. The use of a filter that allows the passage of radiation only has the wavelength equivalent to that of the light in use, which also improves the process. A lighting requirement includes:

- Sufficient intensity
- Homogenous lighting of the scene
- Consistency of the light intensity over time (to guide against depletion).

Due to these lighting requirements, experimental design and implementation of the multi-biometric scanner in this paper adopted an industrial type of lighting for the hardware component of the scanning device. This was needed to achieve the required homogeneity associated with further image processing. For the same reason, the proposed scanning device has been equipped with a special camera. This camera will, however, not function based on automatic corrections of the image as compared to commonly available cameras. Table 2 summarizes the approximated prices of some components of the proposed scanner.

Table 2. Estimated prices of the proposed scanner components.

Component	Seller/Manufacturer	Amount (in €)
CCD camera GuppyPRO F-031B	Allied Vision Technologies GmbH	600
Objective VCN 1.4/4.5 f = 4.5 mm	Vision & Control GmbH	150
IR filter	Heliopan Lichtfilter-Technik Summer GmbH & Co KG	40
Lighting SFD 42/12 IR	Vision & Control GmbH	700
Total		1490

3.4.2. Components of the Scanning Device

A digital camera with resolution of 640 × 480 pixels and 8 bits was used for image scanning. This suggests the realization of grey-toned images, with 255 shades of grey. Firewire interface was used to transfer stored data from the camera. A stable focal distance of 4.5 mm from the manufacturing class VCN was used. The IR filter plays the important role of scanning the IR part of the spectrum [63–65]. This was needed since sunlight impact on the scanned image is a limitation. Another part of the scanning device is the source of IR radiation. Therefore, tests with different sources of IR radiation were conducted. The lighting tests include: direct circular lighting, direct linion lighting, diffused DOM lighting, and background lighting. On the basis of the test results, the most suitable lighting, DOM, was selected. This lighting uses LED lights with the wavelength of 850 nm and allows for generation of images with homogenous lighting, with a well recognizable structure of veins on the top of the hand, as well as a sufficient contrast of the hand against the background. This is good for further evaluation of the hand contour.

Lights and the camera configuration was put on the construction from aluminum profiles, which was mounted on an adjustable tripod (based on the configuration). The background comprises of a black matte surface that helps to get the contrast between the background and the scanned image of the user hand. The whole configuration of the camera, optics, and the lighting is displayed on the images in Figures 10 and 11. From these estimates, the average cost of the scanner is about €1490, which is relatively expensive. Nonetheless, selected sellers/manufacturers are the most revered in terms of sales of most durable spares, which is why the prices are so high. The components may be purchased from other sellers at cheaper rates, which will most likely reduce the average possible cost of the proposed scanner [66].

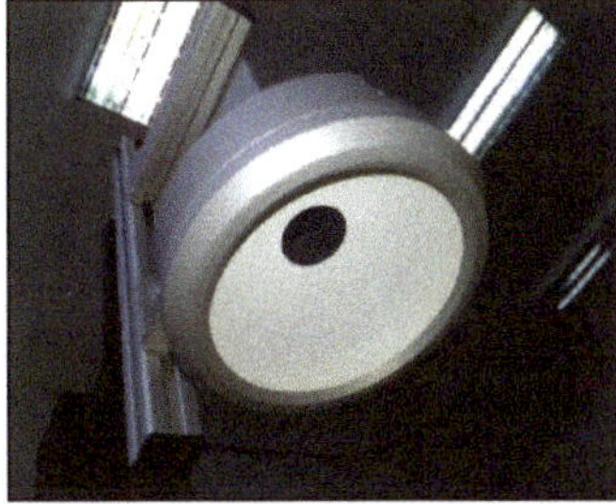

Figure 10. Configuration of the camera and lighting.

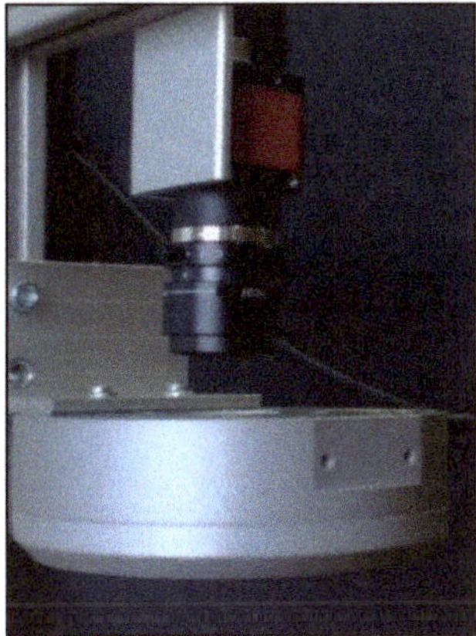

Figure 11. Configuration of the camera and lighting.

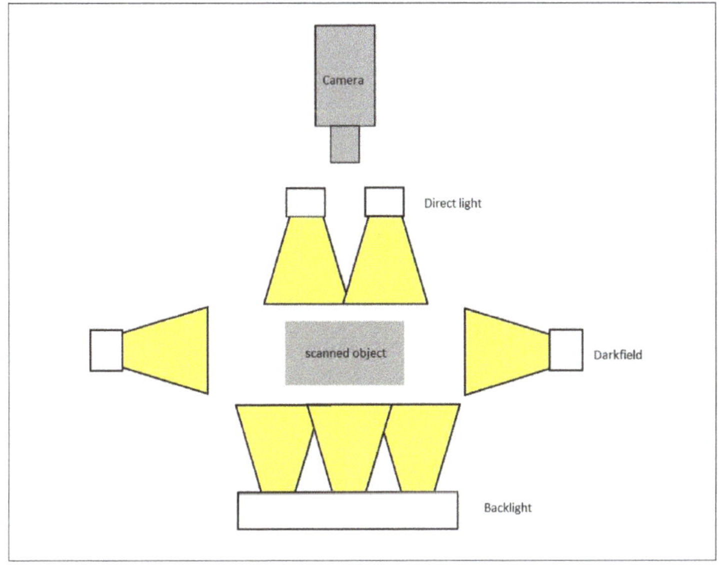

Figure 12. Possible types of lighting within the scene.

4. Results: Testing of Developed Application

4.1. Testing Methodology

For the purpose of testing the biometric verification system proposed in this paper, 280 (40 subjects, where each provided seven images) images of the hands of different persons was used. Although this value is small compared to existing works by References [27,29–34,67], this was mainly due to the unavailability of the scanning system as the ones used were gotten on a friendly loan. Participants consist of males and females between the ages of 25 to 60 years old. Characteristics of the tested persons is summarized in Tables 3 and 4. Two unique experimental design pathways were followed to check the behavior and evaluate how the proposed method progresses. First, the hand geometry was verified by Hamming distance, and then the bloodstream using MHV.

To establish a scientific experimental backing for the current study, it was vital to check the proposed experimental set-up to those in some existing works. As seen in the experimental design from Reference [28], based on the number of test subjects within this experiment, observations show that data associated with the bloodstream are not very sensitive to temperature when the human body is at rest. While Reference [28] noted that perfusion of blood was better in their study, ours yielded a different result since hand geometry seems to be better in terms of the rate of recognition. In Reference [34], the authors explained in their paper that matching scores are derived from the process of triangulation and binarization using vein structures, and from the distance of knuckle points. The current experimental procedure also follows the route of binarization, but with a convex contour polygon for point location detection. A huge part of this experimental procedure follows the ideas of Reference [33]. The pair worked on finger vein and geometry of the image characteristics. Just like in the case of our experiment, segmentation was done through the location of lines at different finger valleys and linking it to the center of the palm through a convex polygon [33]. This was closely followed by locating feature points through which extraction takes place. By adopting the calculation of the Hamming distance, we distinguish between the feature points of an enrolled and input image respectively.

Table 3. Division of tested persons based on gender.

Men	Women
80%	20%

Table 4. Division of tested persons based on age.

20–29	30–39	40–49	50–60
20%	50%	20%	10%

Forty persons were selected for the testing due to the limited number of available hardware components for image scanning. Each had a hand image tested seven times. The scanners were borrowed only for a limited period of time. As such, only a few testing data was generated. The testing of the system was done on a personal computer with the following parameters (only the parameters that may influence the running of the multi-biometric software are mentioned).

- OS—Windows 10, 64 bit
- CPU—Intel Core i5-527U with frequency of 2.7 GHz (maximum turbo frequency of 3.1 GHz)
- RAM—8 GB

4.2. Tests of the Experimental System Speed

Time taken to evaluate a tested image is between 0.6 and 0.9 seconds. In terms of calculations, it was somewhat difficult to derive the algorithm for the extraction of biometric characteristics.

Table 4 summarizes the processing speed of the individual program steps and the corresponding image evaluation.

4.3. Results

FMR (False Match Rate), FNMR (False Non-Match Rate), and EER functions were utilized in evaluating the biometric system. These functions specify the faults frequency of the system. First, the comparison of the individual metrics for the data gathered from participating persons were compared for single-biometric and multi-biometric systems. The threshold value was chosen to be 0.5 [-] and the boundary of the minimum interval varied from 0 to 0.5 [-] (Figure 13). For the metrics normalization, the previously min-max method was applied. The graphs presented in Figures 14 and 15 capture the test results of the two different metrics for hand geometry. Graphs in Figures 16 and 17 capture the test results of the two different metrics for the bloodstream.

In the next part of the testing, the best metric for hand geometry (Hamming metric) as well as the best metric for the bloodstream (MHV) are selected. With these metrics, the multi-biometric system was created and the performance was tested against the data described above (Table 5). The results are shown on the graph in Figure 13. As expected, results showed that the multi-biometric system performed better in comparison to a single biometric system. The EER value for the multi-biometric system was found to be half of what is obtained from a single bloodstream biometry. Moreover, the progress of the FNMR fault was smoothened and partially reduced.

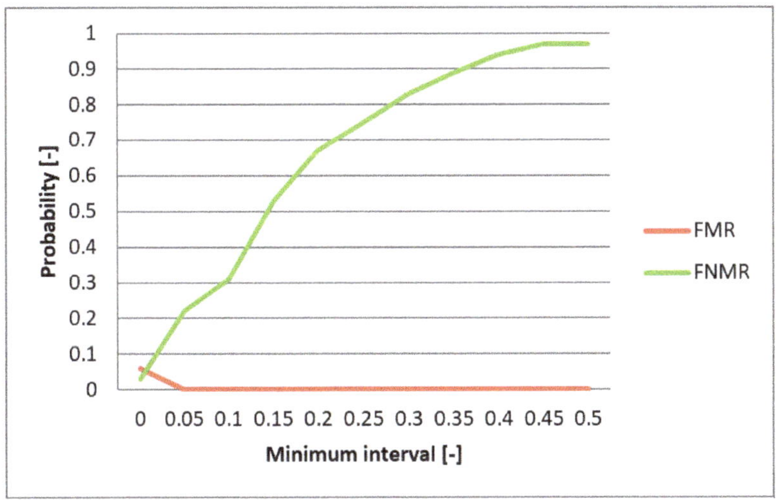

Figure 13. EER for united metrics of Hamming and MHV.

Table 5. Times needed for the execution of individual processing steps and the image evaluation.

Program Step	Hand Geometry		Blood Stream	
	Average Time [ms]	Maximum Deviation [ms]	Average Time [ms]	Maximum Deviation [ms]
Image pre-adaptation	1.25	14.75	3.5	8.5
Segmentation	1.08	8.9	152.7	70.3
Extraction of characteristics	667.5	168.5	8.83	3.26
Calculation of the degree of differentiation and allocation	0.97	3	1	1.9

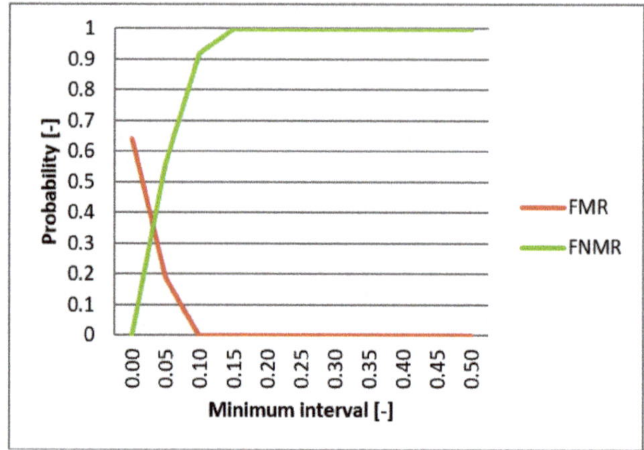

Figure 14. EER for Euclidean distance.

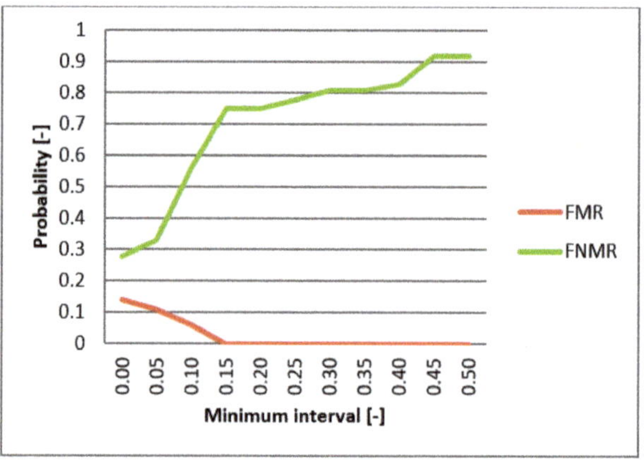

Figure 15. EER for Hamming distance.

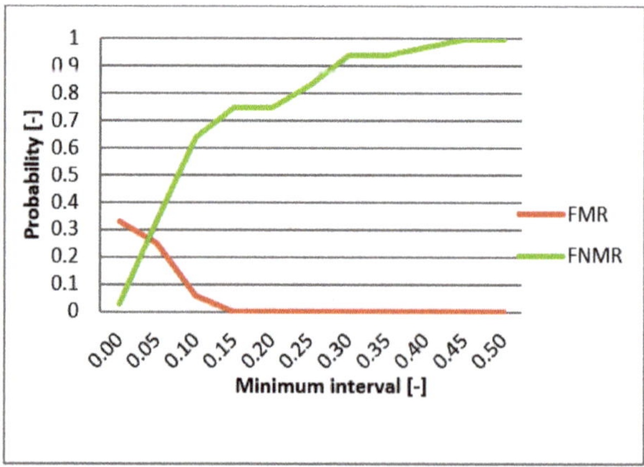

Figure 16. EER for HV.

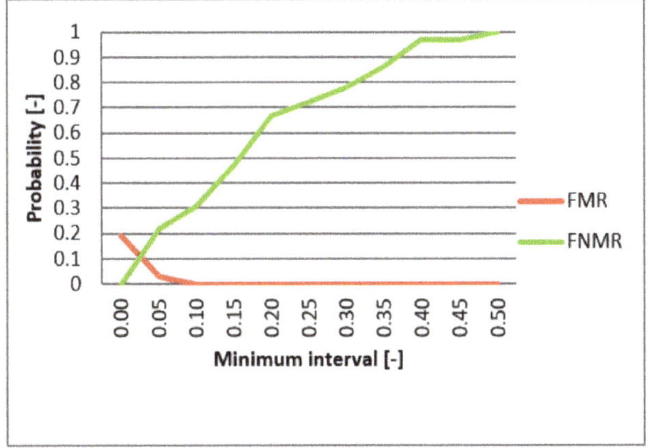

Figure 17. EER for MHV.

5. Discussion

The database developed for this study was built using 280 images (finger view and palm view, vascular images and back of the hand) from 40 individuals, i.e., seven images per subject. Two of the images generated per person (Total of 80 images) was used to train available researchers, so that they get familiar with the systems. Nevertheless, these images were carefully derived so that they were also useful for analysis. In testing our verification technique, each of the generated biometric data is technically subdivided (partitioned) into 2×50, 1×80, and 1×100 samples. Individual scores (from each uni-modal) were then summed up to arrive at the overall score. The equal error rate of the blood stream (11%) and hand geometry consolidated with the blood stream (5%). This approach is dependent on score-level fusion with the individual. Score results were normalized to achieve the stated percentages.

In addition to the samples used for training, extracted characteristics per scanned hand is related to how the proposed multimodal system works. The more the number of scanned hands, the more the extracted features and the better the behavior of the proposed multi-modal systems in terms of its performance. The implication of this is that the system witnesses some form of increased computational cost, which is similar to the finding of Reference [68]. As such, the number of extracted characteristics per scanned hand is checked against the EER. As noted by Reference [68], there is a possibility for an increment in the cost of computation in terms of the overall number of extracted characteristics. For example, this cost covers the time needed to input the first set of features into the database, which, thereby, trains the system, in addition to taking it for verification. Verification time is dependent on the speed of the computer (specification) used, as well as the algorithm for the experiment. There are a number of systems with rapid verification, segmentation, and feature extraction times. Some of these features utilize multi-dimensional vectors with more than two biometric approaches. In general, a high number of extracted characteristics imply longer processing times.

A number of challenges were noticed in the course of the experimentation with the proposed multi-biometric device. However, many of these challenges can be resolved at the point of use except for the issue posed by the extraction algorithm, which causes a lengthy time of verification. Algorithms of the individual metrics (hand geometry and blood stream analysis) demonstrates specific speed and precision issues that were noticed during experimentation, and may have significant effects on results. For instance, the low contrast produced by the bloodstream image during the investigation proved to be a limiting factor. There are also problems of discontinuation of veins and shadow produced by the human skin. This shadow image is confused to be another vein. Such small contrasts are most likely caused by a large distance between the lighting of the scene and the scanned hand. Camera resolution of 640 × 480 pixels is another factor that seems to affect results negatively. This produced very small differences in measured widths of the fingers among persons. This problem can, however, be resolved by using cameras of higher resolution to produce more precise evaluation and differentiation results. However, this was not verified in this study due to time constraint. Differentiation and evaluation of individual metrics is done via frequency comparison of the double type faults (false identification) with the EER values. The lower the EER value, the higher the level of precision of the given metrics. The least obtained value was measured for modified Hausdorff distance metric (MHV), which yielded an EER value that was 11%. Furthermore, the best metrics for the given biometric characteristics (Hamming distance and MHV) were chosen and consolidated into the final evaluation, which results in a multi-biometric model with an EER value of 5%. This implies that, as expected, metrics consolidation meant improved system precision.

The biggest problem of the proposed multi-biometric system could be the unwanted influences from the surrounding. For instance, direct sunlight would influence the intensity of the scanned image. A change in features of the platform on which the scanned hand is laid can also influence the results. Lastly, the user-friendliness of the hardware for hand scanning would need to be improved in order to make the usage of the system faster and more comfortable.

6. Conclusions, Limitations, and Future Work

This paper proposes the use of a multi-biometric system, which is useful for identification and verification of a person through his/her hand geometry and bloodstream behavior (at the upper part of the hand). The study experimentally tested the effectiveness of the system using a number of research participants. The system proved to be able to verify users with high precision by demonstrating good differentiating abilities. The ERR value of the system is estimated as 5%, which is higher than most existing systems (Table 1). This can, however, be optimized. The proposed system is flawed on the lengthy time taken to carry out the verification. As such, attaining commercial usage would mean optimizing an extraction algorithm for the biometric characteristics. This aspect paves way for future research, with goals of speeding up user identification and verification time.

Author Contributions: Conceptualization, L.K., O.K., A.S. and K.K.; methodology, L.K., O.K., O.F. and A.S.; software, L.K. and O.K.; validation, A.S. and O.K.; formal analysis, L.K. and O.K.; investigation, L.K.; resources, O.K., K.K. and A.S.; data curation, L.K., A.S. and O.K.; writing—original draft preparation, L.K. and O.K.; writing—review and editing, O.F. and O.K.; visualization, O.F. and O.K.; supervision, A.S., K.K. and O.K.; project administration, O.K. and K.K.; funding acquisition, O.K. and K.K.

Funding: The work and the contribution were mainly supported by project of excellence 2019/2205, Faculty of Informatics and Management, University of Hradec Kralove. The work was also partially funded by the: (1) the Ministry of Education, Youth and Sports of Czech Republic (project ERDF no. CZ.02.1.01/0.0/0.0/18_069/0010054), (2) Universiti Teknologi Malaysia (UTM) under Research University Grant Vot-20H04, Malaysia Research University Network (MRUN) Vot 4L876 and (3) the Fundamental Research Grant Scheme (FRGS) Vot 5F073 supported under Ministry of Education Malaysia for the completion of the research.

Acknowledgments: We are grateful for the support of student Sebastien Mambou and Michal Dobrovolny in consultations regarding application aspects. The APC was funded by project of excellence 2019/2205, Faculty of Informatics and Management, University of Hradec Kralove.

Conflicts of Interest: The authors declare no conflict of interest.

References

1. Abo-Zahhad, M.; Ahmed, S.M.; Abbas, S.N. A Novel Biometric Approach for Human Identification and Verification Using Eye Blinking Signal. *IEEE Signal Process. Lett.* **2015**, *22*, 876–880. [CrossRef]
2. Zhu, H.; Zhang, Y.; Wang, X. A Novel one-time identity-password authenticated scheme based on biometrics for e-coupon system. *Int. J. Netw. Secur.* **2016**, *18*, 401–409.
3. Luque-Baena, R.M.; Elizondo, D.; López-Rubio, E.; Palomo, E.J.; Watson, T.; Baena, R.M.L. Assessment of geometric features for individual identification and verification in biometric hand systems. *Expert Syst. Appl.* **2013**, *40*, 3580–3594. [CrossRef]

4.   Gupta, S.; Buriro, A.; Crispo, B. Demystifying Authentication Concepts in Smartphones: Ways and Types to Secure Access. *Mob. Inf. Syst.* **2018**, *2018*, 1–16. [CrossRef]

5.   Alpar, O.; Krejcar, O. Frequency and Time Localization in Biometrics: STFT vs. CWT. In Proceedings of the International Conference on Industrial, Engineering and Other Applications of Applied Intelligent Systems, Montreal, QC, Canada, 25–28 June 2018; Volume 10868 LNAI, pp. 722–728.

6.   Sanchez-Reillo, R.; Sanchez-Avila, C.; González-Marcos, A. Biometric identification through hand geometry measurements. *IEEE Trans. Pattern Anal. Mach. Intell.* **2000**, *22*, 1168–1171. [CrossRef]

7.   Gaikwad, A.N.; Pasalkar, N.B. Biometric Person Identification—Methods, Advances and Performance Evaluation. *IETE Tech. Rev.* **2004**, *21*, 211–217. [CrossRef]

8.   Zhang, J.; Cai, L.; Zhang, S. Malicious Cognitive User Identification Algorithm in Centralized Spectrum Sensing System. *Future Internet* **2017**, *9*, 79. [CrossRef]

9.   Zheng, H.; Van Hulle, C.A.; Rathouz, P.J. Comparing Alternative Biometric Models with and without Gene-by-Measured Environment Interaction in Behavior Genetic Designs: Statistical Operating Characteristics. *Behav. Genet.* **2015**, *45*, 480–491. [CrossRef]

10.   Hand-based biometrics. *Biom. Technol. Today* **2003**, *11*, 9–11. [CrossRef]

11.   Alpar, O.; Krejcar, O. A Comparative Study on Chrominance Based Methods in Dorsal Hand Recognition: Single Image Case. In Proceedings of the International Conference on Industrial, Engineering and Other Applications of Applied Intelligent Systems, Montreal, QC, Canada, 25–28 June 2018; Volume 10868 LNAI, pp. 711–721.

12.   Yörük, E.; Dutağaci, H.; Sankur, B. Hand biometrics. *Image Vis. Comput.* **2006**, *24*, 483–497. [CrossRef]

13.   Asem, M.M.; Oveisi, I.S.; Janbozorgi, M. Blood vessel segmentation in modern wide-field retinal images in the presence of additive Gaussian noise. *J. Med Imaging* **2018**, *5*, 1. [CrossRef]

14.   Tan, G.J.; Sulong, G.; Rahim, M.S.M. Writer identification: A comparative study across three world major languages. *Forensic Sci. Int.* **2017**, *279*, 41–52. [CrossRef]

15.   Renukalatha, S.; Suresh, K.V. A review on biomedical image analysis. *Biomed. Eng. Appl. Basis Commun.* **2018**, *30*, 1830001. [CrossRef]

16.   O'Gorman, L.; Schuckers, S.; Derakhshani, R.; Hornak, L.; Xia, X.; D'Amour, M. Spoof Detection for Biometric Sensing Systems. WO0124700 (A1), 12 April 2001.

17.   Waluś, M.; Bernacki, K.; Konopacki, J. Impact of NIR wavelength lighting in image acquisition on finger vein biometric system effectiveness. *Opto Electron. Rev.* **2017**, *25*, 263–268. [CrossRef]

18.   Guennouni, S.; Mansouri, A.; Ahaitouf, A. Biometric Systems and Their Applications. In *Eye Tracking and New Trends*; IntechOpen: London, UK, 2019.

19.   Rinaldi, A. Biometrics' new identity—measuring more physical and biological traits. *EMBO Rep.* **2016**, *17*, 22–26. [CrossRef]

20.   Wang, L.; Leedham, G.; Cho, D.S.-Y. Minutiae feature analysis for infrared hand vein pattern biometrics. *Pattern Recognit.* **2008**, *41*, 920–929. [CrossRef]

21.   Jain, A.; Hong, L.; Kulkarni, Y. A Multimodal Biometric System Using Fingerprint, Face, and Speech. In Proceedings of the 2nd Int'l Conference on Audio-and Video-based Biometric Person Authentication, Washington, DC, USA, 22–24 March 1999; pp. 182–187.

22.   Abhishek, K.; Yogi, A. A Minutiae Count Based Method for Fake Fingerprint Detection. *Procedia Comput. Sci.* **2015**, *58*, 447–452. [CrossRef]

23.   Han, J.; Bhanu, B. Gait Recognition by Combining Classifiers Based on Environmental Contexts. In Proceedings of the International Conference on Audio- and Video-Based Biometric Person Authentication, Rye Brook, NY, USA, 20–22 July 2005; pp. 416–425.

24.   Zhang, J.; Liu, H.; Ding, D.; Xiao, J. A robust probabilistic collaborative representation based classification for multimodal biometrics. In Proceedings of the Ninth International Conference on Graphic and Image Processing (ICGIP 2017), Qingdao, China, 14–16 October 2017; Volume 10615, p. 106151F.

25.   Barra, S.; de Marsico, M.; Nappi, M.; Narducci, F.; Riccio, D. A hand-based biometric system in visible light for mobile environments. *Inf. Sci.* **2019**, *479*, 472–485. [CrossRef]

26.   Engel, E.; Kovalev, I.V.; Ermoshkina, A. The biometric-based module of smart grid system. In *IOP Conference Series: Materials Science and Engineering*; IOP Publishing: Bristol, UK, 2015; Volume 94, p. 012007.

27.   Park, G.; Kim, S. Hand Biometric Recognition Based on Fused Hand Geometry and Vascular Patterns. *Sensors* **2013**, *13*, 2895–2910. [CrossRef]

28. Wu, S.-Q.; Song, W.; Jiang, L.-J.; Xie, S.-L.; Pan, F.; Yau, W.-Y.; Ranganath, S. Infrared Face Recognition by Using Blood Perfusion Data. In Proceedings of the International Conference on Audio- and Video-Based Biometric Person Authentication, Rye Brook, NY, USA, 20–22 July 2005; pp. 320–328.

29. Ribaric, S.; Fratric, I. A biometric identification system based on eigenpalm and eigenfinger features. *IEEE Trans. Pattern Anal. Mach. Intell.* **2005**, *27*, 1698–1709. [CrossRef]

30. Wang, Y.; Shark, L.-K.; Zhang, K. Personal identification based on multiple keypoint sets of dorsal hand vein images. *IET Biom.* **2014**, *3*, 234–245. [CrossRef]

31. Saadat, F.; Nasri, M. A multibiometric finger vein verification system based on score level fusion strategy. In Proceedings of the 2015 International Congress on Technology, Communication and Knowledge (ICTCK), Mashhad, Iran, 11–12 November 2015; pp. 501–507.

32. Veluchamy, S.; Karlmarx, L. System for multimodal biometric recognition based on finger knuckle and finger vein using feature-level fusion and k-support vector machine classifier. *IET Biom.* **2017**, *6*, 232–242. [CrossRef]

33. Kang, B.J.; Park, K.R. Multimodal biometric method based on vein and geometry of a single finger. *IET Comput. Vis.* **2010**, *4*, 209–217. [CrossRef]

34. Kumar, A.; Prathyusha, K.V. Personal Authentication Using Hand Vein Triangulation and Knuckle Shape. *IEEE Trans. Image Process.* **2009**, *18*, 2127–2136. [CrossRef]

35. Jain, A.K.; Duta, N. Deformable matching of hand shapes for user verification. In Proceedings of the 1999 International Conference on Image Processing (Cat. 99CH36348), Kobe, Japan, 24–28 October 1999; Volume 2, pp. 857–861.

36. Wong, R.L.N.; Shi, P. Peg-free hand geometry recognition using hierarchical geometry and shape matching. In Proceedings of the IAPR Workshop on Machine Vision Applications, Nara, Japan, 11–13 December 2002; pp. 281–284.

37. Kang, W.; Liu, Y.; Wu, Q.; Yue, X. Contact-free palm-vein recognition based on local invariant features. *PLoS ONE* **2014**, *9*, e97548. [CrossRef]

38. Van Tilborg, H.C.A.; Jajodia, S. *Encyclopedia of Cryptography and Security*, 2nd ed.; Springer: Boston, MA, USA, 2011.

39. Kim, W.; Song, J.M.; Park, K.R. Multimodal biometric recognition based on convolutional neural network by the fusion of finger-vein and finger shape using near-infrared (NIR) camera sensor. *Sensors* **2018**, *18*, 2296. [CrossRef]

40. Kirimtat, A.; Krejcar, O. Parametric Variations of Anisotropic Diffusion and Gaussian High-Pass Filter for NIR Image Preprocessing in Vein Identification. In Proceedings of the International Conference on Bioinformatics and Biomedical Engineering, Granada, Spain, 25–27 April 2018; Volume 10814, pp. 212–220.

41. Robles, F.E.; Chowdhury, S.; Wax, A. Assessing hemoglobin concentration using spectroscopic optical coherence tomography for feasibility of tissue diagnostics. *Biomed. Opt. Express* **2010**, *1*, 310–317. [CrossRef]

42. Mesicek, J.; Krejcar, O.; Selamat, A.; Kuca, K. A recent study on hardware accelerated Monte Carlo modeling of light propagation in biological tissues. In Proceedings of the International Conference on Industrial, Engineering and Other Applications of Applied Intelligent Systems, Morioka, Japan, 2–4 August 2016; Volume 9799, pp. 493–502.

43. Mesicek, J.; Zdarsky, J.; Dolezal, R.; Krejcar, O.; Kuca, K. Simulations of light propagation and thermal response in biological tissues accelerated by graphics processing unit. In Proceedings of the International Conference on Computational Collective Intelligence, Halkidiki, Greece, 28–30 September 2016; Volume 9876 LNCS, pp. 242–251.

44. Jain, A.; Hong, L. Biometrics: Techniques for personal identification. In Proceedings of the SPIE—The International Society for Optical Engineering, Wuhan, China, 21–23 October 1998; Volume 3545, pp. 2–3.

45. Wayman, J.L. Biometric Verification/Identification/Authentication/Recognition: The Terminology. In *Encyclopedia of Biometrics*; Li, S.Z., Jain, A., Eds.; Springer: Boston, MA, USA, 2009; pp. 153–157.

46. Mayhew, S. Explainer: Verification vs. Identification Systems. Biometric Update. 1 June 2012. Available online: https://www.biometricupdate.com/201206/explainer-verification-vs-identification-systems (accessed on 26 July 2019).

47. Biometric Security Devices. Biometric Verification Vs Biometric Identification Systems. Biometric Security Devices. 2019. Available online: https://www.biometric-security-devices.com/biometric-verification.html (accessed on 26 July 2019).

48. Alpar, O.; Krejcar, O. Detection of Irregular Thermoregulation in Hand Thermography by Fuzzy C-Means. In Proceedings of the International Conference on Bioinformatics and Biomedical Engineering, Granada, Spain, 25–27 April 2018; Volume 10814, pp. 255–265.

49. Kolda, L.; Krejcar, O. Biometrie hand vein estimation using bloodstream filtration and fuzzy e-means. In Proceedings of the IEEE International Conference on Fuzzy Systems, Naples, Italy, 9–12 July 2017.

50. Manjón, J.V.; Thacker, N.A.; Lull, J.J.; Garcia-Martí, G.; Martí-Bonmatí, L.; Robles, M. Multicomponent MR Image Denoising. *Int. J. Biomed. Imaging* **2009**, *2009*. [CrossRef]

51. Blaschke, T.; Hay, G.J.; Kelly, M.; Lang, S.; Hofmann, P.; Addink, E.; Feitosa, R.Q.; van der Meer, F.; van der Werff, H.; van Coillie, F.; et al. Geographic Object-Based Image Analysis—Towards a new paradigm. *ISPRS J. Photogramm. Remote Sens.* **2014**, *87*, 180–191. [CrossRef]

52. Oyebode, K.O. Improved thresholding method for cell image segmentation based on global homogeneity information. *J. Telecommun. Electron. Comput. Eng.* **2018**, *10*, 13–16.

53. Yang, Y.; Stafford, P.; Kim, Y. Segmentation and intensity estimation for microarray images with saturated pixels. *BMC Bioinform.* **2011**, *12*, 462. [CrossRef]

54. Liu, L.; Xie, Z.; Yang, C. A novel iterative thresholding algorithm based on plug-and-play priors for compressive sampling. *Future Internet* **2017**, *9*, 24.

55. Manish, R.; Venkatesh, A.; Ashok, S.D. Machine Vision Based Image Processing Techniques for Surface Finish and Defect Inspection in a Grinding Process. *Mater. Today Proc.* **2018**, *5*, 12792–12802. [CrossRef]

56. Zhang, T.Y.; Suen, C.Y. A fast parallel algorithm for thinning digital patterns. *Commun. ACM* **1984**, *27*, 236–239. [CrossRef]

57. Guo, C.; Ngo, D.; Ahadi, S.; Doub, W.H. Evaluation of an Abbreviated Impactor for Fine Particle Fraction (FPF) Determination of Metered Dose Inhalers (MDI). *AAPS PharmSciTech* **2013**, *14*, 1004–1011. [CrossRef]

58. Zhou, Z.-Q.; Wang, B. A modified Hausdorff distance using edge gradient for robust object matching. In Proceedings of the 2009 International Conference on Image Analysis and Signal Processing (IASP 2009), Linhai, China, 11–12 April 2009; pp. 250–254.

59. Ross, A.; Jain, A.K. Fusion Techniques in Multibiometric Systems. In *Face Biometrics for Personal Identification*; Hammoud, R.I., Abidi, B.R., Abidi, M.A., Eds.; Springer: Berlin/Heidelberg, Germany, 2007; pp. 185–212.

60. Ríos-Sánchez, B.; Arriaga-Gómez, M.F.; Guerra-Casanova, J.; de Santos-Sierra, D.; de Mendizábal-Vázquez, I.; Bailador, G.; Sanchez-Avila, C. gb2s μ MOD: A MUltiMODal biometric video database using visible and IR light. *Inf. Fusion* **2016**, *32*, 64–79. [CrossRef]

61. Moini, A.; Madni, A.M. Leveraging Biometrics for User Authentication in Online Learning: A Systems Perspective. *IEEE Syst. J.* **2009**, *3*, 469–476. [CrossRef]

62. Marasco, E.; Sansone, C. An Experimental Comparison of Different Methods for Combining Biometric Identification Systems. *Inf. Secur. Appl.* **2011**, *6979*, 255–2641.

63. Kirimtat, A.; Krejcar, O.; Selamat, A. A Mini-review of Biomedical Infrared Thermography (B-IRT). In Proceedings of the International Work-Conference on Bioinformatics and Biomedical Engineering Granada, Spain, 8–10 May 2019; Volume 11466, pp. 99–110.

64. Zurek, P.; Cerny, M.; Prauzek, M.; Krejcar, O.; Penhaker, M. New approaches for continuous non invasive blood pressure monitoring. In Proceedings of the XII Mediterranean Conference on Medical and Biological Engineering and Computing 2010, Chalkidiki, Greece, 27–30 May 2010; Volume 29, pp. 228–231.

65. Mambou, S.; Krejcar, O.; Kuca, K.; Selamat, A. Novel Human Action Recognition in RGB-D Videos Based on Powerful View Invariant Features Technique. *Stud. Comput. Intell.* **2018**, *769*, 343–353.

66. Maresova, P.; Sobeslav, V.; Krejcar, O. Cost–benefit analysis–evaluation model of cloud computing deployment for use in companies. *Appl. Econ.* **2017**, *49*, 521–533. [CrossRef]

67. Isah, S.S.; Selamat, A.; Ibrahim, R.; Krejcar, O. An investigation of information granulation techniques in cybersecurity. *Stud. Comput. Intell.* **2020**, *830*, 151–163.

68. De-Santos-Sierra, A.; Sánchez-Ávila, C.; Del Pozo, G.B.; Guerra-Casanova, J. Unconstrained and Contactless Hand Geometry Biometrics. *Sensors* **2011**, *11*, 10143–10164. [CrossRef]

Article

Blind Quality Assessment of Iris Images Acquired in Visible Light for Biometric Recognition [†]

Mohsen Jenadeleh [1,*], Marius Pedersen [2] and Dietmar Saupe [1]

[1] Department of Computer and Information Science, University of Konstanz, 78457 Konstanz, Germany; dietmar.saupe@uni-konstanz.de

[2] Department of Computer Science, Norwegian University of Science and Technology, N-2802 Gjøvik, Norway; marius.pedersen@ntnu.no

* Correspondence: mohsen.jenadeleh@uni-konstanz.de

[†] This paper is an extended version of the conference paper: Jenadeleh, M.; Pedersen, M.; Saupe, D. Realtime quality assessment of iris biometrics in visible light. In Proceedings of the 2018 IEEE/CVF Conference on Computer Vision and Pattern Recognition Workshops (CVPRW), Salt Lake City, UT, USA, 18–22 June 2018.

Received: 4 January 2020; Accepted: 25 February 2020; Published: 28 February 2020

Abstract: Image quality is a key issue affecting the performance of biometric systems. Ensuring the quality of iris images acquired in unconstrained imaging conditions in visible light poses many challenges to iris recognition systems. Poor-quality iris images increase the false rejection rate and decrease the performance of the systems by quality filtering. Methods that can accurately predict iris image quality can improve the efficiency of quality-control protocols in iris recognition systems. We propose a fast blind/no-reference metric for predicting iris image quality. The proposed metric is based on statistical features of the sign and the magnitude of local image intensities. The experiments, conducted with a reference iris recognition system and three datasets of iris images acquired in visible light, showed that the quality of iris images strongly affects the recognition performance and is highly correlated with the iris matching scores. Rejecting poor-quality iris images improved the performance of the iris recognition system. In addition, we analyzed the effect of iris image quality on the accuracy of the iris segmentation module in the iris recognition system.

Keywords: biometric recognition; visible light iris images; image quality assessment; image covariates; quality filtering

---

1. Introduction

The stability of iris patterns over the human lifespan and their uniqueness was first noticed in 1987 [1]. Since then, biometric iris recognition has been extensively investigated for accurate and automatic personal identification and authentication [2]. Most commercial iris recognition systems use near-infrared (NIR) images. However, due to the popularity of smartphones and similar handheld devices with digital cameras, iris recognition systems using images taken in visible light have recently been developed [3–5].

Image quality is a key factor affecting the performance of iris recognition systems [6–8]. In the biometric recognition literature, a biometric quality measure is a covariate that is measurable, influences performance, and is actionable [9–11]. Quality measurement can include subject and image covariates. Subject covariates are attributes of a person, which may be properties of subjects such as eyelid occlusion, glare, iris deformation, or wearing of glasses. Image covariates depend on sensor and acquisition conditions, such as focus, noise, resolution, compression artifacts, and illumination effects. In this work, we develop a real-time quality measure for image covariates as an actionable quality score, e.g., to decide whether an input iris image sample should be enrolled into a dataset or rejected and a new sample should be captured.

The performance of an iris recognition system in visible light suffers from all of the image quality factors mentioned above. To overcome this problem, some researchers have considered image quality in different ways for iris recognition systems [12–17]. However, these systems fall short in two ways:

- The considered image covariates and distortions are limited. Only distortions are taken into account that are often seen, such as Gaussian blur, noise, motion blur, and defocus. However, authentic iris images, especially those taken by handheld devices, may additionally suffer from other types of distortion.
- Typically, quality assessment is applied to accurately segmented iris images. However, image distortion also affects the performance of the segmentation module of iris recognition systems. Thus, poor image quality can lead to poorly segmented irises and increase in the false rejection rate.

In this paper, we propose a general-purpose and fast image quality method that aims to assess the distortion of iris images acquired in unconstrained environments. This method can be used for real-time quality prediction of iris images to rapidly filter image samples with poor quality. Iris images with insufficient quality could lead to high dissimilarity scores for matching pairs and increase the false rejection rate of an iris recognition system. We investigate the effect of iris image quality on the recognition performance of a reference iris recognition system for three challenging iris image datasets acquired in visible light.

This paper is an extended version of our conference paper [18] and mostly a part of the Ph.D. thesis of the first author [19]. The remainder of the paper is organized as follows: Section 2 surveys the literature on iris image quality assessment and iris recognition systems. Section 3 presents the proposed metric for iris image quality assessment. In Section 4, experiments are conducted to study the effect of image quality on the accuracy of iris segmentation. In Section 5, the improvements achieved by filtering poor-quality iris images are discussed using three performance measures on three large iris image datasets acquired in visible light. The paper concludes with suggestions for future research in Section 6.

2. Related Work

In this section, we review the literature on iris image quality assessment, followed by a brief overview of some state-of-the-art iris recognition systems.

Recently, research has been reported to improve the performance of iris recognition systems by considering image quality, but with certain limitations. In some studies, image quality has been examined by considering only certain quality factors, such as sharpness [20], out-of-focus [21], and JPEG compression [22]. These metrics alone cannot be expected to produce reliable quality assessments of authentic in-the-wild iris images.

In some other work, iris image quality metrics are applied after segmentation of the iris. In [23], the result of the iris segmentation module is used to form a quality score. Happold et al. [24] proposed a method for predicting the iris matching scores of an iris image pair based on their quality features. They calculated these features for precisely segmented iris images. They labeled a dataset of iris image pairs with the corresponding matching scores. They trained their method for predicting the matching score of an image pair based on their quality features. Therefore, these methods cannot be used to measure iris image quality in the iris recognition system pipeline before segmentation.

Several metrics for iris image quality were developed based on a fusion of several quality measures of image and subject covariates. The authors of [25,26] combined quality measures relating to motion blur, angular deviation, occlusion, and defocus into an overall quality value of an input iris image. These quality metrics were developed for NIR-based images and compared to traditional NIR-controlled iris image acquisition settings. However, images in visible light and under uncontrolled lighting conditions result in notorious differences in the appearance of the acquired images [3]. Therefore, this method may not be used directly to evaluate the quality of iris images in visible light. Li et al. [27] proposed a method for predicting an iris matching score based on iris quality factors such as motion blur, illumination, off-angle, occlusions, and dilation. This method requires segmented irises to compute some of these quality factors (dilation and occlusions).

The authors of [10] used combined subject and image covariates, such as the degree of defocusing, occlusion, reflection, and illumination, to form an overall quality score. They focused on the evaluation of iris images after iris segmentation, which allows the systems to process images of poor and good quality in the acquisition phase. They considered only a few image covariates for quality estimation.

Proença [3] proposed a metric for the quality assessment of iris images taken in visible light. This metric measures six image quality attributes such as focus score, off-angle score, motion score, occlusion score, iris pigmentation level, and pupil dilation. Then, the impact of image quality on feature matching was analyzed. The results showed a significant performance improvement of the iris recognition system by avoiding low-quality images. However, this method requires precisely segmented iris images, and only the motion-blur score is combined with some quality factors related to the subject's covariates.

The authors in [12] proposed an approach that automatically selects the regions of an iris image with the most distinguishably changing patterns between the reference iris image and the distorted version to compute the feature. The measured occlusion and dilation are combined to form a total image quality score to study the correlation between iris image quality and iris recognition accuracy.

In the approach of [28], the image quality is assessed locally, based on a fusion schema at the pixel level using a Gaussian mixture model, which gives a probabilistic measure of the quality of local regions of the iris image. The local quality measure is used to detect the poorly segmented pixels and remove them from the fusion process of a sequence of iris images.

Recently, many image quality methods have been proposed for perceptual quality assessment of natural images [29–35]. Some of these models use statistics of completed local binary patterns (CLBP) as a part of their feature vectors. In [33], joint statistics of local binary patterns (LBP) and CLBP patterns produced quality-aware features, and a regression function was trained to map the feature space to the perceived quality scores. In [32], features based on several local image descriptors such as CLBP, local configuration patterns (LCP), and local phase quantization (LPQ) were extracted, and then a support vector regressor was used to predict the quality scores. These models are trained to predict the perceptual quality of natural images. Liu et al. [36,37] studied some of these methods for filtering low-quality iris images. This study showed inconsistencies for the predicted quality, e.g., removing more low-quality images did not always increase the performance of the iris recognition system. In addition, they removed the low-quality images for each subject separately. Therefore, the filtered images do not have the same range of quality, and there is no global quality-filtering threshold.

In summary, some of the methods for iris quality assessment, such as [25,26], are proposed for NIR images, and only a few types of distortion are considered. Some other quality metrics, like those in [3,23,24], require a segmented iris image to calculate their quality features. They also take limited distortion types into account and are not expected to work well for quality assessment of authentic iris images taken in visible light in arbitrary environmental conditions. Iris recognition systems based on authentic images will broaden the scope of iris recognition systems, and require more research to develop robust metrics for quality assessment of authentically distorted iris images.

Since we used an iris recognition system as a reference system in this paper, in the following, we briefly review some state-of-the-art iris recognition systems.

The fast iris recognition (FIRE) system for images acquired by mobile phones in visible light was proposed by Galdi et al. [38]. It is based on the combination of three classifiers by exploiting iris color and texture information. Raja et al. [39] proposed a recognition system for iris images captured in visible light. This method extracts deep sparse features from image blocks and the whole iris image in different color channels to form the feature vector for an input iris image. Minaee et al. [40] proposed an iris feature extraction method based on textural and scattering transform features. The principal component analysis (PCA) technique is used to reduce the extracted feature dimension.

Recently, OSIRIS version 4.1, an open-source iris detection system, was proposed by Othman et al. [41]. This system follows the classic Daugman method [42] with some improvements in segmentation, normalization, coding, and matching modules. For iris and pupil segmentation, the Viterbi algorithm is used for optimal contour detection. For normalization, a non-circular iris normalization is performed using the coarse contours detected by the Viterbi algorithm. The coding module is based on 2-D Gabor filters, which are calculated in different scales and resolutions. Finally, the matching module calculates the global dissimilarity score between two iris codes using the Hamming distance. We used this system as a reference iris recognition system.

3. Proposed Method

In this section, we present our fast and general-purpose method for assessing the quality of iris images acquired in visible light.

Earlier works on iris recognition [42,43] employed block-based operations to obtain iris features. Therefore, we can infer that the most distinctive information in the iris pattern comes from the local patterns of an iris image rather than from global features. Local binary patterns (LBP) and their derivatives have been successfully used in many pattern recognition applications, including texture classification [44–46], image retrieval [47,48], object recognition [49,50], action recognition [51,52], and biometric recognition [53–56].

Most of the LBP-based biometric recognition methods use statistical analysis of local patterns for their feature extraction. Wu et al. [29] showed that image distortions could change the statistics of LBPs. They then examined the statistics of the LBPs to suggest an index for evaluating natural image quality. However, this index does not accurately predict image quality for some common image distortions, such as Gaussian blur and impulse noise.

In the proposed differential sign–magnitude statistics index (DSMI), sign and magnitude patterns are first derived. Then, the statistical characteristics of these patterns are analyzed for their sensitivity to iris image distortion. Statistical features of specific coincidence patterns with high sensitivity to image distortion are identified. A weighted nonlinear mapping is applied to the features to form the iris image quality score. This metric takes advantage of the observation that low-quality iris images have fewer of these patterns compared with those in high-quality iris images.

3.1. Proposed Quality Metric

Our iris image quality metric uses statistical features extracted from patterns of signs and magnitudes of local intensity differences. Then, certain locally weighted statistics of specific sign–magnitude coincidence patterns are used to define the quality score. Guo et al. [46] suggested a completed local binary pattern (CLBP) to represent the local difference information that is missed in the LBP representation of an image [57]. We investigate how common distortions in iris images could alter the statistics of the CLBP. Then, a quality metric based on a specific coincidence of sign and magnitude patterns of the CLBP is proposed.

In CLBP, a local grayscale image patch is represented by its central pixel, and the local differences are given by $d_p = x_p - x_c$, where $x_c = I(c)$ is the gray value of the central pixel of the given patch and x_p is the gray value of a pixel in the neighborhood. A local difference d_p can be decomposed into two components, its sign and its magnitude. These signs and magnitudes of local differences are combined into corresponding patterns, CLBP-S and CLBP-M, as follows.

Let $C = \{(i,j)|i = 0, \cdots, M-1, j = 0, \cdots, N-1\}$ be the set of pixels of a normalized grayscale image I of N pixels width by M pixels height. For a given pixel $c \in C$, let x_c and $x_p, p = 0, \cdots, P-1$, denote the gray values of the center pixel c and the P points on a circle of radius R about x_c. For example, suppose the coordinates of x_c are (0,0); then, the coordinates of x_p are $(R\cos(2\pi p/P), R\sin(2\pi p/P))$. The grayscale value x_p is estimated by interpolation if its coordinates do not coincide with the center of a pixel. Then, the CLBP-S patterns are defined by

$$\text{CLBP-S}_{P,R}(c) = \sum_{p=0}^{P-1} b_p \cdot 2^p, \qquad b_p = \begin{cases} 1 & x_p \geq x_c \\ 0 & \text{otherwise} \end{cases}. \tag{1}$$

The CLBP-S operator generates the same code as that of the original LBP operator. The CLBP magnitude patterns are defined similarly by

$$\text{CLBP-M}_{P,R}(c) = \sum_{p=0}^{P-1} b_p \cdot 2^p, \qquad b_p = \begin{cases} 1 & m_p \geq z \\ 0 & \text{otherwise,} \end{cases} \tag{2}$$

where $m_p = |x_p - x_c|$ is the magnitude of the local difference d_p. Furthermore, the threshold value z is the average local difference in the P-neighborhoods of all center pixels together, i.e.,

$$z = \frac{1}{|C|P} \sum_{c \in C} \sum_{p=0}^{P-1} |x_p - x_c|. \tag{3}$$

For each pixel $c \in C$, we consider the P-bit binary representation of the sums in Equations (1) and (2) as binary codes of CLBP-S$_{P,R}$ and CLBP-M$_{P,R}$. Using these binary representations, we define rotation invariant indices or patterns for CLBP-S and CLBP-M in a manner similar to that proposed by Ojala et al. [57] for LBP codes. Equation (4) gives the rotation invariant indices of CLBP-S,

$$\text{CLBP-S}_{P,R}^{riu2}(c) = G(\text{CLBP-S}_{P,R}(c)) = G\left(\sum_{p=0}^{P-1} b_p 2^p\right) = \begin{cases} \sum_{p=0}^{P-1} b_p & U(\sum_{p=0}^{P-1} b_p 2^p) \leq 2 \\ P+1 & \text{otherwise} \end{cases}. \tag{4}$$

Here, U gives the number of bit changes (0 to 1 or 1 to 0) of the P-bit binary representation of a number (including circular shift),

$$U\left(\sum_{p=0}^{P-1} b_p 2^p\right) = \sum_{p=0}^{P-1} |b_p - b_{\text{mod }(p+1,P)}|.$$

Similarly, Equation (5) gives the uniform rotation invariant patterns of CLBP-M.

$$\text{CLBP-M}_{P,R}^{riu2}(c) = G(\text{CLBP-M}_{P,R}(c))). \tag{5}$$

Note that these indices, CLBP-S$_{P,R}^{riu2}$ and CLBP-M$_{P,R}^{riu2}$, range over the set $\{0, ..., P+1\}$. The first indices from 0 up to P correspond to local sign and magnitude patterns with only, at most, two bit changes and, thus, denote uniform local patterns. All non-uniform patterns are assigned to the remaining index $P+1$.

CLBP-$S_{P,R}^{riu2}$ generates fewer codes than the basic CLBP-S. It carries less textural information by simplifying the local structure. CLBP-$M_{P,R}^{riu2}$ provides a compact representation of textural information derived from local magnitude patterns.

For an illustration for the case of $P = 4$ neighbors at distance $R = 1$ from the central pixel of a patch, we provide Figure 1. We obtain six indices k and l for sign and magnitude patterns, corresponding to five rotation invariant uniform patterns ($k, l = 0, ..., 4$) and one index ($k, l = 5$) that represents all non-uniform patterns.

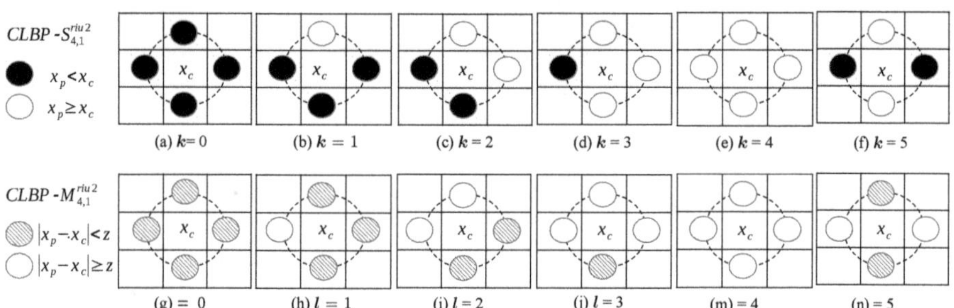

Figure 1. The patterns in the upper row correspond to CLBP-$S_{4,1}^{riu2}$, which compares the gray value of the central pixel of a patch (x_c) with the gray values of its four neighbors (x_p). The black and white disks denote smaller and greater values than those of the central pixel value, respectively. In the lower row, CLBP-$M_{4,1}^{riu2}$ compares the absolute values of the differences of the gray values of the central pixel and its neighbors ($|x_c - x_p|$) with the threshold z from Equation (3). The hatched and white disks denote smaller and greater absolute values than those of the threshold, respectively. Note that the patterns are rotation invariant. Thus, in the case of $P = 4$ shown here, the patterns for $k, l = 1, 2, 3, 5$ may be rotated by multiples of 90 degrees without changing the values of CLBP-$S_{4,1}^{riu2}$ and CLBP-$M_{4,1}^{riu2}$.

Finally, the local indices for sign and magnitude have to be combined to give a quality indicator for an iris image as a whole. We first join the two types of indices into a set of bitmaps $V_{k,l}(c)$, indexed by k, l,

$$V_{k,l}(c) = \begin{cases} 1 & \text{CLBP-}S_{P,R}^{riu2}(c) = k \text{ and CLBP-}M_{P,R}^{riu2}(c) = l \\ 0 & \text{otherwise} \end{cases}. \tag{6}$$

For each pair k, l of indices, we form a weighted sum of $V_{k,l}(c)$ over all pixels c, which is nonlinearly scaled to the unit interval by $r(x) = 1 - e^{-ax}$ as follows:

$$Q_{k,l} = r\left(\frac{1}{|C|} \sum_{c \in C} \frac{V_{k,l}(c)}{\hat{\sigma}^2(c) + \delta^2} \right). \tag{7}$$

Here, $\hat{\sigma}^2(c)$ is the local variance of the P-neighboring pixels of the center pixel c, and δ^2 is a small constant value to prevent division by zero. The parameters δ^2 and a are empirically set to 0.00025 and 0.01, respectively.

In Equation (7), the normalization by the local variance emphasizes local minima and maxima, and normalizing the scores to the range $[0, 1)$ is only for ease of interpretation of the quality scores. The value of $Q_{k,l}$ is considered as an image quality score derived from the sign pattern k and the magnitude pattern l. In our experiments, we used four neighbors ($P = 4$) with unit distance ($R = 1$) from the central pixel c of a local patch.

Our experiments showed that $Q_{k,l}$ with the specific coincidence of the sign pattern $k = 0$ and magnitude pattern $l = 0$ has a high correlation with iris image quality. Therefore, we used $Q_{0,0}$ as our proposed DSMI quality score. We had summarized the proposed DSMI metric in our conference paper [18], considering, however, only the selected coincidence sign–magnitude patterns.

3.2. Empirical Justification

Inspired by Wu et al. [29], we examine the distinctiveness of each pattern of CLBP-$S_{4,1}^{riu2}$, which coincides with patterns of CLBP-$M_{4,1}^{riu2}$ for separating high-quality iris images from distorted versions. To that end, we generated an artificially distorted iris image dataset from 600 pristine high-quality references taken from the Warsaw-BioBase-Smartphone-Iris v1.0 [4], UTIRIS [58], and GC^2 multi-modal [36] datasets. A total of 3 to 12 samples per eye from 75 individuals were selected. This dataset was used only to justify our choice of specific sign–magnitude patterns and also to investigate how filtering out the low-quality iris images using the DSMI metric could affect the performance of the segmentation module of the reference iris recognition system. The reference iris images have no content-dependent deformations such as eyelid occlusion, and were selected from individuals with high, medium, and low degrees of iris pigmentation. The irises of all of these reference iris images were segmented accurately by the reference iris recognition system.

Five common image distortions with different levels and multiple distortions were used to distort the reference iris images. These distortions are Gaussian blur (GB), motion blur (MB), white Gaussian noise (WGN), salt and pepper noise (IN), and overexposure (OE). The parameters of each function and the number of the distorted versions of each reference image are listed in Table 1. In addition to the individual types of distortions, we generated multiple distorted iris images (GB+WGN). First, we distorted the images with GB and then with WGN. Since GB tends to occur during the acquisition phase due to the different working conditions of the image sensors, we applied it first. WGN is a noise model that can be used to mimic the effects of random processes, such as sensor noise due to poor illumination and thermal noise in the imaging device. For simplicity, the recommendation of [59] was followed, and WGN was introduced in the end.

Table 1. Summary of the artificially distorted iris image dataset.

Reference Iris Images		
Degree of Iris Pigmentation	**Number of Individuals**	**Number of All Iris Images**
High	25	200
Medium	25	200
Low	25	200

Distorted Iris Images				
Distortion Type	**MATLAB Function**	**Parameters Interval**	**Distorted Versions**	**All Distorted Iris Images**
GB	imgaussfilt(I, sigma)	0.5–5	10	6000
IN	imnoise(I,'salt &pepper',density)	0.05–0.6	12	7200
OE	I+t	10–100	10	6000
MB	H=fspecial('motion',len, theta); imfilter(I,H,'replicate')	10–60; 10–60	36	21,600
WGN	imnoise(I,'gaussian',0,V)	0.002–0.02	10	6000
GB+WGN	imgaussfilt(I, sigma); imnoise(I,'gaussian',0,V)	0.5–5; 0.002–0.02	100	60,000

To analyze the discrimination power of the scores $Q_{k,l}$ for separating the high-quality reference images from their distorted versions, we show the distributions of the corresponding scores $Q_{k,l}$ for some selected combinations of k and l in Figure 2. Visual inspection clearly shows that the coincidence of sign–magnitude patterns with $k = 0$ and $l = 0$ gives the greatest discrimination power. The predicted quality scores for the reference iris images are mostly between 0.8 and 1, and the scores for the distorted versions are mostly less than 0.8. Therefore, we chose this coincidence pattern to form our DSMI quality metric (DSMI = $Q_{0,0}$).

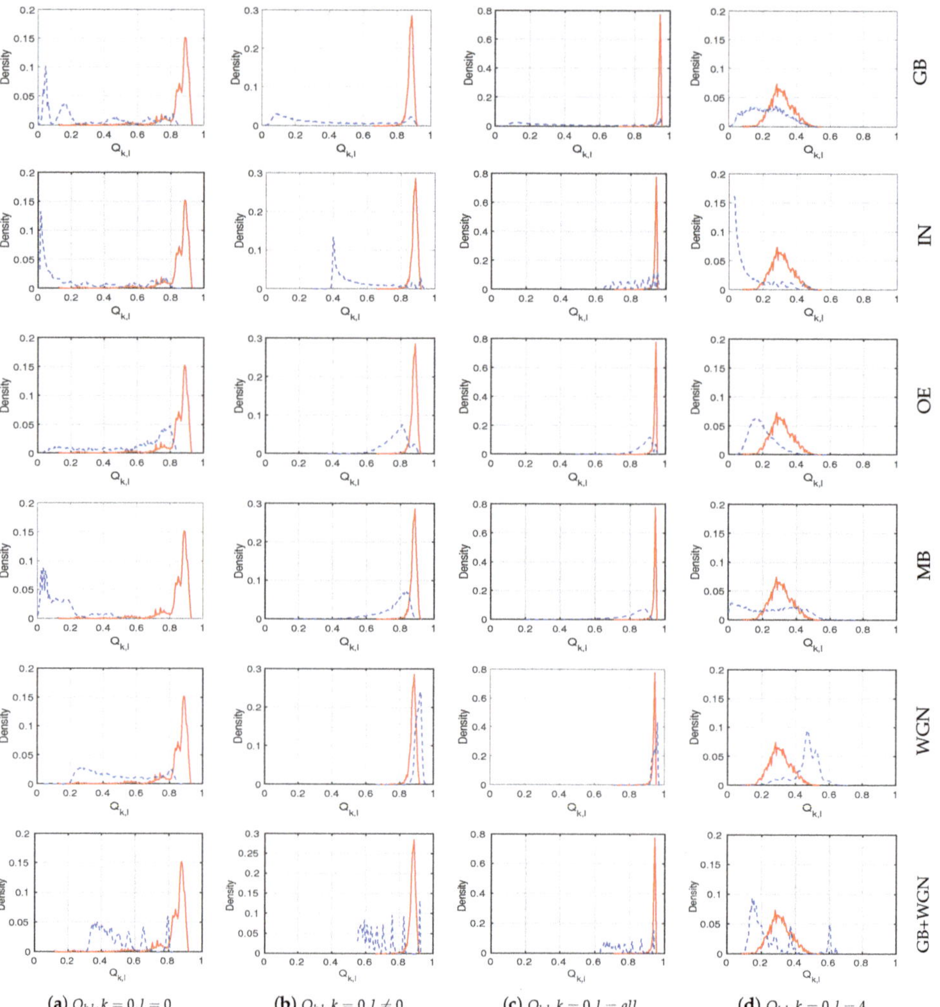

Figure 2. The solid red lines show the distributions of the quality scores of the high-quality iris images, and the dotted blue lines show the distributions for the distorted versions with different distortion types, which are shown on the right side of each row. The quality scores $Q_{k,l}$ are formed based on four different coincidences of sign (k) and magnitude (l) patterns, shown at the bottom of each column. The first column shows the histograms of the quality score $Q_{0,0}$, and the second, third, and fourth columns show the histograms of the coincidence patterns $Q_{0,l}$ with $l \neq 0$, $l = $ all, and $l = 4$.

4. Iris Segmentation Accuracy

The performance of iris segmentation in a classical iris recognition system has a significant impact on the overall performance. In this section, we analyze how image distortions affect the performance of the segmentation module and how quality filtering could improve the segmentation.

Most of the state-of-the-art iris recognition systems for iris imaging acquired in visible light, such as FIRE [38], Raja et al. [39], and OSIRIS, version 4.1 [41], can be used as reference iris recognition systems. We have chosen OSIRIS version 4.1 because (1) OSIRIS is an open-source iris recognition system that facilitates reproducible experiments, (2) it shows high recognition performance [41], and (3) it was used as the reference iris recognition system in some recent biometric recognition studies [4,60–64]. The segmentation module of OSIRIS version 4.1 uses the Viterbi algorithm to detect the iris and pupil contours [65]. The outputs are contours of the iris, which represent the inner boundary between the pupil and iris and the outer boundary between the iris and sclera, resulting in a binary mask for the iris.

For our experiments, we used the artificially distorted dataset from the previous section, which is summarized in Table 1. We segmented all iris images using the OSIRIS segmentation module. The mask of the segmented iris of each reference image was taken as the ground truth for comparison with the segmentation results for the distorted versions. The iris segmentation error is computed by the fraction of mislabeled pixels,

$$e = \frac{1}{|C|} \sum_{c \in C} T(c) \oplus M(c),$$

where $|C|$ is the cardinality of the pixel set C of an iris image, and T and M represent the ground truth and the generated iris masks, respectively. The symbol $\oplus$ represents the exclusive OR operation to identify the segmentation error. If the error e was below the threshold 0.05, the iris segmentation was assumed to be correct. The threshold value was set manually by the authors.

In Figure 3, we show the fractions of incorrectly segmented irises for the different types of distortion and for low, medium, and high degrees of iris pigmentation. The fractions are given as functions of the percentage of low-quality images that were filtered out using the proposed DSMI quality metric.

The results shown indicate a clear correlation between the DSMI quality of iris images and segmentation accuracy. Therefore, filtering out poor-quality images before segmentation will improve the performance by reducing the number of incorrectly segmented images, as indicated by the negative slopes of the plots.

In summary, the experiments performed in this section show that the accuracy of the segmentation module varies for iris images with different pigmentations and different distortions. Highly pigmented iris images present a greater challenge for the reference iris recognition system, while the system is more robust for the segmentation of low-pigmented iris images. However, filtering out poor-quality iris images using the proposed DSMI metric increases the accuracy of iris segmentation.

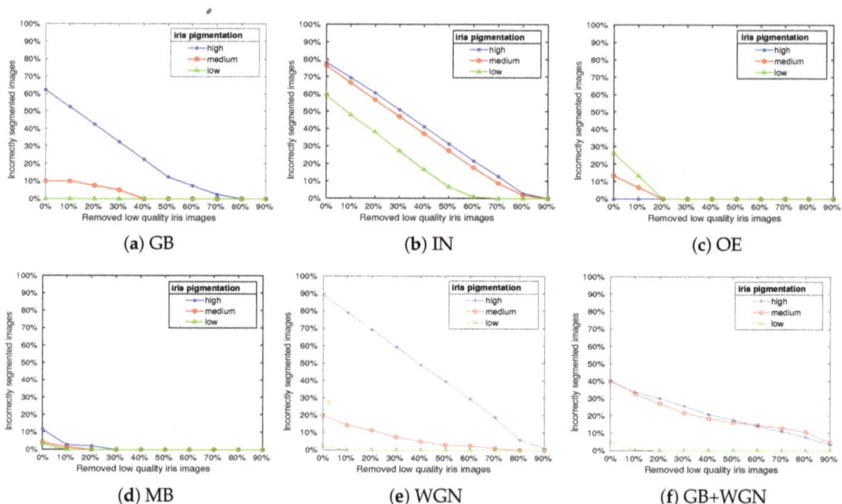

Figure 3. The segmentation performance of the reference iris recognition system is shown for segmenting iris images with high, medium, and low pigmentation, and distorted in different ways. The fraction of incorrectly segmented images is plotted versus the percentage of filtered low-quality images, based on the differential sign–magnitude statistics index (DSMI) metric.

5. Experimental Results

In this section, we investigate to what extent filtering out poor-quality iris images with the proposed quality metric improves the performance of the reference iris recognition system. We also compare our DSMI quality metric with the BRISQUE [66] and WAV1 [67] image quality metrics. BRISQUE uses statistical features extracted from pixel intensities to train a support vector machine for predicting image quality. Pertuz et al. [67] compared 15 metrics to estimate the blur of an image. In their study, WAV1 performed better than the others. WAV1 uses statistical properties of the discrete wavelet transform coefficients. Since blur is a common distortion of iris images taken by handheld imaging devices such as smartphones, we also compare our method with the WAV1 metric. Our experiments were conducted on three large authentic iris image datasets acquired in visible light.

5.1. Iris Image Datasets

There are many iris image datasets recorded with near-infrared cameras such as CASIA V4 [68], CASIA-Iris-Mobile-V1 [69], IIT Delhi [70], and ND CrossSensor Iris 2013 [71]. However, there are just a few iris image datasets acquired in visible light. Four are widely used in iris recognition research: UTIRIS [58], UBIRIS [72], MICHE [73], and VISOB [74].

An optometric framework in a controlled environment was used for capturing the irises of the UTIRIS dataset, resulting in high-quality iris images. UBIRIS iris images were taken from moving subjects and at different distances, resulting in more heterogeneous images compared to UTIRIS. Nevertheless, the pictures have good quality, better than the expected quality of iris images captured by handheld devices. The MICHE and VISOB datasets are challenging datasets for iris recognition systems, including images with varying degrees of iris pigmentation and eye make-up. In addition, the quality of the images is impaired by lack of focus, gaze deviations, specular reflections, eye occlusion, different lighting conditions, and motion blur.

Instead, we chose three datasets of the GC^2 multi-modal biometric dataset [36] because they contain authentically distorted iris images typically seen when capturing iris images with handheld devices such as smartphones. In addition, the iris images were taken from many subjects with different handheld cameras in uncontrolled environments at different distances. Iris pigmentation varied, from European subjects with bright iris textures to Asian subjects with very dark iris textures. In addition to the various authentic distortions corresponding to the image covariates, the iris images are subject to a variety of quality losses related to the subject's covariates, such as gaze deviation, off-angle, reflections, eye closure, and make-up. Also, the datasets contain 12–15 iris images of varying quality per eye and person, which is useful for studying the effect of quality filtering. The iris images have more than 30 different resolutions.

- The first dataset of GC^2, REFLEX, was taken with a Canon D700 camera using a Canon EF 100 mm f/2.8 L macro lens (18 megapixels). It contains 1422 irises of 48 subjects. A total of 12 to 15 samples were taken per eye (left and right).
- The second dataset, LFC, contains iris images taken by a light field camera. The LFC dataset contains 1454 iris images from the right and left eyes of 49 subjects. For each eye, 13 to 15 samples were taken.
- The third dataset, PHONE, was taken by a smartphone (Google Nexus 5, 8 megapixel camera). It contains 1379 iris images from the right and left eyes of 50 subjects, and 12 to 15 samples were taken per eye.

We compare an iris image with all iris images from the same dataset. Table 2 summarizes these datasets and shows the number of matching and non-matching iris pairs. Figure 4 shows some samples from these datasets, and Figure 5 shows the histograms of the quality scores of the datasets, estimated by the proposed DSMI metric.

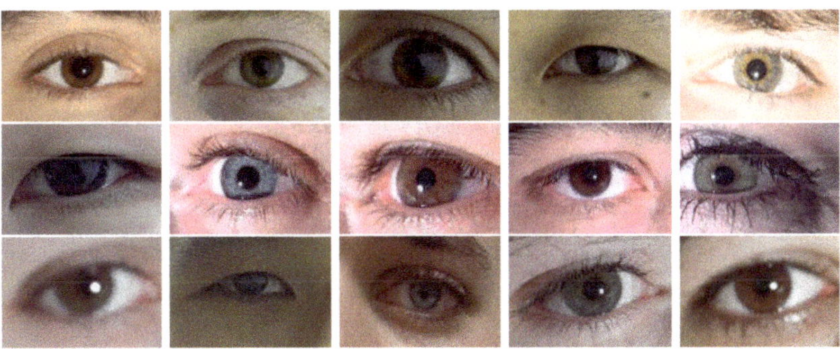

Figure 4. Some iris image samples with high, medium, and low pigmentation from the multi-modal biometric dataset GC^2 [36]. The first, second, and third rows show some images from the REFLEX, LFC, and PHONE datasets, respectively.

Table 2. Summary of the GC^2 dataset.

Datasets	REFLEX	LFC	PHONE
Number of subjects	48	49	50
Total images	1422	1454	1379
Samples per eye	12–15	13–15	12–15
Matching pairs	9457	10,045	9092
Non-matching pairs	975,450	1,056,485	941,039
Camera	Canon D700	Light field camera	Phone Nexus
Lowest resolution	1085×724	327×218	450×300
Highest resolution	2813×1876	1080×1080	1811×1208

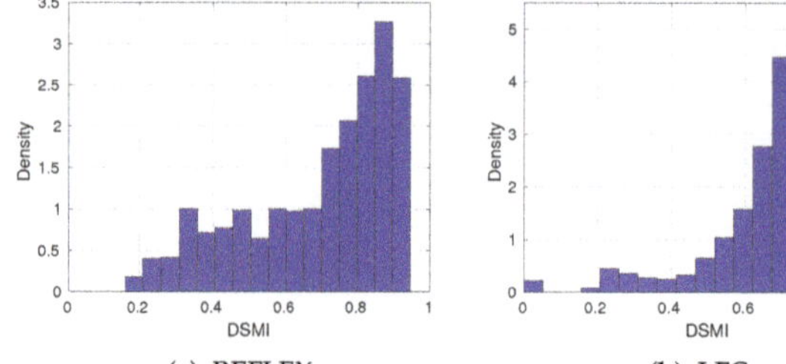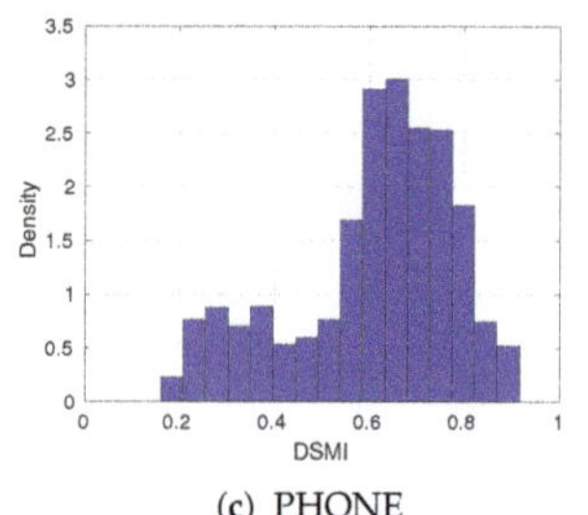

(a) REFLEX   (b) LFC   (c) PHONE

Figure 5. Normalized histograms of the quality scores according to the DSMI metric on three test iris datasets.

5.2. Iris Recognition Performance Analysis

To evaluate the performance improvement of iris recognition achieved by quality filtering using an image quality metric, we used three performance methods, namely the Daugman's decidability index [75], the area under the receiver operating characteristic curves (AUC), and the equal error rates (EER). We compared the performance of three image quality metrics when used for quality filtering. Given a threshold for a metric, we rejected those images that exhibited a quality lower than the threshold. The thresholds for each of the three metrics were chosen such that $1/4$, $1/2$, and $3/4$ of the images were rejected. In our experiments, OSIRIS version 4.1 was used as a reference iris recognition system.

5.2.1. Daugman's Decidability Index

Daugman's decidability index [75] is a widely used method for assessing the performance of iris recognition systems [3,36,75]. In an iris recognition system like OSIRIS, a binary phase code is derived for each presented iris image. Then, the fractional Hamming distance to the phase code of a reference iris image is computed. The distributions of these Hamming distances are compared between a set of matching and a set of non-matching iris image pairs from a test dataset. The larger the overlap between the distributions, the more likely recognition errors become. The Daugman index (d') measures the separation of these distributions by

$$d' = \frac{|\mu_E - \mu_I|}{\sqrt{\frac{1}{2}(\sigma_E^2 + \sigma_I^2)}},$$

where μ_E and μ_I are the means and σ_E and σ_I are the standard deviations of the distributions. Larger values correspond to better discrimination. We follow this procedure using the GC^2 multi-modal biometric dataset and plot the histograms of the Hamming distances for the matching and the non-matching iris pairs in Figure 6. For visualization, normal distributions were fitted to the histograms.

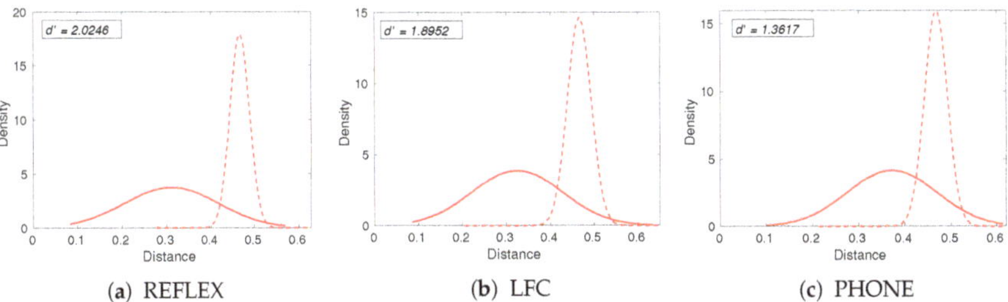

(a) REFLEX   (b) LFC   (c) PHONE

Figure 6. Normal distributions fitted to the normalized histograms of Hamming distances of matching (solid lines) and non-matching (dash lines) iris pairs are shown for three test image datasets.

We can now study the effect of quality filtering on the performance of the iris recognition system. In Figure 7, we show Daugman's decidability index as a function of the fraction of removed poor-quality images. DSMI, BRISQUE, and WAV1 image quality metrics were used for quality filtering. Filtering out low-quality iris images using the DSMI metric leads to the largest performance improvement in the REFLEX dataset, while quality filtering in the PHONE dataset leads only to small improvements. This could be due to the DSMI metric performing better in quality assessment on iris images in the REFLEX dataset or to the PHONE dataset posing a greater challenge to the reference iris recognition system. The Daugman index for the PHONE dataset is only 1.36, compared to 2.02 and 1.90 for REFLEX and LFC, respectively (see Figure 6).

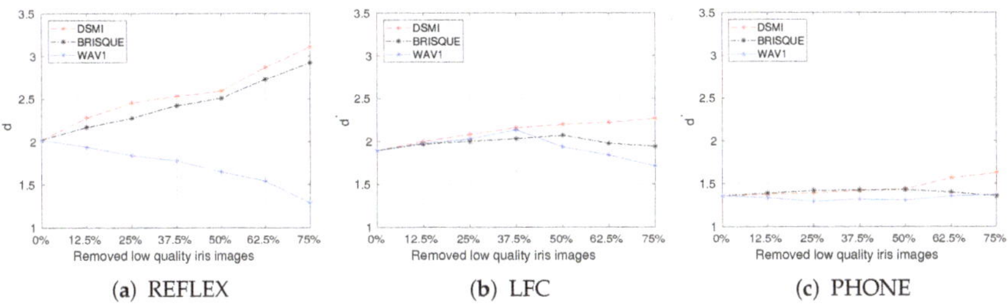

(a) REFLEX   (b) LFC   (c) PHONE

Figure 7. Daugman's decidability index for all iris images, after filtering different parts of the iris images with the poorest quality using three image quality metrics on three test datasets.

From the Daugman's decidability index values in the three test datasets, as shown in Figure 7, we can conclude that filtering out the iris images with the poorest quality using the proposed DSMI metric improves the recognition accuracy of the reference iris recognition system. The BRISQUE metric also performs well in the REFLEX dataset, but it is not consistent for quality filtering in the LFC and PHONE datasets. WAV1 is not consistent with quality filtering on all three test datasets.

5.2.2. Receiver Operating Characteristic Curve

The area under the curve (AUC) of the receiver operating characteristic (ROC) is a widely used performance metric for comparing the accuracy of iris recognition systems. The iris recognition system with the larger AUC is considered to be a more accurate system.

To visualize and measure the improvements of the performance of the reference iris recognition system by filtering out the poor quality iris images, the ROC curves were generated for each dataset by plotting the true positive rate against the false positive rate at various fractional Hamming distances (see Figure 8).

Figure 8 shows the ROC curves for the three test datasets with different quality filtering thresholds using our DSMI metric, BRISQUE, and WAV1 metrics. The solid red lines in Figure 8 show the performance of the reference iris recognition system without quality filtering. Without quality filtering, the corresponding AUC value for the REFLEX dataset is 0.9065, for the LFC dataset it is 0.8861, and for the PHONE dataset it is 0.8226. The AUC values show again that the PHONE dataset is the most challenging one for the reference iris recognition system.

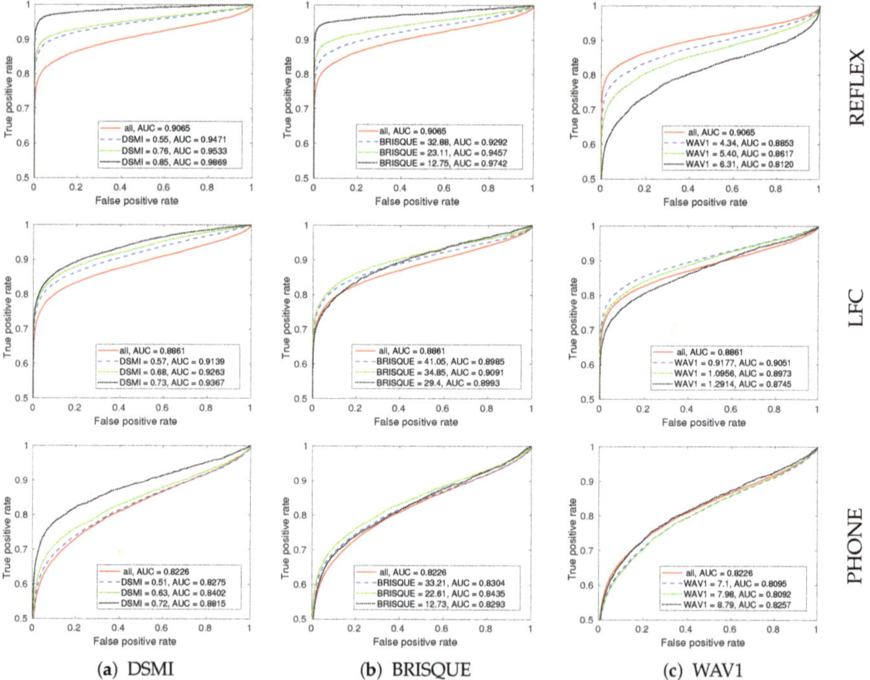

Figure 8. The receiver operating characteristic (ROC) curves for the three test datasets (REFLEX, LFC, and PHONE) with different quality filtering thresholds using our DSMI metric, BRISQUE, and WAV1. The solid red, dashed blue, dot-dashed green, and dotted black lines were plotted without quality filtering, after filtering out one-quarter, half, and three-quarters of the poorest-quality images, respectively.

We also computed the AUC values after removing 1/4, 1/2, and 3/4 of the iris images with the poorest quality from each test dataset. The AUC values are listed in the figure legends for all of the test datasets. Using the proposed DSMI metric for quality filtering increased the AUC value in all test datasets.

In the REFLEX dataset, filtering out a quarter of the iris images with the poorest quality using the DSMI metric greatly improves the performance of the reference iris recognition system in terms of AUC by 0.0406 (4.5%). However, filtering out the second quarter only increases AUC by 0.0062 (0.65%). This indicates that the middle two quarters of the iris images have a small quality deviation, and filtering a part of these images does not result in a considerable improvement in the performance of the iris recognition system. However, filtering the third quarter of the iris images with the poorest quality improves the AUC significantly by 0.0336 (3.5%).

The performance improvements for the LFC dataset after filtering out the first, second, and third quarters of the iris images with the poorest quality using the DSMI metric are 0.0278 (3.1%), 0.0124 (1.4%), and 0.0104 (1.1%), respectively. The values for performance improvement on the PHONE dataset are 0.0049 (0.6%), 0.0127 (1.5%), and 0.0413 (4.9%). Filtering out the first quarter of the iris images with the poorest quality using the DSMI metric only slightly improves the AUC value, but filtering out three quarters of the iris images with the poorest quality improves the performance significantly by 7.2%. We visualized these performance improvements in Figure 9.

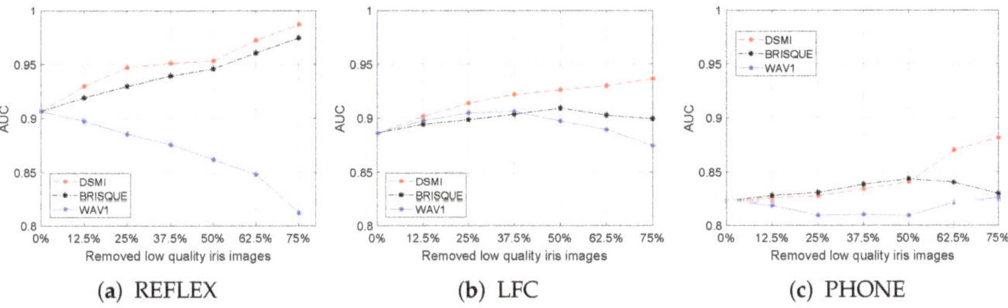

| (a) REFLEX | (b) LFC | (c) PHONE |

Figure 9. Area under the curve (AUC) values for all iris images after removing different parts of the iris images with the poorest quality.

The analysis of the AUC values shows that the performance of the reference iris recognition system has improved by quality filtering in all test datasets when using the DSMI metric for quality assessment. In contrast, BRISQUE is consistent for quality filtering for the REFLEX dataset, but not for the other two test datasets. WAV1 shows inconsistent performance in all test datasets.

The reason for this could be that the DSMI metric is optimized for assessing the image quality of iris images and BRISQUE for the perceptual quality of natural images. Both, however, can assess image quality for different image distortions. The WAV1 metric is optimized for blur assessment. Since blur is common in iris images taken with handheld devices, we compare our method with the WAV1 metric. However, the iris images in test datasets have more complicated authentic in-the-wild image distortions, and these distortions degrade the performance of WAV1 in all test datasets.

5.2.3. Equal Error Rate

The equal error rate (EER) is the rate at which both accept and reject errors are equal. The EER is used for comparing the accuracy of classification systems with different receiver operating characteristic (ROC) curves. With the EER approach, the system with the lowest EER is considered the most accurate.

In Table 3, we calculated the EER values when three image quality metrics were used to filter out the poor-quality iris images from the test datasets. The greatest performance improvement is achieved by filtering out poor-quality iris images using the DSMI metric on the REFLEX dataset. The PHONE dataset is the more challenging dataset for the reference iris recognition system, resulting in higher EER values.

The results confirm that rejecting poor-quality images using the proposed DSMI metric improves the iris recognition performance consistently, while this observation does not hold for BRISQUE and WAV1 metrics.

Table 3. The equal error rate (EER) values are calculated after filtering different parts of the iris images with the poorest quality from each test dataset. This table shows the EER values when all iris images are passed to the iris recognition system and after filtering out one quarter, half, and three quarters of the iris images with the poorest quality from the REFLEX, LFC, and PHONE datasets using the DSMI, BRISQUE, and WAV1 quality metrics.

Removed Part	REFLEX			LFC			PHONE		
	DSMI	BRISQUE	WAV1	DSMI	BRISQUE	WAV1	DSMI	BRISQUE	WAV1
0%	0.1469	0.1469	0.1469	0.1770	0.1770	0.1770	0.2466	0.2466	0.2466
25%	0.0987	0.1202	0.1714	0.1500	0.1604	0.1562	0.2418	0.2374	0.2594
50%	0.0878	0.978	0.1963	0.1376	0.1528	0.1692	0.2293	0.2276	0.2595
75%	0.0382	0.0520	0.2443	0.1287	0.1724	0.1955	0.1845	0.2412	0.2434

In summary, for all of the test iris image datasets (REFLEX, LFC, PHONE) and all of the performance evaluation methods (Daugman's decidability index, AUC, EER), the performance of the reference iris recognition system (OSIRIS, Version 4.1) increased consistently by filtering out iris images with the poor quality using the proposed DSMI quality metric. In contrast, for the other two image quality metrics (BRISQUE, WAV1), the experiments showed inconsistencies, i.e., removing more low-quality images did not always increase the performance of the reference iris recognition system.

Figure 10 shows some iris samples from the test datasets with poor quality scores predicted by the proposed DSMI metric. These samples will be filtered out when we remove a quarter of the iris images with the poorest quality from each test dataset. If we pass these samples to the reference iris detection system for iris recognition, all of them will be falsely rejected. Thus, the proposed DSMI metric can be used to decide whether an input iris sample should be enrolled in a dataset or rejected, and a new sample should be captured based on the quality score. Although our method is designed to consider only image covariates, some subject covariates, such as eyelid occlusion due to blinking, may also result in motion blur or other image quality distortions that can be measured by our proposed quality metric, as shown in Figure 10c. All iris samples shown in Figure 10 suffer from authentic image distortion and other quality degradation due to subject covariates.

Figure 11 shows some iris samples with DSMI scores that are higher than the threshold for filtering out one quarter of the iris samples with the poorest quality from each test dataset. Our proposed framework passes these images for iris segmentation and identification when only a quarter of the iris images with the poorest quality are filtered out from the test datasets. However, all of these samples will be falsely rejected by the reference iris recognition system. Some of these images have quality degradation related to subject covariates, such as eyelashes obscuring the iris or closed eyes.

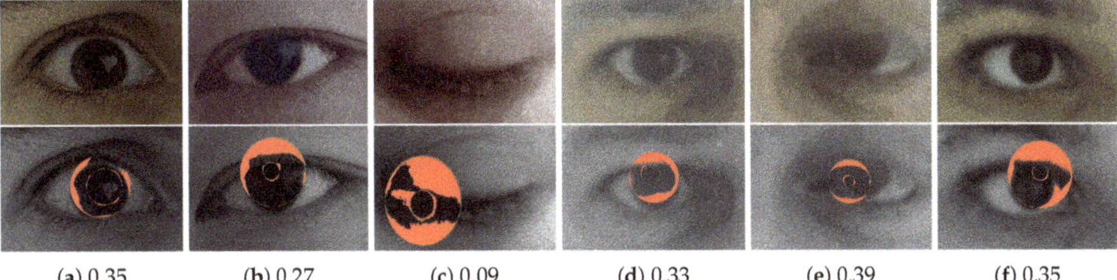

(**a**) 0.35   (**b**) 0.27   (**c**) 0.09   (**d**) 0.33   (**e**) 0.39   (**f**) 0.35

Figure 10. The first row shows some iris samples from the multi-modal biometric dataset GC^2 [36], which are classified as low-quality samples by our DSMI metric. All of these samples would be falsely rejected with high dissimilarity scores (>0.47) by the reference iris detection system. However, if we filter out a quarter of the iris images with the poorest quality from each test dataset, these samples will be removed and not passed to the iris recognition system. The second row shows the segmentation result of the segmentation module of the reference iris recognition system. The DSMI scores are listed below the iris samples.

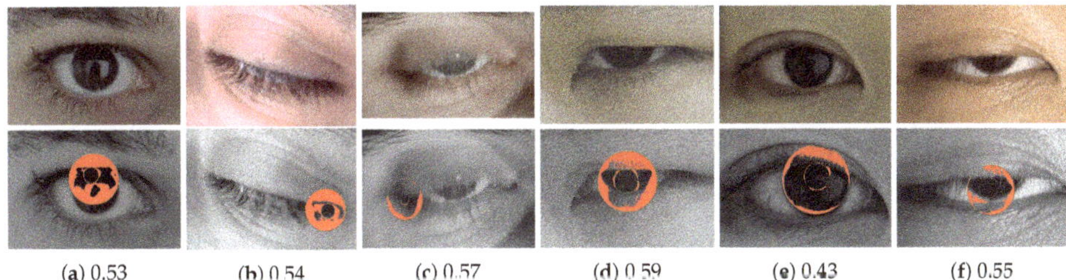

(**a**) 0.53   (**b**) 0.54   (**c**) 0.57   (**d**) 0.59   (**e**) 0.43   (**f**) 0.55

Figure 11. The first row shows some iris samples from the multi-modal biometric dataset GC^2 [36], which are classified by our DSMI metric as iris samples of sufficient quality if only one quarter of the iris images with the poorest quality are filtered out. Therefore, these images are passed to the iris recognition pipeline for further processing. However, all of these samples would be falsely rejected by the reference iris recognition system with high dissimilarity values (>0.47). The second row shows the segmentation result of the segmentation module of the reference iris recognition system. The DSMI scores are listed below the iris samples.

The iris samples that are shown in Figure 11 have fewer image distortions compared to the sample shown in Figure 10. Therefore, our quality metric predicts higher quality scores for these iris images. Some of these images have quality degradations related to subject covariates, such as eyelashes obscuring the iris or closed eyes. If we filter out half of the iris samples with the poorest quality, these samples will be filtered. However, by setting a higher quality filtering threshold, some iris samples may be rejected unnecessarily.

5.3. Computational Complexity

It is straightforward to assess the computational complexity of the DSMI quality metric by checking the algorithmic steps, outlined in Section 3.1, one by one. The result is a time complexity, linear in the size of the input image. More precisely, it is $O(N \times M \times P)$, where $N \times M$ is the image size in pixels, and P is the number of points checked in the neighborhood of each pixel for deriving the sign and magnitude patterns.

We also recorded the actual speed of the quality metric using our implementation, running on an MSI GP60 laptop with an Intel Core i7 processor and 16GB RAM with MATLAB version 2018b in Ubuntu 18.04.3 LTS. We computed the DSMI quality scores on four parts of the test datasets, each containing iris images of the same size in pixels, ranging from 596×397 up to 2036×1358 (see Table 4). The table confirms the linear time complexity, amounting to roughly 0.06×10^{-6} seconds per pixel. At that processing speed, a throughput of 66 frames per second (FPS) can be achieved at resolution 596×397. For the higher resolutions, 625×537, 1233×810, and 2036×1358, the speed is 40, 16, and 6 FPS, respectively. Therefore, the proposed method can be used to assess the quality of iris images in interactive applications, such as iris recognition systems based on handheld imaging devices.

Table 4. Comparison of the average running time (seconds) on four sets of iris images with different resolutions.

Image Resolutions	596×397	625×537	1233×810	2036×1358
Average running time per image (seconds)	0.015	0.026	0.061	0.181
Average running time per pixel (microseconds)	0.065	0.062	0.062	0.064
Frames per second (FPS)	66	40	16	6

6. Conclusions and Future Work

In this paper, we presented a fast image quality metric, based on statistical features of the sign–magnitude transform to estimate the quality of iris images acquired by handheld devices in visible light. We suggest that this method can be used to decide whether an input iris sample should be enrolled in a dataset or rejected, and a new sample should be captured based on the quality score to improve the speed and the recognition rate of the reference iris recognition system.

We conducted extensive experiments to demonstrate these improvements using three performance methods for measuring the iris recognition accuracy on three large datasets acquired in unconstrained environments in visible light. The experiments showed that the proposed approach improved the accuracy of the reference iris recognition system.

However, we would like to point out that the inclusion of quality filtering in an iris recognition system can increase the computational costs of iris image recognition, and some iris images may be rejected unnecessarily. This could be caused by an error in the quality metric, by too conservative of a setting of the quality threshold, or by quality factors related to the subject covariates. In our future work, we will propose a metric for iris image quality assessment that takes into account all of these factors. Furthermore, another future work is to develop an algorithm to monitor criteria, such as recognition performance, time and number of photos required per person, and customer satisfaction, in order to dynamically adapt the threshold for quality filtering to achieve optimal performance.

It may also be promising to examine the use of the proposed quality metric to assess the quality of other biometric images, such as facial image, and NIR biometric images.

Author Contributions: Conceptualization, M.J. and M.P.; Investigation, M.J.; methodology M.J., M.P., D.S.; validation M.J. and D.S.; Writing—original draft preparation, M.J.; Writing—review and editing, M.J., M.P., D.S.; Visualization, M.J., M.P., D.S.; Supervision, M.P. and D.S. All authors have read and agreed to the published version of the manuscript.

Funding: This research was partially funded by the Exzellenzstrategie des Bundes und der Länder (the Excellence Strategy of the German Federal and State Governments), the Deutsche Forschungsgemeinschaft (DFG, German Research Foundation)—Project-ID 251654672—TRR 161 and the Research Council of Norway within project no. 221073 HyPerCept–Color and quality in higher dimensions.

Acknowledgments: The authors thank Jon Yngve Hardeberg, Katrin Franke, and Sokratis Katsikas for their helpful discussions.

Conflicts of Interest: The authors declare no conflict of interest.

References

1. Flom, L.; Safir, A. Iris Recognition System. U.S. Patent 4,641,349, 3 February 1987.
2. Daugman, J. New methods in iris recognition. *IEEE Trans. Syst. Man Cybern. Part B Cybern.* **2007**, *37*, 1167–1175. [CrossRef] [PubMed]
3. Proença, H. Quality assessment of degraded iris images acquired in the visible wavelength. *IEEE Trans. Inf. Forensics Secur.* **2011**, *6*, 82–95. [CrossRef]
4. Trokielewicz, M. Iris recognition with a database of iris images obtained in visible light using smartphone camera. In Proceedings of the 2016 IEEE International Conference on Identity, Security and Behavior Analysis (ISBA), Sendai, Japan, 29 February–2 March 2016; pp. 1–6.
5. Raja, K.B.; Raghavendra, R.; Vemuri, V.K.; Busch, C. Smartphone based visible iris recognition using deep sparse filtering. *Pattern Recognit. Lett.* **2015**, *57*, 33–42. [CrossRef]
6. Thavalengal, S.; Bigioi, P.; Corcoran, P. Iris authentication in handheld devices-considerations for constraint-free acquisition. *IEEE Trans. Consum. Electron.* **2015**, *61*, 245–253. [CrossRef]
7. Thavalengal, S.; Bigioi, P.; Corcoran, P. Evaluation of combined visible/NIR camera for iris authentication on smartphones. In Proceedings of the IEEE Conference on Computer Vision and Pattern Recognition Workshops, Boston, MA, USA, 7–12 June 2015; pp. 42–49. doi:10.1109/CVPRW.2015.7301318. [CrossRef]
8. Bharadwaj, S.; Vatsa, M.; Singh, R. Biometric quality: A review of fingerprint, iris, and face. *EURASIP J. Image Video Process.* **2014**, *2014*, 34. [CrossRef]
9. Phillips, P.J.; Beveridge, J.R. An introduction to biometric-completeness: The equivalence of matching and quality. In Proceedings of the 2009 IEEE 3rd International Conference on Biometrics: Theory, Applications, and Systems, Washington, DC, USA, 28–30 September 2009; pp. 1–5.
10. Daugman, J.; Downing, C. Iris Image Quality Metrics with Veto Power and Nonlinear Importance Tailoring. Available online: https://pdfs.semanticscholar.org/60a3/a6f3e3e047fa1602b735f0682d2a01c84953.pdf (accessed on 12 January 2017).
11. Beveridge, J.R.; Givens, G.H.; Phillips, P.J.; Draper, B.A. Factors that influence algorithm performance in the face recognition grand challenge. *Comput. Vis. Image Underst.* **2009**, *113*, 750–762. [CrossRef]
12. Belcher, C.; Du, Y. A selective feature information approach for iris image-quality measure. *IEEE Trans. Inf. Forensics Secur.* **2008**, *3*, 572–577. [CrossRef]
13. Pillai, J.K.; Patel, V.M.; Chellappa, R.; Ratha, N.K. Secure and robust iris recognition using random projections and sparse representations. *IEEE Trans. Pattern Anal. Mach. Intell.* **2011**, *33*, 1877–1893. [CrossRef]
14. Zhou, Z.; Du, E.Y.; Belcher, C.; Thomas, N.L.; Delp, E.J. Quality fusion based multimodal eye recognition. In Proceedings of the IEEE International Conference on Systems, Man, and Cybernetics, Seoul, Korea, 14–17 October 2012; pp. 1297–1302.
15. Shi, C.; Jin, L. A fast and efficient multiple step algorithm of iris image quality assessment. In Proceedings of the Second International Conference on Future Computer and Communication, Wuhan, China, 21–24 May 2010; pp. 589–593.
16. Dong, W.; Sun, Z.; Tan, T.; Wei, Z. Quality-based dynamic threshold for iris matching. In Proceedings of the 16th IEEE International Conference on Image Processing (ICIP), Cairo, Egypt, 7–10 November 2009; pp. 1949–1952.
17. Makinana, S.; Van Der Merwe, J.J.; Malumedzha, T. A fourier transform quality measure for iris images. In Proceedings of the International Symposium on Biometrics and Security Technologies, Kuala Lumpur, Malaysia, 26–27 August 2014; pp. 51–56.
18. Jenadeleh, M.; Pedersen, M.; Saupe, D. Realtime quality assessment of iris biometrics under visible light. In Proceedings of the IEEE Conference on Computer Vision and Pattern Recognition Workshops, Salt Lake City, UT, USA, 18–22 June 2018; pp. 556–565.
19. Jenadeleh, M. Blind Image and Video Quality Assessment. Ph.D. Thesis, Universität Konstanz, Konstanz, Germany, October 2018.

20. Chen, L.; Han, M.; Wan, H. The fast iris image clarity evaluation based on Brenner. In Proceedings of the 2nd International Symposium on Instrumentation and Measurement, Sensor Network and Automation (IMSNA), Toronto, ON, Canada, 23–24 December 2013; pp. 300–302.

21. Starovoitov, V.; Golińska, A.K.; Predko-Maliszewska, A.; Goliński, M. No-Reference Image Quality Assessment for Iris Biometrics. In *Image Processing and Communications Challenges 4*; Springer: Berlin, Germany, 2013; pp. 95–100.

22. Bergmüller, T.; Christopoulos, E.; Fehrenbach, K.; Schnöll, M.; Uhl, A. Recompression effects in iris recognition. *Image Vis. Comput.* **2017**, *58*, 142–157. [CrossRef]

23. Mottalli, M.; Mejail, M.; Jacobo-Berlles, J. Flexible image segmentation and quality assessment for real-time iris recognition. In Proceedings of the 16th IEEE International Conference on Image Processing (ICIP), Cairo, Egypt, 7–10 November 2009; pp. 1941–1944.

24. Happold, M. Learning to predict match scores for iris image quality assessment. In Proceedings of the IEEE International Joint Conference on Biometrics (IJCB), Clearwater, FL, USA, 29 September–2 October 2014; pp. 1–8.

25. Kalka, N.D.; Zuo, J.; Schmid, N.A.; Cukic, B. Estimating and fusing quality factors for iris biometric images. *IEEE Trans. Syst. Man Cybern. Part A Syst. Hum.* **2010**, *40*, 509–524. [CrossRef]

26. Li, X.; Sun, Z.; Tan, T. Comprehensive assessment of iris image quality. In Proceedings of the 18th IEEE International Conference on Image Processing (ICIP), Brussels, Belgium, 11–14 September 2011; pp. 3117–3120.

27. Li, X.; Sun, Z.; Tan, T. Predict and improve iris recognition performance based on pairwise image quality assessment. In Proceedings of the International Conference on Biometrics (ICB), Madrid, Spain, 4–7 June 2013; pp. 1–6.

28. Othman, N.; Dorizzi, B. Impact of quality-based fusion techniques for video-based iris recognition at a distance. *IEEE Trans. Inf. Forensics Secur.* **2015**, *10*, 1590–1602. [CrossRef]

29. Wu, Q.; Wang, Z.; Li, H. A highly efficient method for blind image quality assessment. In Proceedings of the IEEE International Conference on Image Processing (ICIP), Quebec City, QC, Canada, 27–30 September 2015; pp. 339–343.

30. Ma, K.; Liu, W.; Liu, T.; Wang, Z.; Tao, D. dipIQ: Blind image quality assessment by learning-to-rank discriminable image pairs. *IEEE Trans. Image Process.* **2017**, *26*, 3951–3964. [CrossRef] [PubMed]

31. Jenadeleh, M.; Moghaddam, M.E. BIQWS: efficient Wakeby modeling of natural scene statistics for blind image quality assessment. *Multimed. Tools Appl.* **2017**, *76*, 13859–13880. [CrossRef]

32. Freitas, P.G.; da Eira, L.P.; Santos, S.S.; Farias, M.C. Image quality assessment using BSIF, CLBP, LCP, and LPQ operators. *Theor. Comput. Sci.* **2020**, *805*, 37–61. [CrossRef]

33. Wu, Q.; Li, H.; Wang, Z.; Meng, F.; Luo, B.; Li, W.; Ngan, K.N. Blind image quality assessment based on rank-order regularized regression. *IEEE Trans. Multimed.* **2017**, *19*, 2490–2504. [CrossRef]

34. Liu, L.; Liu, B.; Huang, H.; Bovik, A.C. No-reference image quality assessment based on spatial and spectral entropies. *Signal Process. Image Commun.* **2014**, *29*, 856–863. [CrossRef]

35. Gu, J.; Meng, G.; Redi, J.A.; Xiang, S.; Pan, C. Blind image quality assessment via vector regression and object oriented pooling. *IEEE Trans. Multimed.* **2018**, *20*, 1140–1153. [CrossRef]

36. Liu, X.; Pedersen, M.; Charrier, C.; Bours, P. Can no-reference image quality metrics assess visible wavelength iris sample quality? In Proceedings of the IEEE International Conference on Image Processing, Beijing, China, 17–20 September 2017; pp. 3530–3534.

37. Xinwei, L.; Christophe, C.; Marius, P.; Patrick, B. Performance Evaluation of no-reference image quality metrics for visible wavelength iris biometric images. In Proceedings of the 26th European Signal Processing Conference (EUSIPCO 2018), Rome, Italy, 3–7 September 2018.

38. Galdi, C.; Dugelay, J.L. FIRE: fast iris recognition on mobile phones by combining colour and texture features. *Pattern Recognit. Lett.* **2017**, *91*, 44–51. [CrossRef]

39. Raja, K.B.; Raghavendra, R.; Venkatesh, S.; Busch, C. Multi-patch deep sparse histograms for iris recognition in visible spectrum using collaborative subspace for robust verification. *Pattern Recognit. Lett.* **2017**, *91*, 27–36. [CrossRef]

40. Minaee, S.; Abdolrashidi, A.; Wang, Y. Iris recognition using scattering transform and textural features. In Proceedings of the 2015 IEEE Signal Processing and Signal processing Education Workshop (SP/SPE), Salt Lake City, UT, USA, 9–12 August 2015; pp. 37–42.

41. Othman, N.; Dorizzi, B.; Garcia-Salicetti, S. OSIRIS: An open source iris recognition software. *Pattern Recognit. Lett.* **2016**, *82*, 124–131. [CrossRef]

42. Daugman, J. How iris recognition works. *IEEE Trans. Circuits Syst. Video Technol.* **2004**, *14*, 21–30. [CrossRef]

43. Miyazawa, K.; Ito, K.; Aoki, T.; Kobayashi, K.; Nakajima, H. An effective approach for iris recognition using phase-based image matching. *IEEE Trans. Pattern Anal. Mach. Intell.* **2008**, *30*, 1741–1756. [CrossRef] [PubMed]

44. Nguyen, V.D.; Nguyen, D.D.; Nguyen, T.T.; Dinh, V.Q.; Jeon, J.W. Support local pattern and its application to disparity improvement and texture classification. *IEEE Trans. Circuits Syst. Video Technol.* **2014**, *24*, 263–276. [CrossRef]

45. Liu, L.; Lao, S.; Fieguth, P.W.; Guo, Y.; Wang, X.; Pietikäinen, M. Median robust extended local binary pattern for texture classification. *IEEE Trans. Image Process.* **2016**, *25*, 1368–1381. [CrossRef] [PubMed]

46. Guo, Z.; Zhang, L.; Zhang, D. A completed modeling of local binary pattern operator for texture classification. *IEEE Trans. Image Process.* **2010**, *19*, 1657–1663.

47. Dubey, S.R.; Singh, S.K.; Singh, R.K. Multichannel decoded local binary patterns for content-based image retrieval. *IEEE Trans. Image Process.* **2016**, *25*, 4018–4032. [CrossRef]

48. Murala, S.; Wu, Q.J. Local mesh patterns versus local binary patterns: biomedical image indexing and retrieval. *IEEE J. Biomed. Health Inf.* **2014**, *18*, 929–938. [CrossRef]

49. Satpathy, A.; Jiang, X.; Eng, H.L. LBP-based edge-texture features for object recognition. *IEEE Trans. Image Process.* **2014**, *23*, 1953–1964. [CrossRef]

50. Shang, J.; Chen, C.; Pei, X.; Liang, H.; Tang, H.; Sarem, M. A novel local derivative quantized binary pattern for object recognition. *Visual Comput.* **2017**, *33*, 221–233. [CrossRef]

51. Yu, M.; Liu, L.; Shao, L. Structure-preserving binary representations for RGB-D action recognition. *IEEE Trans. Pattern Anal. Mach. Intell.* **2016**, *38*, 1651–1664. [CrossRef]

52. Chen, C.; Liu, M.; Liu, H.; Zhang, B.; Han, J.; Kehtarnavaz, N. Multi-Temporal Depth Motion Maps-Based Local Binary Patterns for 3-D Human Action Recognition. *IEEE Access* **2017**, *5*, 22590–22604. [CrossRef]

53. Kang, W.; Wu, Q. Contactless palm vein recognition using a mutual foreground-based local binary pattern. *IEEE Trans. Inf. Forensics Secur.* **2014**, *9*, 1974–1985. [CrossRef]

54. Popplewell, K.; Roy, K.; Ahmad, F.; Shelton, J. Multispectral iris recognition utilizing hough transform and modified LBP. In Proceedings of the 2014 IEEE International Conference on Systems, Man, and Cybernetics (SMC), San Diego, CA, USA, 5–8 October 2014; pp. 1396–1399.

55. Hezil, N.; Boukrouche, A. Multimodal biometric recognition using human ear and palmprint. *IET Biom.* **2017**, *6*, 351–359. [CrossRef]

56. Piciucco, E.; Maiorana, E.; Campisi, P. Palm vein recognition using a high dynamic range approach. *IET Biom.* **2018**, *7*, 439–446. [CrossRef]

57. Ojala, T.; Pietikainen, M.; Maenpaa, T. Multiresolution gray-scale and rotation invariant texture classification with local binary patterns. *IEEE Trans. Pattern Anal. Mach. Intell.* **2002**, *24*, 971–987. [CrossRef]

58. Hosseini, M.S.; Araabi, B.N.; Soltanian-Zadeh, H. Pigment melanin: Pattern for iris recognition. *IEEE Trans. Instrum. Meas.* **2010**, *59*, 792–804. [CrossRef]

59. Jayaraman, D.; Mittal, A.; Moorthy, A.K.; Bovik, A.C. Objective quality assessment of multiply distorted images. In Proceedings of the 2012 Conference Record of the Forty Sixth Asilomar Conference on Signals, Systems and Computers (ASILOMAR), Pacific Grove, CA, USA, 4–7 November 2012; pp. 1693–1697.

60. Czajka, A.; Bowyer, K.W.; Krumdick, M.; VidalMata, R.G. Recognition of image-orientation-based iris spoofing. *IEEE Trans. Inf. Forensics Secur.* **2017**, *12*, 2184–2196. [CrossRef]

61. Raghavendra, R.; Raja, K.B.; Busch, C. Exploring the usefulness of light field cameras for biometrics: An empirical study on face and iris recognition. *IEEE Trans. Inf. Forensics Secur.* **2016**, *11*, 922–936. [CrossRef]

62. Talreja, V.; Ferrett, T.; Valenti, M.C.; Ross, A. Biometrics-as-a-service: A framework to promote innovative biometric recognition in the cloud. In Proceedings of the 2018 IEEE International Conference on Consumer Electronics (ICCE), Las Vegas, NV, USA, 12–14 January 2018; pp. 1–6.
63. Zhao, D.; Fang, S.; Xiang, J.; Tian, J.; Xiong, S. Iris template protection based on local ranking. *Secur. Commun. Netw.* **2018**, *2018*, 1–9. [CrossRef]
64. Thavalengal, S. Contributions to Practical Iris Biometrics on Smartphones. Ph.D. Thesis, National University of Ireland, Galway, Ireland, May 2016.
65. Sutra, G.; Garcia-Salicetti, S.; Dorizzi, B. The Viterbi algorithm at different resolutions for enhanced iris segmentation. In Proceedings of the Fifth IAPR International Conference on Biometrics (ICB), New Delhi, India, 29 March–1 April 2012; pp.310–316.
66. Mittal, A.; Moorthy, A.K.; Bovik, A.C. No-reference image quality assessment in the spatial domain. *IEEE Trans. Image Process.* **2012**, *21*, 4695–4708. [CrossRef]
67. Pertuz, S.; Puig, D.; Garcia, M.A. Analysis of focus measure operators for shape-from-focus. *Pattern Recognit.* **2013**, *46*, 1415–1432. [CrossRef]
68. CASIA V4. Available online: http://biometrics.idealtest.org/dbDetailForUser.do?id=4 (accessed on 2 May 2016).
69. CASIA-Iris-Mobile-V1. Available online: http://biometrics.idealtest.org/dbDetailForUser.do?id=13 (accessed on 25 May 2016).
70. Kumar, A.; Passi, A. Comparison and combination of iris matchers for reliable personal authentication. *Pattern Recognit.* **2010**, *43*, 1016–1026. [CrossRef]
71. ND-CrossSensor-Iris-2013 Dataset. Available online: https://cse.nd.edu/labs/cvrl/data-sets/biometrics-data-sets (accessed on 12 June 2016).
72. Proença, H.; Filipe, S.; Santos, R.; Oliveira, J.; Alexandre, L.A. The UBIRIS. v2: A database of visible wavelength iris images captured on-the-move and at-a-distance. *IEEE Trans. Pattern Anal. Mach. Intell.* **2010**, *32*, 1529. [CrossRef] [PubMed]
73. De Marsico, M.; Nappi, M.; Riccio, D.; Wechsler, H. Mobile Iris Challenge Evaluation (MICHE)-I, biometric iris dataset and protocols. *Pattern Recognit. Lett.* **2015**, *57*, 17–23. [CrossRef]
74. Rattani, A.; Derakhshani, R.; Saripalle, S.K.; Gottemukkula, V. ICIP 2016 competition on mobile ocular biometric recognition. In Proceedings of the IEEE International Conference on Image Processing (ICIP), Phoenix, AZ, USA, 25–28 September 2016; pp. 320–324.
75. Daugman, J. Biometric Decision Landscapes. Technical Report 482, University of Cambridge, Computer Laboratory. 2000. Available online: https://www.cl.cam.ac.uk/techreports/UCAM-CL-TR-482.pdf (accessed on 10 July 2016).

Article

Enhancing Security on Touch-Screen Sensors with Augmented Handwritten Signatures

Majd Abazid, Nesma Houmani * and Sonia Garcia-Salicetti

SAMOVAR, Telecom SudParis, Institut Polytechnique de Paris, 9 rue Charles Fourier, 91011 Evry, France; majd.abazid@telecom-sudparis.eu (M.A.); sonia.garcia@telecom-sudparis.eu (S.G.-S.)
* Correspondence: nesma.houmani@telecom-sudparis.eu

Received: 6 January 2020; Accepted: 4 February 2020; Published: 10 February 2020

Abstract: We aim at enhancing personal identity security on mobile touch-screen sensors by augmenting handwritten signatures with specific additional information at the enrollment phase. Our former works on several available and private data sets acquired on different sensors demonstrated that there are different categories of signatures that emerge automatically with clustering techniques, based on an entropy-based data quality measure. The behavior of such categories is totally different when confronted to automatic verification systems in terms of vulnerability to attacks. In this paper, we propose a novel and original strategy to reinforce identity security by enhancing signature resistance to attacks, assessed per signature category, both in terms of data quality and verification performance. This strategy operates upstream from the verification system, at the sensor level, by enriching the information content of signatures with personal handwritten inputs of different types. We study this strategy on different signature types of 74 users, acquired in uncontrolled mobile conditions on a largely deployed mobile touch-screen sensor. Our analysis per writer category revealed that adding alphanumeric (date) and handwriting (place) information to the usual signature is the most powerful augmented signature type in terms of verification performance. The relative improvement for all user categories is of at least 93% compared to the usual signature.

Keywords: automatic signature verification; touch-screen sensor; data quality; enrollment phase; performance assessment; augmented signature; security enhancement; mobile conditions

---

1. Introduction

The handwritten signature has been for a long time a usual mean to establish personal consent, with legal value for administrative and financial institutions. With the impressive proliferation of mobile devices having embedded sensors (smartphones, tablets), added to the development of online services, signing on digital platforms has become a reality in different sectors for identity security (banking, legal transactions, e-commerce among other). This reality has signified a turning point in the field of online signature biometrics.

In the last forty years, research studies were focused on online signatures captured on high quality sensors such as Wacom digitizing tablets, in controlled office-like scenarios, with a devoted ad-hoc pen stylus. Impedovo and Pirlo [1] published an article giving a detailed overview on the state-of-the-art techniques. Diaz et al. [2] presented a recent update on automatic signature verification (ASV). The research community made significant efforts for acquiring several online signature corpora [1–11] and conducting international evaluations of ASV systems [2,6,12–16].

Recent research studies have focused on signature verification in mobile conditions, using touch-screen sensors largely deployed nowadays. Nevertheless, the mobile scenario implies much more variability of acquisition conditions, like posture, writing tool (stylus or finger), screen size, sensor technology, interoperability, setting several new challenging issues that impact verification performance [2,17].

Usually, for improving verification performance, different strategies were exploited in the literature: (i) acquiring signatures in controlled conditions [1–16]; (ii) using a high quality sensor (such as a Wacom tablet) with high temporal and spatial resolution, and able to capture other time functions than pen coordinates, as pen pressure and pen inclination angles [18]; (iii) selecting reference signatures in order to control intra-personal variability [19–21]; (iv) extracting several features for signature description (as pressure, speed, and acceleration, etc.) [1–16] or by means of a deep neural network [2,22–25].

However, some of these strategies are no longer possible in the mobile scenario: as pointed out by [17], the sensors are not of the same quality, in terms of temporal resolution in particular, acquisition conditions are highly variable, and some sensors are limited to the capture of only pen coordinates. In the so-called "cloud scenario" [17], users acquire their signatures as they want, standing, sitting or moving, handling the device on the hand at different angles or orientations, or placing it on any support. A smartphone is usually handheld, while a tablet may be placed on the desktop or sustained by the left arm if the writer is right-handed. The consequence is that verification performance is strongly degraded in mobile conditions [2,15,16,26–39].

In the present paper, we study the online signature biometrics in the framework of uncontrolled mobile conditions. The challenging question then is how to improve verification performance in uncontrolled mobile conditions? To respond to this question, we propose a novel and original scheme for enhancing signature information content at the enrollment phase and reinforce its resistance to attacks, on a largely deployed touch-screen sensor technology. To this end, we propose different enrollment strategies for signature enrichment and assess them in terms of data quality and verification system performance.

The enrollment phase is critical for any biometric system since it determines the genuine signatures that will represent the user at the verification step. These signatures are called "Reference signatures". In our previous works on signature quality assessment, we have shown that a signature's resistance to attacks depends on its information content, quantified by an entropy-based measure, called personal entropy [28,40–43]. We identified automatically different risk levels in signatures related to three user categories, and in particular a "problematic" population, characterized by simple and highly variable signatures, very vulnerable to attacks.

Based on these findings, we propose in this paper, since the enrollment phase on a touch screen sensor, a novel strategy that turns any signature with a "high risk" into a "low risk" one. For signature enrichment, we use complementary personal handwritten information, as initials, name-surname, date and place of birth. We choose these information since a person is familiarized to append it for expressing her consent in administrative or legal frameworks. For this study, we consider different types of signatures (the usual signature, initials, name-surname, date and place of birth) and hybrid types as well (some combinations of the already mentioned types), and analyze the impact of each in terms of information content and resistance to attacks (skilled forgeries).

This paper is organized as follows: in Section 2, we present previous works of the literature related to online signature analysis on mobile devices. In Section 3, we describe the signature database and recall the personal entropy concept and the verification system used. In Section 4, we report the obtained results later summarized and discussed in Section 5. Finally, Section 6 presents the conclusions and future perspectives of our study.

2. Related Works

Different works of the literature pointed out the diversity of acquisition conditions and the subsequent degradation of verification performance in the mobile context [2,15,16,26–39]. The majority of these works focused on assessing and improving verification performance in several ways.

Martinez-Diaz et al. [26] indicated that ASV systems traditionally used on signatures acquired with Wacom digitizing tablets in an office-like scenario should be adapted in the context of handheld touch-screen devices. Indeed, the Wacom digitizing tablet used needs to be connected to a personal computer and thus leads to an office-like scenario. The authors exploited the BioSecure datasets DS2 and DS3 that contain signatures of the same 120 persons acquired, respectively, on a Wacom digitizing tablet following an office-like scenario and on a mobile touch-screen sensor (PDA) while holding it in the hand. Based on a feature selection algorithm and a hidden markov model (HMM) classifier, they observed the low discriminative power of dynamic features and the high consistency of geometric features in the mobile scenario.

Houmani et al. [27] evaluated an HMM-based classifier on two different databases both acquired on a PDA, namely PDA-64 containing online signatures of 64 persons, and BioSecure DS3 dataset (DS3-210) containing online signatures of 210 persons. Experiments showed significant performance degradation in mobile conditions: an average Equal Error Rate (EER) of 3.5% is obtained on the Wacom digitizer with skilled forgeries, while on PDA-64 and DS3-210, the EER is of 16.02% and 9.95% respectively.

In the context of the international online signature competition BSEC'2009 [15], different ASV systems were assessed on two large BioSecure datasets containing signatures of the same 382 persons acquired in a controlled scenario on a Wacom digitizer, namely DS2-382, and on a mobile device (PDA), namely DS3-382. Results showed a clear degradation of systems' performance when signing in mobile conditions.

Another work on BioSecure databases was conducted by Houmani and Garcia-Salicetti [28] to quantify the quality of signatures of the same 104 persons when captured in office-like conditions (DS2-104, captured on a Wacom digitizer) and in the mobile context (DS3-104, captured on a PDA). Results showed that signatures' quality degrades in mobile conditions and especially signature complexity decreases.

Blanco-Gonzalo et al. [29] evaluated an ASV system based on dynamic time warping (DTW), on seven mobile devices (using stylus and finger): two Wacom tablets, a tactile laptop, a Samsung Galaxy Note, an iPad, a Samsung Galaxy Tab and a Blackberry Playbook. They confronted the system with different acquisition conditions, such as sensor technology, interoperability, visual feedback, and screen size. Experiments showed that the best system performance is obtained when signing on a small screen, using a stylus instead of the finger. However, this study is of limited scope because only 11 writers were considered. In [31], Blanco-Gonzalo et al. exploited a DTW-based classifier on a database containing signatures of 43 users acquired on different mobile platforms. However, performance assessment was carried out only on random forgeries.

Martinez-Diaz et al. [32] presented the first publicly available database collected on a touch-screen sensor embedded in a mobile phone, namely the DooDB database. The database contains finger-drawn doodles and pseudo-signatures from 100 persons and skilled forgeries for all of them. To create pseudo-signatures, participants were asked to draw a simplified version of their signature, for example signing with their initials or part of their signature flourish, which could be used as a graphical password. Using a DTW-based classifier, the authors obtained an EER of around 26.9% on skilled forgeries considering time variability and 19.8% when the system was evaluated on only one session.

Sae-Bae and Memon [34] collected a new database that contains finger-drawn signatures from 180 persons captured in uncontrolled mobile conditions, on user owned iOS mobile devices. The authors proposed a histogram-based feature set for representing an online signature. They pointed out the importance of updating reference signatures to reduce the intra-class variability and thus improve systems' performance. The authors claimed that personalized feature selection is necessary to attain an acceptable performance level; they obtained an EER of 3.18% on random forgeries.

Antal et al. [36] introduced the MOBISIG database that contains finger-drawn pseudo-signatures from 83 persons, captured on a capacitive touch-screen sensor embedded in a mobile device. Participants were asked to create a signature for a given family name and were instructed on how to produce signatures with their finger. For performance assessment, the authors used a personalized threshold; they obtained an EER of 8.56% with a DTW classifier (vs. 25.45% with a global threshold), considering skilled forgeries and five reference signatures in the enrollment phase. Then, when considering 15 samples for enrollment, they improved verification performance, reaching an EER of 5.81% with a DTW classifier and a personalized threshold (vs. 20.82% with a global threshold).

Tolosana et al. [37] showed that performance is much better with the stylus than with the finger, on 65 persons of the e-BioSign database, using signatures captured with both Wacom tablets and Samsung mobile devices. Based on a DTW classifier, the authors obtained an EER of 22.1% with the finger versus roughly 7.9% with the stylus, considering skilled forgeries. This performance is obtained on 35 persons of the evaluation dataset; the 30 remaining persons were used as a development dataset to select 23 relevant parameters among the whole set of 117 features.

Zareen and Jabin [38] used a publicly available database [33] that contains 500 signatures from 25 persons, acquired on a Samsung Galaxy Note. As no skilled forgeries were acquired, the verification system based on a feed-forward multilayer neural network was evaluated only on random forgeries. The authors obtained an EER of 0.12% on random forgeries. These results seem preliminary since only 25 persons were considered and no skilled forgeries were used for performance assessment.

Nam et al. [39] used a private database that contains real finger-drawn signatures of only 20 persons, collected on a Samsung Galaxy S3. The authors proposed convolutional neural networks (CNN) for feature extraction, trained with genuine and forged signatures. Then, using an autoencoder for classification, they obtained an EER of 4.4% on skilled forgeries. However, this study is of limited scope because only 20 writers were considered.

All the above-mentioned works pointed out the degradation of systems' performance in the mobile context. However, in most of them, the data corpora presented do not contain signatures acquired in totally uncontrolled mobile conditions. This specific point is sometimes not even mentioned in the description of the acquisition protocol used, mainly focused on the sensor characteristics (technology, resolution, sampling rate), the writing tool, the design of the interface for acquisition, and the number of captured genuine signatures and forgeries. In addition, some works [32,36] evaluated the impact of mobile conditions based on pseudo-signatures, which are not exploited in real-world usages. Other studies evaluated ASV systems only on random forgeries because of the burden of acquiring skilled forgeries [34,38]. Finally, some works considered new challenging scenarios for ASV system assessment in terms of interoperability, as enrolling the person with a given writing tool and testing the system with another one [37].

In conclusion, we notice significant efforts in the literature for assessing verification performance in mobile conditions with different sensors, scenarios and classification strategies. Most works focused on the development of algorithms for biometric verification to enhance user authentication: DTW, HMM, neural networks and more recently some deep architectures. However, none of these works addressed quality-driven signature verification since the enrollment step, by quantifying and enhancing the information content of input data given to the sensor.

3. Materials and Methods

In this section, we first describe the signature database acquired for this study in mobile conditions and the signature types considered for enhancing the information content of the data given as input to the sensor. Then, we recall how Personal Entropy is quantified considering each signature type, and the classifier used for assessing the impact of our strategy in terms of verification performance.

3.1. Signature Data Acquisition

For this study, we captured online signatures from 74 persons on an iPad tablet with a capacitive touch-screen of 2048 × 1536 pixels. The signatures were sampled at 63 Hz and stored as a sequence of discrete values $[x_t, y_t]$, where x_t and y_t are the coordinate values and t is the time stamp.

Each person signed 25 times with their usual signatures. No instructions were given to participants when they signed, letting them acquire their signatures naturally, freely in terms of posture and position of the device, so that they would feel comfortable with the mobile device when signing. This leads to different acquisition conditions according to persons, exactly like it would be in real mobile usages.

Additionally to their usual signatures, we asked participants to append other types of signatures separately: name-surname, initials, date and place of birth. We considered these signature types because: (i) in terms of usages, they are traditionally reported by persons in legal and administrative documents; (ii) they convey complementary handwritten information on the user's identity. Each type of signature was done by the person 25 times. This dataset thus contains 9250 (74 × 25 × 5) genuine signatures of different types.

Figure 1 displays an example of one person's usual signature and the associated place of birth. We plot below in Figure 2 the velocity temporal function for both handwritten information.

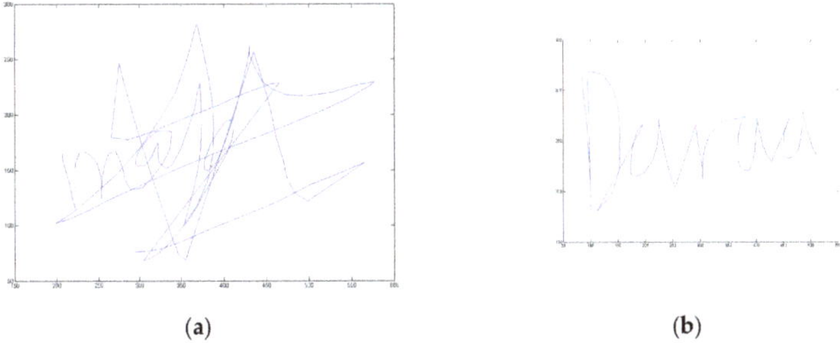

(a)   (b)

Figure 1. Examples of (**a**) a usual signature and (**b**) the associated place information of a user who authorized their publication.

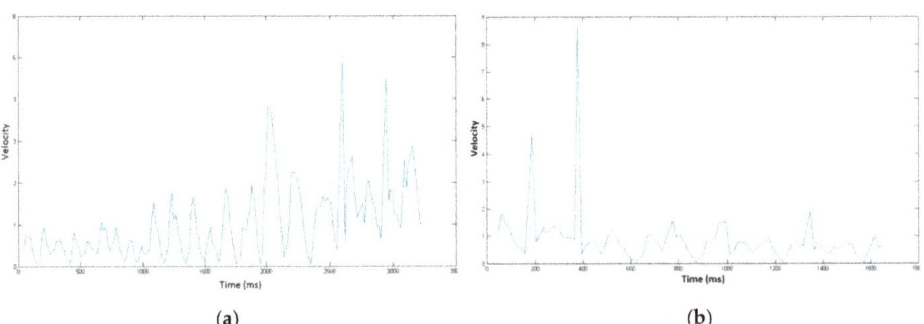

(a)   (b)

Figure 2. Velocity profile of (**a**) the usual signature and (**b**) the associated place information displayed in Figure 1.

In order to assess signature vulnerability to attacks, we acquired 10 skilled forgeries per signature type after displaying on the screen the shape and kinematics of the target signature. This type of forgery is considered in the literature as being the best attacks [3,43,44]. We thus obtain 3700 skilled forgeries (74 × 10 × 5) done by different forgers. Figure 3 shows an example of skilled forgeries of the usual signature and the associated place of birth displayed in Figure 1. We also display in Figure 4 the velocity temporal function for both handwritten information forgeries.

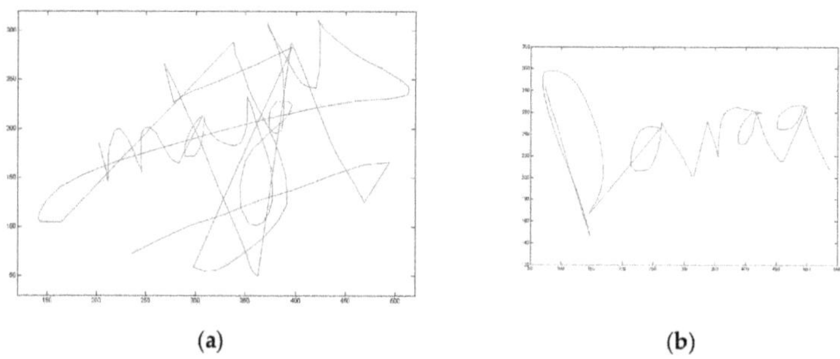

(a)   (b)

Figure 3. Examples of skilled forgeries of (**a**) the usual signature and (**b**) the associated place information displayed in Figure 1.

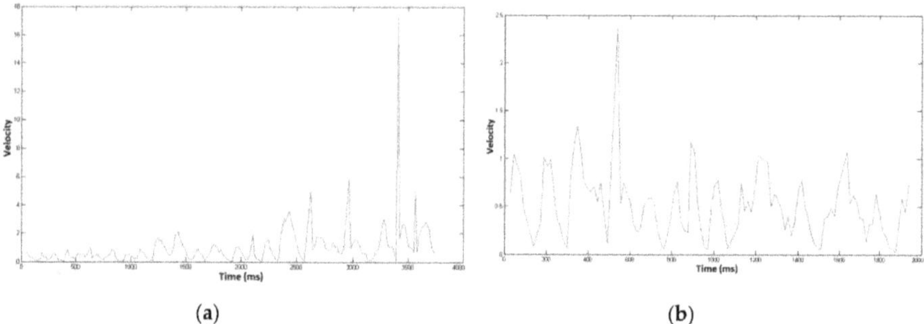

(a)   (b)

Figure 4. Velocity profile of (**a**) the forged usual signature and (**b**) the forged place information displayed in Figure 3.

3.2. Signature Types

We considered the five types of signatures separately: the usual signature (S), the initials (I), the name-surname (N), the date of birth (D), and the place of birth (P). From these five simple types, we constructed 7 hybrid signature types by combining:

- the usual signature with initials (SI);
- the usual signature with name-surname (SN);
- the usual signature with date (SD);
- the usual signature with place (SP);
- the usual signature with date and place (SDP);
- the usual signature with initials, date and place (SIDP);
- the name-surname with date and place (NDP).

These instances of hybrid types were constructed by concatenating the sequences of the corresponding simple signature types, resulting in a single time sequence. The identity of the user is thus expressed through several signature types of different length, which convey different complementary information to strengthen the user's identity.

3.3. Quantifying Quality of Signature Types

To assess information enrichment at the enrollment phase, we quantify the information content of other simple signature types than the usual signature (initials, name-surname, date and place), and also of the 7 hybrid types mentioned in Section 3.2.

The concept of entropy is a good alternative for quantifying the information content or the disorder in signatures. In [28,40–43], we proposed the concept of personal entropy (PE), an entropy-based quality measure that quantifies simultaneously both the complexity and variability of a person's signatures. In fact, complexity and variability are related to disorder at two different levels: complexity corresponds to the intrinsic disorder in a signature sample; variability corresponds to the intra-class disorder in a set of signatures belonging to a given user.

A user's PE is measured by exploiting the local probability densities estimated when training the user's HMM on a set of 10 genuine signatures described only by x and y attributes. Indeed, the HMM automatically generates portions by the Viterbi algorithm and estimates a mixture of Gaussian densities on each portion [28]. Figure 5 illustrates how PE is computed locally, on the segments generated by the user's HMM.

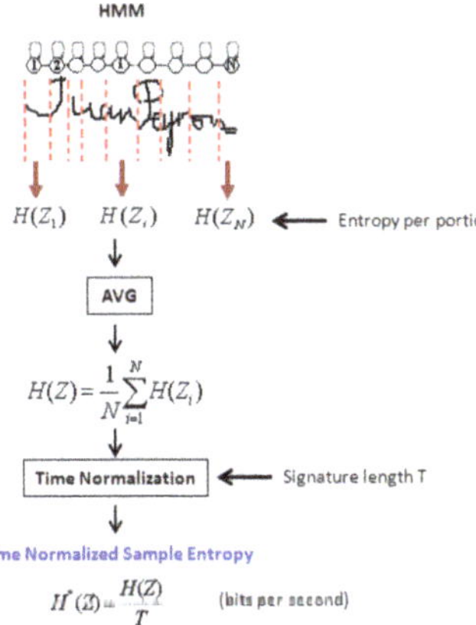

Figure 5. Personal entropy computation on portions of a signature.

Therefore, a random variable Z can be associated to each stationary portion i of the signature, generated by the Viterbi algorithm by the user's HMM. The number of portions N is the number of states of the HMM. The entropy $H(Z_i)$ of a portion i is computed as follows:

$$H(Z_i) = -\sum_{z \in S_i} p(z).log_2(p(z)), \tag{1}$$

where z corresponds to a given point in the signature described by its coordinates (x,y), belonging to the current portion i, and $p(z)$ is the probability of observing z.

We studied the number of genuine samples necessary for a good HMM estimation and showed that 10 instances lead to stable PE values [28]. The local probability distribution function is estimated using all the sample points belonging to each portion, across the 10 genuine samples. After that, the entropy of each genuine signature $H^*(Z)$ is the average of entropy values $H(Z_i)$ on all the N portions of the signature, divided by the signing time T:

$$H^*(Z) = \frac{1}{N*T}\sum_{i=1}^{N} H(Z_i) , \tag{2}$$

Finally, by averaging $H^*(Z)$ across the 10 user's genuine samples, we obtain a user's PE for each signature type. We demonstrated that PE allows obtaining three categories of signatures, coherent across several databases, spanning from short and highly variable signatures (high PE category) to stable, longer and complex signatures (low PE category). Moreover, we showed that for different classifiers, persons with low PE are the most robust to skilled forgeries. Persons with high PE are considered being "problematic" users in the literature [28,43,45]. These results were obtained considering the usual signature of each person [28,40–43].

3.4. Signature Verification System

As our aim is to assess the impact of our strategy in a mobile scenario, we used a statistical verification system that has already been evaluated on large databases acquired on mobile sensors [9,12, 15,16,27,46], and has shown to maintain good performance on well-known databases in interoperability scenarios [47], as reported in Table 1. Indeed, Table 1 presents our system's performance on several online signature databases, some acquired in an office-like scenario using a Wacom digitizer with an inking pen, and other in a mobile scenario on different touch-screen sensors (PDA, iPad, iPhone). We report the EER values on skilled forgeries only, since it is the most challenging configuration for signature verification. The system has been evaluated in BSEC'2009 and ESRA'2011 competitions on very large databases of 382 persons [15,16] that signed both on a Wacom digitizer and on a PDA device. We observe that in the mobile context, the verification performance of our system is clearly better on recent capacitive touch-screen sensors (iPad and iPhone) compared to the results obtained on the PDA device (DS3-210, PDA-64, DS3-382).

Table 1. Performance of our HMM-based ASV system on several online signature databases acquired in office-like (Wacom device) and mobile scenarios (touch-screen sensors), considering skilled forgeries.

Databases	Evaluation Campaigns	Year	Devices	Users	EER (in %)
DS2-382 [15]	BSEC'2009	2012	Wacom tablet	382	4.47
DS3-382 [15]	BSEC'2009	2012	PDA (stylus)	382	11.27
DS2-382 [16]	ESRA'2011	2011	Wacom tablet	382	2.73–4.04
DS3-382 [16]	ESRA'2011	2011	PDA (stylus)	382	8.13–10.92
DS3-210 [27]	-	2010	PDA (stylus)	210	9.95
PDA-64 [27]	-	2010	PDA (stylus)	64	16.02
iPad-74 [46]	-	2019	iPad (stylus)	74	7.04
iPhone-74 [46]	-	2019	iPhone (stylus)	74	4.95
BIOMET [47]	-	2007	Wacom tablet	84	2.33
PHILIPS [47]	-	2007	Digitizing tablet	51	3.25
SVC2004 [47]	-	2007	Digitizing tablet	40	4.83
MCYT-100 [47]	-	2007	Wacom tablet	100	3.37
MCYT-330 [48]	-	2009	Wacom tablet	330	3.91

Table 2 summarizes the state-of-the-art of online signature verification systems on mobile sensors, when considering skilled forgeries. We observe that in some publications, performance is not reported on skilled forgeries, which is the most challenging case for ASV systems. When comparing the results on mobile sensors in Tables 1 and 2, we note that our system shows good performance compared to the state-of-the art. Indeed, on iPad and iPhone mobile sensors, an EER of 7.04% and 4.95% respectively is reached on signatures of the same 74 users. Compared to e-Biosign database containing real signatures of 65 users acquired with a stylus on two mobile devices, we notice that our HMM-based system shows slightly better performance on the iPad device (7.04% vs. 7.9% in the best case, or vs. 10.7% on the other mobile device) and much better performance on the iPhone device (4.95% vs. 7.9% in the best case, or vs. 10.7% on the other mobile device).

Our system behaves well in mobile conditions because it is based on a statistical model, namely a continuous left-to-right HMM with four Gaussian components per state [47–49]. In other words, each writer's signature is modeled through a double stochastic process, characterized by a given number of states with an associated set of transition probabilities among them, and in each state, a continuous density, a multivariate Gaussian mixture is used to model the emission probability density. This model has the advantage of absorbing the intra-personal variability of signatures [47], which increases significantly in mobile conditions.

Table 2. Performance of ASV systems of the literature on several online signature databases acquired in mobile scenarios with touch-screen sensors, considering skilled forgeries.

Databases	Year	Users	Sensor	EER (in %)
ATVS-DooDB [32]	2013	100	HTC Touch HD (Pseudo-signatures)	Finger: 26.9
Blanco-Gonzalo et al. [31]	2013	43	Asus Eee PC Touch (stylus) Samsung Gal. Note (stylus/finger) BlackBerry Playbook (finger) Apple Ipad2 (finger) Samsung Gal. Tab (finger)	-
e-Biosign [37]	2016	65	Samsung ATIV7 Samsung Gal. Note	Stylus: 7.9 Finger: 22.1 Stylus: 10.7 Finger: 26.4
Zareen and Jabin [38]	2016	25	Samsung Gal. Note	-
MOBISIG [36]	2018	83	Nexus 9 tablet capacitive (Finger-drawn pseudo-signatures)	Personalized vs. global threshold: 8.56% vs. 25.45%
Nam et al. [39]	2018	20	Samsung Gal. S3	Finger: 4.4%

A personalized number of states is determined according to the total number T_{total} of sampled points available in the genuine signatures of the HMM's training set. We consider that in average 30 sampled points are enough to estimate the mean vector and the covariance matrix of each Gaussian [47]. The number of states N is computed as:

$$N = \left[\frac{T_{total}}{M * 30} \right], \tag{3}$$

where $M = 4$ is the number of Gaussian densities per state and brackets denote the integer part.

Nineteen dynamic features are extracted point-wise for all signature types. These features are described in detail in the Appendix A. The usual information extracted from an HMM is the likelihood of the input signature given the user's model. We have noticed that the information coming from the segmentation of the test signature by the target user's model is complementary to that of the likelihood, especially for forgery detection. Indeed, we have shown in [47] that the segmentations made by the target model on forgeries differ from those obtained on genuine signatures. For this reason, in the verification phase, the classifier performs a score fusion combining two levels of signature analysis: one based on a local point-wise analysis of each signature by the HMM (log-Likelihood score), the other on the analysis of the signature at the level of portions, automatically segmented by the same HMM (Viterbi score) [47–49]. At the first level (log-Likelihood score), on a particular test signature, a distance is computed between its log-Likelihood and the average log-Likelihood obtained on the training signatures; then it is shifted to a similarity value—called "Log-Likelihood score"—between 0 and 1, by the use of an exponential function [47]. At the second level of analysis (Viterbi score), the user's HMM automatically performs by the Viterbi algorithm, a segmentation of each training signature into portions, according to the most likely path displayed in Figure 6. A "segmentation vector" can then be associated to each signature: the N-components segmentation vector, N being the number of states in the claimed identity's HMM has in the i-th position the number of points (observations) associated to state i by the Viterbi path, as illustrated in Figure 6. Each training signature is then characterized by a *Reference segmentation vector*. In the verification phase, on a particular test signature, a distance between its corresponding segmentation vector and each *Reference segmentation vector* is computed, and such distances are averaged to compute the final distance. It is then shifted to a similarity measure between 0 and 1 (Viterbi score) by an exponential function [47].

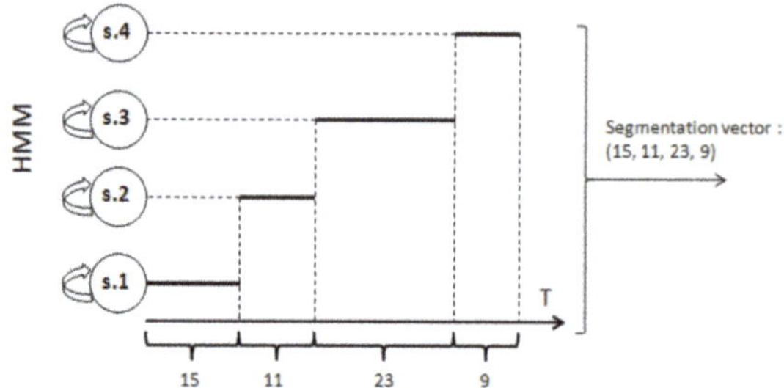

Figure 6. Computation of a signature's segmentation vector generated by the user's HMM. Feature vectors describing the signature are on the *x*-axis and the left-to-right HMM on the *y*-axis.

Finally, the similarity score for a given test signature is thus the fusion by a simple arithmetic mean of the log-Likelihood score and the Viterbi score. If the final score is higher than the value of the decision threshold the claimed identity is accepted, otherwise it is rejected.

In this work, for simple signature types, we train an HMM per person and per signature type. For hybrid types, we train an HMM for each person considering the whole time sequence constructed by concatenating the time sequences of the concerned simple types. As example, for the SDP type, we train an HMM on the complete sequence composed of the usual signature, the date and the place. Note that according to signature types, the length of the complete signature sequence will vary: for simple signature types, it will tend to increase when considering name-surname and to decrease when considering initials. For hybrid types, the length of the signature will be even higher. This will impact the number of states of the user's HMM.

4. Results

4.1. Quality Measure of Usual Signatures

In a first step, we quantify the quality of usual signatures of the 74 persons available in our dataset. To this end, we trained for each person, an HMM on 10 genuine signatures to measure the user's PE. Then, a hierarchical clustering was performed on the obtained PE values, resulting in three user categories displayed in Figure 7.

Figure 7. Examples of signatures captured in uncontrolled mobile conditions with (**a**) high, (**b**) medium and (**c**) low PE.

In Figure 7a, we observe three examples of signatures with high PE: they are the shortest and the simplest signatures, having the aspect of a flourish, and are the most variable (see Figure 8). Such signatures are considered as being "problematic" in the literature [28,43,45]. On the other hand, Figure 7c shows three examples of signatures with low PE: they are longer, the most complex and the most stable (see Figure 8). In between, there is a transition category in terms of complexity and stability, the category of medium PE (see Figures 7b and 8).

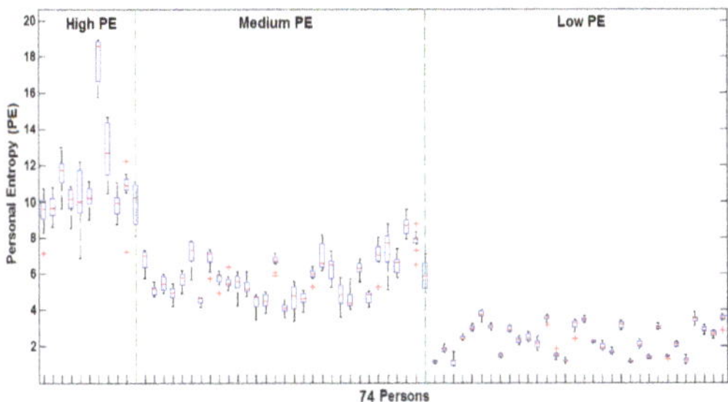

Figure 8. Boxplots of PE values for all 74 persons clustered into high, medium and low PE.

4.2. Quality Measure of All Signature Types

Figure 9 displays two examples of each simple signature type, captured separately on the iPad: usual signature, initials, name-surname, date and place of birth.

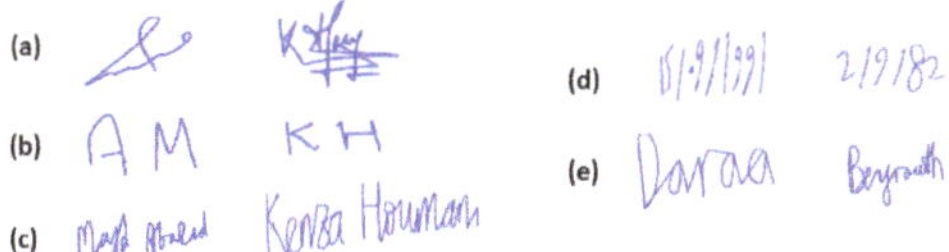

Figure 9. Examples of the simple signature types: (**a**) usual signature; (**b**) initials; (**c**) name-surname, (**d**) date of birth, and (**e**) place of birth. These signatures belong to persons who have authorized their publication.

For each person, we compute PE values for the five simple signature types separately, and for the seven hybrid signature types: SI, SD, SP, SN, SDP, NDP and SIDP. Figure 10 presents the boxplots of the obtained PE values for the 12 signature types.

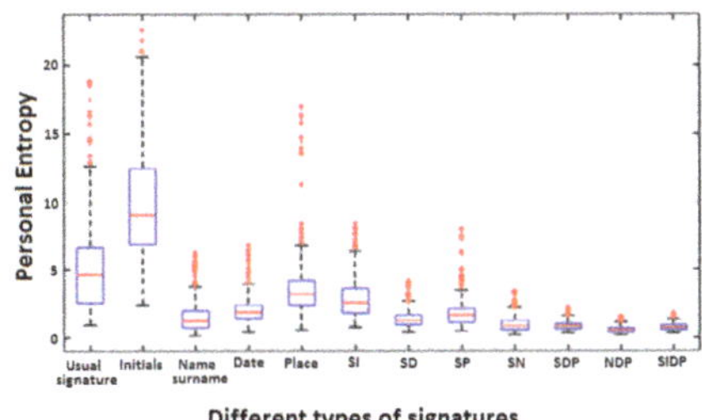

Figure 10. Boxplots of PE values for the 74 persons per signature type.

We first notice that initials have the highest PE values. This result is coherent since initials are the most simple and variable type of signature. We also notice a strong spread out of their boxplot in Figure 10. Moreover, we observe that some initials have a comparable PE to that of usual signatures: indeed, for initials, some entropy values are below the first quartile. This can be explained by the fact that some persons appended their initials into two, three and even four letters, sometimes linking them as usually done when producing a short signature. In this case, the initials show a higher complexity and stability.

Furthermore, we notice that the more the signature is enriched (name-surname, SI, SD, SP, SN, SDP, NDP, SIDP), the lower PE becomes: the complexity of signatures is higher and variability is lower. The hybrid types SDP, NDP and SIDP are those showing the lowest PE values and the lowest variance between persons in the boxplots displayed in Figure 10.

In the sequel, we study the relationship between information content quantified by PE and verification performance. Our objective is to identify which types of signatures are more resistant to attacks in uncontrolled mobile conditions.

4.3. Evaluation of the Proposed Scheme

As we have proven in former works the significant difference in verification performance between PE categories [28,40–43], we naturally adopted a methodology assessing the impact of our strategy on each PE category separately. This methodology consists in the following steps: first we computed PE values of the 74 persons considering only their usual signatures. Then, we generated three user categories based on the obtained PE values, by a Hierarchical Clustering (as explained in Section 4.1). Finally, we assessed, per user category, verification performance on usual signatures, and compared it to performance when considering the other signature types: initials, name-surname, SI, SN, NDP, SDP, and SIDP.

For performance assessment, we considered, for each person and each signature type, the remaining 15 genuine signatures (the other 10 genuine signatures were used for PE computation) and the 10 available skilled forgeries. For each person, the HMM classifier was trained on five genuine signatures among the 15, and tested on the remaining 10 genuine instances and the 10 skilled forgeries. The same signature type is considered in the training and testing phases.

Five random samplings were carried out on the training signatures. The false acceptance rate (FAR) and false rejection rate (FRR) are computed relying on the total number of false rejections and false acceptances obtained on all the five random samplings.

4.3.1. Results on High PE Category

We analyze in this section the results obtained on the so-called "problematic" users in the literature [28,43,45], which are the main target of our strategy for enhancing signature security in uncontrolled mobile conditions.

Figure 11 and Table 3 display the system performance on problematic users, those with the highest PE. The EER reaches 7.17% when considering their usual signature (see Table 3 and the blue curve in Figure 11). We first notice a significant degradation of performance when persons sign with their initials (green curve in Figure 11). A relative degradation of 93% at the EER is observed even if the usual signature is already simple and variable. This highlights the importance of the ballistic aspect of the signing process in terms of resistance to attacks. Note that the vulnerability of initials is also predictable by their very high PE values observed in Figure 12.

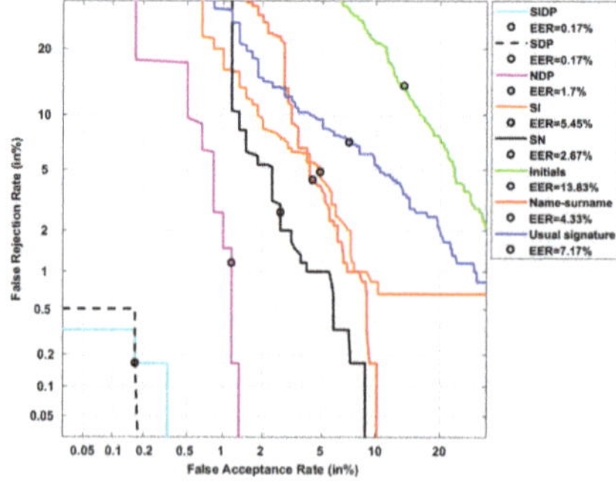

Figure 11. System performance on users of the highest PE category considering the 8 signature types.

Table 3. System performance on users of the highest PE category in terms of EER.

Type of Signatures	EER
Usual signature	7.17%
Initials	13.83%
Name-surname	4.33%
SN	2.67%
SI	5.45%
NDP	1.7%
SDP	0.17%
SIDP	0.17%

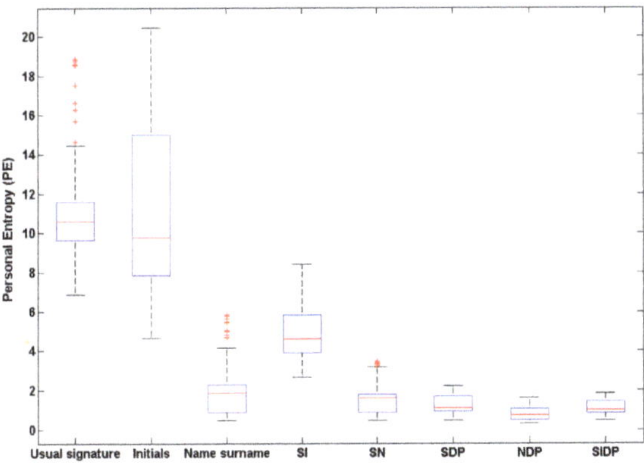

Figure 12. Boxplots of PE values per signature type for users of the highest PE category.

Moreover, we notice a significant improvement in performance when persons sign with their name-surname (red curve in Figure 11). The FAR is in this case bounded around 10%. Also, the hybrid type SN, which combines the usual signature and name-surname, improves significantly performance (black curve in Figure 11): at the EER, the relative improvement is of 63% compared to the usual signature. This result confirms the robustness of this hybrid type to attacks, predictable by its low PE values displayed in its corresponding boxplot in Figure 12.

Besides, adding the date and place information clearly enhances performance. Indeed, the NDP type (magenta curve in Figure 11) improves performance of 83.68% at the EER when compared to the usual signature. But the SDP type outperforms the NDP: the relative improvement is of 98% at the EER, when compared to the usual signature (see Table 3 and black dotted curve in Figure 11). Moreover, it leads to a bounded FAR at 0.2% and a bounded FRR at 0.5%. Interestingly, we notice that this could not be predicted by PE since SDP type has higher PE values than NDP (see Figure 12). This result shows that the ballistic gesture inherent to the usual signature remains more discriminant than the name-surname, when being combined to an alphanumeric information (the date) and handwriting (the place), even in the case of a very simple problematic signature.

Finally, the SIDP type does not perform significantly better than the SDP type. This may be explained by the fact that in this particular category of users, the usual signature is simple and variable, and thus close to initials in terms of information content.

4.3.2. Results on Low PE Category

Figure 13 and Table 4 show system performance on persons with low PE, whose signatures are the most complex and stable, and the most robust to attacks. The EER reaches 6.93% (see Table 4) when considering their usual signature (blue curve in Figure 13).

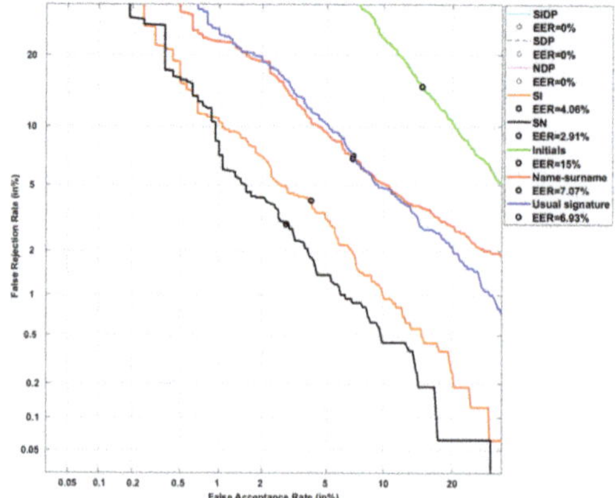

Figure 13. System performance on users of the lowest PE category considering the 8 signature types.

Table 4. System performance on users of the lowest PE category in terms of EER.

Type of Signatures	EER
Usual signature	6.93%
Initials	15%
Name-surname	7.07%
SN	2.91%
SI	4.06%
NDP	0%
SDP	0%
SIDP	0%

Some trends observed on problematic users in the previous section are here confirmed. First, a significant degradation of 116% is obtained at the EER with initials relatively to the usual signatures. PE predicts this trend in Figure 14 (higher PE values for initials). Besides, as expected, this relative degradation of 116% is higher in the case of complex signatures of this category, compared to problematic users (relative degradation of 98% as reported in Section 4.3.1). Figure 14 shows the significant gap between initials and usual signatures for the low PE category, compared to that obtained on problematic users (high PE category).

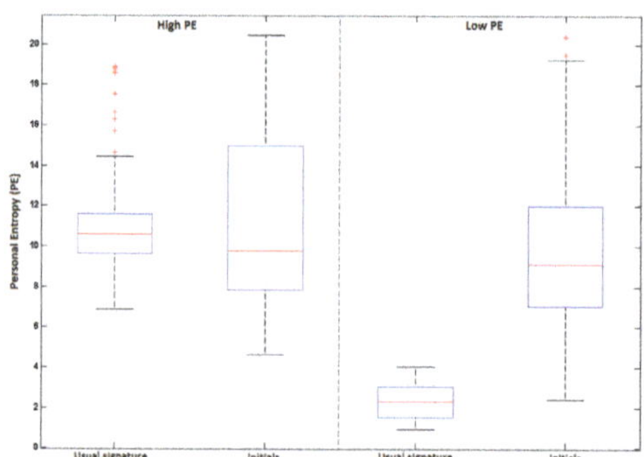

Figure 14. Boxplots of PE for users with highest (left) and lowest (right) PE values, considering their usual signature and initials.

Moreover, the hybrid types SI, SN, NDP, SDP and SIDP outperform significantly the usual signature; we note a relative improvement of 100% at the EER for NDP, SDP and SIDP types. For this reason, the three associated DET-curves are not visible in Figure 13. This confirms again their resistance to attacks, predictable by their low PE values, as shown in Figure 14.

On the other hand, some trends differ from those observed on problematic users. We notice that the name-surname type (red curve in Figure 13) gives similar performance to that of the usual signature (blue curve in Figure 13); while for problematic users, the name-surname outperforms by 40% the usual signature (see Figure 11 and Table 3). This means that in this category of persons, if we consider the name-surname as a possible signature for identity verification, although it has higher complexity (low PE in Figure 15), performance would not be improved relatively to the usual signature.

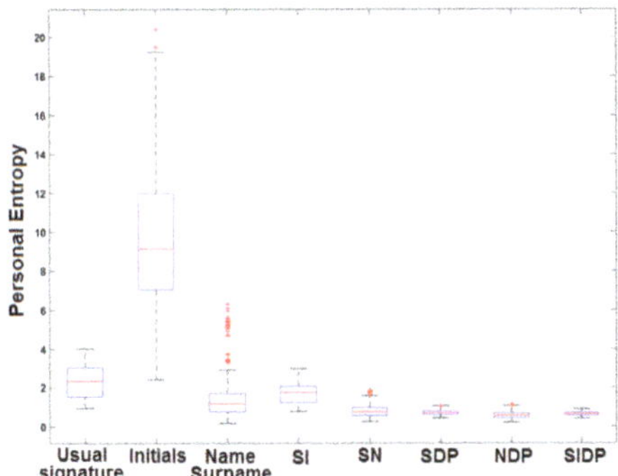

Figure 15. Boxplots of PE values per signature type for users of the lowest PE category.

In conclusion, this result shows on one hand that the complexity criterion is not sufficient to enhance the security of a signature. On the other hand, it highlights the importance of the ballistic process for identity verification.

4.3.3. Results on Medium PE Category

Figure 16 and Table 5 display system performance on persons belonging to the category of medium PE. The obtained results on this category show an intermediate behavior compared to that observed on high and low PE categories. More precisely, the initials are the worst type in terms of performance.

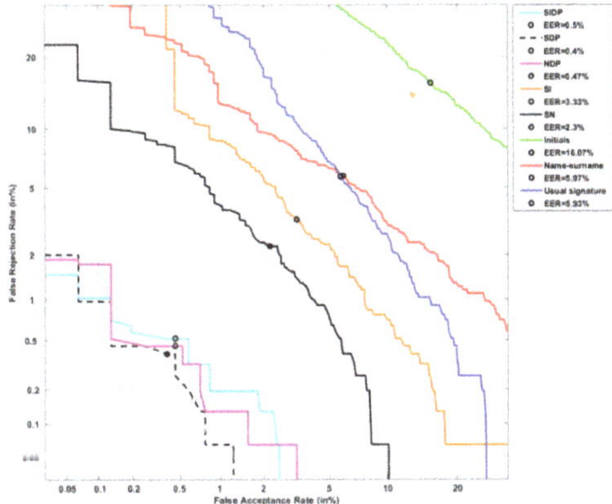

Figure 16. System performance on users of the medium PE category considering the 8 signature types.

Table 5. System performance on users of the medium PE category in terms of EER.

Type of Signatures	EER
Usual signature	5.93%
Initials	16.07%
Name-surname	5.97%
SN	2.3%
SI	3.33%
NDP	0.47%
SDP	0.4%
SIDP	0.5%

Moreover, the DET-curves corresponding to name-surname and usual signature types intersect at the EER value and get closer compared to problematic users. Also, the hybrid types including alphanumeric and handwriting information enhance significantly verification performance in uncontrolled mobile conditions. The SDP type in particular leads to a significant relative improvement of 93% at the EER compared to the usual signature; besides we note that the FAR and FRR are both bounded at 1.2% and 2%, respectively.

5. Discussion

In this paper, we have proposed a novel strategy for securing personal identity on a touch-screen sensor embedded in a mobile device, largely used nowadays. This strategy operates upstream from the verification system, at the sensor level, by enriching the information content of handwritten inputs. Specific additional inputs then reinforce the usual signature with alphanumeric and handwritten personal information, frequently used in public and legal usages.

We quantified information enrichment with PE measure that characterizes both signature complexity and stability. Several simple and hybrid signature types were proposed for our experimental study.

We assessed the effectiveness of our proposal across three well-established user categories in terms of signature complexity, signature stability and verification performance. This allowed highlighting inside categories, subtle differences in terms of relative performance enhancement, depending on the signature type. This methodology allows understanding which characteristics are relevant in the signing process to reinforce the digital identity for all persons.

Experiments were performed on 74 writers that signed on a tablet with a stylus in uncontrolled mobile conditions. Our analysis per writer category revealed a common trend to all: adding alphanumeric (date) and handwriting (place) information to the usual signature is the most powerful hybrid type in terms of verification performance. This can be explained by the fact that this hybrid type combines complementary information, and keeps the ballistic aspect of the signature, so important for identity verification. The relative improvement for all user categories is of at least 93% compared to the usual signature.

6. Conclusions and Future Work

The important outcome of our study is the possibility of extending the concept of handwritten identity to other personal information than the usual signature. Actually, by combining the usual signature with alphanumeric (date) and handwritten (place) personal information, personal identity security is significantly enhanced for all persons in uncontrolled mobile conditions. Moreover, with our strategy, the concept of user categories even disappears because all persons become very robust to attacks.

Another interesting outcome is that the complexity criterion is not sufficient to enhance the security of a signature. This is clearly observed on persons with the most complex signatures (low PE category): although the name-surname type is more complex than the usual signature, it is not more reliable in terms of resistance to attacks. This is because the usual signature conveys specific ballistic information about identity; this information can be completed by other handwritten information but cannot be removed and replaced for robust identity verification.

The finding of combining signature, date and place for enhancing identity security is in total accordance with public and legal usages in which identity information is requested. This may facilitate the implementation of the proposed enrollment strategy at a large scale.

In future work, we envisage implementing our strategy by developing an application on different mobile devices to study the practical usage of the proposed enrollment strategy, in terms of acquisition time during enrollment, user HMM training when acquiring signature followed by date and place in one shot, accuracy in mobility, and user acceptability and comfort. This will be conducted considering challenging mobile scenarios in terms of interoperability and time variability. Also, since our study demonstrates that augmenting the usual signature with alphanumeric and handwritten personal information enhances significantly verification performance, it would be interesting to study the impact of reducing the number of enrollment inputs. Furthermore, we will investigate the effectiveness of our strategy in terms of relative performance improvement when confronted to other classifiers.

Author Contributions: Formal analysis, M.A., N.H. and S.G.-S.; Investigation, M.A., N.H. and S.G.-S.; Methodology, N.H. and S.G.-S.; Project administration, N.H. and S.G.-S.; Software, M.A. and N.H.; Supervision, N.H. and S.G.-S.; Validation, N.H. and S.G.-S.; Visualization, M.A. and N.H.; Writing—original draft, N.H. and S.G.-S. All authors have read and agreed to the published version of the manuscript.

Funding: This research received no external funding.

Conflicts of Interest: The authors declare no conflict of interest.

Appendix A

Table A1. The 19 dynamic features extracted point-wise on all signature types.

	N°	Feature Name
	1–2	Normalized coordinates ($x(t){-}x_g$, $y(t){-}y_g$) relative to the gravity center (x_g, y_g) of the signature
	3–4	Speed in x and y
Gesture related features	5	Absolute speed
	6	Ratio of the minimum over the maximum speed on a window of 5 points
	7–8	Acceleration in x and y
	9	Absolute acceleration
	10	Tangential acceleration
	11	Angle α between the absolute speed vector and the x axis
	12–13	Sine(α) and Cosine(α)
Local shape related features	14	Variation of the α angle: Φ
	15–16	Sine(Φ) and Cosine(Φ)
	17	Log(1 + r) where r is the curvature radius of the signature at the present point
	18	Length to width ratio on windows of size of 5 points
	19	Length to width ratio on windows of size of 7 points

For computing the derivatives of a parameter $x(t)$, we used the regression equation as follows:

$$x'(t) = reg(x(t), Z) = \frac{\sum_{z=1}^{Z} z * (x(t+z) - x(t-z))}{2 \sum_{z=1}^{Z} z^2}$$

where z = 2, in order to obtain soft derivative curves

Accordingly,

- Speed in x and y (N° 3–4 in Table A1):

$$v_x(t) = reg(x(t), 2) \qquad v_y(t) = reg(y(t), 2)$$

- The absolute speed (N° 5 in Table A1):

$$v(t) = \sqrt{v_x^2(t) + v_y^2(t)}$$

- Acceleration in x and y (N° 7–8 in Table A1):

$$a_x(t) = reg(v_x(t), 2) \qquad a_y(t) = reg(v_y(t), 2)$$

- The absolute acceleration (N° 9 in Table A1):

$$a(t) = \sqrt{a_x^2(t) + a_y^2(t)}$$

- The tangential acceleration (N° 10 in Table A1):

$$a_t(t) = reg(v(t), 2)$$

- Angle α between the absolute speed vector and the x axis (N° 11 in Table A1):

$$\alpha(t) = arcsin\left(\frac{v_y(t)}{v(t)}\right)$$

- Sine and cosine of angle α (N° 12–13 in Table A1):

$$Sin(\alpha(t)) = \left(\frac{v_y(t)}{v(t)}\right) \qquad Cos(\alpha(t)) = \left(\frac{v_x(t)}{v(t)}\right)$$

- Variation ϕ of the angle α angle (N° 14 in Table A1):

$$\phi(t) = reg(\alpha(t), 2)$$

- Sine and cosine of angle ϕ (N° 15–16 in Table A1):

$$sin(\phi(t)) \qquad cos(\phi(t))$$

- $\log(1 + r(t))$ (N° 17 in Table A1), where r is the curvature radius of the signature at the present point t:

$$lr(t) = \log(1 + r(t)) = log\left(1 + \frac{v_t(t)}{\phi(t)}\right)$$

- Length to width ratio on windows of size of 5 points centered on the current point t (N° 18 in Table A1).
- Length to width ratio on windows of size of 7 points centered on the current point t (N° 19 in Table A1).

References

1. Impedovo, D.; Pirlo, G. Automatic Signature Verification: The State of the Art. *IEEE Trans. Syst. Man Cybern. Part C Appl. Rev.* **2008**, *38*, 609–635. [CrossRef]

2.   Diaz, M.; Ferrer, M.A.; Impedovo, D.; Malik, M.I.; Pirlo, G.; Plamondon, R. A Perspective Analysis of Handwritten Signature Technology. *ACM Comput. Surv.* **2019**, *51*, 1–39. [CrossRef]

3.   Dolfing, J.; Aarts, E.; Van Oosterhout, J. On-line signature verification with hidden Markov models. In Proceedings of the Fourteenth International Conference on Pattern Recognition, Brisbane, Australia, 20 August 1998; pp. 1309–1312.

4.   Garcia-Salicetti, S.; Beumier, C.; Chollet, G.; Dorizzi, B.; Jardins, J.L.L.; Lunter, J.; Ni, Y.; Petrovska-Delacrétaz, D. BIOMET: A Multimodal Person Authentication Database Including Face, Voice, Fingerprint, Hand and Signature Modalities. In Proceedings of the Computer Vision, Guildford, UK, 9–11 June 2003; Springer Science and Business Media LLC: Berlin, Germany; pp. 845–853.

5.   Ortega-Garcia, J.; Fierrez-Aguilar, J.; Simon, D.; González, J.; Faundez-Zanuy, M.; Espinosa, V.; Satue, A.; Hernaez, I.; Igarza, J.J.; Vivaracho, C.; et al. MCYT baseline corpus: A bimodal biometric database. *IEE Proc. Vision Image Signal Process.* **2003**, *150*, 395–401. [CrossRef]

6.   Yeung, D.Y.; Chang, H.; Xiong, Y.; George, S.; Kashi, R.; Matsumoto, T.; Rigoll, G. SVC2004: First International Signature Verification Competition. In Proceedings of the Biometric Authentication, First International Conference ICBA, Hong Kong, China, 15–17 July 2004; pp. 16–22.

7.   Kholmatov, A.; Yanikoglu, B. SUSIG: An on-line signature database, associated protocols and benchmark results. *Pattern Anal. Appl.* **2009**, *12*, 227–236. [CrossRef]

8.   SUSIG On-Line Signature Database. Available online: https://biometrics.sabanciuniv.edu/susig.html (accessed on 16 December 2011).

9.   Ortega-Garcia, J.; Fierrez, J.; Alonso-Fernandez, F.; Galbally, J.; Freire, M.R.; Gonzalez-Rodriguez, J.; Garcia-Mateo, C.; Alba-Castro, J.L.; Gonzalez-Agulla, E.; Otero-Muras, E.; et al. The Multiscenario Multienvironment BioSecure Multimodal Database (BMDB). *IEEE Trans. Pattern Anal. Mach. Intell.* **2010**, *32*, 1097–1111. [CrossRef] [PubMed]

10.   Biometrics in Identity Management: Concepts to Applications. Available online: http://biosecure.itsudparis. eu/AB/ (accessed on 4 October 2017).

11.   Fierrez, J.; Galbally, J.; Ortega-Garcia, J.; Freire, M.R.; Alonso-Fernandez, F.; Ramos, D.; Toledano, D.T.; Gonzalez-Rodriguez, J.; Siguenza, J.A.; Garrido-Salas, J.; et al. BiosecurID: A multimodal biometric database. *Pattern Anal. Appl.* **2010**, *13*, 235–246. [CrossRef]

12.   Mayoue, A.; Dorizzi, B.; Allano, L.; Chollet, G.; Hennebert, J.; Petrovska-Delacr, D.; Verdet, F. Biosecure multimodal evaluation campaign 2007 (BMEC'2007). In *Guide to Biometric Reference Systems and Performance Evaluation*; Springer: London, UK, 2009; pp. 327–369.

13.   Encyclopedia of Biometrics: I–Z. Available online: http://biometrics.itsudparis.eu/BMEC2007/ (accessed on 12 June 2008).

14.   Blankers, V.L.; Heuvel, C.E.V.D.; Franke, K.Y.; Vuurpijl, L.G. ICDAR 2009 Signature Verification Competition. In Proceedings of the 10th International Conference on Document Analysis and Recognition, Barcelona, Spain, 26–29 July 2009; pp. 1403–1407.

15.   Houmani, N.; Mayoue, A.; Garcia-Salicetti, S.; Dorizzi, B.; Khalil, M.I.; Moustafa, M.N.; Abbas, H.; Muramatsu, D.; Yanikoglu, B.; Kholmatov, A.; et al. Biosecure Signature Evaluation Campaign (BSEC'2009): Evaluating Online Signature Verification Systems' Performance. *Pattern Recogn.* **2012**, *45*, 993–1003. [CrossRef]

16.   Houmani, N.; Garcia-Salicetti, S.; Dorizzi, B.; Montalvão, J.; Canuto, J.C.; Andrade, M.V.; Qiao, Y.; Wang, X.; Scheidat, T.; Makrushin, A.; et al. BioSecure Signature Evaluation Campaign (ESRA'2011): Evaluating systems on quality-based categories of skilled forgeries. In Proceedings of the 2011 International Joint Conference on Biometrics (IJCB), Washington, DC, USA, 11–13 October 2011; pp. 1–10.

17.   Impedovo, D.; Pirlo, G. Automatic signature verification in the mobile cloud scenario: Survey and way ahead. *IEEE Trans. Emerg. Top. Comput.* **2018**, 1. [CrossRef]

18.   Houmani, N.; Garcia-Salicetti, S. Digitizing Tablet. In *Encyclopedia of Biometrics*, 2nd ed.; Springer: Boston, MA, USA, 2015; pp. 351–356.

19.   Di Lecce, V.; DiMauro, G.; Guerriero, A.; Impedovo, S.; Pirlo, G.; Salzo, A.; Sarcinella, L. Selection of Reference Signatures for Automatic Signature Verification. In Proceedings of the Fifth International Conference on Document Analysis and Recognition, ICDAR '99 (Cat. No.PR00318), Bangalore, India, 22 September 1999; pp. 597–600.

20. Guest, R.; Fairhurst, M. Sample selection for optimising signature enrolment. In Proceedings of the 10th International Workshop on Frontiers in Handwriting Recognition, La Baule, France, 23–26 October 2006.

21. Kahindo, C.; Garcia-Salicetti, S.; Houmani, N. A Signature Complexity Measure to Select Reference Signatures for Online Signature Verification. In Proceedings of the 2015 International Conference of the Biometrics Special Interest Group (BIOSIG), Darmstadt, Germany, 9–11 September 2015; pp. 1–8.

22. Tolosana, R.; Vera-Rodriguez, R.; Fierrez, J.; Ortega-Garcia, J. Exploring Recurrent Neural Networks for On-Line Handwritten Signature Biometrics. *IEEE Access* **2018**, *6*, 5128–5138. [CrossRef]

23. Lai, S.; Jin, L.; Yang, W. Online Signature Verification Using Recurrent Neural Network and Length-Normalized Path Signature Descriptor. In Proceedings of the 2017 14th IAPR International Conference on Document Analysis and Recognition (ICDAR), Kyoto, Japan, 9–15 November 2017; pp. 400–405.

24. Sekhar, C.; Mukherjee, P.; Guru, D.S.; Pulabaigari, V. OSVNet: Convolutional Siamese Network for Writer Independent Online Signature Verification. *arXiv* **2019**, arXiv:1904.00240. Available online: https://arxiv.org/abs/1904.00240 (accessed on 7 February 2020).

25. Wu, X.; Kimura, A.; Uchida, S.; Kashino, K. Prewarping Siamese Network: Learning Local Representations for Online Signature Verification. In Proceedings of the ICASSP 2019–2019 IEEE International Conference on Acoustics, Speech and Signal Processing (ICASSP), Brighton, UK, 12–17 May 2019; pp. 2467–2471.

26. Martinez-Diaz, M.; Fierrez, J.; Galbally, J.; Ortega-Garcia, J. Towards mobile authentication using dynamic signature verification: Useful features and performance evaluation. In Proceedings of the 19th International Conference on Pattern Recognition, Tampa, FL, USA, 8–11 December 2008; pp. 1–5.

27. Houmani, N.; Garcia-Salicetti, S.; Dorizzi, B.; El-Yacoubi, M. On-line signature verification on a mobile platform. In Proceedings of the International Conference on Mobile Computing Applications, and Services, Santa Clara, CA, USA, 25–28 October 2010; pp. 396–400.

28. Houmani, N.; Garcia-Salicetti, S. Quality criteria for on-line handwritten signature. In *Signal and Image Processing for Biometrics, ser. Lecture Notes in Electrical Engineering*; Scharcanski, J., Proença, H., Du, E., Eds.; Springer: Berlin/Heidelberg, Germany, 2014; pp. 255–283.

29. Blanco-Gonzalo, R.; Miguel-Hurtado, O.; Mendaza-Ormaza, A.; Sanchez-Reillo, R. Handwritten signature recognition in mobile scenarios: Performance evaluation. In Proceedings of the IEEE International Carnahan Conference on Security Technology (ICCST), Boston, MA, USA, 15–18 October 2012; pp. 174–179.

30. Blanco-Gonzalo, R.; Sanchez-Reillo, R.; Hurtado, O.M.; Liu-Jimenez, J. Usability analysis of dynamic signature verification in mobile environments. In Proceedings of the International Conference of the BIOSIG Special Interest Group (BIOSIG), Darmstadt, Germany, 5–6 September 2013; pp. 1–9.

31. Blanco-Gonzalo, R.; Sanchez-Reillo, R.; Liu-Jimenez, J.; Miguel-Hurtado, O. Performance evaluation of handwritten signature recognition in mobile environments. *IET Biom.* **2014**, *3*, 139–146. [CrossRef]

32. Martinez-Diaz, M.; Fierrez, J.; Galbally, J. The DooDB Graphical Password Database: Data Analysis and Benchmark Results. *IEEE Access* **2013**, *1*, 596–605. [CrossRef]

33. Martinez-Diaz, M.; Galbally, J.; Krish, R.P.; Fierrez, J. Mobile signature verification: Feature robustness and performance comparison. *IET Biom.* **2014**, *3*, 267–277. [CrossRef]

34. Sae-Bae, N.; Memon, N. Online Signature Verification on Mobile Devices. *IEEE Trans. Inf. Forensics Secur.* **2014**, *9*, 933–947. [CrossRef]

35. Antal, M.; Bandi, A. Finger or stylus: Their impact on the performance of on-line signature verification systems. *MACRo* **2015**, *2*, 11–22. [CrossRef]

36. Antal, M.; Zsolt Szabó, L.; Tordai, T. Online Signature Verification on MOBISIG Finger-Drawn Signature Corpus. *Mob. Inf. Syst.* **2018**, *2018*, 3127042. [CrossRef]

37. Tolosana, R.; Vera-Rodriguez, R.; Fierrez, J.; Morales, A.; Ortega-Garcia, J. Benchmarking desktop and mobile handwriting across COTS devices: The e-BioSign biometric database. *PLOS ONE* **2017**, *12*, e0176792. [CrossRef]

38. Zareen, F.J.; Jabin, S. Authentic mobile-biometric signature verification system. *IET Biom.* **2016**, *5*, 13–19. [CrossRef]

39. Nam, S.; Park, H.; Seo, C.; Choi, D. Forged Signature Distinction Using Convolutional Neural Network for Feature Extraction. *Appl. Sci.* **2018**, *8*, 153. [CrossRef]

40. Garcia-Salicetti, S.; Houmani, N.; Dorizzi, B. A Novel Criterion for Writer Enrolment Based on a Time-Normalized Signature Sample Entropy Measure. *EURASIP J. Adv. Signal Process.* **2009**, *2009*, 964746. [CrossRef]

41. Salicetti, S.G.; Houmani, N.; Dorizzi, B. A client-entropy measure for On-line Signatures. In Proceedings of the Biometrics Symposium, Tampa, FL, USA, 23–25 September 2008; pp. 83–88.

42. Houmani, N.; Garcia-Salicetti, S.; Dorizzi, B. A Novel Personal Entropy Measure confronted with Online Signature Verification Systems' Performance. In Proceedings of the IEEE 2nd International Conference on Biometrics: Theory, Applications and Systems, Arlington, VA, USA, 29 September–1 October 2008; pp. 1–6.

43. Houmani, N.; Garcia-Salicetti, S. On Hunting Animals of the Biometric Menagerie for Online Signature. *PLoS ONE* **2016**, *11*, e0151691. [CrossRef] [PubMed]

44. Houmani, N.; Garcia-Salicetti, S.; Dorizzi, B. On measuring forgery quality in online signatures. *Pattern Recognit.* **2012**, *45*, 1004–1018. [CrossRef]

45. Brault, J.J.; Plamondon, R. How to detect problematic signers for automatic signature verification. In Proceedings of the International Canadian Conference on Security Technology (ICCST), Zurich, Switzerland, 3–5 October 1989; pp. 127–132.

46. Abazid, M.; Houmani, N.; Garcia-Salicetti, S. Impact of Spatial Constraints when Signing in Uncontrolled Mobile Conditions. In Proceedings of the ACM Workshop on Information Hiding and Multimedia Security, Paris, France, 3–5 July 2019; pp. 89–94.

47. Ly Van, B.; Garcia-Salicetti, S.; Dorizzi, B. On using the Viterbi Path along with HMM Likelihood Information for On-line Signature Verification. *IEEE Trans. Syst. Man Cybern. Part B Cybern.* **2007**, *37*, 1237–1247. [CrossRef] [PubMed]

48. Garcia-Salicetti, S.; Houmani, N.; Ly Van, B.; Dorizzi, B.; Alonso-Fernandez, F.; Fierrez, J.; Ortega-GarciaClaus, J.; Vielhauer, C.; Scheidat, T. On-line Handwritten Signature Verification. In *Guide to Biometric Reference Systems and Performance Evaluation*; Petrovska-Delacrétaz, D., Chollet, G., Dorizzi, B., Eds.; Springer: London, UK, 2009; pp. 125–164.

49. Garcia-Salicetti, S.; Fierrez-Aguilar, J.; Alonso-Fernandez, F.; Vielhauer, C.; Guest, R.; Allano, L.; Trung, T.D.; Scheidat, T.; Ly Van, B.; Dittmann, J.; et al. Biosecure Reference Systems for On-Line Signature Verification: A Study of Complementarity. *Ann. Telecommun.* **2007**, *62*, 36–61.

Article

A Two-Stage Method for Online Signature Verification Using Shape Contexts and Function Features [†]

Yu Jia, Linlin Huang * and Houjin Chen

School of Electronic and Information Engineering, Beijing Jiaotong University, No. 3 Shangyuancun Haidian District, Beijing 100044, China; 16120010@bjtu.edu.cn (Y.J.); hjchen@bjtu.edu.cn (H.C.)
* Correspondence: huangll@bjtu.edu.cn; Tel.: +86-010-5168-8206
† This paper is an extended version of our paper published in PRCV 2018: Chinese Conference on Pattern Recognition and Computer Vision, Guangzhou, China, 23–26 November 2018.

Received: 18 March 2019; Accepted: 13 April 2019; Published: 16 April 2019

Abstract: As a behavioral biometric trait, an online signature is extensively used to verify a person's identity in many applications. In this paper, we present a method using shape contexts and function features as well as a two-stage strategy for accurate online signature verification. Specifically, in the first stage, features of shape contexts are extracted from the input and classification is made based on distance metric. Only the inputs passing by the first stage are represented by a set of function features and verified. To improve the matching accuracy and efficiency, we propose shape context-dynamic time warping (SC-DTW) to compare the test signature with the enrolled reference ones based on the extracted function features. Then, classification based on interval-valued symbolic representation is employed to decide if the test signature is a genuine one. The proposed method is evaluated on SVC2004 Task 2 achieving an Equal Error Rate of 2.39% which is competitive to the state-of-the-art approaches. The experiment results demonstrate the effectiveness of the proposed method.

Keywords: online signature verification; shape contexts; function features; SC-DTW; symbolic representation; two-stage method

---

1. Introduction

Biometric verification technology has aroused a lot of interest due to its reliability, effectiveness, and convenience in verifying personal identity [1]. Verification techniques based on face [2], fingerprint, and some such physiological biometric attributes have brought extra convenience and changed our lifestyle [3]. Although behavioral biometric attributes are slightly inferior to physiological ones in stability and uniqueness, they are more accessible and less intrusive to users. Voice, signature, gait, etc. are all typical behavioral attributes. Among them, signature remains the most widespread and recognized socially and legally verification approach in our day-to-day life [4]. Signing is a customary and fast movement driven by long-term nervous system and writing habit. Therefore, signature verification techniques can have more potential applications in the real world.

Depending on the different methods of signature acquisition, signature verification technique can be split into two categories: offline and online. In the system of offline signature verification [5], images containing signatures are collected after finishing the signing process. For online signature verification, signatures are captured by sensor-based devices while the user is signing and represented by a set of temporal functions, from which both static and dynamic features are extracted and then used to make a decision on whether the signature belongs to its claimed user. Compared with offline signature verification, the dynamic information collection of online signature ensures its uniqueness and higher difficulty to forge, so online signature verification technique usually owns better performance in accuracy and security.

There are two parts of an online signature verification system: enrolment and verification. Several signatures are provided as reference signatures by the users during enrolment and their extracted features along with calculated thresholds would be stored in the knowledge base. In verification, the authenticity of a test signature is evaluated by matching its features with those from reference signatures of its claimed user [6].

Online signatures are collected by electronic devices such as tablets, smart phones, and so on. Most of them use sensors to capture various real-time data such as coordinates, pressure, timestamp, etc. during signing. After collection, the signatures are represented as time series and then undergo preprocessing and feature extraction modules successively.

Online signature verification methods can be categorized based on the feature extraction process and matching strategy [7]. According to the employed features, there are broadly two groups: parametric and function features-based approaches. In the framework of parametric features-based methods, a signature is characterized as a vector of elements and each one is a representative of the value of one feature [8]. Examples of such attributes are width, height, average speed, etc. The dimensions of parametric features of signatures are all equal. In the function features-based method, a signature is represented by a multi-dimension feature set constituted by several time functions. Coordinate, timestamp, pressure, etc. are commonly used function features. Generally, the function features-based approaches perform better due to more dynamic information application, but these kinds of method consume more computational time and memory.

With regards to the matching methods, distance-based and model-based approaches are two main techniques [9]. Dynamic time warping (DTW) has been often adopted in distance-based methods [10]. DTW is a well-known approach for aligning vectors of different lengths. For application in signature verification, a set of features at each sample point is extracted and the similarity between the test signatures and enrolled reference signatures is then computed using dynamic programming. Point-based warping technique is a variant of DTW, wherein only selective points are warped. Extreme point [11] and stroke point [12] are often used. In addition, some works make a fusion of DTW with other methods. Sharma and Sundaram [9] propose a method that uses the information from DTW cost matrix and warping paths alignments. The decision is made by the conjunction of warping path score and DTW score. Yanikoglu and Kholmatov [13] fuse the Fast Fourier Transform with DTW and the fusion system lowers the error rate by up to about 25%. Chen and Xia [14] extract a set of function features for comparing the dissimilarity-based DTW between the test signature and the template database. In addition, the nearest template and majority vote are proposed to classify. Model-based approaches employ either generative-based classifiers such as hidden Markov model (HMM) [15–17] or discriminative ones such as neural network (NN) [18–20] and support vector machine (SVM) [21,22]. Also, there are some hybrid methods that combine different methods mentioned above. Multi-stage cascade framework [23], multi-stage decision-level score fusion [24,25] or a multi-expert system for signature verification [26,27] have been reported in the literature. Recently, inspired by the great success of recurrent neural networks (RNNs) in sequential modeling, several verification methods based on RNNs are proposed. Lai et al. [28] propose a novel descriptor called the length-normalized path signature (LNPS) for feature representation and then features are fed into the GRU (Gated recurrent unit) network. Triplet loss and center loss were used to train the network with the BP algorithm. The method proposed in [29] extracts 23 hand-crafted time function features and uses the bidirectional LSTM (Long short-term memory) and GRU networks with Siamese architecture to learn a dissimilarity metric from the pairs of signatures.

Although it is not that easy for a forger to fake a signature that is exactly the same as the genuine one, due to the large intra-class variations from one person and small inter-class variations between forgeries and genuine ones, accurate online signature verification still remains a challenging problem.

In real applications, the forgeries are usually classified to be two types, named skilled forgery and random one. A skilled forgery is signed by a person who had access to the genuine signatures and practiced for a while. A random forgery is signed without with any information about the signature, or even the name of the person whose signature is forged [30]. Compared with skilled forgeries, the random forgeries are more common in our daily life. Obviously, the skilled forgeries are more difficult to verify. In addition, the loss brought by accepting forgeries is higher than that by rejecting genuine signatures, which means accepting a signature as genuine should be stricter. Considering these factors, we propose a two-stage method using shape contexts and function features for accurate online signature verification. Features of shape contexts are extracted from the input firstly and classification of this stage is based on shape distance metric. Only the inputs passing by the first stage are represented by a set of function features and verified. To improve the matching accuracy and efficiency, we employ a shape context-dynamic time warping (SC-DTW) to compare the test signature with the enrolled reference ones based on the extracted function features. An interval-valued symbolic representation-based classifier is proposed to decide if the test signature is a genuine one.

The contributions of this paper are as follows:

- Based on the fact of unbalanced occurrence probability of skilled signature forgeries and random ones, a fast and accurate two-stage verification method is proposed.
- Shape context feature extractor is designed to describe global shape characteristics of signature for fast classification of random forgeries.
- SC-DTW is applied to fulfill comparison task and interval-valued-based representation classifier is proposed for final decision-making to achieve state-of-the-art verification performance.

This paper is an extended version of the one published in proceedings of PRCV2018 [31]. In this paper, more details on feature extraction and matching methods are given. Moreover, to further improve the performance method in the paper of PRCV2018, more effective features are extracted. Instead of distance metric classification, an interval-valued symbolic representation-based classifier is employed to enhance classification ability. Besides, more detailed experimental results are reported.

The rest of this paper is organized as follows. Section 2 details the methodology we proposed. Signature preprocessing is presented in Section 2.1. Section 2.2 presents the shape context descriptor and online signature verification method based on it. The function features extraction, feature alignment, and symbolic classifier are showed in Section 2.3. Section 2.4 discusses the two-stage verification protocol. The database used in our experiment, experimental results, and performance analysis are provided in Section 3. The conclusion is offered finally in Section 4.

2. Methodology

The diagram of the proposed method is shown in Figure 1. The input signature is first preprocessed for smoothing and normalization, and then it is fed into the shape context-based verification module, which does well in quickly distinguishing the random forgeries owing to their manifest differences in shape. Most obvious forgeries can be rule out in this stage. The signature passed through the first module is verified by function features-based verification module. This module achieves more accurate verification results due to the application of details in signature and decision fusion by interval-valued symbolic representation-based classifier.

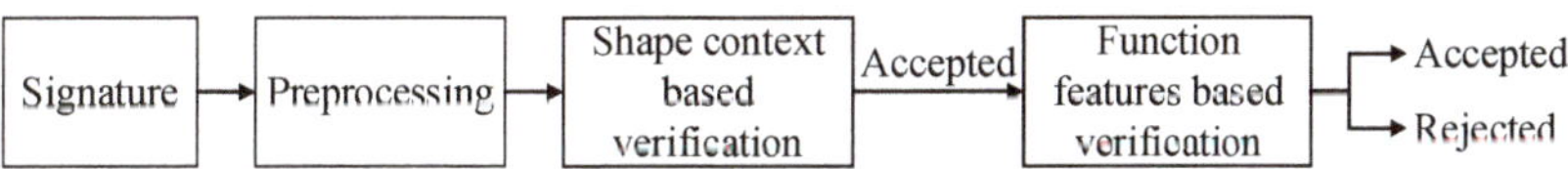

Figure 1. Diagram of proposed verification system.

2.1. Preprocessing

Captured by electronic devices, the time series of a signature are mixed with noises and fluctuations unavoidably. In addition, the acquired signatures of one individual vary with time or places, with the result that there are differences in size and location between signatures. Therefore, we firstly let the acquired signatures pass the preprocessing module to address those issues. The preprocessing module includes smoothing and normalization. Gaussian smoothing is employed to filter the artifacts and smooth the data. Then we adopt moment normalization technique [32] to standardize the size and location of acquired signatures.

Set the signature as $S = (s_1, s_2, ..., s_i, ..., s_N)$, $s_i = (x_i, y_i)$. N is the number of sample points, (x_i, y_i) is x and y coordinates information.

In the moment normalization technique, the size of a signature is not the difference between maximum and minimum in horizontal and vertical directions, but the width and height of the window derived from its moment, as is show in Figure 2. Denote the width and height of window as W and H, given by

$$W = 4\sqrt{\mu_{20}}, H = 4\sqrt{\mu_{02}} \tag{1}$$

μ_{pq} denotes the center moment, and (x_c, y_c) denotes the signature's centroid.

$$\mu_{pq} = \sum_x \sum_y (x - x_c)^p (y - y_c)^q \tag{2}$$

After window calculation, the size normalization technique is implemented as follows. The heights of the signatures are normalized to a predetermined value that in this paper is 300. Moreover, the aspect ratio of before and after preprocessing remains consistent to keep the signature shape unchanged.

$$x' = \alpha \times (x - x_c) + x'_c$$
$$y' = \beta \times (y - y_c) + y'_c \tag{3}$$

where x and y are smoothed originate coordinates. x' and y' are normalized coordinates. x'_c and y'_c are the centroid of normalized signature. α and β are the ratio of the normalized signature size to its original size, given by

$$\alpha = \frac{W_{norm}}{W},$$
$$\beta = \frac{H_{norm}}{H}, \tag{4}$$
$$\frac{W_{norm}}{H_{norm}} = \frac{W}{H}$$

where W_{norm} and H_{norm} denote the normalized width and height.

Signatures are centered at $(0, 0)$ to normalize their locations. After preprocessing, the signatures have the same size and location. In this paper, we did not employ translation normalization since we believe signature's angle is an out-of-habit feature. Figure 2 shows some examples of original signatures and corresponding preprocessed signatures.

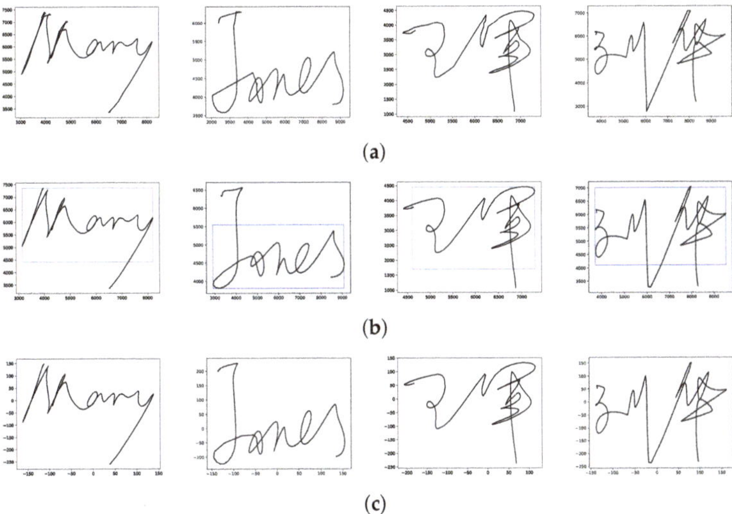

Figure 2. Examples of signature preprocessing. (**a**) Four English or Chinese examples of original signatures. (**b**) Window calculated by moment of signatures. (**c**) Preprocessed results of corresponding signatures.

2.2. Shape Context-Based Online Signature Verification

In the methods proposed for online signature verification, the dynamics properties of the signatures, for example, velocity, pressure, acceleration, etc. are widely applied. However, the shape of signature contains very useful details, which is critical for distinguish a signature between forgery and genuine one. The method proposed by Gupta and Joyce [33] extracts the dynamics properties of position extreme points of signatures and achieved better performance. Features based on shape also have been successively applied in offline signature verification [34].

In this paper, we propose a verification method based on shape context features. Specifically, shape context descriptor [34,35] is used to extract features of a signature and a cost matrix is computed. After finding the best one-to-one matching between two signatures' shape and modeling transformation, the measurable shape distance is used for classification. To further improve the efficiency, only trend-transition-points (TTPs) that can represent the shape of a signature roughly are used for calculating distance.

2.2.1. Shape Context Feature Extraction

Shape context descriptor captures the distribution over relative positions of shape points and the connectivity properties between features points along curves. Therefore, shape context features not only provide global characterization of shape but also contain more contextual information within a certain range of a signature. Besides, shape context descriptor is designed in a way of describing shapes that allows for measuring shape similarity and the recovering of point correspondences. Traditionally, the first step is to randomly select a set of points that lie on the edges of two shapes separately. Here the shape of an online signature is represented by a set of sampled points which in this work is (x_i, y_i), $i = 1, 2, ..., N$. N is the number of sampled points.

Figure 3 shows the shapes and shape context histograms of a reference signature, a genuine signature and a skilled forgery of one user. Because the writing speed is a kind of relatively fixed and unique information, the number and distribution of sample points between genuine signature and reference signature are more similar. Taking one point as the origin of polar coordinate, the shape contexts of this point can be represented using log-polar histogram. We set five bins for *logr* and

12 bins for θ. The number of neighboring points that fall into the very bin is just the histogram value. Figure 3d–f present the corresponding histograms for certain points. We can see that the difference in shape context histograms between genuine signature and reference signature is relatively small, while the histogram of skilled forgeries is quite dissimilar to the reference ones.

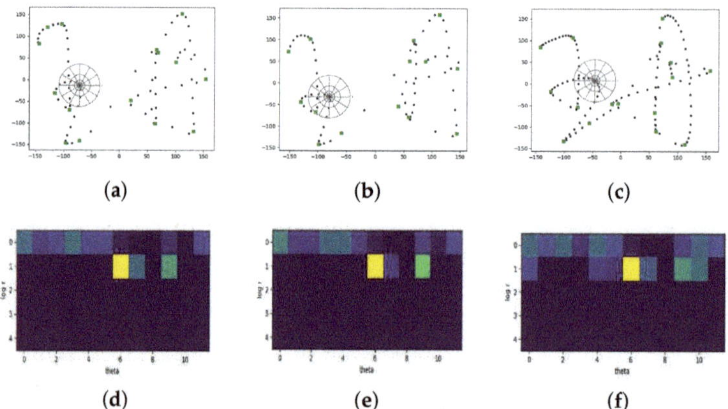

(a)   (b)   (c)

(d)   (e)   (f)

Figure 3. Examples of shape context feature extraction. (**a**) A reference signature of one user. (**b**) A corresponding genuine signature from the same user with reference one. (**c**) A skilled corresponding forgery from the same user with reference one. The green square points represent selected trend-transition-points. (**d–f**) Shape context histograms for chosen trend-transition-point in the signatures of (**a–c**), respectively.

Considering a point p_i on the first shape and a point q_j on the second shape, denote $C_{ij} = C(p_i, q_j)$ as the matching cost of these two points, given by

$$C_{ij} = C(p_i, q_j) = \frac{1}{2} \sum_{k=1}^{K} \frac{[h_i[k] - h_j[k]]^2}{h_i[k] + h_j[k]} \tag{5}$$

where $h_i[k]$ and $h_j[k]$ denote the k_{th} bin histogram at p_i and q_j respectively.

For all pairs of points p_i on the first shape and q_j on the second shape, calculate the cost as Equation (5) and then we got a cost matrix. The next step is to find the optimal alignment between two shapes that minimizes total cost. This can be done by the Hungarian method with time complexity of $O(N^3)$. The cost between shape contexts is based on the chi-square test statistic that is not a suitable distance metric. Thin plate spline (TPS) model is adopted for modeling transformation [35]. After that, we get the measurable distance of two shapes. The smaller the distance, the more similar these two shapes, or vice versa. So, if the average distance between test signature and reference signatures is lower than a threshold, it would be accepted as a genuine signature of its claimed user.

2.2.2. Trend-Transition-Point Selection

The shape context representation of signature should not only capture specific shape features but also allow considerable variations. Besides, the computational load of distance calculation is closely related to the number of points. Therefore, with the hope of efficiency improvement and variances tolerance, a few representative points are selected. Only selected points can participate in shape distance calculation. In this paper, we propose the TTP selection method.

Trend-transition-points are the points where the curve trends before and after them are completely different while the trends between two successive TTPs do not have obvious change so that the curve shape of the segment approximates to a straight line. So, the signature could be re-constructed with these points. In our method, local extreme points and corner points are all defined as TTPs. The local extreme points are selected depending on its value greater or smaller than its neighborhood. The corner points selection we adopted is proposed in [36,37], which makes use of the smaller eigenvalues of covariance matrices of regions of support.

Let $S_k(s_i)$ denotes the region of support (ROS) of point s_i, a small curve segment containing itself and k points in its left and right neighborhoods. That is

$$S_k(s_i) = [s_j = (x_j, y_j) | j = i - k, i - k + 1, \cdots, i + k - 1, i + k]$$

where (x_j, y_j) are the Cartesian coordinates of s_i.

Therefore, the 2×2 covariance matrix for points in the segment $S_k(s_i)$ is calculated. λ_L and λ_S are two eigenvalues corresponding to the covariance matrix. The smaller eigenvalues λ_S can be used to measure prominence of corners over its ROS. In other words, sharper corner points have the large λ_S and weaker corners have small one. When the points are on a straight line or on a flat curve, the λ_S will be very small, even approximate to zero. So, corners can be determined if its λ_S exceeds a predetermined threshold.

Shape contexts are calculated on every point, but only TTPs are used to distance calculation. For every sample point of signatures, the algorithm is implemented as follows.

Step 1:   If the point is a start point, add it to TTP dataset. Else, go to Step 2;
Step 2:   If the point is an end point, add it to TTP dataset and go to Step 5. Else, go to Step 3;
Step 3:   If the point is an extreme point, add it to TTP dataset. Else, go to Step 4;
Step 4:   If the point is a corner point, add it to TTP dataset. Else, head to next sample point and return to Step 2;
Step 5:   For all points in TTP dataset, the point with smaller λ_S would be deleted when the distance of two successive points is lower than a threshold. The process repeats several times until the distances between points are long enough.

2.3. Function Features-Based Online Signature Verification

One of the advantages of online signature verification is that signature is captured by specialized sensors-based devices. So dynamic information can be recorded and used for verification, which makes verification more accurate and reliable. In function features-based methods, a set of function features, such as position, pressure, velocity, acceleration, etc., is firstly captured. Then matching between features of the test and the reference and decision-making are implemented.

2.3.1. Function Features Extraction

Usually, lots of features can be obtained directly from the specialized electronic devices. Horizontal and vertical position, pressure and timestamp of each sample point are the basic measurements. Let x, y, p, t be the mentioned basic measurements, $n = 1, 2, 3, \ldots N$ be the discrete time index of the temporal functions and N be the time duration of a signature in sampling units [14]. Based on them, various features can be derived. Among them, 20 frequently used function features are selected. The features are grouped according to their properties, such as position-related, pressure-related, velocity-related, acceleration-related, and angle-related. The features are listed in Table 1.

Table 1. Function features extracted for online signature verification.

Category	Description	Symbols
Position-related	x coordinate	$x(n)$
	y coordinate	$y(n)$
	Displacement	$S(n) = \sqrt{x(n)^2 + y(n)^2}$
	Change of x coordinate	$\Delta x_n = x(n+1) - x(n)$
	Change of y coordinate	$\Delta y_n = y(n+1) - y(n)$
	Change of displacement	$\Delta S(n) = \sqrt{(\Delta x(n))^2 + (\Delta y(n))^2}$
Pressure-related	Pressure	$p(n)$
	Change of pressure	$\Delta p_n = p(n+1) - p(n)$
Velocity-related	x velocity	$v_x[n] = \frac{x(n+1)-x(n)}{t(n+1)-t(n)}$
	y velocity	$v_y[n] = \frac{y(n+1)-y(n)}{t(n+1)-t(n)}$
	Total velocity	$v(n) = \sqrt{v_x^2(n) + v_y^2(n)}$
Acceleration-related	x acceleration	$a_x[n] = \frac{v_x(n+1)-v_x(n)}{t(n+1)-t(n-1)}$
	y acceleration	$a_y[n] = \frac{v_y(n+1)-v_y(n)}{t(n+1)-t(n-1)}$
	Total acceleration	$a(n) = \sqrt{a_x^2(n) + a_y^2(n)}$
	Centripetal acceleration	$a_c(n) = [v_x(n) \cdot a_y(n) - v_y(n) \cdot a_x(n)]/v(n)$
Angle-related	Cosine of the angle between x-axis and signature curve	$\cos \alpha = \dfrac{x(n+1)-x(n)}{\sqrt{(x(n+1)-x(n))^2+(y(n+1)-y(n))^2}}$
	Sine of the angle between x-axis and signature curve	$\sin \alpha = \dfrac{y(n+1)-y(n)}{\sqrt{(x(n+1)-x(n))^2+(y(n+1)-y(n))^2}}$
	Cosine of the angle between x velocity and total velocity	$\cos \beta = v_x(n)/v(n)$
	Angle between x-axis and signature curve	$\theta(n) = \tan^{-1} \frac{y(n+1)-y(n)}{x(n+1)-x(n)}$
	Angle velocity	$v_\theta(n) = \frac{\theta(n+1)-\theta(n)}{t(n+1)-t(n)}$

2.3.2. Matching Based on Shape Context-Dynamic Time Warping (SC-DTW)

Feature matching is very critical for function features-based verification. In recent years, DTW has been widely applied as the matching technique in online signature verification. The DTW method compress or expand the time axis of two temporal functions locally to make them aligned.

Here are two time series $T = (t_1, t_2, ..., t_N)$ and $R = (r_1, r_2, ..., r_M)$ and their lengths are N and M respectively. The similarity between the n_{th} point of T and the m_{th} point of R are calculated according to defined similarity rule. All the similarity values constitute a DTW cost matrix denoted by $d(m, n)$ defined as:

$$d(m, n) = \| r_m - t_n \| \tag{6}$$

The overall distance is calculated as following equation:

$$D(m, n) = d(m, n) + \min \begin{cases} D(m, n-1) + C \\ D(m-1, n-1) \\ D(m-1, n) + C \end{cases} \tag{7}$$

where $D(n, m)$ is the cumulative distance up to the current element and C is gap cost. To alleviate the situation of signature at different length, the distance is normalized by Equation (8).

$$d = \frac{D}{\sqrt{M \times N}} \tag{8}$$

DTW has been an effective method of finding the alignment between two signatures with different length. However, a time series has both numerical nature and shape nature. DTW warps time series depending on the similarity of their numerical characteristics as Equation (6) but ignores the shape properties. It may lead to abnormal alignment sometimes. Zhang and Tang [38] propose a novel variant of DTW, named SC-DTW. The SC-DTW employs shape context to replace the raw observed values used in conventional DTW, getting ahead in time series data mining. In this paper, we adopted the SC-DTW for function features-based verification to further improve the accuracy.

Specifically, the alignment of two point is decided by their shape matching cost of shape contexts, which means

$$d(n,m) = C_{nm} \tag{9}$$

where C_{nm} is defined in Equation (5).

Under this circumstance, a function feature is considered to be a 1-D array and a 2-D shape. The problem of measuring the similarity of two function features can be translate into how similar these two shapes. Figure 4 shows the process of SC-DTW. Figure 4a,b are the time series of 11_{th} function feature v listed in Table 1 of two signatures from the same user. Figure 4c,d are the corresponding shape context histograms. That the shape context is similar means the sample points in time series are well matched. Please note that the application of shape contexts is only used to find the alignment between two time series. The measurable cumulative distance of them is still obtained by the original cost matrix for the convenience of following classification.

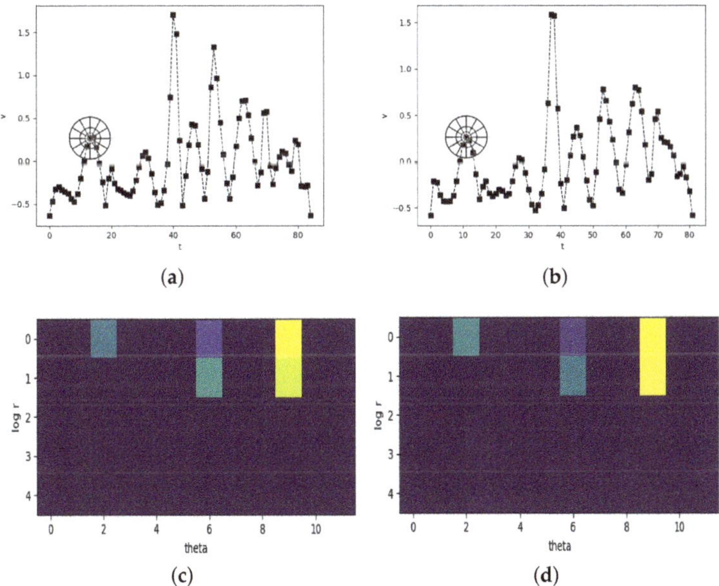

Figure 4. SC-DTW. (**a,b**) Time series of total velocity v from two signatures and a pair of corresponding points found by shape context. (**c,d**) show the shape context histograms of the points marked in (**a,b**), respectively.

Given a $N^{(k)} \times D$ feature set $X^{(k)}$, extracted from a reference signature and a $N^{(q)} \times D$ feature set $X^{(q)}$, extracted from a signature which is claimed to belong to the same user, a D-dimensional vector $z^{(k,q)}$ called 'similarity feature vector' can be derived by calculating the similarity between each pair of corresponding function features using SC-DTW mentioned above.

2.3.3. Classification Based on Interval-Valued Symbolic Representation

The concept of symbolic data analysis has been applied in the field of document image analysis and cluster analysis. Interval-valued and histogram-valued symbolic representation can represent the variability and distribution of feature values. Guru and Prakash [39] extract global features of signature to form an interval-valued feature vectors and proposed a method for verification and recognition based on the symbolic representation. Pal and Alaei [5] also propose an interval-valued symbolic representation-based method for offline verification. In this paper, we first use the interval-valued symbolic representation to model the similarity features derived from SC-DTW and then build a classifier for verification.

Let $[ref_1, ref_2, \cdots, ref_n]$ be a set of n enrolled reference signatures of user. In addition, denote D as the similarity feature vector of an user, where D_{ij}^r is the SC-DTW distance of feature r between signature ref_i and ref_j, as is showed in Table 2. Each user has a feature vector like that. For the k_{th} feature, we compute the statistical mean μ_k and standard deviation σ_k and the lower and upper bound of interval value can be computed as Equation (10).

$$
\begin{aligned}
sf_k &= ([f_k^-, f_k^+], \mu_k, \sigma_k) \\
f_k^- &= \mu_k - \alpha\sigma_k \\
f_k^+ &= \mu_k + \alpha\sigma_k \\
\mu_k &= mean(f_k) \\
\sigma_k &= std(f_k)
\end{aligned}
\tag{10}
$$

where sf_k is the symbolic representation of k_{th} feature of a user and includes an interval value and two continuous values. α is a scalar to control the upper and lower limit of each feature. In addition, the symbolic feature vectors are computed for all users and stored in the template base for future verification.

Table 2. Similarity feature vector of each individual.

Fea. Ref.	f_1	f_2	$\cdots$	f_r	$\cdots$	f_D
ref_1/ref_2	D_{12}^1	D_{12}^2	$\cdots$	D_{12}^r	$\cdots$	D_{12}^D
ref_1/ref_3	D_{13}^1	D_{13}^2	$\cdots$	D_{13}^r	$\cdots$	D_{13}^D
$\cdots$	$\cdots$	$\cdots$	$\cdots$	$\cdots$	$\cdots$	$\cdots$
ref_2/ref_3	D_{23}^1	D_{23}^2	$\cdots$	D_{23}^r	$\cdots$	D_{23}^D
ref_2/ref_4	D_{24}^1	D_{24}^2	$\cdots$	D_{24}^r	$\cdots$	D_{24}^D
$\cdots$	$\cdots$	$\cdots$	$\cdots$	$\cdots$	$\cdots$	$\cdots$
ref_i/ref_j	D_{ij}^1	D_{ij}^2	$\cdots$	D_{ij}^2	$\cdots$	D_{ij}^D
$\cdots$	$\cdots$	$\cdots$	$\cdots$	$\cdots$	$\cdots$	$\cdots$

For signature verification problem, the signature is compared with all the reference signatures belonging to the claimed ID. Let $F_t = [f_{t1}, f_{t2}, \cdots, f_{tD}]$ denote a D-dimensional feature vector representing the average SC-DTW distance with reference signatures. In addition, denote $sf = [[f_{r1}^-, f_{r1}^+], [f_{r2}^-, f_{r2}^+], \cdots, [f_{rD}^-, f_{rD}^+]]$ as the reference signatures of the claimed identity described by an interval-valued symbolic feature vector. Each feature value of the test signature is compared with corresponding interval in sf to examine whether it lies within the interval. The feature value represents the dissimilarity of two signatures. That is, the more similar the two signatures, the smaller the value and the closer to 0. The total value of features of a test signature which fall inside the interval value decides how this test signatures is similar to genuine ones, as is showed in Equations (11) and (12). Define A as the measure of measure of degree of authenticity:

$$
A = \sum_{i=1}^{D} C(f_{ti}, [f_{ri}^-, f_{ri}^+])
\tag{11}
$$

where

$$C(f_{ti}, [f_{ri}^-, f_{ri}^+]) = \begin{cases} 2 & \text{if } 0 \le f_{ti} \le f_{ri}^- \\ 1 & \text{if } f_{ri}^- < f_{ti} \le f_{ri}^+ \\ 0 & \text{else} \end{cases} \tag{12}$$

If the acceptance count A is greater that a threshold th, the test signature will be classified as a genuine signature of its claimed user. In the user-dependent scenario, every person has its own A which is computed using those training samples. For each training signature, there is an A we got. For each person, we compute several A and then average them thus getting A_m. The threshold th equals to $\beta \times A_m$.

2.4. Two-Stage Online Signature Verification

The forgery can be classified to be two types, named skilled forgery and random one. In real applications, random forgeries appear more frequently while skilled forgeries occur less. On the other hand, skilled forgeries are much more difficult to be verified correctly. In this paper, we propose a method using shape contexts and function features as well as a two-stage strategy for accurate online signature verification. The shape context-based verification module is firstly used to reject obvious random forgeries quickly while the function features-based verification module is applied to re-check the signatures survived from the previous module. In this way, the whole system can achieve higher accuracy and consume less computation cost at the mean time.

Two metrics named FRR (False Reject Rate) and FAR (False Accept Rate) have been widely used to evaluate signature verification system. For cascade structure applied in our method, the relationship of FRR and FAR between the sub-verification modules and the whole system are showed in Table 3, where p denotes the reject percentage of first sub-verification module. Obviously, p takes the value smaller than 1.

$$\begin{aligned} r_1 < r_2 &\Rightarrow pr_1 + (1-p)r_2 < r_2 \\ p < 1 &\Rightarrow (1-p)a_2 < a_2 \end{aligned} \tag{13}$$

Table 3. FRR and FAR of individual verification modules and cascade system.

System Framework	FRR	FAR
Shape Context Module	r_1	a_1
Feature Function Module	r_2	a_2
Cascade of two Modules	$pr_1 + (1-p)r_2$	$(1-p)a_2$

It can be seen that the performance of the cascade system depends on the thresholds of two sub-verification modules. If $p < 1$ and r_1 is set to be smaller than r_2, the cascade system can achieve better performance than the sub-verification modules in terms of false acceptance rate, which is illustrated in Equation (13).

3. Experimental Results

3.1. Database and Evaluation Measurement

To evaluate the effectiveness of the proposed method, we run a set of experiments on public database SVC 2004 Task2. There are 40 users and each user has 20 genuine signatures and 20 skilled forgeries. These genuine signatures are collected in two sessions, spaced apart by at least one week. The skilled forgeries are contributed by who could replay the writing sequence of the signatures on the computer screen and practice the forgeries for a few times until they were confident to proceed to the actual data collection. The signatures are mostly in either English or Chinese [40]. In our experiments, for each of the users, we randomly select five male/female genuine signatures for enrolment as reference signatures. The signatures are chosen from the first or second session. The remaining 15 genuine signatures (not selected for enrolment) and 20 skilled forgeries are considered for testing the performance of our proposal. As for the random-forgeries scenario, corresponding to any user, we randomly select 20 signatures from other users. The trial is conducted ten times for each user.

We evaluate the performance primarily using the Equal Error Rate (EER): which is the error when false acceptance rate is equal to false rejection rate. We considered two forms of calculating EER: EER-*commonThreshold* and EER-*userThreshold*. EER-*commonThreshold* is calculated using a global decision threshold. In this case, all the feature values from all training signatures are used to find an optimal value based on minimum EER. The same threshold is shared by all users. EER-*userThreshold* is using user-specific decision threshold. It is derived from feature values of training samples of each user. For the respective user, the best threshold corresponds to his/her lowest EER. Since there are multiple users in the database SVC 2004, the average of EER across all users is applied as overall performance of the method when using user threshold in our experiments.

3.2. Experiment Results

Performance evaluations of shape context (SC)-based verification and function features (FF)-based verification method is firstly conducted. Here Skilled and Random denotes skilled forgeries and random forgeries. In the case of common threshold, the Receiver Operating Characteristic (ROC) curves are given to evaluate the performance. As for user-dependent threshold set-up, EER of every user are expressed as histograms.

The results of these two methods are shown in Figures 5 and 6. From the results, we can see that both the SC and FF method perform well, and the better results are achieved using user thresholds on random forgery verification. It is a general statement that the usage of user threshold usually can yield better performance than common threshold, as is proved by the results. For common threshold, it is difficult to use one value to cover the differences of different individuals. For user threshold, the value is user-specific, varying from one user to another.

As descried in the previous section, 20 frequently used features are categorized into 5 groups according to their properties. To achieve best performance and to investigate contributions of different features, we run a series of experiments. Since only single feature or single feature group cannot provide enough classification ability for online signature verification, we test several combinations of feature groups. For clear illustration, we use $G1 - G5$ to represent the 5 groups: position-related, pressure-related, velocity-related, acceleration-related, and angle-related. The symbol $\cup$ denotes combination of different groups. The experimental results are given in Table 4. From the results, we can see that using all 20 features performs the best. It is also shown that when velocity-related FF are removed, verification performance deteriorates a lot.

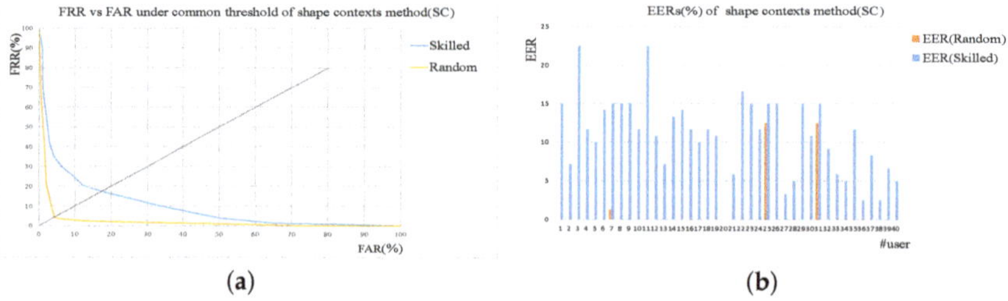

Figure 5. Results of shape context-based verification method (SC). (a) ROC curves under common threshold. (b) EER of each user under user threshold.

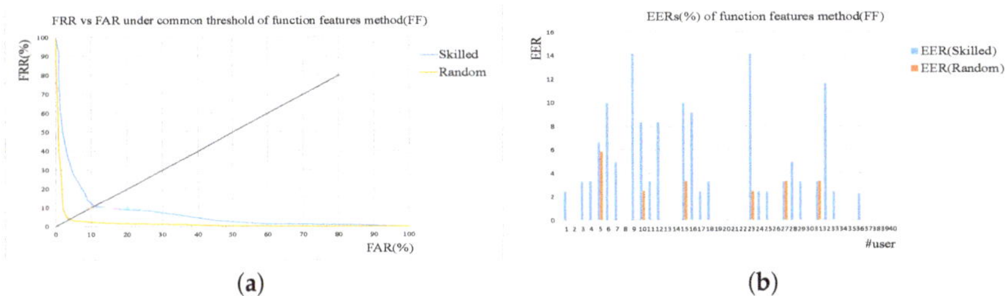

(a)   (b)

Figure 6. Results of function features-based verification method (FF). (**a**) ROC curves under common threshold. (**b**) EER of each user under user threshold.

Table 4. Comparisons between different group of function features.

Group	Not Included Feature Group	EER(SF)	EER(RF)
G1 ∪ G2 ∪ G3 ∪ G4 ∪ G5	None	10.8	4.5
G1 ∪ G2 ∪ G3 ∪ G4	Angle-related	11.5	4.8
G1 ∪ G2 ∪ G3 ∪ G5	Acceleration-related	12.2	6.3
G1 ∪ G2 ∪ G4 ∪ G5	Velocity-related	13.5	7.2
G1 ∪ G3 ∪ G4 ∪ G5	Pressure-related	11.8	6.8
G2 ∪ G3 ∪ G4 ∪ G5	Position-related	11.2	5.9

To compare the performances of SC and FF method more clearly, the experimental results of two methods are shown together. Figure 7 gives the results of SC and FF method on skilled forgery while Figure 8 on random one. From the figures, it can be seen that for random forgery verification the performances of SC method and Feature Function method (FF) are similar while FF method outperforms SC method much more on skilled forgery verification. As described in the previous section, SC method is good at extracting global features from signatures with low computation cost, which are quite effective and sufficient for random forgery verification. FF method extracts more detailed features, thus achieving better performance than SC method on skilled forgery verification.

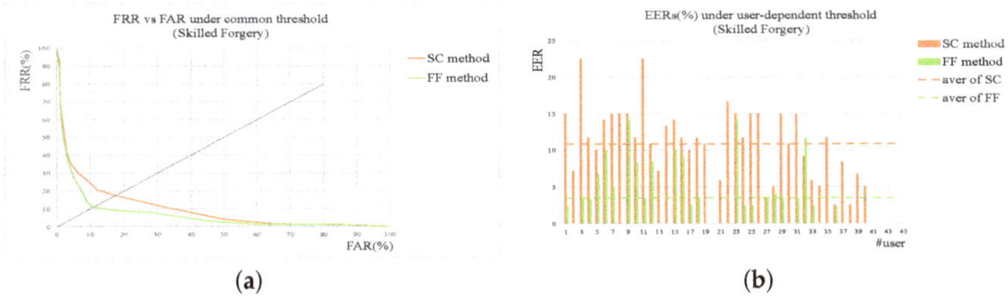

(a)   (b)

Figure 7. Results of SC and FF method on skilled forgery. (**a**) ROC under common threshold. (**b**) EER of each user under user threshold. In addition, those dotted lines are the average levels of corresponding methods.

In real applications, random forgeries occur much more frequently that skilled ones. Based on the experimental results, a cascade verification method is designed and tested. The shape context-based verification method is firstly used to reject obvious random forgeries quickly while the function features-based verification method is applied to re-check the signatures survived from the previous module. As illustrated in Section 2.4, FRR of SC method should be smaller than function features-based verification to achieve higher accuracy with lower computation cost. In case of common threshold, FRR of the SC method is set to be 1% and 65% skilled forgeries and 25% random forgeries can be accepted by the second module for re-verification. Figures 9 and 10 give the detailed results on SC method, function feature method, and the two-stage method. Table 5 shows the detailed results on EERs. From the results, it can be seen that the two-stage method achieves the best performance with tolerable computation cost.

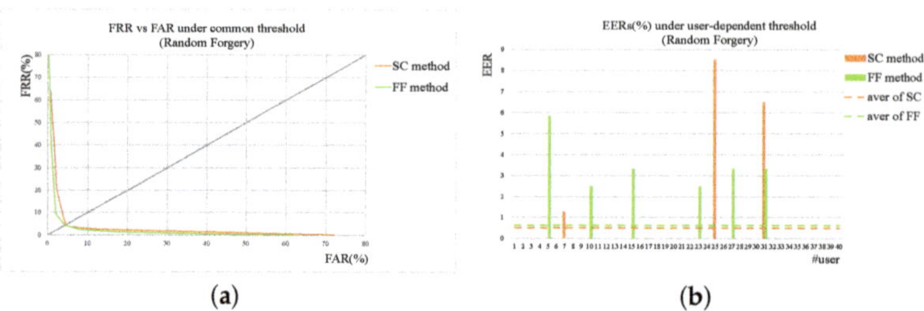

Figure 8. Results of SC and FF method on random forgery. (**a**) ROC under common threshold. (**b**) EER of each user under user threshold. In addition, those dotted lines are the average levels of corresponding methods.

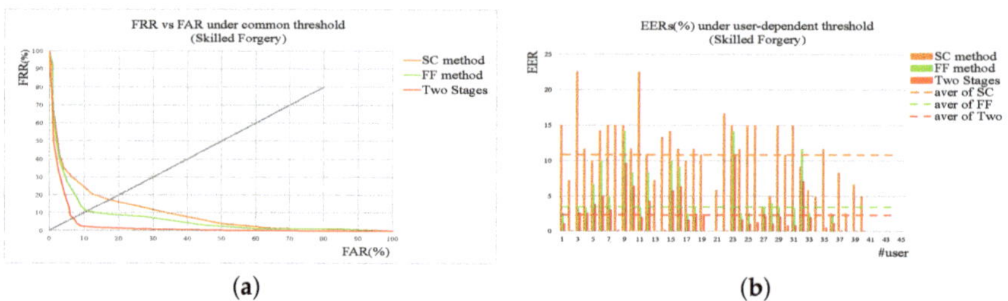

Figure 9. Results of SC, FF, and two-stage method on skilled forgery. (**a**) ROC under common threshold. (**b**) EER of each user under user threshold. In addition, those dotted lines are the average levels of corresponding methods.

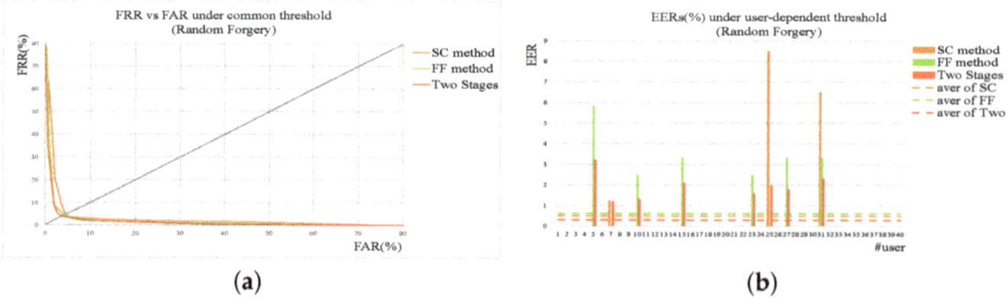

Figure 10. Results of SC, FF, and two-stage method on random forgery. (**a**) ROC under common threshold. (**b**) EER of each user under user threshold. In addition, those dotted lines are the average levels of corresponding methods.

Comparisons with the state of the art on database SVC2004 are given in Table 6. It is not easy to make fair comparisons of online signature verification methods due to different databases, training, testing, etc. We select several recently published works which use the same database (SVC2004) with us. The method proposed by Lai et al. [28] based on GRU network obtained slightly higher EER than our method. However, it needs more training samples and consumes more computation costs.

Table 5. Verification results (EER%) of different methods with common threshold and user threshold.

Method	Time(s)	Common Threshold		User Threshold	
		EER(SF)	EER(RF)	EER(SF)	EER(RF)
Shape context-based verification	0.47	17.4	4.5	10.45	0.5
Function features-based verification	1.26	10.8	4.3	3.5	0.62
Two-stage verification	1.04	6.92	3.8	2.39	0.3

Table 6. Comparisons with the state-of-the-art works on database SVC2004.

Works	Method	EER(%)
Song et al., 2016, [41]	DTW with SCC	2.89
Liu et al., 2017, [42]	Spare representation	2.98
Xia et al., 2018, [6]	GMM+DTW with SCC	2.63
Sharma et al., 2018, [9]	DTW+warping path alignment	2.53
Lai et.al., 2017, [28]	LNPS+GRU	2.37
Proposed method	Two-stage verification	2.39

4. Conclusions

In this paper, we propose a two-stage method using SCs and FF for accurate online signature verification. Features of SCs are extracted from the input firstly and classification of this stage is based on shape distance metric. Only the inputs passing by the first stage are represented by a set of FF and verified. To improve the matching accuracy and efficiency, we propose a SC-DTW to compare the test signature with the enrolled reference ones based on the extracted FF. Then an interval-valued symbolic representation-based classifier is proposed to decide if the test signature is a genuine one. The proposed method is evaluated on SVC2004 Task 2 database achieving an EER of 2.39% which is competitive to the state-of-the-art approaches. The experiment results demonstrate the effectiveness of the proposed method.

Author Contributions: L.H., Y.J. and H.C. contributed to algorithm and system design; Y.J. conducted the experiments; L.H. and Y.J. contributed to experiment results analysis and manuscripts.

Funding: This research was funded by National Natural Science Foundation of China (NSFC) grant number 61271306.

Acknowledgments: The authors thank for the help of reviewers and editors.

References

1. Plamondon, R.; Pirlo, G.; Impedovo, D. *Online Signature Verification*; Springer: London, UK, 2014; pp. 156–161.
2. Seal, A.; Bhattacharjee, D.; Nasipuri, M.; Gonzalo-Martin, C.; Menasalvas, E. A-trous wavelet transform-based hybrid image fusion for face recognition using region classifiers. *Expert Syst.* **2018**, *35*, 12307. [CrossRef]
3. Jain, A.K.; Griess, F.D.; Connell, S.D. On-line signature verification. *Pattern Recognit.* **2007**, *35*, 2963–2972. [CrossRef]
4. Mohammed, R.A.; Nabi, R.M.; Mahmood, M.R.; Nabi, R.M. State-of-the-Art in Handwritten Signature Verification System. In Proceedings of the International Conference on Computational Science and Computational Intelligence, Las Vegas, NV, USA, 7–9 December 2015; pp. 519–525.
5. Pal, S.; Alaei, A.; Pal, U.; Blumenstein, M. Interval-valued symbolic representation based method for off-line signature verification. In Proceedings of the International Joint Conference on Neural Networks, Killarney, Ireland, 12–17 July 2015; pp. 1–6.
6. Xia, X.; Song, X.; Luan, F.; Zheng, J.; Chen, Z.; Ma, X. Discriminative feature selection for on-line signature verification. *Pattern Recognit.* **2018**, *74*, 422–433. [CrossRef]
7. Rua, E.A.; Castro, J.L.A. Online Signature Verification Based on Generative Models. *IEEE Trans. Syst. Man Cybern.* **2012**, *42*, 1231–1242.

8. Impedovo, D.; Pirlo, G. Automatic Signature Verification: The State of the Art. *IEEE Trans. Syst. Man Cybern.* **2008**, *38*, 609–635. [CrossRef]

9. Sharma, A.; Sundaram, S. On the Exploration of Information From the DTW Cost Matrix for Online Signature Verification. *IEEE Trans. Cybern.* **2018**, *48*, 611–624. [CrossRef] [PubMed]

10. Griechisch, E.; Malik, M.I.; Liwicki, M. Online Signature Verification Based on Kolmogorov-Smirnov Distribution Distance. In Proceedings of the International Conference on Frontiers in Handwriting Recognition, Heraklion, Greece, 1–4 September 2014; pp. 738–742.

11. Feng, H.; Wah, C.C. Online signature verification using a new extreme points warping technique. *Pattern Recognit. Lett.* **2003**, *24*, 2943–2951. [CrossRef]

12. Kar, B.; Mukherjee, A.; Dutta, P.K. Stroke Point Warping-Based Reference Selection and Verification of Online Signature. *IEEE Trans. Instrum. Meas.* **2018**, *67*, 2–11. [CrossRef]

13. Yanikoglu, B.; Kholmatov, A. Online Signature Verification Using Fourier Descriptors. *Eurasip J. Adv. Signal Process.* **2009**, *2009*, 260516. [CrossRef]

14. Chen, Z.; Xia, X.; Luan, F. Automatic online signature verification based on dynamic function features. In Proceedings of the IEEE International Conference on Software Engineering and Service Science, Beijing, China, 26–28 Auguest 2016; pp. 964–968.

15. Bao, L.V.; Garcia-Salicetti, S.; Dorizzi, B. On Using the Viterbi Path Along With HMM Likelihood Information for Online Signature Verification. *IEEE Trans. Syst. Man Cybern.* **2007**, *37*, 1237–1247.

16. Muramatsu, D.; Kondo, M.; Sasaki, M.; Tachibana, S.; Matsumoto, T. A Markov chain Monte Carlo algorithm for bayesian dynamic signature verification. *IEEE Trans. Inform. Forensics Secur.* **2006**, *1*, 22–34. [CrossRef]

17. Fierrez, J.; Ortega-Garcia, J.; Ramos, D.; Gonzalez-Rodriguez, J. HMM-based on-line signature verification: Feature extraction and signature modeling. *Pattern Recognit. Lett.* **2007**, *28*, 2325–2334. [CrossRef]

18. Fuentes, M.; Garciasalicetti, S.; Dorizzi, B. On-Line Signature Verification: Fusion of a Hidden Markov Model and a Neural Network via a Support Vector Machine. In Proceedings of the International Workshop on Frontiers in Handwriting Recognition, Niagara on the Lake, ON, Canada, 6–8 Auguest 2002; pp. 253–258.

19. Lejtman, D.Z.; George, S.E. On-line Handwritten Signature Verification Using Wavelets and Back-propagation Neural Networks. In Proceedings of the International Conference on Document Analysis and Recognition, Seattle, WA, USA, 10–13 September 2001; pp. 992–996.

20. Rashidi, S.; Fallah, A.; Towhidkhah, F. Feature extraction based DCT on dynamic signature verification. *Sci. Iran.* **2012**, *19*, 1810–1819. [CrossRef]

21. Gruber, C.; Gruber, T.; Krinninger, S.; Sick, B. Online Signature Verification With Support Vector Machines Based on LCSS Kernel Functions. *IEEE Trans. Syst. Man Cybern.* **2010**, *40*, 1088–1100. [CrossRef]

22. Swanepoel, J.; Coetzer, J. Feature Weighted Support Vector Machines for Writer-Independent On-Line Signature Verification. In Proceedings of the International Conference on Frontiers in Handwriting Recognition, Heraklion, Greece, 1–4 Sept. 2014; pp. 434–439.

23. Liu, N.N.; Wang, Y.H. Fusion of global and local information for an on-line Signature Verification system. In Proceedings of the International Conference on Machine Learning and Cybernetics, Kunming, China, 12–15 July 2008; pp. 57–61.

24. Fierrez-Aguilar, J.; Nanni, L.; Lopez-Peñalba, J.; Ortega-Garcia, J.; Maltoni, D. *An On-Line Signature Verification System Based on Fusion of Local and Global Information*; Springer: Berlin/Heidelberg, Germany, 2005; pp. 523–532.

25. Nanni, L.; Maiorana, E.; Lumini, A.; Campisi, P. Combining local, regional and global matchers for a template protected on-line signature verification system. *Expert Syst. Appl.* **2010**, *37*, 3676–3684. [CrossRef]

26. Bovino, L.; Impedovo, S.; Pirlo, G.; Sarcinella, L. Multi-Expert Verification of Hand-Written Signatures. In Proceedings of the Seventh International Conference on Document Analysis and Recognition, Edinburgh, UK, 6 Auguest 2003; pp. 932–936.

27. Wan, L.; Wan, B.; Lin, Z.C. On-line signature verification with two-stage statistical models. In Proceedings of the International Conference on Document Analysis and Recognition, Seoul, Korea, 31 August–1 September 2005; Volume 1, pp. 282–286.

28. Lai, S.; Jin, L.; Yang, W. Online Signature Verification Using Recurrent Neural Network and Length-Normalized Path Signature Descriptor. In Proceedings of the International Conference on Document Analysis and Recognition, Kyoto, Japan, 9–15 November 2017; pp. 400–405.

29. Tolosana, R.; Verarodriguez, R.; Fierrez, J.; Ortegagarcia, J. Exploring Recurrent Neural Networks for On-Line Handwritten Signature Biometrics. *IEEE Access* **2018**, *6*, 5128–5138. [CrossRef]

30. Kholmatov, A.; Yanikoglu, B. Identity authentication using improved online signature verification method. *Pattern Recognit. Lett.* **2005**, *26*, 2400–2408. [CrossRef]

31. Jia, Y.; Huang, L. Online Signature Verification Based on Shape Context and Function Features. In Proceedings of the Chinese Conference on Pattern Recognition and Computer Vision, Guangzhou, China, 23–26 November 2018; pp. 62–73.

32. Liu, C.L.; Nakashima, K.; Sako, H.; Fujisawa, H. Handwritten digit recognition: Investigation of normalization and feature extraction techniques. *Pattern Recognit.* **2004**, *37*, 265–279. [CrossRef]

33. Gupta, G.K.; Joyce, R.C. Using position extrema points to capture shape in on-line handwritten signature verification. *Pattern Recognit.* **2007**, *40*, 2811–2817. [CrossRef]

34. Agam, G.; Suresh, S. Shape matching through particle dynamics warping. In Proceedings of the IEEE Conference on Computer Vision And Pattern Recognition, Minneapolis, MN, USA, 17–22 June 2007; pp. 1–7.

35. Belongie, S.J.; Malik, J.; Puzicha, J. Shape matching and object recognition using shape contexts. *IEEE Trans. Pattern Anal. Mach. Intell.* **2002**, *24*, 509–522. [CrossRef]

36. Tsai, D.M.; Hou, H.T.; Su, H.J. *Boundary-Based Corner Detection Using Eigenvalues of Covariance Matrices*; Elsevier Science Inc.: Amsterdam, The Netherlands, 1999; pp. 31–40.

37. Horng, W.B.; Chen, C.W. Revision of Using Eigenvalues of Covariance Matrices in Boundary-Based Corner Detection. *IEICE Trans. Inf. Syst.* **2009**, *92*, 1692–1701. [CrossRef]

38. Zhang, Z.; Tang, P.; Duan, R. Dynamic time warping under pointwise shape context. *Inf. Sci.* **2015**, *315*, 88–101. [CrossRef]

39. Guru, D.S.; Prakash, H.N. Online Signature Verification and Recognition: An Approach Based on Symbolic Representation. *IEEE Trans. Pattern Anal. Mach. Intell.* **2008**, *31*, 1059–1073. [CrossRef]

40. Yeung, D.Y.; Chang, H.; Xiong, Y.; George, S.E.; Kashi, R.S.; Matsumoto, T.; Rigoll, G. SVC2004: First International Signature Verification Competition. In *Biometric Authentication*; Lecture Notes in Computer Science; Springer: Cham, Switzerland, 2004; pp. 16–22.

41. Song, X.; Xia, X.; Luan, F. Online Signature Verification Based on Stable Features Extracted Dynamically. *IEEE Trans. Syst. Man Cybern. Syst.* **2017**, *47*, 2663–2676. [CrossRef]

42. Liu, Y.; Yang, Z.; Yang, L. Online Signature Verification Based on DCT and Sparse Representation. *IEEE Trans. Cybern.* **2017**, *45*, 2498–2511. [CrossRef]

Article

Online Signature Verification Based on a Single Template via Elastic Curve Matching

Huacheng Hu [1], Jianbin Zheng [1], Enqi Zhan [1,*] and Jing Tang [2]

[1] School of Information Engineering, Wuhan University of Technology, Wuhan 430070, China; honyeal@whut.edu.cn (H.H.); jbzheng@whut.edu.cn (J.Z.)
[2] Hubei Collaborative Innovation Center for High-Efficient Utilization of Solar Energy, Hubei University of Technology, Wuhan 430068, China; mimitang85119@163.com
* Correspondence: eqzhan@whut.edu.cn; Tel.: +86-1530-7109-170

Received: 29 August 2019; Accepted: 29 October 2019; Published: 7 November 2019

Abstract: Person verification using online handwritten signatures is one of the most widely researched behavior-biometrics. Many signature verification systems typically require five, ten, or even more signatures for an enrolled user to provide an accurate verification of the claimed identity. To mitigate this drawback, this paper proposes a new elastic curve matching using only one reference signature, which we have named the curve similarity model (CSM). In the CSM, we give a new definition of curve similarity and its calculation method. We use evolutionary computation (EC) to search for the optimal matching between two curves under different similarity transformations, so as to obtain the similarity distance between two curves. Referring to the geometric similarity property, curve similarity can realize translation, stretching and rotation transformation between curves, thus adapting to the inconsistency of signature size, position and rotation angle in signature curves. In the matching process of signature curves, we design a sectional optimal matching algorithm. On this basis, for each section, we develop a new consistent and discriminative fusion feature extraction for identifying the similarity of signature curves. The experimental results show that our system achieves the same performance with five samples assessed with multiple state-of-the-art automatic signature verifiers and multiple datasets. Furthermore, it suggests that our system, with a single reference signature, is capable of achieving a similar performance to other systems with up to five signatures trained.

Keywords: curve similarity; curve similarity model; curve similarity transformation; similarity distance; segmentation matching; evolutionary computation

---

1. Introduction

Biometric authentication has always been a field of primary concern in the security application field [1,2]. Person authentication or verification using handwritten signatures is one of the most widely researched behavior-biometrics and the most popular method for identity verification [3]. Usually, signature verification systems can be divided into two categories, namely, off-line and on-line systems, which have a significant difference. Dynamic signatures are too difficult to imitate and forge, even for skilled forgers [4] because they are unique and consistent for a given period. Compared with off-line signatures [5], online signatures are more robust and gain a higher level of security by monitoring dynamic features like time series of position trajectories, pressure, altitude, and azimuth. There is a tendency to recover online signatures from offline signature images [6]

Online signature verification can basically be viewed as a problem of similarity discrimination, whereby a decision must be made about whether a given online signature corresponds to the claimed identity or not. In a signature verification system, we compare the features of a test signature against those from a set of genuine signatures of an enrolled user, which can be called reference signatures or template signatures. By stable and discriminative feature extraction and selection, there are two approaches to identify the authenticity of a signature, which can be called the function approach and the parameter approach.

Signature verification methods based on the parameter approach include the statistical classification [7,8], neural network [9], support vector machine (SVM) [10], Bayesian decision [11] and features cluster [12], where some global or local features derived from the original signature signal, e.g., average speed, pressure, the number of strokes, etc., constitute signature feature patterns or feature vectors.

The signature verification system based on a function considers each signature signal as a function of time and verifies the signer by comparing the reference signature with the test signature directly. Usually, matching procedures or special function parameter calculations are a need between signatures, requiring more time and space. The common approaches include Dynamic Time Wrapping (DTW) [13], its improved version [14–19] and the hidden Markov model (HMM) [20,21].

Bonus template matching approaches are considered, and a longest common subsequences (LCSS) combined elastic distance metrics is also used [22]. A discrete cosine transform (DCT) [23] has been applied to 44 time signals. A multi-section vector quantization (VQ) approach [24] has been suggested where all signatures are represented by vectors of the same length. Similar methods such as the Fourier description [25], wavelet packet and discrete wavelet transform (DWT) [26] have been presented too.

A function-based system utilizes all original information about the signature, and shows better performance than parameter-based systems. Nowadays, fusion for improving verification accuracy has become a promising trend, and a combination of parametric approaches and functional approaches is often adopted in literature [16–18,27].

1.1. Related Work

Although a signature can show individual behavior features, it is more unstable and diverse than other biometric verification technologies such as fingerprint recognition, iris recognition, face recognition and so on. Due to changes in the internal and external environment, there are fluctuations in the size, location and rotation angle of signatures with the same signer at different input times. In addition, signatures will not maintain high consistency for a long time as writing habits and the external environment change. As a result, two repetitions of a signature from the same writer never have an identical appearance [28]. Each person can even have several signatures of diverse styles, and one style of signature is obviously not suitable for the verification of another style of signature. Of course, the style variability also makes signature verification better for privacy protection than fixed biometric recognition technology.

In a traditional signature verification system, a large number of samples are mandatory when building a reliable statistical classifier and many algorithms even also require skilled forgery samples [29]. Figuring out a stable signature region is also a hot topic in recent research. Similarly [19,29], extraction of a stable signature region also depends on availabiliy of a large number of training signatures. In practice, it is often impractical to obtain various signature samples from a signer, which limits the applicability of a signature verification system.

How to reduce the enrollment signature size is a crucial issue. Another problem is how to reduce the differences between different signatures, that is, the problem of signature alignment is also a key problem to be solved.

It is the most widely used and recommended method for size alignment by max-min normalization [30]. Some template matching methods, such as DTW [14], LCSS [22] and so on, also apply for alignment. Recently, alignment methods based on Gaussian mixture model (GMM) have been developed [17], but training a Gauss model requires a large number of samples.

For selecting effective reference signatures, the intra-class variation of genuine signatures can be quantified with a correlation-based criterion which detects and recovers non-linear time distortions in different specimens as described in [31].

A single reference signature system (SRSS) for training with only a single reference signature has been proposed in [28], which followed the strategy of duplicating the reference signature to enlarge the training set. In this work, the strategy consists of duplicating the given signature a number of times and training an automatic signature verifier with each of the resulting signatures and the duplication scheme is based on a sigma lognormal decomposition of the reference signature.

Nevertheless, in a real situation, it is sometimes difficult to obtain enough signatures from a signer, especially in commercial applications and forensic covers. Therefore, this paper discusses the model and method of designing an automatic signature verification system using only one real reference signature per enrolled signer. Moreover, in this study, it is vitally important to effectively align the test signature to the reference signature for verification in order to cut down the influence of fluctuations caused by variances of size, location and rotation angle, which may deteriorate the performance of verification.

The signature trajectory can be viewed as a 2D/3D curve. The similarity between two signatures can be measured by curve similarity [32]. Curve similarity is a major category of similarity measure and a large number of similarity problems can be transformed or abstract into curve similarity problems.

Measuring curve similarity is a common method for curve matching. The curves are usually assumed to be represented as polygonal chains in the plane and to be measured by distance such as DTW or Fréchet distance as in [33]. The Fréchet distance, which relies on fewer features, can be applied for signature verification as proposed in [33].

By computing cumulative distance, DTW provides normalization and alignment as a computational technique to determine the best match between two curves, which might produce different sample points. The Fréchet distance belongs to a general class of distance measures that are sometimes called "dog-man" distances and is a max measure which is outlined in terms of the maximum leash length over a parameterization. However, two classical curve similarity measures are sensitive to data anomaly points and cannot adapt to changes in the translation and scaling of the curve.

1.2. This Paper

Our goal of this paper is to design an automatic signature verification system for a SRSS. To this end, a new curve similarity measure model and calculation method has been established, which we call the curve similarity model (CSM). The curve similarity draw lessons from geometric similarity, and can be adapted to various transformations such as translation, scaling and rotation, and can better be adapted to the inconsistency of signatures such as signature size, position and rotation angle in the signature curve.

The procedure presented in this paper considers a rigorous and adaptive CSM to build a robust SRSS. Therefore, we completed the exploratory work reported in [32], proposed a continuous and discrete curve similarity model based on transformation, and accomplished the curve optimal matching calculation based on evolutionary computation (EC) [32]. Based on the characteristics of the SRSS, a differentiated fusion feature named local similarity score (LSC) is designed for the difference calculation between two signatures.

The paper is organized as follows: Section 2 introduces the relevant definitions of CSM. Section 3 describes the proposed curve similarity calculation method and process. The fourth section describes the optimal sectional matching of signature curves and local matching feature extraction for SRSS. The experimental results will be presented and discussed in Section 3. Conclusions are drawn in the last section.

2. Model and Method

2.1. Curve Similarity Model

2.1.1. Original Definition

One curve is typically represented by a function. A definition must be provided to study the problem of the curve similarity. In geometry, there is a strict definition for shape similarity, which is an accurate similarity. In engineering applications, due to the large number of error factors, the definition of fuzzy curve similarity is adopted. Taking a 2D plane curve as an example, a kind of curve similarity is defined as follows:

Definition 1 ([32]). *Given functions $f_1(x)$ and $f_2(x)$, $d(f_1, f_2) = \int_{C_1}^{C_2} |f_1(x) - f_2(x)| dx$ is the distance between two functions, and also known as the function similarity distance or the curve similarity distance, where $[C_1, C_2]$ is the function definition domain or the definition domain.*

Definition 2 ([32]). *For a given threshold ε, if $d(f_1, f_2) < \varepsilon$, then $f_1(x)$ and $f_2(x)$ are similar, otherwise they are not.*

As mentioned above in the definitions of curve similarity distance (CSD) and curve similarity, the two functions have the same definition domains, that is to say, two curves must be aligned first, which is very limited in practical application. In most cases, the definition domains of two functions are different, and it is necessary to perform a truncation, translation, stretching, or even rotation transformation on a function to calculate the similarity distance.

As shown in Figure 1, given a curve L, the curves L_1 to L_3 are new curves obtained after applying different transformations such as translation or stretching. If the above calculation method is adopted, the distances between L_1 to L_3 and L are different. Similar problems are available in the calculation of DTW and Fréchet distances of the curves.

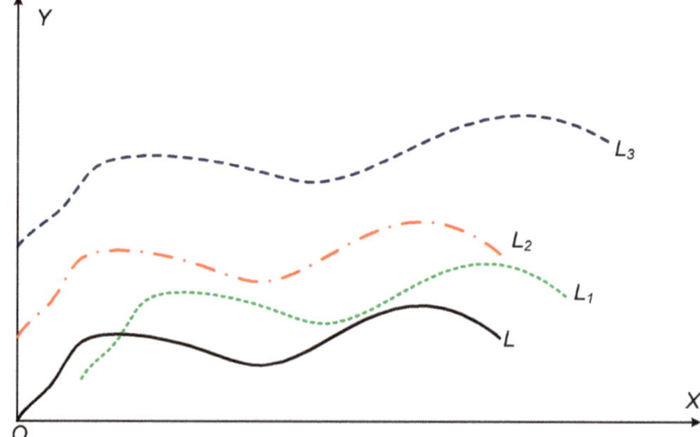

Figure 1. Translation and stretching transformation of curves.

From the perspective of transformation, the above several curves are similar, and the distance is 0. Therefore, when to measure the similarity between curves, the transformation of the curve should be taken into consideration. Firstly, a curve similarity distance definition based on curve transformation may be suggested.

2.1.2. Improvement Definition

Definition 3. *Given functions $f_1(x)$ defined on $[R_1, R_2]$ and $f_2(x)$ defined on $[C_1, C_2]$, and make transform $T \rightarrow f'_1(x) = k \cdot f_1((x-b)/a) - h$, then the minimum distance of all the matching distances between $f'_1(x)$ and $f_2(x)$ with different transformation T is as follows:*

$$
\begin{aligned}
dis(a,b,k,h) &= \int_{aR_1+b}^{aR_2+b} \left| T(f_1(x)) - f_2(x) \right| dx \\
&= \int_{aR_1+b}^{aR_2+b} \left| f'_1(x) - f_2(x) \right| dx \\
&= \int_{aR_1+b}^{aR_2+b} \left| k \cdot f_1((x-b)/a) - h - f_2(x) \right| dx \\
&= \int_{R_1}^{R_2} a \cdot \left| k \cdot f_1(t) - h - f_2(a \cdot t + b) \right| dt \quad t = (x-b)/a \\
&= \int_{R_1}^{R_2} a \cdot \left| k \cdot f_1(x) - h - f_2(a \cdot x + b) \right| dx
\end{aligned}
\tag{1}
$$

$$
d(f_1, f_2) = d(f'_1, f_2) = \min\{dis(a,b,k,h)\}
\tag{2}
$$

which is called the curve similarity distance of $f_2(x)$ to $f_1(x)$ under the transformation T, where $f_1(x)$ is called the reference function or the reference curve, $f_2(x)$ is called the comparison function or the comparison curve, $f'_1(x)$ is called the transform function or the transform curve, T is called a function similarity transformation or a curve similarity transformation (CST), and dis(a,b,k,h) is called the distance of the two curves under the curve similarity transformation T.

Obviously, the distance between the two curves is different under different similarity transformations T, and the curve similarity distance (CSD) is the distance after the optimal matching of the reference curve for the comparison curve. After the curve similarity transformation, the curve similarity distance is denoted by min{*dis(a,b,k,h)*}.

Once the optimal matching of the curves is obtained, the corresponding curve similarity distance can be obtained, as shown in Figure 2.

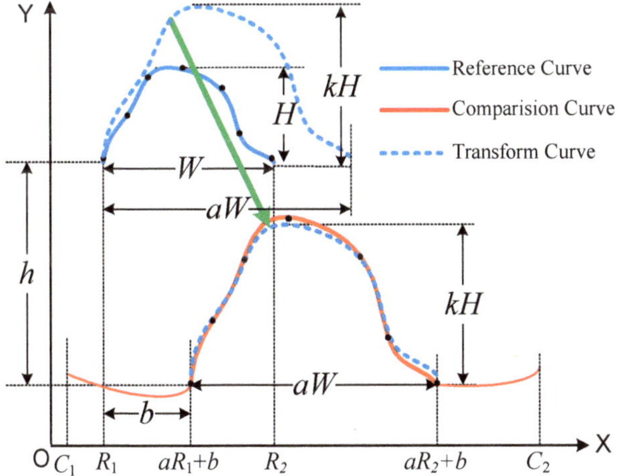

Figure 2. The curve similarity transformation between the two curves.

From Figure 2, a, k are translational transformations in the horizontal and vertical directions, respectively, and b, h are scaling transformations in the horizontal and vertical directions, respectively. Therefore, the similarity of curves based on curve similarity transformation is defined as follows:

Definition 4. *Given a reference curve $f_1(x)$ and a comparison curve $f_2(x)$, for a given threshold ε, if $d(f_1 f_2) \leq \varepsilon$, then the curve $f_2(x)$ is called similar to curve $f_1(x)$, and vice versa.*

2.1.3. Discrete Definition

In engineering applications, the expression of the curve is therefore difficult to obtain, and can only be represented by an implicit function. By sampling, a continuous curve can be represented by a set of discrete ordered points. In this way, it is a common problem to measure the similarity of the two curves, which is to determine the similarity between two discrete ordered sets. As a result, a discrete curve similarity definition is needed. Similarly, the definition of discrete curve similarity transformation is given by reference to planar image transformation.

Definition 5. *Given one discrete curve $F_A = \{(x_1, y_1), (x_2, y_2) \dots (x_m, y_m)\}$, and*

$$T = \begin{bmatrix} a & 0 & 0 \\ 0 & k & 0 \\ b & h & 1 \end{bmatrix} \tag{3}$$

is the curve similarity transformation matrix, where the meanings of a, b, k, h are the same as those described in Formula (1), corresponding to the translation and scale transformation of the horizontal and translation directions, respectively. $F'_A = T \cdot F_A = \left\{(x_1^*, y_1^*), (x_2^*, y_2^*), \cdots, (x_m^*, y_m^*)\right\}$ is called the similarity transformation curve of F_A, where:

$$\begin{cases} x_i^* = a \cdot x_i + b \\ y_i^* = k \cdot y_i + h \end{cases} \tag{4}$$

Without any doubt, the transformation matrix T can be more complicated, such as rotation, mirroring and shearing transformation, and various combinations thereof. Moreover, the definition of discrete curve similarity distance under the condition of curve similarity transformation can be given.

Definition 6. *Given a discrete curve $F_A = \{(x_1, y_1), (x_2, y_2) \dots (x_m, y_m)\}$ as a reference curve, another curve $F_B = \{(x'_1, y'_1), (x'_2, y'_2) \dots (x'_n, y'_n)\}$ is considered as a comparison curve. $F'_A = T \bullet F_A$ is transformed by a similarity transformation matrix T from F_A. Among all similarity transformation matrices T, the minimum distance of all the matching distances between F'_A and F_B is as follows:*

$$\begin{aligned}
dis(t, a, b, k, h) &= \frac{1}{m} \sum_{i=1, j=i+t}^{m} \left| T(x_i, y_i) - (x'_j, y'_j) \right| \\
&= \frac{1}{m} \sum_{i=1, j=i+t}^{m} \left| (x_i^*, y_i^*) - (x'_j, y'_j) \right| \\
&= \frac{1}{m} \sum_{i=1, j=i+t}^{m} \left| (a \cdot x_i + b, k \cdot y_i + h) - (x'_j, y'_j) \right| \\
&= \frac{1}{m} \sum_{i=1, j=i+t}^{m} \left| (a \cdot x_i + b - x'_j, k \cdot y_i + h - y'_j) \right| \\
&= \frac{1}{m} \sum_{i=1, j=i+t}^{m} \sqrt{(a \cdot x_i + b - x'_j)^2 + (k y_i + h - y'_j)^2}
\end{aligned} \tag{5}$$

$$D(F_A, F_B) = D(F'_A, F_B) = \min\{dis(t, a, b, k, h)\}, \tag{6}$$

which is called the curve similarity distance of F_B to F_A under the transformation T. The meanings of a, b, k, h are the same as those described in Formula (1) and Figure 2, corresponding to the translation and scale *transformation of the horizontal and translation directions, respectively. Meanwhile, t is the starting point position of the optimal matching of the reference curve on the comparison curve.*

Likewise, the definition of discrete curve similarity is as follows:

Definition 7. *Given a discrete reference curve F_A and a comparison curve F_B, for a given threshold ε, if $D(F_A, F_B) \leq \varepsilon$, then the curve F_B is called similar to curve F_A, and vice versa.*

Finally, calculating the similarity distance of the comparison curve to the reference curve is equivalent to calculating the optimal matching of the reference curve in the sense of the average distance on the comparison curve after the transformation. Therefore, performing a similarity measure between two curves requires two steps, one is to calculate the optimal matching, and the other is to perform threshold discrimination.

In the classic curve matching algorithm DTW, once a curve of two curves is translated or stretched, the distance measurement between them will be modified. In the curve similarity measure process, the curve similarity distance calculation depends on a transformation matrix. The characteristics of the transformation matrix can well prevent the deformation process of the translation, stretching and rotation of the curve. However, the transformation matrix is sometimes difficult to solve directly.

Accordingly, an evolutionary computation (EC) [32] algorithm has been introduced to obtain the transformation matrix and the optimal matching interval between two curves by minimizing the matching distance. The specific search algorithm is introduced later.

2.1.4. Matching Calculation

It can be observed from the definition that the curve similarity distance is the optimal matching of the reference curve with the comparison curve under different similarity transformations. Therefore, the matching distance between two curves under different CSTs can be obtained by random search algorithm. EC is a very effective algorithm for intelligent random search, where the similarity transformation matrix T is the parameter space of random search.

The whole idea of EC is to generate S random populations $POP = \{POP(0), POP(1), \dots , POP(S - 1)\}$, each of which corresponds to a set of parameters of CST T and a fitness value *fitness* $= D(F_A, F_B)$ from Formula (6), where the *fitness* is smaller, the population is better. At the same time in each iteration search, each individual will generate new descendants near itself, and the best individuals can be chosen to enter the next iteration search.

The above process is repeated until the iterative search reaches the maximum number of iterations, at which point the optimal individual parameters and fitness will be regarded as CST and CSD, respectively.

At the same time, the parameters (t, a, b, k, h) should meet certain constraint conditions and adopt real coding for the signature curve matching. The boundary condition of the following system is $0 \leq t < 2n$, $0.90 \leq a \leq 1.10$, $0.90 \leq k \leq 1.10$, $-100 \leq b < 100$, and $-100 \leq h \leq 100$, where a and k are elastic scales in the horizontal and vertical directions, and b and h are translations.

The Algorithm 1 process is as follows:

Algorithm 1. Optimal matching calculation by EC

$i = 0$
For $j = 0$: $S - 1$ **Do**
//random generation of parameters *t, a, b, k, h*
initial a population $POP_i(j) = (t, a, b, k, h)$
calculate fitness of each member in $POP_i(j)$
End For
While $i < Iterations$ **Do**
  $i = i + 1$
  //sorting and classification
  order POP_{i-1} by fitness in ascending and divide them into 4 levels
  //acceleration search
  For $j = 0:S - 1$ **Do**
    If $POP_{i-1}(j)$ is at level *kind*
      //where, *kind* = 1, 2, 3, 4, indicating the classification level of each population
      //random generation of new parameters *t, a, b, k, h* in the neighborhood
      //i.e., $a = $ rand($POP_{i-1}(j).a$, *kind*)
      generate new *kind*+1 subpopulations *nPOP* from $POP_{i-1}(j)$
      calculate fitness of each member in *nPOP*
      //sorting subpopulations
      order *nPOP* by fitness in ascending
      //select best subpopulation *nPOP* as $POP_{i-1}(j + S)$
        $POP_{i-1}(j + S) = nPOP(0)$
    End If
  End For
  //global selection
  order POP_{i-1} by fitness in ascending, where there are $2S$ populations
  select the top S from POP_{i-1} as POP_i
End While

When generating new *kind* + 1 subpopulations from one population, each subpopulation has different parameter generation range at different level. Generally speaking, the *fitness* is smaller, the parameter generation range of each subpopulation is narrower.

Suppose that the *j*-th population $POP(j)$ is at the *kind*-th level, on which parameter generation range is $GR_{kind} = kind \bullet GR_{max}/4$, *kind* + 1 subpopulations *nPOP* can be randomly generated, here:

$$nPOP = \left\{ nPOP(k) \middle| \left| nPOP(k) - POP(j) \right| < GR_{kind}, k = 0 \sim kind \right\} \tag{7}$$

For example, for parameter *a*, $GR_{max} = \max(a) - \min(a)$. When *kind* = 1 and a new random number $r = -1$ or 1 is generated, $a^{new} = a^{old} + r \bullet GR_{kind} = a^{old} + r \bullet GR_{max}/4$. At the same time, check whether a^{new} satisfies the boundary condition.

The other parameters *t, b, k, h* all perform similar operations. Therefore, from the new subpopulation process generated by the parameters represented by each original population, the parameters of the generated subpopulations are determined according to the classification level of the original population. Although this parameter is also random, it varies randomly to the original population by its classification level.

2.2. Proposed System

Online signature verification system can be discriminated by the similarity distance of two signature curves, where one signature can be called as the reference signature and the other can be called as the comparison signature. For instance, two signature trajectories can be considered as plane curves which can be also divided into *X* curves and *Y* curves, as shown in Figure 3.

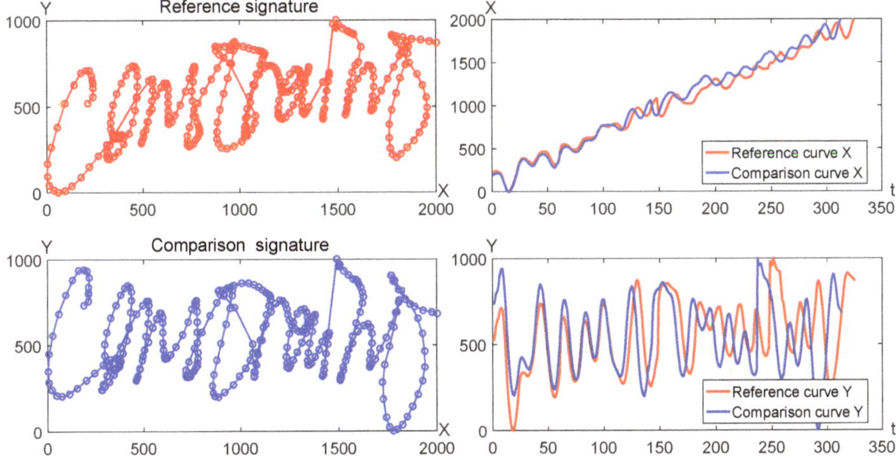

Figure 3. Two signature curves and their corresponding X and Y curves.

For a reference signature curve and a comparison signature curve, if the similarity distance between them is sufficiently small, it can be considered that the comparison signature is genuine one, otherwise it is a forged signature. A block diagram of the proposed system is illustrated in Figure 4.

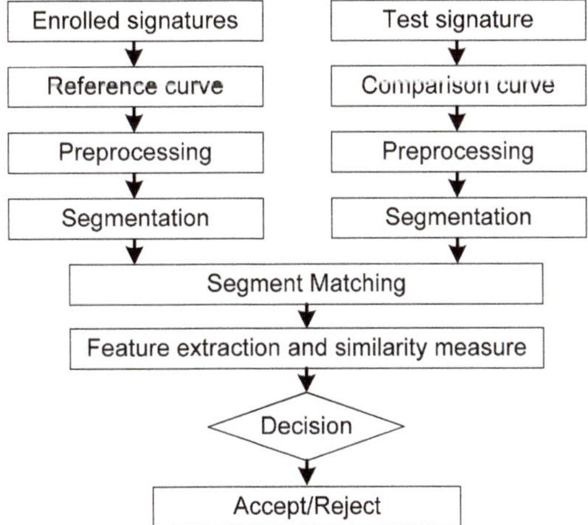

Figure 4. Block diagram of the proposed system.

Given $SignR = \{(x_1, y_1), (x_2, y_2) \ldots (x_M, y_M)\}$ and $SignC = \{(x'_1, y'_1), (x'_2, y'_2) \ldots (x'_N, y'_N)\}$ as the reference curve and the comparison curve, respectively. For the calculation of the similarity distance of the signature curves, if the reference curve is calculated as a whole with the comparison signature, it will lead to greater errors. Segmentation curve matching can be used to better measure local differences between curves, which could be common in complex curve similarity measures.

2.2.1. Preprocessing

As the variations in different signatures have different dynamic ranges, min–max normalization is implemented in their X and Y coordinates. One signature should be reprocessed as follows:

$$x_i = 2000 \times \frac{x - \min(x)}{\max(x) - \min(x)}, y_i = 1000 \times \frac{y - \min(y)}{\max(y) - \min(y)} \tag{8}$$

where x and y are the original coordinates, and x_i and y_i are the normalized coordinates. The normalization scales of horizontal and vertical directions are different, and 2000 and 1000 are taken respectively to keep the original scale of the signature curve as much as possible.

2.2.2. Segmentation

SignR is divided into K sections and there are m data points of each the reference segmentation curve. Each the reference segmentation curve can be defined as

$$\begin{cases} (SignR)_i = \left\{(x_t, y_t) \middle| t \in [R_i, R_i'], R_i' - R_i = m\right\} \\ m = \mathrm{INT}(M/K) \\ R_0 = (M - m{\cdot}K)/2 \\ R_i = R_0 + (i-1){\cdot}m \\ R_i' = R_0 + i{\cdot}m \\ i \in [1, K] \end{cases} \tag{9}$$

where, $\mathrm{INT}(x)$ is the integral function.

Likewise for the comparison curve, each possible matching interval corresponding to the reference segmentation curve can be defined as

$$\begin{cases} (SignC)_i = \left\{(x'_t, y'_t) \middle| t \in [C_i, C_i'], C_i' - C_i = 2n + m\right\} \\ n = \mathrm{INT}(N/K) \\ C_0 = (N - n{\cdot}K)/2 \\ C_i = C_0 + (i-1){\cdot}n - n/2 - m/2 \\ C_i' = C_0 + (i+1){\cdot}n - n/2 + m/2 \\ 0 \le C_i < C_i' \le N, i \in [1, K] \end{cases} \tag{10}$$

Here, considering the positional correlation and the deviation between the signature curves, when the reference signature length of each segmentation is m, the interval to be matched of the comparison signature is at least m in length, and is offset by n before and after the corresponding segmentation position. Equivalently, the reference curve swims within the interval to be matched on the comparison curve to obtain the optimal matching position.

2.2.3. Segmentation Matching

The process of the optimal segmentation matching is as follows:

Step 1: Take the reference signature curve as a template, and divide it into K segments according to Equation (9).

Step 2: The comparison curve should be divided into K segmentations according to Equation (10).

Step 3: For the i-th segmentation of the comparison curve, search the optimal matching with the corresponding the i-th segmentation of the reference curve by EC algorithm, and get the similarity distance d_i. Meanwhile, the matching distance dx_i and dy_i of the corresponding X, Y curves can be calculated based on the current matching result.

Step 4: Set σx_i, σy_i which are the standard deviation of X, Y data points in i-th segmentation of the reference curve, compare it with the matching distance dx_i and dy_i, and calculate the similarity score sx_i and sy_i of this segmentation, respectively, as in Equations (11) and (12).

$$\begin{cases} sx_i = h(dx_i, \sigma x_i, \alpha) + h(dx_i, \sigma x_i, \beta) + h(dx_i, \sigma x_i, \gamma) \\ sy_i = h(dy_i, \sigma y_i, \alpha) + h(dy_i, \sigma y_i, \beta) + h(dy_i, \sigma y_i, \gamma) \end{cases} \tag{11}$$

$$\begin{cases} h(d, \sigma, \delta) = 100 \cdot \exp(-0.5\delta \cdot d^2 / (\sigma/4 + 10)^2) \\ \alpha = 0.25 \sim 0.5, \beta = 1 \sim 2, \gamma = 2 \sim 5 \end{cases} \tag{12}$$

Step 5: Repeat step 3 until all segmentation curves parameters are calculated.

Step 6: Calculate the average of sx, sy as the result outputs of X, Y curves' similarity measure, and use the weighted average as the result output of the similarity measure *Score* of the two curves, as shown in later Formula (14).

It should be noted that when performing the optimal matching segmentation calculation, the similarity distance between the two curves can be obtained, which are considered as 2D curves. Next, the matching distances of the corresponding 1D curves X and Y can be separately calculated. Obviously, the calculated matching distances are absolute values, and if a similarity evaluation is to be performed, one threshold is needed to discriminate. For this reason, the similarity average is calculated using Gaussian functions of three different widths and the absolute distance measure is converted to a relative measure between 0 and 100. Thus, the discrimination threshold can be unified to a value between 0 and 100.

A pair of genuine signature curves is adopted for the optimal segmentation matching as seen in Figure 5. The matching results of X, Y curves are shown in Figure 6, respectively.

A pair of genuine and forged signature curves is adopted for the segmentation matching as seen in Figure 7. The matching results of X, Y curves are shown in Figure 8, respectively.

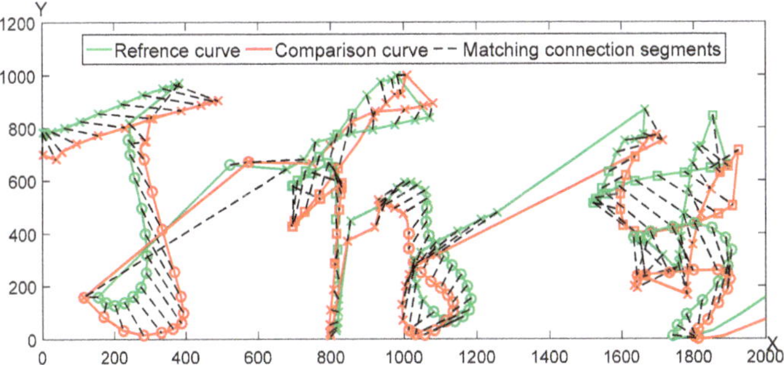

Figure 5. The matching results of two genuine signatures.

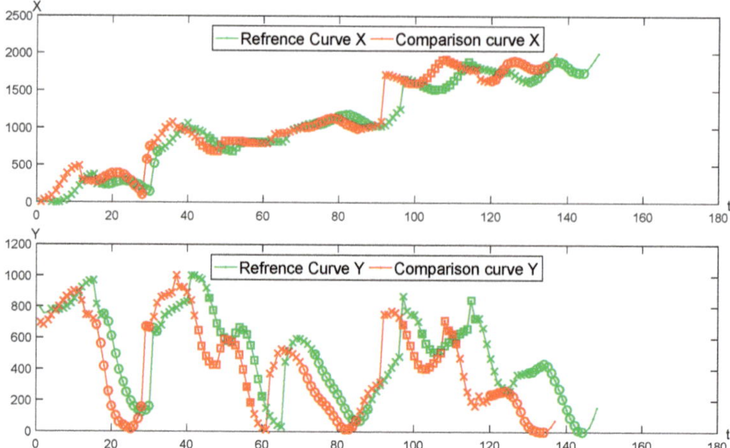

Figure 6. The matching result of X and Y curves between two genuine signatures.

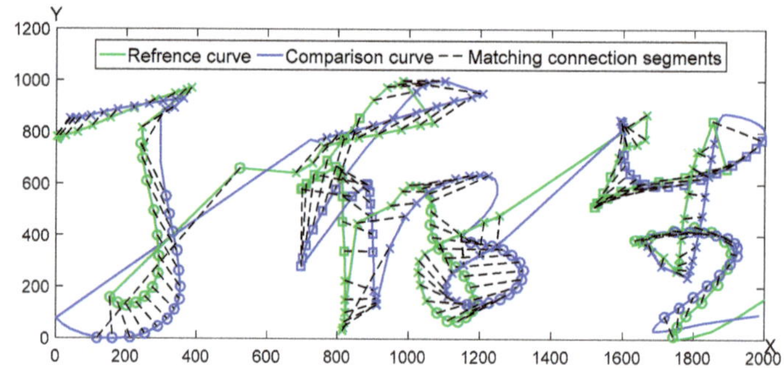

Figure 7. The matching results of the genuine and the forged signatures.

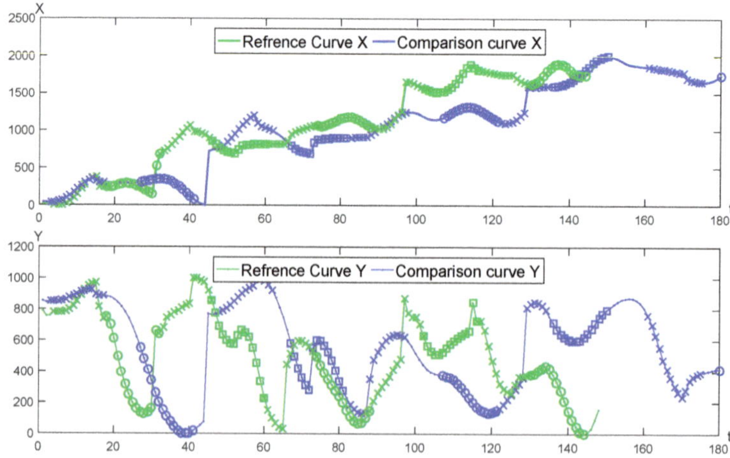

Figure 8. The matching result of X, Y curve between the genuine and forged signatures.

The optimal segmentation matching results between three curves are shown in Tables 1 and 2.

Table 1. Results of segmentation matching between a pair of genuine signature curves ($K = 10$).

No.	$[R_i, R_i']$	Matching	t	a	b	k	h	d	dx	dy	ox	oy	sx	sy
1	[3,17]	[0,14]	0	1.03	36	0.93	10	44.6	22.4	27.8	70.5	92.3	77.9	69.2
2	[17,31]	[15,29]	15	0.99	9	0.92	95	61.6	11.7	42.4	79.7	220.1	94.1	81.6
3	[31,45]	[27,41]	25	1.05	−23	0.90	29	55.4	20.7	34.3	54.7	107.9	73.8	65.8
4	[45,59]	[42,56]	27	0.99	−11	0.90	61	55.4	13.9	47.9	23.5	130.2	67.8	57.8
5	[59,73]	[55,69]	27	0.90	6	0.90	61	26.3	3.9	21.8	66.9	234.3	99.1	95.0
6	[73,87]	[70,84]	29	1.06	−86	0.90	68	18.5	6.6	8.6	20.9	112.1	89.5	97.1
7	[87,101]	[82,96]	28	1.06	−61	1.09	17	66.3	24.7	40.4	151.5	251.8	91.3	86.3
8	[101,115]	[96,110]	29	1.02	31	0.91	62	66.3	16.6	47.4	77.2	108.4	88.2	50.9
9	[115,129]	[108,122]	28	0.99	8	1.01	37	32.6	24.8	14.7	35.3	137.7	52.2	94.2
10	[129,143]	[119,133]	26	0.92	66	0.92	100	44.7	11.0	43.8	42.5	154.2	87.6	68.6
		Average											82.1	76.7

Table 2. Results of segmentation matching between a pair of genuine and forged signature curves ($K = 10$).

No.	$[R_i, R_i']$	Matching	t	a	b	k	h	d	dx	dy	ox	oy	sx	sy
1	[3,17]	[2,16]	2	0.96	32	0.94	−95	56.4	15.1	30.2	70.5	92.3	88.6	65.6
2	[17,31]	[26,40]	26	0.91	53	0.9	94	112.0	96.7	172.2	79.7	220.1	20.9	23.6
3	[31,45]	[47,61]	31	1.04	30	0.93	−63	38.3	9.0	16.6	54.7	107.9	93.8	89.2
4	[45,59]	[66,80]	30	1.07	13	0.91	53	97.0	26.6	76.9	23.5	130.2	38.3	36.4
5	[59,73]	[82,96]	26	0.94	82	0.9	−40	50.2	10.7	59.3	66.9	234.3	93.5	71.8
6	[73,87]	[106,120]	30	1.00	82	0.9	−12	28.6	7.9	18.7	20.9	112.1	85.3	87.4
7	[87,101]	[119,133]	23	0.90	76	1.09	−20	61.7	16.4	39.9	151.5	251.8	96.0	86.6
8	[101,115]	[135,149]	19	0.95	96	0.91	−30	37.1	12.3	29.1	77.2	108.4	93.2	72.9
9	[115,129]	[160,174]	24	0.95	40	0.9	41	19.6	4.5	10.4	35.3	137.7	97.1	97.0
10	[129,143]	[179,193]	23	1.00	3	0.9	24	39.6	17.3	16.6	42.5	154.2	73.9	93.9
		Average											78.1	72.4

It can be seen from Tables 1 and 2 that for the similarity of X, Y curves is calculated by the same template signature and segmentations, the similarity between two genuine signatures is usually higher than that between the genuine and forged signatures. On each segmentation, the similarity calculation of the X, Y curves depends on the standard deviations of the respective segmentations in the template signature, as shown in Equation (11). To accurately estimate this deviation, a large number of genuine and forged signatures are needed for matching calculations and statistics. Obviously, this is difficult to obtain in practical applications. Here we use the intra-segmentation standard deviation of the signature segmentation itself and a deviation as the empirical value. In addition, the three control parameters α, β, and γ are equivalent to controlling the width of the Gaussian function, and are also an empirical value. Here, $\alpha = 0.4$, $\beta = 1.6$, $\gamma = 3.2$ are selected. The changes in these parameters are not sensitive to the correct rate of the final evaluation results. Due to space limitations, this article will not discuss them.

2.2.4. Feature Extraction

In the i-th matching interval $[R_i, R_{i+1}]$, the interval velocity ratio (*IVR*) of corresponding points is calculated as follows:

$$\begin{cases} v_j = \sqrt{(x_j - x_{j-1})^2 + (y_j - y_{j-1})^2} \\ v'_j = \sqrt{(x'_j - x'_{j-1})^2 + (y'_j - y'_{j-1})^2} \\ IVR_i = \frac{100}{R_i' - R_{i+1}} \sum_{j=R_i+1}^{R_i'} \min\left(\frac{v_j+1}{v'_j+1}, \frac{v'_j+1}{v_j+1}\right) \end{cases} \tag{13}$$

2.2.5. Similarity Measure

The similarity measure *Score* of two signature curves is calculated as follows:

$$\begin{cases} Score = w_a \cdot LSC + w_b \cdot GSC \\ LSC = \frac{1}{K} \sum_{i=1}^{K} (0.2sx_i + 0.3sy_i + 0.5IVR_i) \\ GSC = g(M/N) \\ w_a + w_b = 1, w_a \geq 0, w_b \geq 0 \end{cases} \tag{14}$$

where *LSC* and *GSC* are local similarity score and global similarity score, respectively, while w_a and w_b are the corresponding weights. M and N are the lengths of the reference signature and the comparison signature, respectively.

Here:

$$g(x) = \begin{cases} 0 & x < 0.5 \\ 100 \times \exp(-2(x-1)^2) & 0.5 \leq x \leq 2 \\ 0 & x > 2 \end{cases} \tag{15}$$

is used to calculate score of the writing time ratio of two origin signatures.

The calculation of the weight is calculated by enumeration, where $w_a = 0.85$ and $w_b = 0.15$, and the detailed process is shown in the next Section 3.4.

It is considered that threshold ε of the signature verification system is 60 in many cases, and when *Score* is greater than 60, it may be distinguished into a genuine signature, and below 60 may be considered as a forged signature.

This is similar to the 100 point test. Passing more than 60 points is a pass, and below 60 is a failure. Of course, accurately determining the threshold is also a problem that needs to be studied in depth. For each user's signature threshold determination, some other real registration signatures or even skilled forged signatures are needed. As a single template signature authentication system, only a reasonable threshold is given here, which is one of the key issues that need to be studied in the future.

3. Experiments

In this section, experiments to evaluate the efficacy of the four datasets are described and signature verification performances are reported.

3.1. Dataset and Evaluation Protocol

The efficacy of the proposal is demonstrated on the publicly available SUSIG, SVC2004 Task1&Task2 and MCYT datasets. The main differences among the four datasets are the acquisition protocol, device, and signer. In the following, the datasets used in this paper are briefly described:

(1) SUSIG Visual Subcorpus [34]: This dataset consists of 2820 western signatures from 94 signers with 20 genuine signatures collected in two sessions and 10 skilled forgery signatures (half are highly skilled) with an LCD touch device. For convenience, this subcorpus is called SUSIG for short in this paper. The data in SUSIG consists of X, Y, pressure, and timestamp, collected at 100 Hz.

(2) SVC2004 Task1 Subcorpus [35]: This datasets are acquired with a Wacom graphic tablet. It consists of 800 English and Chinese signatures from 40 signers with 20 genuine signatures collected in two sessions and 20 skilled forgeries per signer. For convenience, this subcorpus is called SVC1 for short in this paper. The data in SVC1 consists of X, Y and timestamp, collected at 100 Hz.

(3) SVC2004 Task2 Subcorpus [35]: It is also composed of 40 signers with the same number of genuine and forged signatures as in Task1. For convenience, this subcorpus is called SVC2 for short in this paper. The data in SVC2 consists of X, Y, pressure, azimuth, altitude, timestamp, and button status, collected at 100 Hz.

(4) MCYT-100 Subcorpus [36]: It is also composed of 100 signers with 25 genuine and 25 forged signatures. For convenience, this subcorpus is called MCYT for short in this paper. The data in MCYT consists of X, Y, pressure, azimuth, altitude, timestamp, and button status, collected at 100 Hz.

Out of these, one genuine signature is selected randomly to be used as reference sample (template), and the other genuine signatures and all skilled forgeries are used as test samples. Thus, in our work there are 9480 reference and test signatures from 274 signers to be verified in total.

We adopt EER, i.e., the error rate at which false acceptance rate (FAR) and false rejection rate (FRR) are equal, as a measure for characterizing verification performance. In order to obtain reliable results for independent test data, this process of random selection of reference signatures and performance evaluation is repeated ten repetitions.

3.2. Parameter Determination for EC

Taking the template signature of Figure 5 as an example, the signature is from SVC2004 Task1 signer #1, with 147 data points. The total number of segmentations K of different data points of the reference signature is shown in Table 3.

Table 3. Proposal parameter K with different data points of the reference signature.

Order	1	2	3	4	5	6	7	8	9	10
Data Points Count M	≤100	≤150	≤200	≤250	≤300	≤350	≤400	≤500	≤600	Other
Proposal K	8	10	12	14	16	18	20	22	24	30

The optimal segmentation matching calculation between the template signature and itself was executed. The theoretically similarity distance of each segmentation is theoretically zero. The parameters S and *Iterations* of the EC are determined by enumeration calculation as seen in Figure 9, where S = 4, 8, 12, 16, 20 and *Iterations* = 100, 200, 400, 800. In the abovementioned optimal segmentation matching calculation, although the global optimal parameters are not obtained, the different S and *Iterations* can quickly reach the local minimum. Comprehensive calculation of speed and accuracy requirements, choose S = 20 and *Iterations* = 400 as the control parameters for EC.

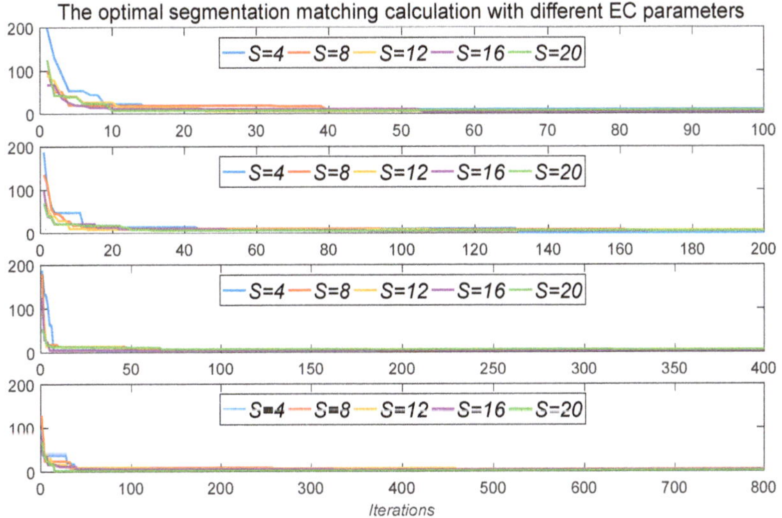

Figure 9. The optimal segmentation matching calculation with different EC parameters.

3.3. Feature Validity Test

Select 20 genuine signatures and 20 skilled forged signatures of the first signer on SVC1, take one of the first 10 genuine signatures as a template in turn, and the remaining 19 genuine signatures and 20 skilled forged signatures respectively perform matching calculation, and then calculate sx, sy, IVR, GSC, LSC, and final *Score*.

Figure 10a–d show the distribution of similarity differences between genuine and skilled forged signatures. Signature verification can be regarded as a two-category problem. It can be seen from the distribution process of the comparison results of the signatures in Figure 10a that only the similarity of X and Y curves are used, and it is difficult to distinguish the authenticity of each test signature. From Figure 10b, it can be seen that the IVR has a certain degree of discrimination, but there are many indistinguishable confusing signatures. However it can be clearly seen in Figure 10c that the fusion feature LSC and the global feature GSC have a high degree of discrimination. Only a very small number of test signatures are misidentified. In the Formula (11), the above several features are merged, and the similarity of the two curves can be used as a one-dimensional index. The discriminant threshold can be used to directly identify or classify the signature authenticity.

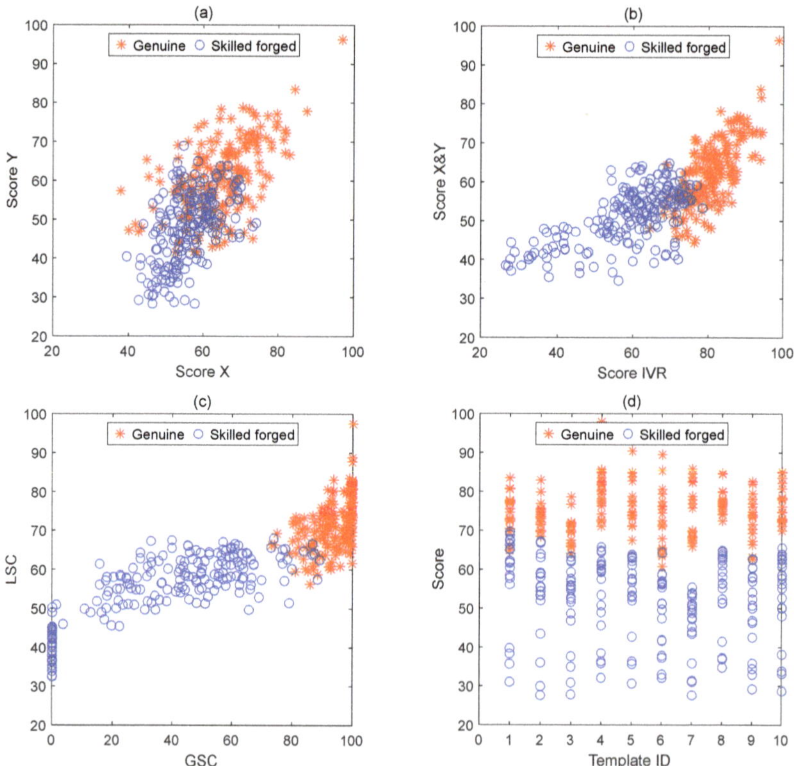

Figure 10. One of the first 10 genuine signatures is used as the template signature in turn, and the remaining genuine and skilled forged signatures are used as the comparison result of the test signatures. (**a**) Score X and score Y distribution; (**b**) Score IVR distribution; (**c**) GSC and LSC distribution; (**d**) 10 templates in turn *Score* distribution.

In Figure 10d, it can also be clearly seen that different template signatures have different discriminating thresholds, and the degree of discrimination between genuine and forged signatures is also different. Using the #4, #5, and #7 signatures as templates, the signature authenticity can be completely distinguished, and the #7 template has the largest degree of discrimination.

3.4. Feature Weight Calculation

Let the number of signatures of the template equal one and the weight value w_a is increments from 0 to 1 with the interval 0.05. The signatures of the first four signers on SVC1 dataset are selected for training to find the optimal w_a. On the four signature datasets, the respective EERs under different weights are calculated. The results are shown in Table 4.

Table 4. EERs under different weights on four datasets.

| # | Weight | | Average EER (in %) | | | | |
| | | | Train | | Test | | |
	w_a	w_b	SVC1 *	SUSIG	SVC1	SVC2	MCYT *
1	0.00	1.00	17.34	4.49	24.20	23.73	
2	0.05	0.95	16.56	4.32	22.06	21.66	
3	0.10	0.90	15.75	4.13	20.04	19.13	
4	0.15	0.85	15.19	3.98	18.47	18.13	
5	0.20	0.80	14.63	3.83	17.07	16.66	
6	0.25	0.75	14.00	3.68	16.05	15.54	
7	0.30	0.70	13.38	3.54	15.16	14.65	
8	0.35	0.65	12.50	3.41	14.46	13.97	
9	0.40	0.60	12.06	3.26	14.01	13.38	
10	0.45	0.55	11.44	3.15	13.53	12.93	
11	0.50	0.50	11.06	3.06	13.05	12.59	7. 50
12	0.55	0.45	10.50	2.99	12.79	12.43	7.08
13	0.60	0.40	10.25	2.98	12.64	12.25	6.74
14	0.65	0.35	9.81	2.99	12.49	12.22	6.51
15	0.70	0.30	9.31	3.03	12.40	12.16	6.33
16	0.75	0.25	8.94	3.08	12.32	12.13	6.13
17	0.80	0.20	8.75	3.21	12.32	12.15	6.10
18	0.85	0.15	8.63	3.47	12.30	12.25	6.07
19	0.90	0.10	8.64	3.74	12.32	12.42	6.14
20	0.95	0.05	8.81	4.09	12.47	12.66	6.34
21	1.00	0.00	9.88	4.84	12.67	12.91	6.85

* Only choose w_a = 0.5~1.0 as there is a long computation time with the most signatures.

From Table 4, it can be seen on the training samples that the minimum EER is obtained when w_a = 0.85, and the EERs on the other datasets are 3.47%, 12.30%, 12.25%, and 6.07%, respectively. At this point, the same minimums are obtained on SVC1 and MCYT datasets. In Figure 11, we can see that when w_a is incremented, the EER value changes to a convex function, and the EER on different datasets has only a minimum value.

In fact, we can also see that the EER is not much different when $w_a \in [0.6, 0.95]$, which means that our weight selection has better robustness, and when $w_a \leq 0.3$, The EER has increased dramatically. At the same time, when w_a = 0, signature verification actually only depends on GSC the signature writing time ratio. On the SVC1 and SVC2 datasets, the EER of each is over 20%, and on the SUSIG dataset, the EER is still less than 5%. It can be seen that the writing time ratio is a better feature distinguishing the test signature. In addition, SVC1&SVC2 is much higher than the SUSIG dataset in the difficulty of signature verification of four signature datasets. This can also be obtained from the whole experimental results.

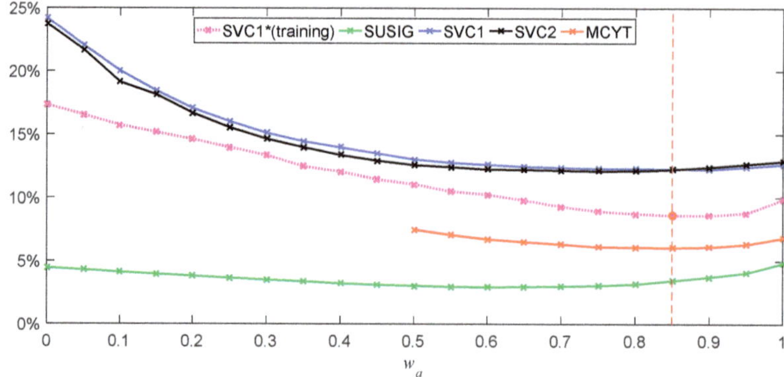

Figure 11. EERs and the mean on four signature datasets with different weight w_a.

3.5. Experimental Results

Performance of the system with the maximum EER, the minimum EER, the average EER and the standard deviation of EERs measured in percentage for different number reference signatures of similarity metrics are shown in Table 5.

Table 5. Performances of the system with different dataset.

DataSet	# of Samples	EER (in %)		
		Average	Minimum	Maximum
SUSIG	1	3.47	2.27%	4.32%
SVC1	1	12.30	9.62%	14.94%
SVC2	1	12.25	9.53%	14.58%
MCYT	1	6.07	3.98%	7.92%

From the results described above, when experiments are implemented on SUSIG, EER = 3.47% can be the best result based on CSM with five reference samples. As for SVC1&SVC2&MCYT, it can be provided EER = 12.30%, EER = 12.25% and EER = 6.07%, respectively. For four different datasets and different number of reference samples, the EERs of test results with #1 template repeated 10 times are arranged in ascending order, as shown in Figure 12.

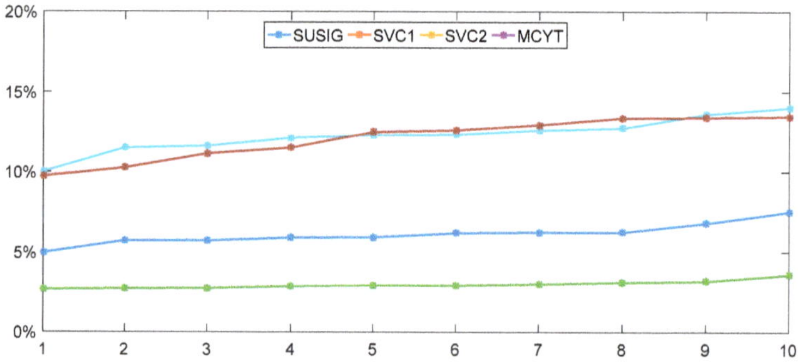

Figure 12. EERs of test results with #1 template repeated 10 times for SUSIG, SVC1, SVC2 and MCYT, respectively.

At the same time, it should be emphasized that the deviation of the maximum and minimum values of EER is more than double almost when #1 sample is randomly selected as templates in ten repetitions as seen in Table 5 and Figure 12. It demonstrates that the selection of template samples is also essential and representative template samples can effectively improve the accuracy of signature verification.

In order to demonstrate the effectiveness of our proposed method, we compare the results of our proposed method with other state-of-the-art methods. It is to be mentioned that each of these methods have different features and classifiers, and it is difficult to make comparisons between them based on different datasets. Hence, we just compare the performance of methods which are carried out on SUSIG, SVC1, SVC2 and MCYT datasets. Nevertheless, the best results methods carried out both on them are taken for comparative studies, which use five genuine reference signature of a signer for enrollment. The best EERs reported from the reference works on SUSIG, SVC1, SVC2 and MCYT are given in Tables 6–9 with one and five genuine reference signatures.

Table 6. Comparative studies of state-of-the-art methods implemented on SUSIG.

Method	# of Samples	Average EER (in %)
Fuzzy modeling [37]	5	5.38
Histogram + Manhattan [38]	5	4.37
FFT + DTW [25]	5	3.03
DTW_Linear C [34]	5	2.10
35 global feature + FLD [39]	5	1.59
Parzen window + DCT [23]	5	1.49
TASS + RLCSS [26]	5	0.52
Target-wise [28]	1	6.67
Proposed method	1	3.47

Table 7. Comparative studies of state-of-the-art methods implemented on SVC1.

Method	# of Samples	Average EER (in %)
DTW [40]	5	6.96
DTW + HMM [18]	5	6.91
LCSS + SVM [22]	5	6.84
Wavelet Packet [41]	5	6.65
TASS + RLCSS [26]	5	5.33
SVC-competition [35]	5	2.84
Target-wise [28]	1	17.25
Proposed method	1	12.30

Table 8. Comparative studies of state-of-the-art methods implemented on SVC2.

Method	# of Samples	Average EER (in %)
Fuzzy modeling [37]	5	7.57
Function-based + HMM [42]	5	7.14
LCCS-SVM [22]	5	6.84
DTW + HMM [18]	5	6.91
LCSS [26]	5	5.33
Feature selection + DTW [43]	5	3.38
SVC-competition [35]	5	2.89
5 features DTW + VQ [15]	5	2.73
DTW with SCC [17]	5	2.63
Target-wise [28]	1	18.25
Proposed method	1	12.25

Table 9. Comparative studies of state-of-the-art methods implemented on MCYT.

Method	# of Samples	Average EER (in %)
DTW + Fourier descriptors [25]	5	7.22
Symbolic Representation [4]	5	6.12
HMM+Parzen Window [44]	5	5.29
Time function_LDP [45]	5	5.20
Histogram + Manhattan [38]	5	4.02
Neuro-fuzzy system [46]	5	4.02
Fusion matchers [47]	5	3.81
Dynamic programming [48]	5	3.52
HMM + Viterbi Path [49]	5	3.37
Wavelet coefficients [28]	5	3.21
GMM + DTW [16]	5	3.05
Velocity and pressure partition [50]	5	1.09
Target-wise [28]	1	13.56
Proposed method	1	6.07

Furthermore, it should be pointed out that the performance of a signature verification system is related to the number of samples used to build the model. Even in some statistical models, true and false signatures need to be trained. However, in many cases, it is difficult to register a large number of signatures in the actual system, which also limits the practical applications of many excellent methods, but for our signature verification system, the requirement for the number of reference signatures or templates is minimal and even only one signature can be used. Moreover, among the already known single signature systems, our performance is the best, and is very close to that of multi-signature systems.

4. Conclusions

The similarity measurement of curves is an old problem. A lot of pattern recognition problems can be converted into curve similarity problems to study. In this research, we presented a novel signature verification based on the curve similarity model, which is equally competitive when compared to other approaches and leads to much simpler and easier matching procedures. Considering internal and external writing environments being always varied, signatures were effectively aligned to the reference signature curve by CSM and a curve similarity distance was proposed to make an assessment the similarity between test signatures and references. Open access signature datasets SUSIG, SVC2004 Task1&Task2, and MCYT-100 were used in our work, and several experiments were implemented. Experimental results illustrated that the best matching could be obtained by our proposed CSM method with one signature template. The error rates $EER_{SUSIG} = 3.47\%$, $EER_{SVC1} = 12.30\%$, $EER_{SVC2} = 12.25\%$ and $EER_{MCYT} = 6.07\%$ were provided, respectively, which demonstrated the effectiveness and robustness of our proposed method. The most important thing is the case that our method can use one signature to authenticate, and the performance of our method is not much different from that of multi-signature verification systems. Finally, this innovative method opens the door to new competitions on signature verification using a single signature as reference template.

Author Contributions: H.H. and J.Z. proposed the conception of this research and drafted this article. E.Z. and J.T. focused on the collection, analysis and interpretation of data. E.Z. and J.T. revised the whole content of this manuscript. All authors approved and agreed with the final paper version to be published.

Funding: This work was supported by the National Key R&D Program of China "The study on Load-bearing and Moving Support Exoskeleton Robot Key Technology and Typical Application" [grant numbers 2017YFB1300502]; the National Natural Science Fund "The mathematical model of natural computation and research" [grant numbers 61672319, 2017].

Conflicts of Interest: None of these authors has conflict of interest in this research.

References

1. Hammoud, R.I.; Abidi, B.R.; Abidi, M.A. *Biometrics—Personal Identification in Networked Society*; Springer: New York, NY, USA, 2005.
2. Lim, M.H.; Yuen, P.C. Entropy Measurement for Biometric Verification Systems. *IEEE Trans. Cybern.* **2015**, *46*, 1065–1077. [CrossRef] [PubMed]
3. Impedovo, D.; Pirlo, G. Automatic Signature Verification: The State of the Art. *IEEE Trans. Syst. Man Cybern.* **2008**, *38*, 609–635. [CrossRef]
4. Guru, D.S.; Prakash, H.N. Online Signature Verification and Recognition: An Approach Based on Symbolic Representation. *IEEE Trans. Pattern Anal. Mach. Intell.* **2009**, *31*, 1059–1073. [CrossRef] [PubMed]
5. Réjean, P.; Srihari, S.N. On-line and off-line handwriting recognition: A comprehensive survey. *IEEE Trans. Pattern Anal. Mach. Intell.* **2000**, *22*, 63–84.
6. Diaz, M.; Ferrer, M.A.; Parziale, A.; Marcelli, A. Recovering Western On-Line Signatures from Image-Based Specimens. In Proceedings of the International Conference on Document Analysis and Recognition (ICDAR 14), Kyoto, Japan, 9–15 November 2017; pp. 1204–1209.
7. Nelson, W.; Turin, W.; Hastie, T. Statistical methods for on-line signature verification. *Int. J. Pattern Recognit. Artif. Intell.* **1994**, *8*, 749–770. [CrossRef]
8. Kiran, G.V.; Kunte, R.S.R.; Samuel, S. On-line signature verification system using probabilistic feature modeling. In Proceedings of the IEEE International Symposium on Signal Processing & Its Applications, Kuala Lumpur, Malaysia, 13–16 August 2001.
9. Wu, Q.Z.; Jou, I.C.; Lee, S.Y. On-line signature verification using LPC cepstrum and neural networks. *IEEE Trans. Cybern.* **1997**, *27*, 148–153.
10. Gruber, C.; Gruber, T.; Sick, B. Online Signature Verification with New Time Series Kernels for Support Vector Machines. In Proceedings of the International Conference on Biometrics ICB 2006: Advances in Biometrics, Hong Kong, China, 5–7 January 2006.
11. Lv, H.; Wang, W. On-Line Signature Verification Based on Dynamic Bayesian Network. In *International Conference on Advances in Natural Computation*; Springer: Berlin/Heidelberg, Germany, 2006.
12. Guru, D.S.; Prakash, H.N.; Manjunath, S. On-line Signature Verification: An Approach Based on Cluster Representations of Global Features. In Proceedings of the 2009 Seventh International Conference on Advances in Pattern Recognition, Kolkata, India, 4–6 February 2009; IEEE Computer Society: Washington, DC, USA, 2009.
13. Martens, R.; Claesen, L. On-line signature verification by dynamic time-warping. In Proceedings of the 13th International Conference on Pattern Recognition, Vienna, Austria, 25–29 August 1996.
14. Faundez-Zanuy, M. On-line signature recognition based on VQ-DTW. *Pattern Recognit.* **2017**, *40*, 981–992. [CrossRef]
15. Sharma, A.; Sundaram, S. An enhanced contextual DTW based system for online signature verification using Vector Quantization. *Pattern Recognit. Lett.* **2016**, *84*, 22–28. [CrossRef]
16. Sharma, A.; Sundaram, S. A Novel Online Signature Verification System Based on GMM Features in a DTW Framework. *IEEE Trans. Inf. Forensics Secur.* **2016**, *12*, 705–718. [CrossRef]
17. Xia, X.; Chen, Z.; Luan, F.; Song, X. Signature alignment based on GMM for on-line signature verification. *Pattern Recogn.* **2017**, *65*, 188–196. [CrossRef]
18. Fiérrez-Aguilar, J.; Krawczyk, S.; Ortega-Garcia, J.; Jain, A.K. Fusion of Local and Regional Approaches for On-Line Signature Verification. In Proceedings of the Advances in Biometric Person Authentication, International Workshop on Biometric Recognition Systems (IWBRS 2005), Beijing, China, 22–23 October 2005; Springer: Berlin/Heidelberg, Germany, 2005.
19. Parziale, A.; Diaz, M.; Ferrer, M.A.; Marcelli, A. SM-DTW: Stability Modulated Dynamic Time Warping for signature verification. *Pattern Recogn. Lett.* **2019**, *121*, 113–122. [CrossRef]
20. Wada, N.; Hangai, S. HMM Based Signature Identification System Robust to Changes of Signatures with Time. In Proceedings of the 2007 IEEE Workshop on Automatic Identification Advanced Technologies, Alghero, Italy, 7–8 June 2007.
21. Shafiei, M.M.; Rabiee, H.R. A New On-Line Signature Verification Algorithm Using Variable Length Segmentation and Hidden Markov Models. In Proceedings of the International Conference on Document Analysis and Recognition, Edinburgh, UK, 6 August 2003; IEEE Computer Society: Washington, DC, USA, 2003.

22. Gruber, C.; Gruber, T.; Krinninger, S.; Sick, B. Online Signature Verification with Support Vector Machines Based on LCSS Kernel Functions. *IEEE Trans. Syst. Man Cybern.* **2010**, *40*, 1088–1100. [CrossRef] [PubMed]

23. Rashidi, S.; Fallah, A.; Towhidkhah, F. Feature extraction based DCT on dynamic signature verification. *Sci. Iran.* **2012**, *19*, 1810–1819. [CrossRef]

24. Faundez-Zanuy, M.; Pascual-Gaspar, J.M. Efficient on-line signature recognition based on multi-section vector quantization. *Pattern Anal. Appl.* **2011**, *14*, 37–45. [CrossRef]

25. Yanikoglu, B.; Kholmatov, A. Online Signature Verification Using Fourier Descriptors. *EURASIP J. Adv. Signal Process.* **2009**, *2009*, 1–14. [CrossRef]

26. Fahmy, M.M.M. Online handwritten signature verification system based on DWT features extraction and neural network classification. *Ain Shams Eng. J.* **2010**, *1*, 59–70. [CrossRef]

27. Barkoula, K.; Economou, G.; Fotopoulos, S. Online signature verification based on signatures turning angle representation using longest common subsequence matching. *Int. J. Doc. Anal. Recognit.* **2013**, *16*, 261–272. [CrossRef]

28. Diaz, M.; Fischer, A.; Ferrer, M.A.; Plamondon, R. Dynamic Signature Verification System Based on One Real Signature. *IEEE Trans. Cybern.* **2017**, *48*, 228–239. [CrossRef]

29. Parziale, A.; Fuschetto, S.G.; Marcelli, A. Exploiting stability regions for online signature verification. In *New Trends in Image Analysis and Processing—ICIAP 2013*; Springer: Berlin/Heidelberg, Germany, 2013.

30. Thumwarin, P.; Pernwong, J.; Matsuura, T. FIR signature verification system characterizing dynamics of handwriting features. *J. Adv. Signal Process.* **2013**, *2013*, 183. [CrossRef]

31. Di Lecce, V.; Dimauro, G.; Guerriero, A. Selection of reference signatures for automatic signature verification. In Proceedings of the Fifth International Conference on Document Analysis and Recognition, Bangalore, India, 22 September 1999.

32. Zheng, J.B.; Zhu, G.X. A new algorithm for on-line handwriting signature verification based on evolutionary computation. *Wuhan Univ. J. Nat. Sci.* **2006**, *11*, 596–600.

33. Zheng, J.B.; Gao, X.L.; Zhan, E.Q. Algorithm of On-Line Handwriting Signature Verification Based on Discrete Fréchet Distance. In *International Symposium on Advances in Computation and Intelligence*; Springer: Berlin/Heidelberg, Germany, 2008.

34. Kholmatov, A.; Yanikoglu, B. SUSIG: An on-line signature database, associated protocols and benchmark results. *Pattern Anal. Appl.* **2009**, *12*, 227–236. [CrossRef]

35. Yeung, D.Y.; Chang, H.; Xiong, Y.; George, S.E.; Kashi, R.S.; Matsumoto, T.; Rigoll, G. *SVC2004: First International Signature Verification Competition in Biometric Authentication*; Lecture Notes in Computer Science; Springer: Cham, Switzerland, 2004; pp. 16–22.

36. Ortega-Garcia, J.; Fierrez-Aguilar, J.; Simon, D.; Gonzalez, J.; Faundez-Zanuy, M.; Espinosa, V.; Satue, A.; Hernaez, I.; Igarza, I.-I.; Vivaracho, C.; et al. MCYT baseline corpus: A bimodal biometric database. *IEEE Proc. Vis. Image Signal Process.* **2003**, *150*, 395–400. [CrossRef]

37. Ansari, A.Q.; Singh, A.K.; Hanmandlu, M.; Kour, J. Online signature verification using segment-level fuzzy modelling. *IET Biom.* **2014**, *3*, 113–127. [CrossRef]

38. Sae-Bae, N.; Memon, N. Online Signature Verification on Mobile Devices. *IEEE Trans. Inf. Forensics Secur.* **2014**, *9*, 933–947. [CrossRef]

39. Ibrahim, M.T.; Kyan, M.; Guan, L. On-line signature verification using global features. In Proceedings of the IEEE Conference on Electrical & Computer Engineering, St. John's, NL, Canada, 3–6 May 2009.

40. Kholmatov, A.; Yanikoglu, B. Identity authentication using improved online signature verification method. *Pattern Recognit. Lett.* **2005**, *26*, 2400–2408. [CrossRef]

41. Wang, K.; Wang, Y.; Zhang, Z. On-line signature verification using wavelet packet. In Proceedings of the IEEE International Joint Conference on Biometrics, Washington, DC, USA, 11–13 October 2011.

42. Fierrez-Aguilar, J.; Ortega-Garcia, J.; Gonzalez-Rodriguez, J. Target Dependent Score Normalization Techniques and Their Application to Signature Verification. *IEEE Trans. Syst. Man Cybern. Part C Appl. Rev.* **2005**, *35*, 418–425. [CrossRef]

43. Pascual-Gaspar, J.M.; Cardeñoso-Payo, V.; Vivaracho-Pascual, C. Practical On-Line Signature Verification. In Proceedings of the Advances in Biometrics, Third International Conference, Alghero, Italy, 2–5 June 2009.

44. Fierrez-Aguilar, J.; Nanni, L.; Lopez-Peñalba, J.; Ortega-Garcia, J.; Maltoni, D. An On-Line Signature Verification System Based on Fusion of Local and Global Information. In *Audio- and Video-Based Biometric Person Authentication*; Springer: Berlin/Heidelberg, Germany, 2005.

45. Nanni, L.; Lumini, A. A novel local on-line signature verification system. *Pattern Recognit. Lett.* **2008**, *29*, 559–568. [CrossRef]

46. Cpałka, K.; Zalasiński, M.; Rutkowski, L. A new algorithm for identity verification based on the analysis of a handwritten dynamic signature. *Appl. Soft Comput.* **2016**, *43*, 47–56. [CrossRef]

47. Nanni, L.; Maiorana, E.; Lumini, A.; Campisi, P. Combining local, regional and global matchers for a template protected on-line signature verification system. *Expert Syst. Appl.* **2010**, *37*, 3676–3684. [CrossRef]

48. Muramatsu, D.; Matsumoto, T. Effectiveness of Pen Pressure, Azimuth, and Altitude Features for Online Signature Verification. In *International Conference on Biometrics*; Springer: Berlin/Heidelberg, Germany, 2007.

49. Bao, L.V.; Garcia-Salicetti, S.; Dorizzi, B. On Using the Viterbi Path along with HMM Likelihood Information for Online Signature Verification. *IEEE Trans. Cybern.* **2007**, *37*, 1237–1247.

50. Ibrahim, M.T.; Khan, M.A.; Alimgeer, K.S.; Khan, M.L.; Taj, I.A.; Guan, L. Velocity and pressure-based partitions of horizontal and vertical trajectories for on-line signature verification. *Pattern Recognit.* **2010**, *43*, 2817–2832. [CrossRef]

Article

End-to-End Deep Learning Fusion of Fingerprint and Electrocardiogram Signals for Presentation Attack Detection

Rami M. Jomaa [1,*], Hassan Mathkour [1], Yakoub Bazi [2] and Md Saiful Islam [1]

[1]   Computer Science Department, College of Computer and Information Sciences, King Saud University, Riyadh 11543, Saudi Arabia; mathkour@ksu.edu.sa (H.M.); saislam@ksu.edu.sa (M.S.I.)

[2]   Computer Engineering Department, College of Computer and Information Sciences, King Saud University, Riyadh 11543, Saudi Arabia; ybazi@ksu.edu.sa

*   Correspondence: rami@ksu.edu.sa

Received: 29 February 2020; Accepted: 4 April 2020; Published: 7 April 2020

Abstract: Although fingerprint-based systems are the commonly used biometric systems, they suffer from a critical vulnerability to a presentation attack (PA). Therefore, several approaches based on a fingerprint biometrics have been developed to increase the robustness against a PA. We propose an alternative approach based on the combination of fingerprint and electrocardiogram (ECG) signals. An ECG signal has advantageous characteristics that prevent the replication. Combining a fingerprint with an ECG signal is a potentially interesting solution to reduce the impact of PAs in biometric systems. We also propose a novel end-to-end deep learning-based fusion neural architecture between a fingerprint and an ECG signal to improve PA detection in fingerprint biometrics. Our model uses state-of-the-art EfficientNets for generating a fingerprint feature representation. For the ECG, we investigate three different architectures based on fully-connected layers (FC), a 1D-convolutional neural network (1D-CNN), and a 2D-convolutional neural network (2D-CNN). The 2D-CNN converts the ECG signals into an image and uses inverted Mobilenet-v2 layers for feature generation. We evaluated the method on a multimodal dataset, that is, a customized fusion of the LivDet 2015 fingerprint dataset and ECG data from real subjects. Experimental results reveal that this architecture yields a better average classification accuracy compared to a single fingerprint modality.

Keywords: fingerprint; ECG; presentation attack detection; deep learning

---

1. Introduction

Biometric systems in which the physiological or behavioral characteristics of humans, e.g., fingerprints, electrocardiogram (ECG), gait, iris, and face, are captured and utilized for authentication are increasingly used. Fingerprints are among the most extensively employed biometrics owing to their several advantages, such as acceptability, collectability, and high authentication accuracy [1]. The widespread availability of fingerprint-based systems has made them vulnerable to numerous attacks, mainly presentation attacks (PAs). ISO/IEC 30107 defines a PA as the presentation of a fraudulent sample, such as an artefact or a fake biological sample, to an input biometric sensor with the intention of circumventing the system policy [2]. An artefact can be an artificial or synthetic fingerprint presented as a copy of a real fingerprint, which is also known as a spoof [3]. Figure 1 shows examples of numerous artefact fingerprint samples created by different artificial materials, such as gelatin, Play-Doh, and silicon [4]. For fake biological sample-based attacks, a severed or altered finger, or a finger of a cadaver, is presented to deceive a biometric sensor. The automated process used for detecting a PA in a biometric system is called PA detection (PAD) [2]. The aim of PAD is discriminating the bona fide (i.e., real or live) biometric samples from PA (i.e., artefact) samples.

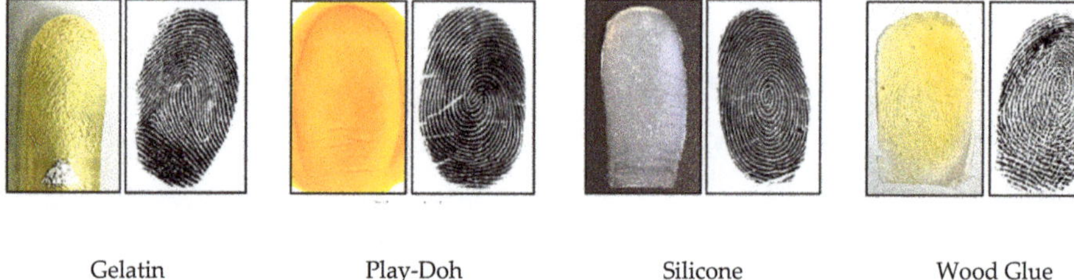

Gelatin   Play-Doh   Silicone   Wood Glue

Figure 1. Examples of fingerprint artefacts fabricated using different materials. A real image of a fabricated fingerprint is shown on the left and a scanned image using a fingerprint sensor is shown on the right [4].

Fingerprint PAD methods can be divided into hardware- and software-based methods [5]. In a hardware-based method, additional hardware devices are added to the biometric system to capture additional characteristics indicating the liveness of the fingerprint, such as blood pressure in the fingers, skin transformation, and skin odor [6–8]. With software-based methods, in contrast, the PAs of the fingerprints are analyzed by applying image processing techniques on fingerprint images. By exploring software-based techniques for fingerprint PAD studied in the literature, these methods can be grouped into handcrafted feature- and deep-learning-based techniques. In handcrafted feature-based techniques, expert knowledge is required to formulate the feature descriptors, whereas in deep-learning-based techniques, no such expert knowledge is required.

The local binary pattern (LBP) is one of the earliest and most common handcrafted techniques that has been investigated for fingerprint liveness detection, in which LBP histograms are applied to extract the texture liveness information using binary coding [9]. Measuring the loss of information while fabricating fake fingerprints is utilized in local phase quantization (LPQ) to differentiate between bona fide and artefact fingerprint images [10]. The Weber local descriptor (WLD) is applied for fingerprint liveness detection, in which 2D histograms representing differential excitation and orientation features are applied [11]. Combining these local descriptors such as WLD with LPQ [11], or WLD with LBP [12], improves the accuracy of detecting the liveness of a fingerprint. A new local contrast phase descriptor is proposed for fingerprint liveness detection as 2D histogram features composed of spatial and phase information [13].

Deep learning techniques have recently proven their superiority over traditional approaches in image classification problems [14,15]. Deep learning techniques have also proven their advantages on 1D signals, including ECG [16–19]. Several studies have investigated the utilization of deep learning techniques in biometrics systems [20–22], and for fingerprint PAD [23–26]. Convolutional neural network (CNN) networks have exhibited continuous improvements for spoof detection compared with handcrafted techniques. An early work that introduced CNN for fingerprint PAD [23] employed transfer learning using a pre-trained CNN model for detecting fake fingerprints, which achieved the best results in the LivDet 2015 competition [27]. Another use of deep learning for fingerprint PAD is presented in [28], where local patches of minutiae have been extracted and processed using a well-known CNN model called Inception-V3, which achieved state-of-the-art accuracy in fingerprint liveness detection. A CNN model with improved residual blocks was proposed to balance between the accuracy and the convergence time in a fingerprint liveness system [29], wherein they extracted local patches using the statistical histogram and center of gravity. This approach won first place in the LivDet 2017 competition. A small CNN network was proposed to overcome the difficulties in the deployment of a fingerprint liveness detection system in mobile systems by utilizing the structure of the SqueezNet fire module and removing the fully connected layers [24].

Recently, a new group of fingerprint PAD methods have also been considered, which fall outside of software- and hardware-based approaches and are based on the fusion of fingerprint with a more secure biometric modality [30,31]. Several researchers investigated the fusion of fingerprints with a variety of biometric modalities, such as face, ECG, and fingerprint dynamics, to improve the accuracy and security of biometric systems [19,32–41]. The fusion of an ECG with other biometric modalities [37,38,42–46] has also received attention because the ECG has certain biometric advantages, such as a natural inherence of the liveness characteristic and a continuous authentication over time [47]. The crucial location of the heart in the body enables this biometric to be used as a secure modality. Moreover, a high-quality ECG can be captured from fingers, which make this modality a convenient candidate for a multimodal fusion with fingerprints [48]. These characteristics render ECG biometrics robust against PAs and provide them with advantages over other traditional biometrics. Several studies have considered the fusion of fingerprints and an ECG for PAD in fingerprint biometrics. A sequential score level fusion between an ECG and a fingerprint was proposed in [37]. Later, this approach was improved to be appropriate for fingerprint PAD in an authentication system [38]. Another study on fusing a fingerprint with an ECG was proposed in [36], in which the fusion is achieved at the score level by applying an automatic updating of the ECG templates. In this study, the authors fused an ECG matching score with the liveness score to evaluate the liveness of the fingerprint sample, demonstrating a good performance.

Several recent studies have proposed utilizing a CNN to deal with two-branch networks for processing video data [49–51]. A CNN network has been introduced into a multimodal biometric system combining an ECG with a fingerprint [19,52], in which the CNN is used for extracting ECG and fingerprint features. Although CNN was used in these studies, they did not achieve an end-to-end fusion in which the CNN is only used as features extractor and the classification carried out by an independent classifier. Furthermore, these studies focused on authentication performance rather than fingerprint PAD.

In this study, we propose a novel architecture for fusing a fingerprint and an ECG to detect and prevent fingerprint PAs. The proposed architecture is learnable end-to-end from the signal level to the final decision. The proposed method is intended to achieve a high degree of robustness against the PA targeting of a fingerprint modality. We evaluated the proposed system using a customized dataset composed of fingerprints and ECG signals.

The main contributions of this paper are listed as follows:

- Proposal of a novel end-to-end neural fusion architecture for fingerprints and ECG signals.
- A novel application of state-of-the-art EfficientNets for fingerprint PAD.
- Proposal of a 2D-convolutional neural network (2D-CNN) architecture for converting 1D ECG features into 2D images, yielding a better representation for ECG features compared to standard models based on fully-connected layers (FC) and 1D-convolutional neural networks (1D-CNNs).

The remainder of this paper is organized as follows. In Section 2, we introduce our proposed end-to-end deep learning approaches. In Section 3, we present the datasets and experimental setup applied. In Section 4, we present experimental results and discussions. Finally, in Section 5, we provide some concluding remarks and suggest areas of future study.

2. Proposed Methodology

Assume a fingerprint dataset $D = \{X_i, y_i\}_{i=1}^{N}$ composed of $N = A + B$ (where A is the number of artefact samples and B is the number of bona fide samples), where X_i represents the input fingerprint image and y_i is a binary label indicating if a fingerprint is an artefact or a bona fide (real). The aim of ordinary fingerprint PAD is to detect whether a fingerprint image is a PA (artefact) and differentiate it from a bona fide fingerprint sample. In this study, we consider ECG signals as an additional input modality to strengthen the fingerprint PAD system. To this end, the dataset becomes triplet $D = \left\{ \left(X_i^f, X_i^e \right), y_i \right\}_{i=1}^{N}$, where X_i^f is the fingerprint image and X_i^e is the ECG signal.

Figure 2 shows the proposed fusion approach, which is composed of three parts, i.e., the fingerprint branch, the ECG branch, and a fusion module. Detailed descriptions for these branches are provided in the next subsections.

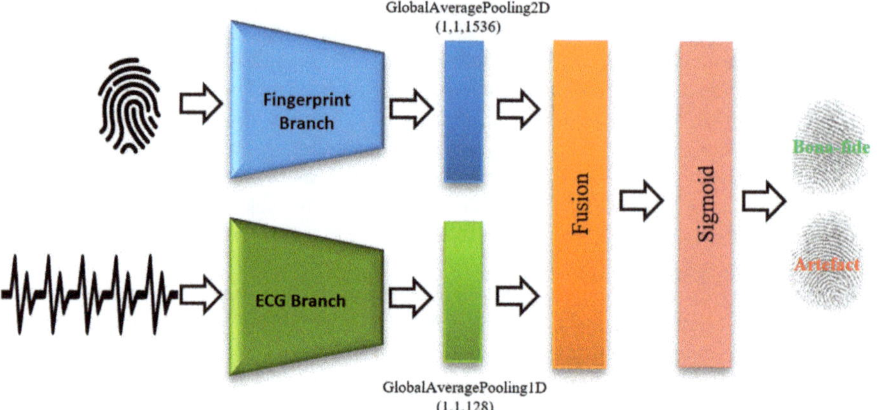

Figure 2. Overall architecture of the proposed end-to-end convolutional neural network-based (CNN) fusion architecture. ECG, electrocardiogram.

2.1. Fingerprint Branch

A fingerprint branch uses state-of-the-art EfficientNets [53] to obtain the feature representations of a fingerprint as shown in Figure 3 EfficientNets are a family of models that were recently developed by the Google Brain team by applying a new model scaling method for balancing the depth, width, and resolution of the CNNs [53]. Their scaling method uniformly scales the dimensions of a network using a simple and efficient compound coefficient. The compound scaling method enables a baseline CNN network to be scaled up with respect to the available resources while maintaining a high efficiency and accuracy. EfficientNets include mobile inverted bottleneck convolution (MBConv) as the basic building block [54]. In addition, this network uses an attention mechanism based on squeeze excitation (SE) to improve feature representations. This attention layer starts by applying a global average pooling (GAP) after each block. This operation is then followed by a fully-connected layer (with weight W_1) to reduce the number of dimensions by $(1/16)$. The resulting feature vector s is then used to calibrate the feature maps of each channel (V) using a channel-wise scale operation with an extra fully-connected layer with weight W_2. SE operates as shown below:

$$s = Sigmoid(W_2(ReLU(W_1(V)))), \tag{1}$$

$$V_{SE} = s \odot V, \tag{2}$$

where s is the scaling factor, $\odot$ refers to the channel-wise multiplication, and V represents the feature maps obtained from a particular layer of the EfficientNet.

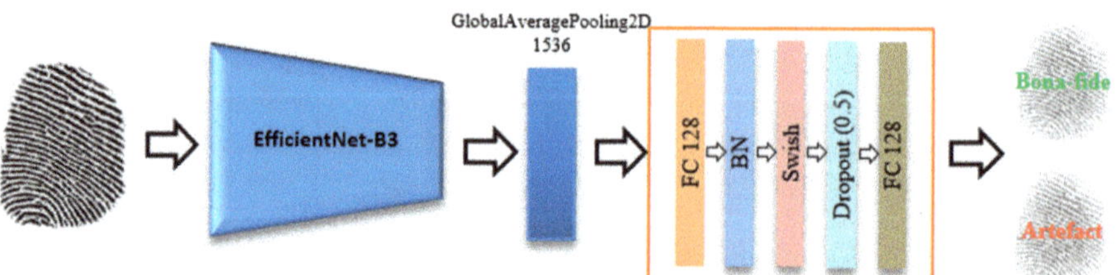

Figure 3. Flowchart of a fingerprint branch.

Furthermore, a novel activation function called Swish is used by an EfficientNet, which is essentially the sigmoid function multiplied by x according to Equation (3). Figure 4 shows the behavior of the following Swish activation function:

$$f(x) = x \cdot Sigmoid(x) \tag{3}$$

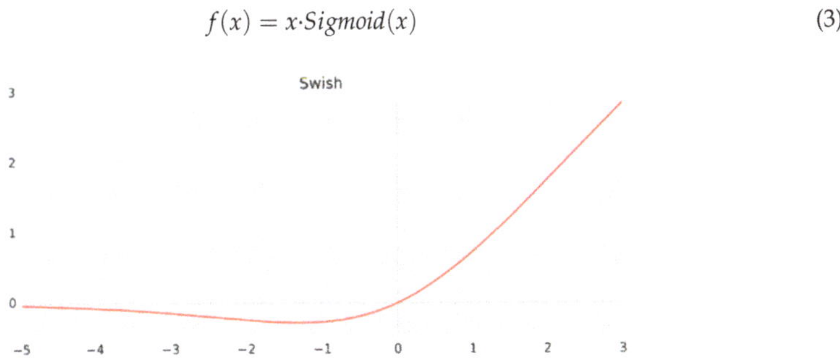

Figure 4. Swish activation function [55].

EfficientNet models surpass the accuracy of state-of-the-art CNN approaches on the ImageNet dataset [56] by minimizing the numbers of parameters and FLOPS, as shown in Figure 5. In this study, we investigate the baseline EfficientNet-B3 in terms of the feature representations of a fingerprint. To the best of our knowledge, this is the first time EfficientNets have been used for fingerprint PAD.

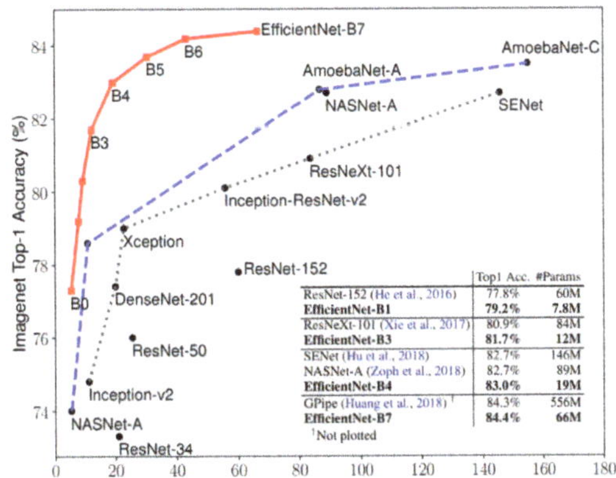

	Top1 Acc.	#Params
ResNet-152 (He et al., 2016)	77.8%	60M
EfficientNet-B1	**79.2%**	**7.8M**
ResNeXt-101 (Xie et al., 2017)	80.9%	84M
EfficientNet-B3	**81.7%**	**12M**
SENet (Hu et al., 2018)	82.7%	146M
NASNet-A (Zoph et al., 2018)	82.7%	89M
EfficientNet-B4	**83.0%**	**19M**
GPipe (Huang et al., 2018) †	84.3%	556M
EfficientNet-B7	**84.4%**	**66M**
†Not plotted		

Figure 5. Comparison among EfficientNet and other popular CNN models in terms of ImageNet accuracy vs. model size [53].

During the experiments, we truncated EfficientNet-B3 by removing its 1000 softmax classification layer and used the output of the "swish_78" layer as an input to a fusion module, which has the task of fusing the fingerprint and ECG features, as described later.

2.2. ECG Branch

Regarding an ECG branch, we propose three different feature representation architectures, FC, 1D-CNN, and 2D-CNN, as shown in Figure 6. The FC architecture is composed of simple fully-connected layers followed by batch normalization (BN), a Swish activation function, and dropout regularization to reduce an overfitting, as presented in Figure 6a. The second architecture, 1D-CNN, is based on the application of 1D convolution operations on the ECG signals. With this architecture, the ECG signals are fed through three 1D consecutive convolutional layers, the first two layers of which are followed by a BN, Swish, dropout (0.25), and 1D average pooling, and the last layer is followed by a BN, Swish, and 1D average pooling, as shown in Figure 6b. The last architecture, which is one of the main contributions of the present paper, is based on the idea of converting ECG feature into a 2D features image using a generator module and then processing the signal using a standard 2D CNN network, as shown in Figure 6c. In particular, this architecture learns and reshapes a 1D ECG feature into a 2D image using fully connected layers. The resulting image is then fed to two consecutives MBConv blocks to obtain the final 2D representations. Transforming a 1D ECG feature into an image can play a significant role in achieving powerful 2D convolution and pooling operations when learning the appropriate ECG features. During the experiments, we show that this architecture allows the generation of a better representation compared to those based on an analysis of 1D features.

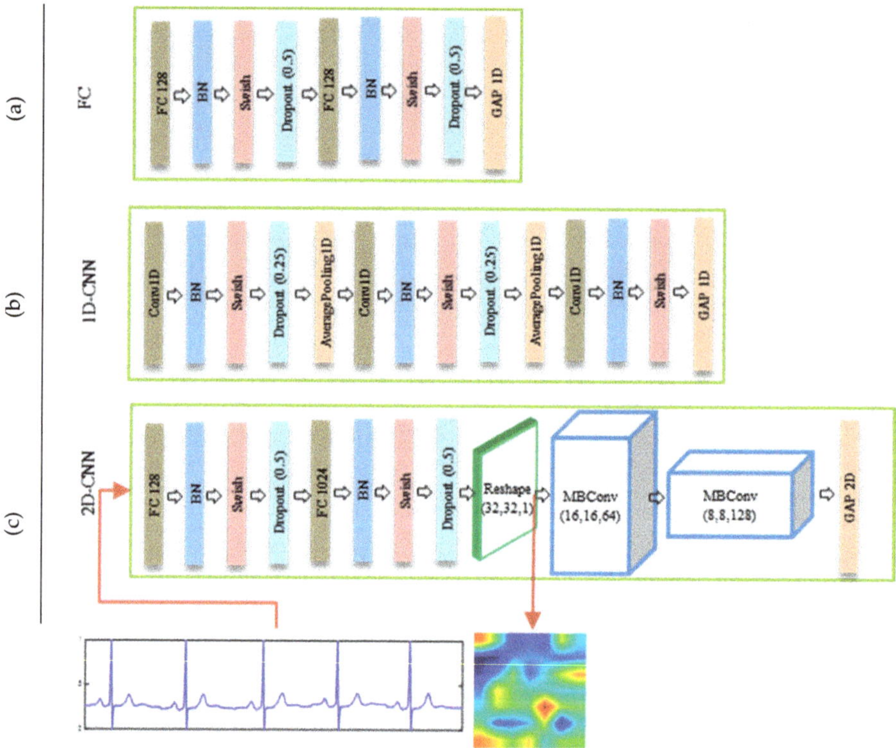

Figure 6. Details of ECG feature extraction architectures in ECG branch: (**a**) FC (**b**) 1D-CNN, and (**c**) 2D-CNN.

2.3. Fusion Module

The feature representations obtained from both fingerprint and ECG branches are further processed using a fusion module. This fusion module is composed of a sequence of layers, as shown in Figure 7. First, the feature vector of an ECG is concatenated into a fingerprint feature vector to produce a single global feature vector. The concatenated feature is fed to an additional fully connected layer followed by a BN, Swish activation function, and dropout (0.5) regularization. Finally, the output of this module is fed to a binary classification layer using a sigmoid activation function to determine the final fingerprint class, i.e., an artefact or bona fide.

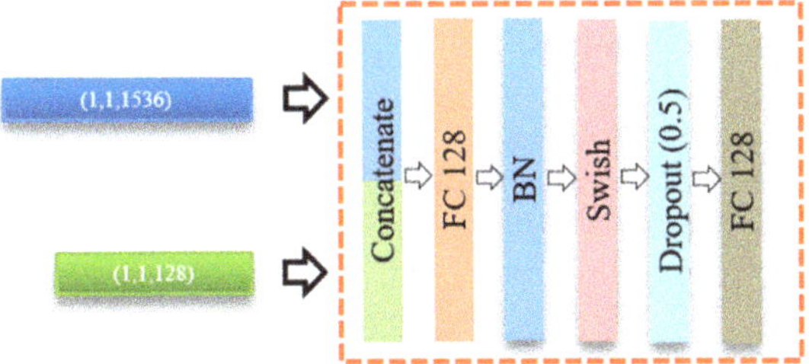

Figure 7. Structure of the fusion module.

2.4. Network Optimization

As mentioned previously, the complete architecture proposed in this study is a learnable end-to-end network using a backpropagation algorithm. If we define the output of the sigmoid function in the final layer of the trained network as $\hat{y}_i$, then the distribution of the network output $\hat{y}_i$ follows a Bernoulli distribution. The determination of the weights W of the network, including those of the fingerprint and ECG branches, can be carried out by maximizing the following likelihood function:

$$L(D, W) = \prod_{i=1}^{N} \hat{y}_i^{y_i} (1 - \hat{y}_i)^{1-y_i},\tag{4}$$

which is equivalent to minimizing the following log-likelihood function:

$$L(D.W) = -\sum_{i=1}^{N} y_i ln(\hat{y}_i) + (1 - y_i) ln(1 - \hat{y}_i).\tag{5}$$

The loss function in (5) is usually called a cross-entropy loss function. To optimize this loss, we use the RMSProp optimization algorithm proposed by Hinton [57], which is considered one of the most common adaptive gradient algorithms, dividing the gradient by averaging the magnitude of its recent movement as follows:

$$E\left[g^2\right]_t = \beta E\left[g^2\right]_{t-1} + (1 - \beta)\left(\frac{\partial L}{\partial W}\right)^2,\tag{6}$$

$$W_t = W_{t-1} - \alpha\left(\frac{\partial L}{\partial W}\right)\frac{1}{\sqrt{E[g^2]_t}},\tag{7}$$

where $E\left[g^2\right]_t$ represents a moving average of the squared gradients during the iteration process (t), and $\frac{\partial L}{\partial W}$ is known as the gradients of the loss function of the weights of the network W. Parameters α and β are the learning rate and moving average, respectively. During the experiments, parameter β is set to its default value ($\beta = 0.9$), whereas α is initially set to 0.0001 and is periodically decreased by a factor of 1/10 for every 20 epochs.

3. Experiments

3.1. Datasets

To evaluate the proposed approach, we used the LivDet 2015 dataset for the fingerprint and real ECG datasets collected in our lab. LivDet 2015 has approximately 19,000 images divided into two parts: training and testing. Each part has bona fide (live) and artefact (fake) images captured using different fingerprint sensors, as shown in Table 1. Numerous materials are used for fabricating the artefact fingerprint samples, e.g., Ecoflex, gelatin, latex, and wood glue. The testing dataset contains artefact samples fabricated using various materials, some of which are not used in the training dataset, such as OOMOO and RTV, as shown in Table 2. Figure 8 shows bona fide and artefact samples for the same subject captured from two different sensors, i.e., Green Bit and Digital Persona sensors.

Table 1. Device and image characteristics of the LivDet 2015 dataset.

Sensor	Resolution (dpi)	Image Size (pixel)	Training		Testing	
			Live	Fake	Live	Fake
Green Bit	500	500×500	1000	1000	1000	1500
Biometrika (Hi Scan)	1000	1000×1000	1000	1000	1000	1500
Digital Persona	500	252×324	1000	1000	1000	1500
Crossmatch	500	640×480	1500	1500	1500	1448

Table 2. Materials used for fabricating fake images in the LivDet 2015 dataset. Some materials in the testing are unknown during training (<u>underlined</u>).

Sensor	Training	Testing
Green Bit		Ecoflex, gelatin, latex, wood glue,
Biometrika	Ecoflex, gelatin, latex, wood glue	Liquid Ecoflex, <u>RTV</u>
Digital Persona		
Crossmatch	Body Double, Ecoflex, PlayDoh	Body Double, Ecoflex, PlayDoh, <u>OOMOO</u>, gelatin

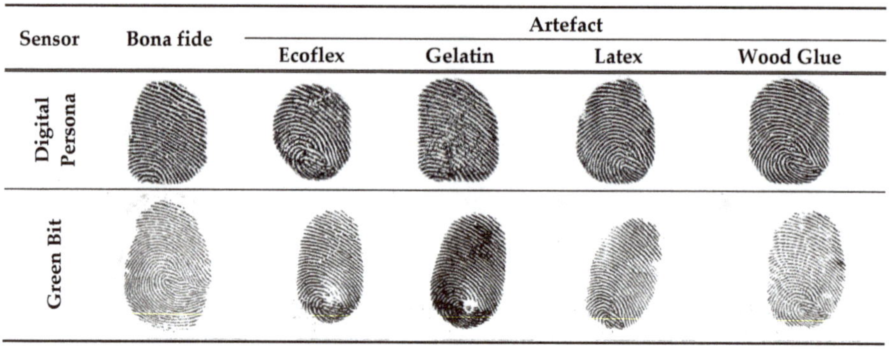

Sensor	Bona fide	Artefact			
		Ecoflex	Gelatin	Latex	Wood Glue
Digital Persona					
Green Bit					

Figure 8. Bona fide and artefact fingerprint samples from the LivDet 2015 dataset captured using Digital Person and Green Bit sensors. Artefact samples were fabricated using different materials.

For the ECG dataset, we used a dataset collected in our lab. We collected this dataset using a commercially available handheld ECG device, i.e., ReadMyHeart by DailyCare BioMedical, Inc. (https://www.dcbiomed.com/webls-en-us/index.html), as shown in Figure 9. We built a database of 656 ECG records captured from 164 individuals collected in two sessions [48,58]. Now, we have extended this database with a third session to have 10 records for most of the users. The device captures a signal for 15 seconds, digitalizes it and exports it to the computer as an ECG record. Generally, such a signal may contain different types of noise, such as power-line interface, baseline wanders, and patient-electrode motion artifacts. In the preprocessing step, we use a band-pass Butterworth filter of order four with cut-off frequencies of 0.25 and 40 Hz to remove the noises. Then an efficient curvature-based method is used to detect heartbeats [59,60] and we take the first 10 beats from each record for this experiment. Figure 10 shows such preprocessed ECG samples from four different subjects.

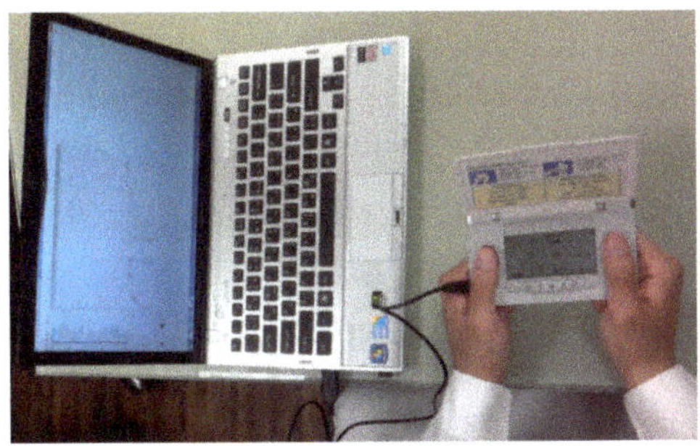

Figure 9. ECG data collection using the ReadMyHeart device.

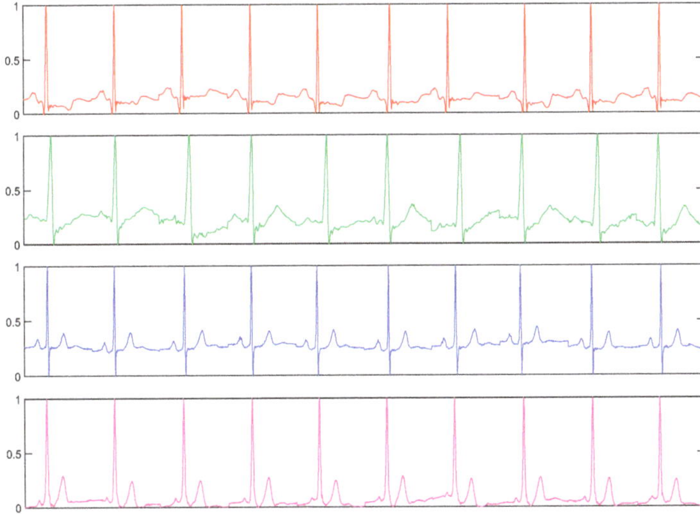

Figure 10. ECG sample of 10 heart beats from four different subjects.

Owing to a lack of availability of public multimodal datasets containing fingerprint and ECG signals, we constructed a multi-model dataset from the LivDet 2015 dataset and an ECG dataset. First, we built a mini fingerprint dataset from the LivDet 2015 dataset, called the mini-livdet2015 dataset, containing images from Digital Persona sensor. This mini-livdet2015 is composed of 70 subjects, each of which has bona fide and artefact samples (10 and 12, respectively). Subsequently, we randomly selected the artefact samples from all available fabricating materials. To form this multimodal dataset, we assigned a random subject from the ECG dataset to each subject from the mini-livdet2015 dataset. Table 3 describes this new dataset, which is comprised of 70 subjects, each of which has 10 bona fide and 12 artefact fingerprint samples, and 10 ECG samples.

During training, we feed the network with batches of input triplets that cover both possible classes. For the bona fide label, we assign a bona fide fingerprint sample with a bona fide ECG sample from the same subject; i.e., X_i^f and X_i^e are bona fide samples belonging to the same subject. Because we do not have artefact ECG signals, we assign an artefact fingerprint sample from one subject with a bona fide ECG sample from another subject (zero-effort ECG sample); i.e., X_i^f and X_i^e are bona fide samples from two different subjects.

Feeding the network with these inputs allows learning the correlations between bona fide fingerprint samples and ECG samples of the same subject to correctly predict which samples are bona fide. Furthermore, this network learns how to correctly predict an artefact by learning the features representing the incoherence between artefact fingerprint sample and a bona fide ECG sample of the same subjects or between a bona fide fingerprint sample of a subject and a bona fide ECG sample belonging to different subjects.

Table 3. Description of the customized multimodal dataset, which contains 70 subjects.

	Fingerprint		ECG
	Bona Fide	Arteact	Bona Fide
Number of samples per subject	10	12	10
Total number of samples	700	840	700

3.2. Experiment Setup

To evaluate the proposed approach, we conducted several experiments. First, we carried out an initial experiment to evaluate the performance of the fingerprint branch net regarding the detection of PAs. We compared our results with previous state-of-the-art methods. For this purpose, we utilized the LivDet 2015 dataset, whereas the fingerprint branch net was trained on the training portion of the LivDet 2015 and tested on the testing portion of the same dataset. In the second experiment, we evaluated the three proposed fusion architectures in detecting and preventing the PAs. We then conducted an experiment to analyze the sensitivity of the highest performing architecture during the second experiment. To this end, we analyzed the effects of increasing the number of subjects during the training on the classification accuracy. Finally, we reported the number of parameters and classification time by the proposed architectures compared with state-of-the-art methods.

For training the network, we use the RMSProp optimizer with the following parameters: β is set to its default value ($\beta = 0.9$), whereas α is initially set to 0.0001 and is periodically decreased by a factor of 1/10 for every 20 iterations (epochs). For compatibility with the LivDet 2015 competition [27], the accuracy was used as the evaluation parameter in all of the experiments. The accuracy is defined as the percentage of correctly classified samples.

All experiments were repeated five times and the average classification accuracy was reported. The experiments were carried out using a workstation with i9 CPU @ 2.9 GHz, 32 GB of RAM, and NVIDIA GeForce GTX 1080 Ti (11 GB GDDR5X).

4. Results and Discussions

4.1. Experiments Using Fingerprint Modality Only

Initially, we examined the performance of the proposed fingerprint network based on EfficientNet. This evaluation allows us to compare the performances of this network in terms of fingerprint PAD with that of the methods proposed in the LivDet 2015 competition [27]. Table 4 shows the results after training the network for 50 iterations.

Table 4. Comparison between the results of the proposed fingerprint branch net and the best methods from the LivDet 2015 competition, where we present the average accuracy %.

Algorithm	Green Bit	Biometrika	Digital Persona	Crossmatch	Overall
Nogueira (first place winner)	95.40	94.36	**93.72**	**98.10**	**95.51**
Proposed	94.68	95.12	91.96	97.29	94.87
Unina (second place winner)	**95.80**	**95.20**	85.44	96.00	93.92

We can see from the results in Table 4 that the proposed fingerprint network achieves an overall classification accuracy of 94.87%. A comparison of the reported accuracy of the proposed network with those reported from the LivDet 2015 competition shows that our method would have been the second-best approach. Moreover, the proposed method follows the same behavior as the other two algorithms in terms of its accuracy for the individual sensors, achieving a high accuracy of 97.29% for the Crossmatch sensor (i.e., an easy to learn sensor) and a relatively lower accuracy of 91.96% for the Digital Persona sensor (i.e., a difficult to learn sensor). Furthermore, the proposed method achieves moderately high accuracies of 94.68% and 95.12% for the Green bit and Biometrika sensors, respectively. Figure 11 shows the progress of the loss function during the training on the LivDet 2015 dataset (training part). Note that the loss converges at a low number of iterations (nearly 25 iterations). The reported results confirm the promising capability of the network in detecting PAs, motivating us to improve it further by proposing a multimodal solution that fuses fingerprints with ECG signals.

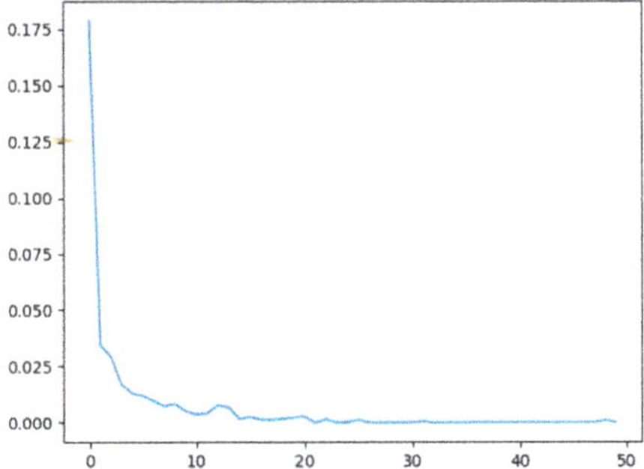

Figure 11. Model loss versus number of epochs (50) by training on LivDet 2015 dataset.

4.2. Fusion of Fingerprints and ECGs

As mentioned previously, owing to the lack of a multimodal dataset containing fingerprints and ECG modalities, we built a mini-livdet2015 dataset and fused it with the ECG dataset. We used a Digital Persona sensor, the most difficult sensor used for the LivDet 2015 dataset, as demonstrated in the previous experiment, and achieved the lowest accuracy 91.96% in comparison to the other sensors. This mini-livdet2015 contains 70 subjects, each of which has 10 bona fide and 12 artefact fingerprint samples. We constructed the multimodal dataset by randomly linking each subject from the mini-livdet2015 dataset to a subject from the ECG dataset, as previously discussed. Before running the fusion network on this multimodal dataset, we first trained the fingerprint network on the mini-livdet2015 dataset to obtain an indication regarding its performance, which is considered a baseline for our fusion mechanism. We obtained an accuracy of 92.98% using 50% of the subjects for training and 50% for testing, i.e., 35 subjects for training and the other 35 subjects for testing.

After this step, we evaluated the complete architecture using the three proposed feature extraction solutions (i.e., FC, 1D-CNN, and 2D-CNN). Table 5 shows the average classification accuracy of the three fusion architectures. Furthermore, the average classification accuracy of the fingerprint network on the mini-livdet2015 dataset is reported.

Table 5. Average accuracy of three proposed fusion architectures and the fingerprint branch net. The reported results are achieved on the customized dataset.

Biometric Modality	ECG Architecture	Average Accuracy %
Fingerprint	(No fusion)	92.98
Fingerprint + ECG	FC	94.99
	1D-CNN	94.84
	2D-CNN	**95.32**

The reported results show that fusing fingerprints with ECG data clearly improves the accuracy of artefact fingerprint detection. The different architectures, namely, FC, 1D-CNN, and 2D-CNN, achieve accuracies of 94.99%, 94.84%, and 95.32%, respectively, thereby outperforming the accuracy achieved by fingerprint net (i.e., without applying a fusion). As the results indicate, the 2D-CNN model achieves the highest accuracy (95.32%) compared with the other two fusion architectures. The high performance of the 2D-CNN model can be attributed to the conversion of ECG signals into images, thus utilizing the power of 2D convolution and pooling operations, in addition to the introduction of MBConv blocks as the main blocks for learning the representative features.

4.3. Sensitivity Analysis of the Number of Training Subjects

During this experiment, we discuss how the number of subjects used for training can affect the level of accuracy. We repeated the above experiment with different percentages of subjects used for training (between 20% and 80%), the average accuracy of which is reported in Table 6.

Table 6. Sensitivity analysis of the proposed architectures against the number of training subjects in terms of the reported testing accuracy (ACC %).

ECG Architecture	Percentage of Subjects Used for Training				
	20%	30%	50%	70%	80%
FC	89.71	93.90	94.49	93.92	96.17
1D-CNN	89.31	92.45	94.26	93.36	96.95
2D-CNN	**90.79**	**94.08**	**95.32**	**95.61**	**97.10**

The reported results show that increasing the number of subjects (70% and 80%) during the training improves the testing accuracy. Although this behavior is the same for the three proposed architectures, we can see that the 2D-CNN model consistently outperforms the other two models with an accuracy of 97.10% when using 80% of the subjects in the dataset for training. In contrast, decreasing the number of subjects during the training degrades the testing accuracy. Despite the decrease in testing accuracy, the level achieved is still acceptable (89.71%, 89.31%, and 90.79%) for the FC, 1D-CNN, and 2D-CNN, respectively when using 20% of the subjects for training.

4.4. Sensitivity with Respect to the Pre-Trained CNN

In order to further assess the sensitivity of the proposed approach with respect to the pre-trained CNN model, we carried out additional experiments using other well-known pre-trained CNN models: Inception-v3 [61], DenseNet [62], and residual network (ResNet) [63]. We used these recent pre-trained models for this experiment as they require a comparatively small number of parameters as shown in Table 7.

Inception-v3 is one of Inception models family developed by Google [61], in which they introduced the concept of factorizing of convolutions. Inceptions models are based on increasing the width and depth of the network, by utilizing a module called inception [64], which contains several convolutional layers with different filter sizes. The utilization of the inception module allows the Inception model for a better deal with scale and spatial variations. DenseNet was proposed by Szegedy et al. [62] for better utilization of computing resources. DeneNet is based on adding connections between each layer and every other layer in feed-forward fashion, in which each layer receives the feature maps of all preceding layers as input and feeds its own feature maps as input into all subsequent layers. ResNet was proposed by He et al. [63] to overcome the difficulties in training deeper networks by learning residual functions. These residual networks achieved better optimization and generalization as the depth increases.

Table 7. Classification Accuracy using different pre-trained CNN models. We used Inception-v3, DenseNet-169, and ResNet-50.

CNN Model	Architecture	#Parameters	Average Accuracy %
EfficientNet-B3	FC	**10 M**	**94.99**
	1D-CNN		**94.84**
	2D-CNN		**95.32**
Inception-v3	FC	21 M	92.80
	1D-CNN		94.32
	2D-CNN		95.20
DenseNet-169	FC	12 M	91.28
	1D-CNN		92.92
	2D-CNN		93.29
ResNet-50	FC	23 M	93.56
	1D-CNN		93.68
	2D-CNN		94.00

We note from the results in Table 7, that EfficientNet-B3 achieves the highest accuracies in the three architectures and outperforms the other pre-trained CNNs. The accuracies of EfficientNet-B3 in the three architecture (94.99%, 94.84%, and 95.32%) consistently exceed the best accuracies of the other models except for Inception-v3 which achieves a comparable result in the case of 2D-CNN (95.20%). However, EfficientNet-B3 requires the minimum number of parameters (10 M after removing the top layers), whereas Inception-v3 and resNet-50 require 21 M and 23 M, respectively (after removing the top layers). Finally, from the reported accuracies, we note that 2D-CNN outperforms the FC and 1D-CNN for all the pre-trained models (i.e., 95.32%, 95.20%, 93.29%, and 94.00% for EffieicentNet-B3, Inception-v3, DenseNet-169, and ResNet-50 respectively).

4.5. Sensitivity of the ECG Network Architecture

In order to further assess the sensitivity of the proposed approach with different configurations, we carried out additional experiments to show the effect of using different configurations of the ECG branch net on the overall accuracy. Considering that the 2D-CNN architecture proves its superiority over the FC and 1D-CNN as shown in the previous sections, we reported the experiments that cover applying different configurations using 2D-CNN architecture. We tested 8 different configurations as described in Table 8. Let us consider the configuration #8, which is shown in Figure 6c: (2 fc = (128, 1024), 2 blocks MBConv (64, 128), fc = 128), this means we use two consecutive fully-connected layers of size 128 and 1024, respectively, in addition to using two consecutive MBConv blocks, of depth 64 and 128, respectively, and finally one fully-connected layer of size 128. The second fully-connected layer fc (1024) means that the feature vector is reshaped into ($32 \times 32 \times 1$) as shown in Figure 6c.

From the reported results in Table 8, we note the following points. Removing the first fully-connected layer fc (128) in configuration #1, degraded the accuracy (91.90%), whereas increasing the feature map in configuration #2 by replacing the second fully-connected layer fc (1024) with fc (4096); i.e., the feature vector is reshaped into ($64 \times 64 \times 1$); will not significantly improve the accuracy (93.56%). Furthermore, changing the number and sizes of MBConv blocks up to 3 (configurations #6 & #7) or down to 1 (configurations #3, #4, & #5), produces better accuracies up to 95.56% in configuration #4. In the proposed configuration #8, we used 2 MBConv blocks, in which the networks achieved the second best accuracy of 95.32%.

Table 8. Classification accuracy of 2D-CNN network by applying three different configurations for ECG architecture.

Configuration	Configuration Description	Accuracy %
1	2 fc = (1024), 2 blocks MBConv (64, 168), fc = 128	91.90
2	2 fc = (128, 4096), 2 blocks MBConv (64), fc = 128	93.56
3	2 fc= (128, 1024), 1 block MBConv (32), fc = 128	94.82
4	2 fc = (128, 1024), 1 block MBConv (64), fc = 128	**95.58**
5	2 fc = (128, 1024), 1 block MBConv (128), fc = 128	95.07
6	2 fc = (128, 1024), 3 blocks MBConv (64), fc = 128	94.68
7	2 fc = (128, 1024), 3 blocks MBConv (64, 128, 128), fc = 128	95.20
8 (Proposed)	2 fc = (128, 1024), 2 blocks MBConv (64, 128), fc = 128	95.32

4.6. Classification Time

In this study, we used an EfficientNet with 12 million parameters as the main building block for the fingerprint branch. This model provides impressive results with a low computational cost. In particular, our models converge using only 50 epochs. The complete architecture provides an average classification time for one subject (i.e., fingerprint image + ECG signal) of 30–35 ms (depending on the architecture), which is faster than previous state-of-the-art approaches (i.e., 128 ms [24] and 800 ms [28]). Recall that the approaches described in [21] and [24] applied solutions using only the fingerprint modality and networks with a larger number of weights.

5. Conclusions

In this paper, we proposed an end-to-end deep learning approach fusing fingerprint and ECG signals for boosting the PAD capabilities of fingerprint biometrics. We also introduced EfficientNet, a state-of-the-art network, for learning efficient fingerprint feature representations. The experimental results prove the superiority of the EfficientNet over other known pre-trained CNNs in terms of accuracy and efficiency. For the ECG signals, we proposed three different architectures and configurations FC, a 1D-CNN, and a 2D-CNN. With the 2D-CNN model, we transformed the ECG features into images using a generator network. The experimental results obtained on a multimodal dataset composed of fingerprint and ECG signals reveal the promising capability of the proposed solution in terms of the classification accuracy and computation time. Although we used a customized database of fingerprint and ECG signal to validate the proposed method, we intend to use a database of fingerprint and ECG signal captured simultaneously using a multimodal sensor. Since this type of sensor is not commercially available, we would like to develop a prototype of such a sensor, create a database of real multimodal data, and use it for the validation of the proposed method as our future work.

Author Contributions: Data curation, M.S.I.; methodology, R.M.J. and Y.B.; supervision, H.M.; writing—original draft, R.M.J.; writing—review and editing, H.M., Y.B., and M.S.I. All authors have read and agreed to the published version of the manuscript.

Funding: The authors would like to thank Deanship of scientific research for funding and supporting this research through the initiative of DSR Graduate Students Research Support (GSR).

Acknowledgments: The authors thank the Deanship of Scientific Research and RSSU at King Saud University for their technical support.

Conflicts of Interest: The authors declare no conflicts of interest.

References

1. Jain, A.K.; Kumar, A. Biometric Recognition: An Overview. In *Second Generation Biometrics: The Ethical, Legal and Social Context*; Mordini, E., Tzovaras, D., Eds.; Springer: Dordrecht, Netherlands, 2012; pp. 49–79, ISBN 978-94-007-3891-1.
2. Standard, I. *Information Technology—Biometric Presentation Attack Detection—Part 1: Framework*; ISO: Geneva, Switzerland, 2016.

3. Schuckers, S. Presentations and attacks, and spoofs, oh my. *Image Vis. Comput.* **2011**, *55 Pt 1*, 26–30. [CrossRef]

4. Chugh, T.; Jain, A.K. Fingerprint Spoof Generalization. *arXiv* **2019**, arXiv:191202710.

5. Coli, P.; Marcialis, G.L.; Roli, F. Vitality detection from fingerprint images: a critical survey. In Proceedings of the International Conference on Biometrics, Berlin/Heidelberg, Germany, 27 August 2007; pp. 722–731.

6. Lapsley, P.D.; Lee, J.A.; Pare, D.F., Jr.; Hoffman, N. Anti-Fraud Biometric Scanner that Accurately Detects Blood Flow. U.S. Patent Application No. US 5,737,439, 7 April 1998.

7. Antonelli, A.; Cappelli, R.; Maio, D.; Maltoni, D. Fake finger detection by skin distortion analysis. *IEEE Trans. Inf. Forensics Secur.* **2006**, *1*, 360–373. [CrossRef]

8. Baldisserra, D.; Franco, A.; Maio, D.; Maltoni, D. Fake fingerprint detection by odor analysis. In Proceedings of the International Conference on Biometrics, Berlin/Heidelberg, Germany, 5 January 2006; pp. 265–272.

9. Nikam, S.B.; Agarwal, S. Texture and wavelet-based spoof fingerprint detection for fingerprint biometric systems. In Proceedings of the 2008 First International Conference on Emerging Trends in Engineering and Technology, Nagpur, Maharashtra, 16 July 2008; pp. 675–680.

10. Ghiani, L.; Marcialis, G.L.; Roli, F. Fingerprint liveness detection by local phase quantization. In Proceedings of the 21st International Conference on Pattern Recognition (ICPR2012), Tsukuba, Japan, 11 November 2012; pp. 537–540.

11. Gragnaniello, D.; Poggi, G.; Sansone, C.; Verdoliva, L. Fingerprint liveness detection based on weber local image descriptor. In Proceedings of the IEEE Workshop on Biometric Measurements and Systems for Security and Medical Applications, Naples, Italy, 9 September 2013; pp. 46–50.

12. Xia, Z.; Yuan, C.; Lv, R.; Sun, X.; Xiong, N.N.; Shi, Y.-Q. A novel weber local binary descriptor for fingerprint liveness detection. *IEEE Trans. Syst. Man Cybern. Syst.* **2018**, *50*, 1526–1536. [CrossRef]

13. Gragnaniello, D.; Poggi, G.; Sansone, C.; Verdoliva, L. Local contrast phase descriptor for fingerprint liveness detection. *Pattern Recognit.* **2015**, *48*, 1050–1058. [CrossRef]

14. Chatfield, K.; Simonyan, K.; Vedaldi, A.; Zisserman, A. Return of the devil in the details: Delving deep into convolutional nets. *arXiv* **2014**, arXiv:14053531.

15. Simonyan, K.; Zisserman, A. Very deep convolutional networks for large-scale image recognition. *arXiv* **2014**, arXiv:14091556

16. Al Rahhal, M.M.; Bazi, Y.; Almubarak, H.; Alajlan, N.; Al Zuair, M. Dense Convolutional Networks with Focal Loss and Image Generation for Electrocardiogram Classification. *IEEE Access* **2019**, *7*, 182225–182237. [CrossRef]

17. Al Rahhal, M.M.; Bazi, Y.; Al Zuair, M.; Othman, E.; BenJdira, B. Convolutional neural networks for electrocardiogram classification. *J. Med. Biol. Eng.* **2018**, *38*, 1014–1025. [CrossRef]

18. Al Rahhal, M.M.; Bazi, Y.; AlHichri, H.; Alajlan, N.; Melgani, F.; Yager, R.R. Deep learning approach for active classification of electrocardiogram signals. *Inf. Sci.* **2016**, *345*, 340–354. [CrossRef]

19. Hammad, M.; Wang, K. Parallel score fusion of ECG and fingerprint for human authentication based on convolution neural network. *Comput. Secur.* **2019**, *81*, 107–122. [CrossRef]

20. Minaee, S.; Abdolrashidi, A.; Su, H.; Bennamoun, M.; Zhang, D. Biometric Recognition Using Deep Learning: A Survey. *arXiv* **2019**, arXiv:191200271.

21. Talreja, V.; Valenti, M.C.; Nasrabadi, N.M. Multibiometric secure system based on deep learning. In Proceedings of the 2017 IEEE Global Conference on Signal and Information Processing (GlobalSIP), Montreal, QC, Canada, 14–16 November 2017; pp. 298–302.

22. Al-Waisy, A.S.; Qahwaji, R.; Ipson, S.; Al-Fahdawi, S.; Nagem, T.A. A multi-biometric iris recognition system based on a deep learning approach. *Pattern Anal. Appl.* **2018**, *21*, 783–802. [CrossRef]

23. Nogueira, R.F.; de Alencar Lotufo, R.; Machado, R.C. Fingerprint Liveness Detection Using Convolutional Neural Networks. *IEEE Trans Inf. Forensics Secur.* **2016**, *11*, 1206–1213. [CrossRef]

24. Park, E.; Cui, X.; Nguyen, T.H.B.; Kim, H. Presentation attack detection using a tiny fully convolutional network. *IEEE Trans. Inf. Forensics Secur.* **2019**, *14*, 3016–3025. [CrossRef]

25. Souza, G.B.; Santos, D.F.; Pires, R.G.; Marana, A.N.; Papa, J.P. Deep Boltzmann Machines for robust fingerprint spoofing attack detection. In Proceedings of the 2017 International Joint Conference on Neural Networks (IJCNN), Anchorage, AK, USA, 3 July 2017; pp. 1863–1870.

26. Tolosana, R.; Gomez-Barrero, M.; Kolberg, J.; Morales, A.; Busch, C.; Ortega-Garcia, J. Towards Fingerprint Presentation Attack Detection Based on Convolutional Neural Networks and Short Wave Infrared Imaging. In Proceedings of the 2018 International Conference of the Biometrics Special Interest Group (BIOSIG), Darmstadt, Germany, 3 December 2018; pp. 1–5.

27. Mura, V.; Ghiani, L.; Marcialis, G.L.; Roli, F.; Yambay, D.A.; Schuckers, S.A. LivDet 2015 fingerprint liveness detection competition 2015. In Proceedings of the 2015 IEEE 7th International Conference on Biometrics Theory, Applications and Systems (BTAS), Arlington, VA, USA, 8–11 September 2015; pp. 1–6.

28. Chugh, T.; Cao, K.; Jain, A.K. Fingerprint spoof detection using minutiae-based local patches. In Proceedings of the 2017 IEEE International Joint Conference on Biometrics (IJCB), Denver, CO, USA, 1 February 2017; pp. 581–589.

29. Zhang, Y.; Shi, D.; Zhan, X.; Cao, D.; Zhu, K.; Li, Z. Slim-ResCNN: A Deep Residual Convolutional Neural Network for Fingerprint Liveness Detection. *IEEE Access* **2019**, *7*, 91476–91487. [CrossRef]

30. Galbally, J.; Fierrez, J.; Cappelli, R. An introduction to fingerprint presentation attack detection. In *Handbook of Biometric Anti-Spoofing. Advances in Computer Vision and Pattern Recognition*; Springer: Cham, Switzerland, 2019.

31. Ross, A.A.; Nandakumar, K.; Jain, A.K. *Handbook of multibiometrics*; Springer: New York, NY, USA, 2006; Volume 6.

32. Huang, Z.; Feng, Z.-H.; Kittler, J.; Liu, Y. Improve the Spoofing Resistance of Multimodal Verification with Representation-Based Measures. In Proceedings of the Chinese Conference on Pattern Recognition and Computer Vision (PRCV), Guangzhou, China, 23–26 November 2018; pp. 388–399.

33. Wild, P.; Radu, P.; Chen, L.; Ferryman, J. Robust multimodal face and fingerprint fusion in the presence of spoofing attacks. *Pattern Recognit.* **2016**, *50*, 17–25. [CrossRef]

34. Marasco, E.; Shehab, M.; Cukic, B. A Methodology for Prevention of Biometric Presentation Attacks. In Proceedings of the 2016 Seventh Latin-American Symposium on Dependable Computing (LADC), Cali, Colombia, 19–21 October 2016; pp. 9–14.

35. Bhardwaj, I.; Londhe, N.D.; Kopparapu, S.K. A spoof resistant multibiometric system based on the physiological and behavioral characteristics of fingerprint. *Pattern Recognit.* **2017**, *62*, 214–224. [CrossRef]

36. Komeili, M.; Armanfard, N.; Hatzinakos, D. Liveness detection and automatic template updating using fusion of ECG and fingerprint. *IEEE Trans. Inf. Forensics Secur.* **2018**, *13*, 1810–1822. [CrossRef]

37. Pouryayevali, S. ECG Biometrics: New Algorithm and Multimodal Biometric System. Master's Thesis, Department of Electrical and Computer Engineering, University of Toronto, Toronto, ON, Canada, 2015.

38. Jomaa, R.M.; Islam, M.S.; Mathkour, H. Improved sequential fusion of heart-signal and fingerprint for anti-spoofing. In Proceedings of the 2018 IEEE 4th International Conference on Identity, Security, and Behavior Analysis (ISBA), Singapore, 12 March 2018; pp. 1–7.

39. Regouid, M.; Touahria, M.; Benouis, M.; Costen, N. Multimodal biometric system for ECG, ear and iris recognition based on local descriptors. *Multimed. Tools Appl.* **2019**, *78*, 22509–22535. [CrossRef]

40. Su, K.; Yang, G.; Wu, B.; Yang, L.; Li, D.; Su, P.; Yin, Y. Human identification using finger vein and ECG signals. *Neurocomputing* **2019**, *332*, 111–118. [CrossRef]

41. Blasco, J.; Peris-Lopez, P. On the feasibility of low-cost wearable sensors for multi-modal biometric verification. *Sensors* **2018**, *18*, 2782. [CrossRef] [PubMed]

42. Jomaa, R.M.; Islam, M.S.; Mathkour, H. Enhancing the information content of fingerprint biometrics with heartbeat signal. In Proceedings of the 2015 World Symposium on Computer Networks and Information Security (WSCNIS), Hammamet, Tunisia, 4 January 2015; pp. 1–5.

43. Alajlan, N.; Islam, M.S.; Ammour, N. Fusion of fingerprint and heartbeat biometrics using fuzzy adaptive genetic algorithm. In Proceedings of the 2013 World Congress on Internet Security (WorldCIS), London, UK, 9 December 2013; pp. 76–81.

44. Singh, Y.N.; Singh, S.K.; Gupta, P. Fusion of electrocardiogram with unobtrusive biometrics: An efficient individual authentication system. *Pattern Recognit. Lett.* **2012**, *33*, 1932–1941. [CrossRef]

45. Pinto, J.R.; Cardoso, J.S.; Lourenço, A. Evolution, current challenges, and future possibilities in ECG biometrics. *IEEE Access* **2018**, *6*, 34746–34776. [CrossRef]

46. Zhao, C.; Wysocki, T.; Agrafioti, F.; Hatzinakos, D. Securing handheld devices and fingerprint readers with ECG biometrics. In Proceedings of the 2012 IEEE Fifth International Conference on Biometrics: Theory, Applications and Systems (BTAS), Arlington, VA, USA, 23–27 September 2012; pp. 150–155.

47. Agrafioti, F.; Hatzinakos, D.; Gao, J. *Heart Biometrics: Theory, Methods and Applications*; INTECH Open Access Publisher: Shanghai, China, 2011.

48. Islam, M.S.; Alajlan, N. Biometric template extraction from a heartbeat signal captured from fingers. *Multimed. Tools Appl.* **2016**. [CrossRef]

49. Minaee, S.; Bouazizi, I.; Kolan, P.; Najafzadeh, H. Ad-Net: Audio-Visual Convolutional Neural Network for Advertisement Detection In Videos. *arXiv* **2018**, arXiv:180608612.

50. Torfi, A.; Iranmanesh, S.M.; Nasrabadi, N.; Dawson, J. 3d convolutional neural networks for cross audio-visual matching recognition. *IEEE Access* **2017**, *5*, 22081–22091. [CrossRef]

51. Zhu, Y.; Lan, Z.; Newsam, S.; Hauptmann, A. Hidden two-stream convolutional networks for action recognition. In Proceedings of the Asian Conference on Computer Vision, Perth, Australia, 2 December 2018; pp. 363–378.

52. Hammad, M.; Liu, Y.; Wang, K. Multimodal biometric authentication systems using convolution neural network based on different level fusion of ECG and fingerprint. *IEEE Access* **2018**, *7*, 26527–26542. [CrossRef]

53. Tan, M.; Le, Q.V. Efficientnet: Rethinking model scaling for convolutional neural networks. *ArXiv* **2019**, arXiv:190511946.

54. Sandler, M.; Howard, A.; Zhu, M.; Zhmoginov, A.; Chen, L.-C. Mobilenetv2: Inverted residuals and linear bottlenecks. In Proceedings of the IEEE Conference on Computer Vision and Pattern Recognition, Salt Lake City, UT, USA, 18–23 June 2018; pp. 4510–4520.

55. Ramachandran, P.; Zoph, B.; Le, Q.V. Searching for activation functions. *arXiv* **2017**, arXiv:171005941.

56. Russakovsky, O.; Deng, J.; Su, H.; Krause, J.; Satheesh, S.; Ma, S.; Huang, Z.; Karpathy, A.; Khosla, A.; Bernstein, M. Imagenet large scale visual recognition challenge. *Int. J. Comput. Vis.* **2015**, *115*, 211–252. [CrossRef]

57. Hinton, G.; Srivastava, N.; Swersky, K. Neural networks for machine learning lecture 6a overview of mini-batch gradient descent. *Cited On* **2012**, *14*.

58. Islam, S.; Ammour, N.; Alajlan, N.; Abdullah-Al-Wadud, M. Selection of heart-biometric templates for fusion. *IEEE Access* **2017**, *5*, 1753–1761. [CrossRef]

59. Islam, M.S.; Alajlan, N. An efficient QRS detection method for ECG signal captured from fingers. In Proceedings of the 2013 IEEE International Conference on Multimedia and Expo Workshops (ICMEW), San Jose, CA, USA, 15–19 July 2013; pp. 1–5.

60. Islam, M.S.; Alajlan, N. Augmented-hilbert transform for detecting peaks of a finger-ECG signal. In Proceedings of the 2014 IEEE Conference on Biomedical Engineering and Sciences (IECBES), Kuala Lumpur, Malaysia, 8–10 December 2014; pp. 864–867.

61. Szegedy, C.; Vanhoucke, V.; Ioffe, S.; Shlens, J.; Wojna, Z. Rethinking the inception architecture for computer vision. In Proceedings of the IEEE Conference on Computer Vision and Pattern Recognition, Las Vegas, NY, USA, 26–30 June 2016; pp. 2818–2826.

62. Huang, G.; Liu, Z.; Van Der Maaten, L.; Weinberger, K.Q. Densely connected convolutional networks. In Proceedings of the IEEE Conference on Computer Vision and Pattern Recognition, Honolulu, HI, USA, 21–26 July 2017; pp. 4700–4708.

63. He, K.; Zhang, X.; Ren, S.; Sun, J. Deep residual learning for image recognition. In Proceedings of the 2016 IEEE Conference on Computer Vision and Pattern Recognition (CVPR), Las Vegas, NV, USA, 27–30 June 2016; pp. 770–778.

64. Szegedy, C.; Liu, W.; Jia, Y.; Sermanet, P.; Reed, S.; Anguelov, D.; Erhan, D.; Vanhoucke, V.; Rabinovich, A. Going deeper with convolutions. In Proceedings of the IEEE Conference on Computer Vision and Pattern Recognition, Boston, MA, USA, 7–12 June 2015; pp. 1–9.

 *sensors*

Article

Novel Local Coding Algorithm for Finger Multimodal Feature Description and Recognition

Shuyi Li [1], **Haigang Zhang** [1], **Yihua Shi** [2] and **Jinfeng Yang** [2,*]

1   Tianjin Key Laboratory for Advanced Signal Processing, Civil Aviation University of China,
    Tianjin 300300, China; shuyili0909@163.com (S.L.); zhang_gang1989@126.com (H.Z.)
2   Shenzhen Polytechnic, Shenzhen 518055, China; yhshi@szpt.edu.cn
*   Correspondence: jfyang@szpt.edu.cn; Tel.: +86-022-2409-2422

Received: 9 April 2019; Accepted: 9 May 2019; Published: 13 May 2019

Abstract: Recently, finger-based biometrics, including fingerprint (FP), finger-vein (FV) and finger-knuckle-print (FKP) with high convenience and user friendliness, have attracted much attention for personal identification. The features expression which is insensitive to illumination and pose variation are beneficial for finger trimodal recognition performance improvement. Therefore, exploring suitable method of reliable feature description is of great significance for developing finger-based biometric recognition system. In this paper, we first propose a correction approach for dealing with the pose inconsistency among the finger trimodal images, and then introduce a novel local coding-based feature expression method to further implement feature fusion of FP, FV, and FKP traits. First, for the coding scheme a bank of oriented Gabor filters is used for direction feature enhancement in finger images. Then, a generalized symmetric local graph structure (GSLGS) is developed to fully express the position and orientation relationships among neighborhood pixels. Experimental results on our own-built finger trimodal database show that the proposed coding-based approach achieves excellent performance in improving the matching accuracy and recognition efficiency.

Keywords: finger features; multimodal recognition; local coding; Gabor filter; LGS

---

1. Introduction

With the arrival of the informational age and the rapid development of computer technology, people have higher requirements for the accuracy of biometric identification technology [1]. Compared with other common biometric traits, finger-based traits (e.g., fingerprint [2], finger-vein [3] and finger-knuckle-print [4],) have some advantages in uniqueness, anti-counterfeit, user acceptance, and high security [5–8]. However, affected by the external environment and the inherent differences of individuals, only relying on finger unimodal biometrics for identity authentication still has many security risks, which can no longer achieve the high-performance requirements of a user. Hence, fusing three traits from a finger together should be beneficial to address the person recognition problem [9,10].

However, the quality of three modal finger images is usually degraded seriously due to illumination variation in skin surfaces, which is unhelpful for reliable feature representation [11–13]. In addition, the finger trimodal images vary with the finger in pose rotation during imaging, which reduces the discriminability of images and further decreases the recognition accuracy rate. Therefore, exploring a robust feature representation method is very favorable for finger-based recognition improvement.

Recently, some researchers have developed some coding-based feature expression methods, which were often considered to be able to solve the above two problems [14–21]. Ojala et al. first proposed the classical local binary pattern (LBP) algorithm for facial recognition, which has great rotation invariant and was insensitive to illumination variation [16]. In 2011, Rosdi et al. proposed a local line binary pattern (LLBP) algorithm to effectively make use of the position relationships among surrounding pixels in horizontal and vertical orientations [17]. Meng et al. [19] proposed a local direction coding

(LDC) algorithm, which utilized the gradient relationships to express a venous feature for finger-vein recognition. In 2013, Peng et al. [20] combined the Gabor wavelet and LBP (GLBP) for feature extraction, which could effectively improve the ability of local and global features representation.

Noteworthy, some methods related to local graph have been presented in succession, and their variants have been successfully applied to many biometric fields [22–29]. Abusham et al. first proposed a local graph structure (LGS) algorithm to extract face features, which was insensitive to illumination [22]. However, the structure was non-symmetric, which led to a feature representation with no-equilibrium in the left and right neighborhoods. In order to balance feature representation of neighbor pixels on both sides, Mohd et al. [23] improved the original LGS operator to the SLGS operator by building a symmetrical structure. In 2015, Dong et al. [24] presented a MOW-SLGS operator for the representation of vein networks, and used the ELM to accomplish finger-vein image classification. On the basis of this information, in 2018, Yang et al. [25] put forward the Weber SLGS, which integrated differential direction features by Weber Law and the local graph structure algorithm. However, these algorithms still have some limitations in the application of representing finger multimodal features. On the one hand, the methods above only describe the relationships between the target pixel and its adjacent ones in a fixed neighborhood, while neglecting the hidden relationships among surrounding pixels. On the other hand, the assignment of different weights for symmetric pixels on left and right sides usually results in an imbalance of the feature expression in images.

To effectively overcome these limitations, we propose a Gabor generalized symmetric local graph structure (Gabor-GSLGS) for finger multimodal fusion recognition, as shown in Figure 1. In the image capture part, first, a finger imaging device is designed, and a pose correction algorithm is proposed to reduce pose variations. A robust finger region of interest (ROI) localization approach is then employed. Secondly, a bank of 6-orientation and single-scale Gabor filters are utilized for finger ROI image enhancement. Thirdly, based on the proposed GSLGS operator, a local coding algorithm is developed for finger features representation. The coded trimodal feature images of a finger are then divided evenly into non-overlapping blocks. Thus, for a finger, we can obtain a feature vector by concatenating the histograms of all blocks. Finally, by computing the similarities between the obtained vectors, the matching results can be reported statistically. Experimental results on our established database demonstrated that the proposed feature description approach exhibits better effects than other traditional approaches in finger multimodal fusion recognition.

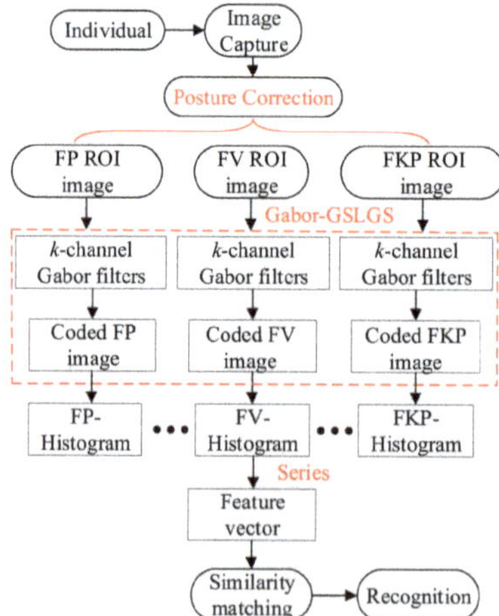

Figure 1. Finger multimodal recognition process based on the Gabor generalized symmetric local graph structure (Gabor-GSLGS).

The reminder of this paper is organized as follows: The finger trimodal imaging device and the proposed posture correction are introduced in Section 2. The enhancement methods used for finger ROI images are described in Section 3. Section 4 details the structure of the proposed local coding algorithm. A feature matching scheme is employed to implement finger multimodal fusion recognition and described in Section 5. Section 6 outlines the extensive experiments conducted and presents the analysis of experimental results in details. Finally, Section 7 presents the summarization.

2. Finger Image Capture and Preprocessing

2.1. Image Acquisition

As shown in Figure 2a, we have developed a homemade image acquisition device to obtain finger trimodal images. The finger imaging device is designed to capture fingerprint (FP), finger-vein (FV), and finger-knuckle-print (FKP) images automatically. It is composed of a binocular camera with two optical filters, a fingerprint acquisition instrument and an array of LEDs at a wavelength of 850 nm. In the imaging device, the FP images are directly obtained through a fingerprint instrument, which has a quick collection speed. Based on the imaging characteristics of a finger, the FV images are collected by using the near infrared (NIR) light to illuminate the palm side of a finger in penetration manner [27]. For FKP modality, we use the principle of reflecting the visible lights source for image acquisition.

For the sake of improving convenient acquisition of images, a collection groove with fixed sizes is designed in the imaging device, which is used to limit the position of a finger for imaging. It can effectively avoid, to a large extent, an image mismatch problem caused by the rotation and translation of a finger. As shown in Figure 2b, the dimensions of our finger imaging device are $10.9 \times 9.8 \times 17.8$ cm (length $\times$ width $\times$ height: cm).

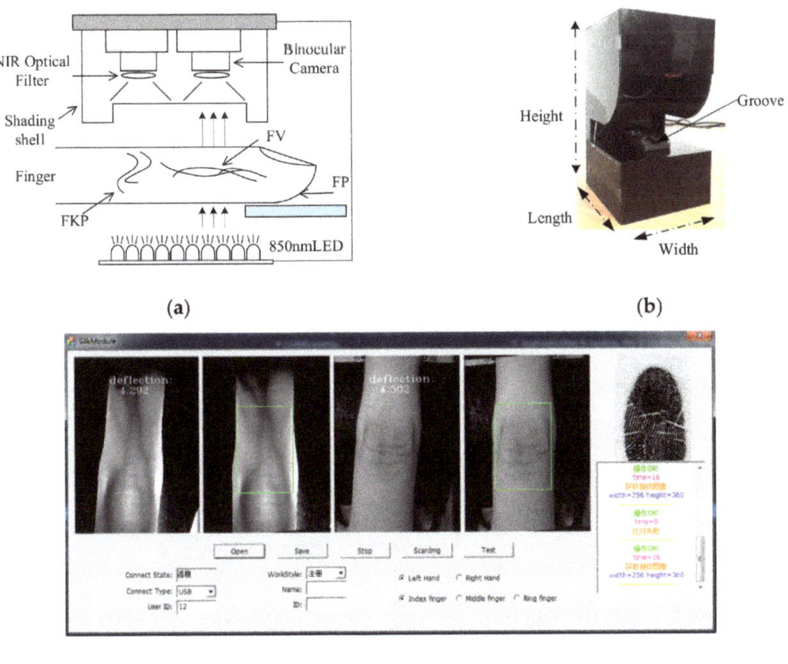

Figure 2. A finger trimodal image acquisition system. (a) the imaging schematic diagram; (b) a homemade image capture device; (c) a system interface of image acquisition.

As shown in Figure 2c, the acquisition program runs on the platform of Windows, and the software interface is built using C++ language. The top of the interface is designed to display the finger trimodal images captured in real time. Considering the friendliness of human-computer interaction, a window on the right side of the system is designed to remind users of the problems in the system operation.

From Figure 2c, it can be clearly seen that the original captured finger trimodal images have, to a small extent, still some posture variation. To solve this problem during the acquisition process, we present the posture correction method for finger trimodal images.

2.2. Posture Correction

Although a collection groove is designed in the acquisition device to fix the position of a finger, the finger still has a plane rotation at a small range. As the finger plane rotates, the edge line of the finger changes steadily. Hence, the rotation angle of a finger posture can be calculated and corrected based on the edge line of the finger. Due to the different illuminations, the edge line of the finger-vein is easier to detect and process than the finger-knuckle-print. Therefore, the finger in the finger-vein imaging space is selected to calculate the rotation angle, and then the three modalities are rotated and corrected together. The calculation process of a finger posture angle is shown in Figure 3.

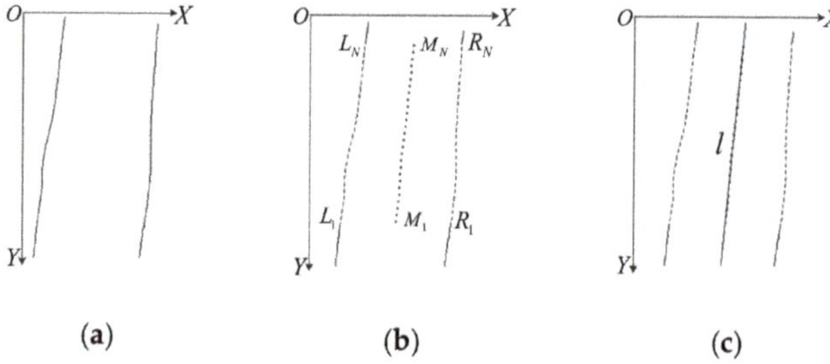

Figure 3. Computing finger posture angle. (**a**) the edge line of the finger; (**b**) the coordinate extraction of the finger edge line; (**c**) finger rotation direction extraction.

At first, the captured finger image is filtered to remove noise, and the edge line of the finger is detected. Then, the point coordinates of two edge lines are extracted to calculate the midpoint coordinates. As shown in Figure 3b, $\{L_n\}$ ($n = 1,2, \ldots , N$) represents the coordinate set in the left edge line of the finger, and $\{R_n\}$ represents the coordinate set in the right edge line of the finger, X and Y represent the row and column coordinates of the midpoint $\{M_n\}$. The calculation is as follows:

$$\begin{cases} X_{M_n} = (X_{L_n} + X_{R_n})/2 \\ Y_{M_n} = Y_{L_n} = Y_{R_n} \end{cases} \tag{1}$$

The linear fitting of the midpoint $\{M_n\}$ by least squares method yields the direction line: $l = kx + b$, where:

$$\begin{cases} k = \dfrac{\sum\limits_{n=1}^{N} (x_{M_n} - \bar{x})(y_{M_n} - \bar{y})}{\sum\limits_{n=1}^{N} (x_{M_n} - \bar{x})^2} \\ b = \bar{y} - k\bar{x} \end{cases} \tag{2}$$

Finally, according to k, the posture angle θ of the finger is calculated as follows:

$$\theta = \arctan\left(\frac{1}{k}\right) \tag{3}$$

Noteworthy, the center of rotation should select the center of the finger direction line l, which can reduce the amplitude of the posture swing of the finger and improve the stability of the correction.

Hence, taking the midpoint $M_{N/2}$ as the center of rotation and θ as the angle of rotation, the finger in the finger-vein imaging space is rotated and corrected. Some original images and corrected images for the same finger are shown in Figure 4.

From Figure 4, we can see that the proposed posture correction algorithm is effective to solve the problem of random plane rotation of the finger. This shows that the selection of the hardware and the posture correction algorithms have achieved better effects in improving the consistency of the finger posture.

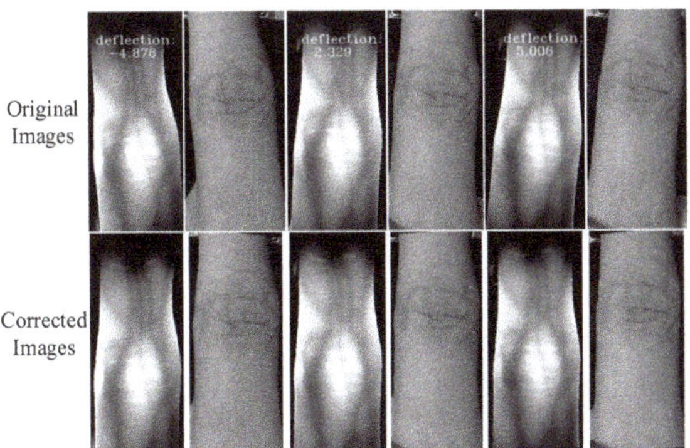

Figure 4. Some corrected image samples after rotation.

However, from the corrected finger images, we can see that they still contain some unnecessary backgrounds and uninformative parts. Hence, the captured finger images need to be processed to implement the regions of interest (ROIs) localization.

2.3. ROI Extraction

Since the imaging principle and acquisition approach of FP, FV, and FKP traits are different, diverse ROI extraction methods are supposed to be adopted accordingly [2]. In this paper, we apply the core point detection method to extract the FP ROI image [30], the convex direction coding method to extract the FKP ROI image [31], and the interphalangeal joint prior method to extract the FV ROI image [3]. Therefore, the FP, FV, and FKP images are cropped into 152×152 pixels, 200×91 pixels and 200×90 pixels, respectively. Some finger trimodal ROI images are shown in Figure 5.

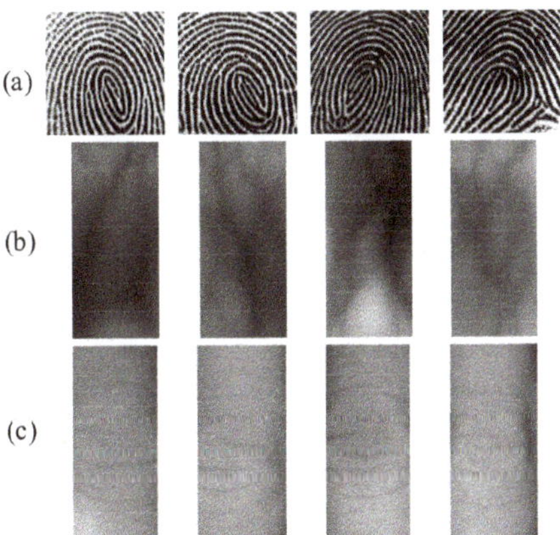

Figure 5. The finger trimodal region of interest (ROI) images of four fingers. (**a**) fingerprint (FP) ROIs samples; (**b**) samples of finger-vein (FV) ROIs; (**c**) finger-knuckle-print (FKP) ROIs samples.

3. Finger Image Enhancement

In recent decades, Gabor filters have been widely applied in many fields since they not only extract the texture information in multiple directions of an image, but also reduce the influence of illumination to some extent [32]. In terms of the abundant texture information of FP, FV, and FKP traits, with respect to direction, oriented Gabor filters are used here to perform image enhancement.

A Gabor filter consists of a real part and an imaginary part. Generally, the real part is also called an even-symmetrical Gabor filter, which is suitable for ridge detection in an image [2]. Since these three modality images of a finger all have their own particular ridge textures, the real part of Gabor filter can be used to extract the flexible feature information effectively [33]. It can be expressed as

$$G(x, y, \theta_k, f_k) = \frac{\gamma}{2\pi\sigma^2} \exp\left\{-\frac{1}{2}\left\{\frac{x_{\theta_k}^2 + \gamma^2 y_{\theta_k}^2}{\sigma^2}\right\}\right\} \cos\left(2\pi f_k x_{\theta_k}\right) \tag{4}$$

where $x_{\theta_k} = x\cos\theta_k + y\sin\theta_k$, $y_{\theta_k} = y\cos\theta_k - x\sin\theta_k$, σ and $k = (1, 2, \ldots, k)$, respectively, represent the scale index and the orientation index, θ_k denotes the orientation of the k-th Gabor filter, and f_k is the central frequency of the sinusoidal plane wave. Assuming $R(x,y)$ is an original ROI image, each k-th Gabor filtered image $I_k(x,y)$ can be obtained by

$$I_k(x, y) = R(x, y) * G(x, y, \theta_k, f_k) \tag{5}$$

where the symbol "$*$" represents two-dimensional convolution.

First, the original ROI image is convoluted with k-channel Gabor filters. Then, the k Gabor filtered images are merged into an image $I(x,y)$ by using the competitive coding method proposed in [15]. Some Gabor filtered images are shown in Figure 6. It can be clearly seen that the texture information of finger images can be effectively enhanced after multichannel Gabor filtering. Based on this, we apply the coding-based theory to obtain more stable and reliable finger features.

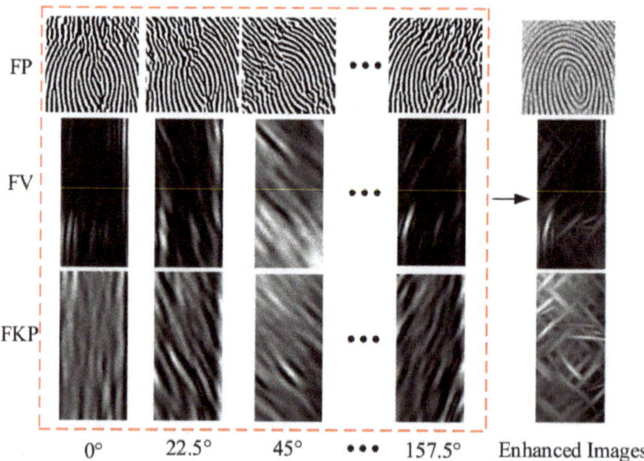

Figure 6. The enhanced images of the finger three modalities.

4. Feature Extraction Based on Local Coding Algorithm

To make full use of local position and gradient features between adjacent pixels in Gabor filtered images, a local coding algorithm based on generalized symmetric LGS is proposed for feature representation. The specific steps are as follows:

Step 1 The finger trimodal images are respectively enhanced by k-channel even symmetric Gabor filters in Section 3, and the Gabor filtered images are obtained.

Step 2 As shown in Figure 7, for each center pixel in the Gabor enhanced images, we respectively select three pixels in the left and right of $n \times n$ neighborhoods (a square area in Figure 7) to constitute the GSLGS operator in the horizontal orientation. In terms of weight distribution, the weights of symmetric pixels in the right and left sides maintain equal weights. More details are shown in Figure 7.

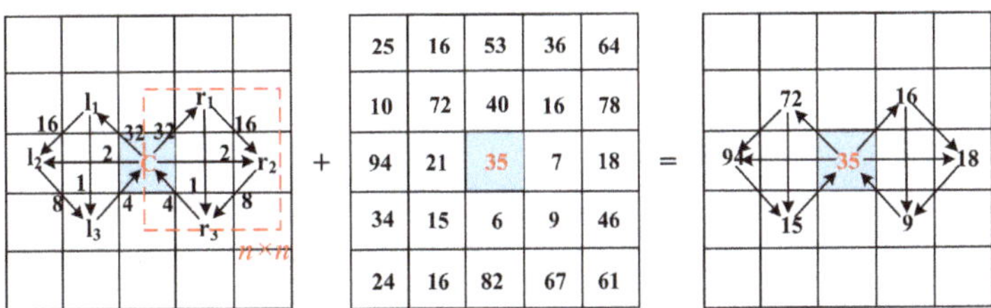

Figure 7. The design of the GSLGS operator ($0°$ direction, $n = 3$).

Since Gabor features of finger trimodal have diverse directions, the information of the surrounding pixels in multiple orientations should be extracted for efficient feature representation. Centered on the target pixel, rotating the GSLGS operator counterclockwise by θ_k (corresponding to Step 1), the structure of GSLGS in an arbitrary orientation can be obtained. For instance, when $k = 4$, the structures of GSLGS are shown in Figure 8.

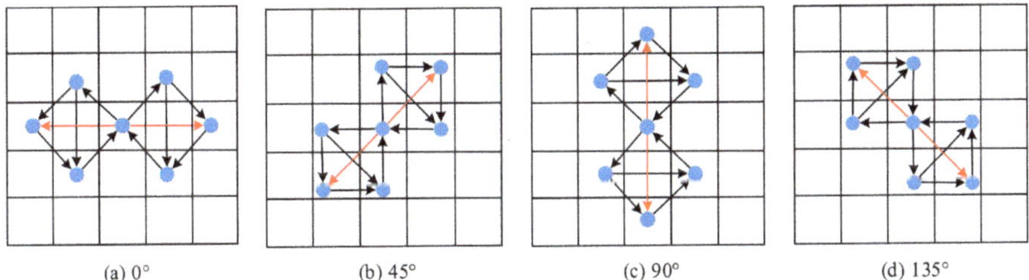

(a) $0°$ (b) $45°$ (c) $90°$ (d) $135°$

Figure 8. The GSLGS operator ($k = 4$).

Step 3 The coding process of the GSLGS operation is shown in Figure 9. In the neighborhood of left and right sides, these gray values of the pixels are respectively compared in succession starting from each target pixel. If the value becomes larger, the relationship between the two pixels to be compared is coded to 1. In contrast, the relationship is coded to 0. The coding calculation process is expressed as

$$F(\theta_k) = \sum_{r=1}^{6} p_r(g_r - f_r)2^{6-r} + \sum_{l=1}^{6} q(g_l - f_l)2^{6-l}, (k = 1, 2, \cdots, K) \tag{6}$$

$$p_r(g_r - f_r) = \begin{cases} 1, g_r - f_r \geq 0, \\ 0, g_r - f_r < 0 \end{cases} \tag{7}$$

$$q_l(g_l - f_l) = \begin{cases} 1, g_l - f_l \geq 0, \\ 0, g_l - f_l < 0 \end{cases} \tag{8}$$

where g_r and f_r (g_l and f_l), respectively, denote values of two adjacent pixels in the right (left) neighborhood, and $F(\theta_k)$ represent the feature coded value in the θ_k orientation.

From Figure 9, we can see that the coded values of the target pixel at $0°$ and $45°$, respectively, can be obtained according to Equations (6)–(8). Similarly, the same calculation process is done at $90°$ and $135°$. Thus, we calculate the coded values of the central pixel in these four directions as follow:

$F(\theta_1) = (010100)_2 + (110110)_2 = (0 \times 32 + 1 \times 16 + 0 \times 8 + 1 \times 4 + 0 \times 2 + 0 \times 1) + (1 \times 32 + 1 \times 16 + 0 \times 8 + 1 \times 4 + 1 \times 2 + 0 \times 1) = 74.$

$F(\theta_2) = (100100)_2 + (000100)_2 = (1 \times 32 + 0 \times 16 + 0 \times 8 + 1 \times 4 + 0 \times 2 + 0 \times 1) + (0 \times 32 + 0 \times 16 + 0 \times 8 + 1 \times 4 + 0 \times 2 + 0 \times 1) = 40.$

$F(\theta_3) = (100110)_2 + (010110)_2 = (1 \times 32 + 0 \times 16 + 0 \times 8 + 1 \times 4 + 1 \times 2 + 0 \times 1) + (0 \times 32 + 1 \times 16 + 0 \times 8 + 1 \times 4 + 1 \times 2 + 0 \times 1) = 60.$

$F(\theta_4) = (010011)_2 + (010101)_2 = (0 \times 32 + 1 \times 16 + 0 \times 8 + 0 \times 4 + 1 \times 2 + 1 \times 1) + (0 \times 32 + 1 \times 16 + 0 \times 8 + 1 \times 4 + 0 \times 2 + 1 \times 1) = 39.$

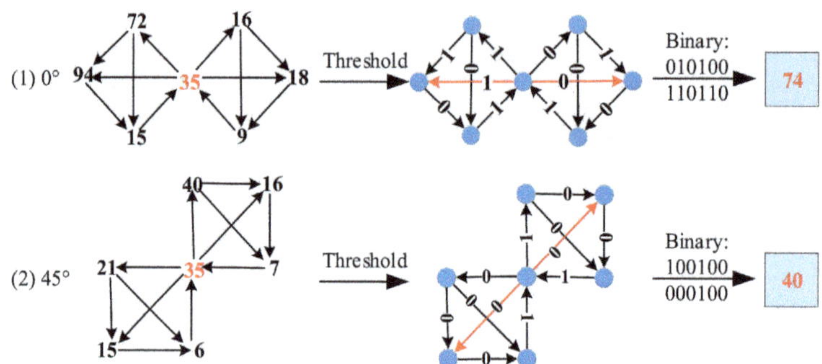

Figure 9. The coding process of GSLGS operator at 0° and 45°.

Step 4 Inspired by the optimal response of Gabor filters in multiple orientations, we choose the maximum value among these coded values as the final coded value, which can be defined as

$$F(x, y) = \arg \max_{\theta_k \in (0°, 180°)} \{F\{\theta_k\}\} \tag{9}$$

As mentioned above, the final coded value of each target pixel in the Gabor enhanced image can be obtained according to the GSLGS operator. For instance, the coded value in Figure 7 is: $F(x,y)$ = argmax $\{F(\theta_1), F(\theta_2), F(\theta_3), F(\theta_4)\} = F(\theta_1) = 74$.

Considering the great capability of a Gabor filter in enhancing texture feature from any orientation, the GSLGS operator is extended into arbitrary orientations, which has superior orientation selectivity. Therefore, it can effectively solve image mismatch problem due to finger pose variation. More importantly, the proposed local coding algorithm can entirely consider the relationships between each target pixel and its surrounding neighborhoods. In addition, the distribution of weights is conformable in the symmetric pixels on both sides. Hence, the finger feature representation of local neighborhoods can maintain balance in the GSLGS operator.

5. Feature Fusion and Matching

In this section, a gray histogram-based feature matching method is used for finger trimodal fusion recognition, as shown in Figure 10. First, the coded finger trimodal images are uniformly separated into M non-overlapping division blocks. Then, the M local histograms corresponding to each sub-block are established, respectively. Assuming that $H_{fv}{}^i (I = 1, 2, \ldots, M)$ represents the histogram of the ith division block in a coded finger-vein image, the global histogram H_{fv} is defined as

$$H_{fv} = \left(H^1, H^2, \cdots, H^M\right) \tag{10}$$

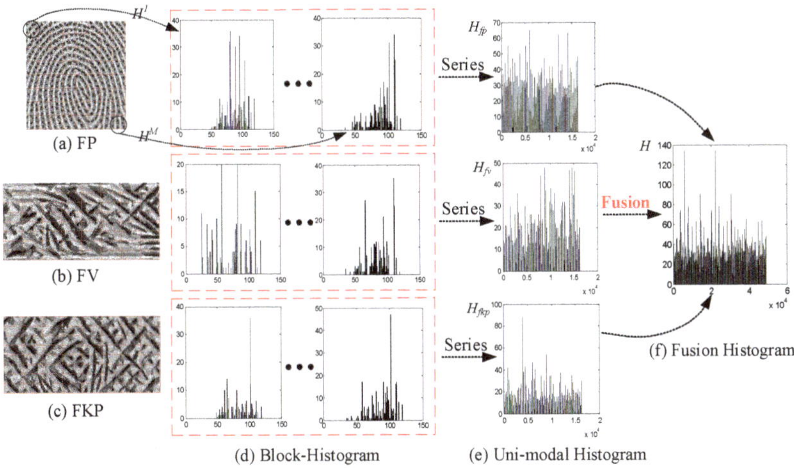

Figure 10. The fusion of finger trimodal features.

Similarly, H_{fp} and H_{fkp} represent the global histogram of a coded fingerprint image and finger-knuckle-print image. Then, the final feature histogram H of a finger trimodal image can be expressed by

$$H = \left(H_{fp}, H_{fv}, H_{fkp} \right) \tag{11}$$

After the above calculation, we can obtain the feature histogram of each finger sample. Here, we can use various classification algorithms, such as SVM, ELM and k-NN [34]. In this section, for convenience, the intersection coefficient between two feature vectors is calculated to determine the similarity of two individuals [29]. Assuming $H_1(i)$ and $H_2(i)$ denote the histograms of two samples to be matched, the similarity can be computed by

$$sim(H_1(i), H_2(i)) = \frac{\sum_{i=1}^{L} \min[H_1(i), H_2(i)]}{\sum_{i=1}^{L} H_1(i)} \tag{12}$$

where L denotes the dimension of a feature vector to be matched. In the matching process, if the intersection coefficient $sim(\cdot)$ is >T (similarity decision threshold), it means that the two samples are similar and are able to be matched. But if the intersection coefficient $sim(\cdot)$ is ≤T, it means that the two samples are not matched. Thus, two samples will tend to be more similar as the intersection coefficient increases. The similarity decision threshold T corresponds to the threshold value when the false rejection rate (FRR) is the same as the false accept rate (FAR).

6. Experimental Results

In order to verify the proposed coding-based method, a finger trimodal database from a homemade image acquisition system is used in our experiments. The database contains a total of 17,550 images from 585 individual fingers (index finger, middle finger, and ring finger) of both hands, and each finger contains 30 images (10 images per modality). Here, we randomly select 3000 images samples from 100 different individuals, each of which, respectively, contains 10 images on the FP, FV, and FKP traits, as the experimental database.

Here, the proposed Gabor-GSLGS algorithm is implemented using MATLAB R2014a on a standard desktop PC which is equipped with Inter Core i5-7400 CPU 3 GHz and 8 GB RAM.

The detailed experiments are as follows: In Section 6.1, we mainly describe the analysis of the influence of different parameter selection on the recognition rate. Section 6.2 presents the detailed comparison of the performance of unimodal and multimodal recognition. The experimental results of different feature extraction methods are compared in Section 6.3.

6.1. Parameter Selection

6.1.1. Selection of k

On the basis of the above introduction in Sections 3 and 4, we can see that the number of orientations in the local coding algorithm corresponds to the number of channels in the Gabor filter. Hence, different k values produce different effects on the performance of the finger multimodal recognition. In order to find the optimal parameters of k, we evaluate it using two recognition indicators, equal error rate (EER) and the time cost of feature extraction. EER listed in Table 1 is the error rate where FRR and FAR are equal. Here, FAR indicates the identification result of incorrect acceptation for an individual, while FRR demonstrates the result of incorrect rejection. The ROC (receiver operating characteristic) curves for intersection coefficient measures are plotted in Figure 11, where FAR and FRR are shown in the same plot at different thresholds.

Table 1. Comparisons on equal error rate (EER) (%) and time cost (single individual).

k	4	6	8
EER (%)	0.042	0.029	0.038
Time cost (s)	0.012	0.017	0.031

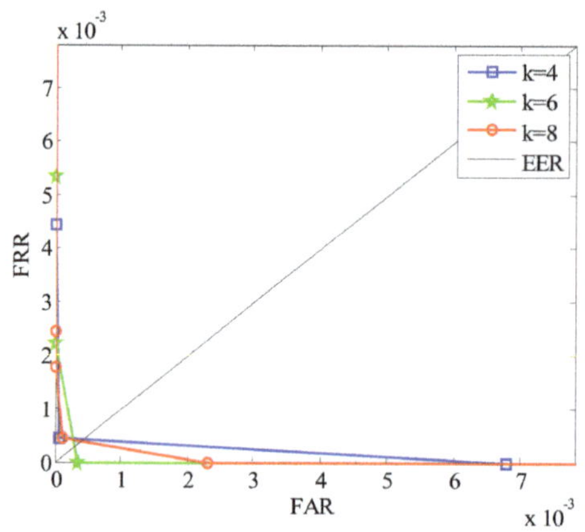

Figure 11. Receiver operating characteristic (ROC) of different k.

From Figure 11, we can see that the EER is lowest when k is 6. However, as the value of k increases, the time cost of finger feature extraction also increases. Considering recognition efficiency and accuracy, the parameter k corresponding to 6 is selected in following experiments.

6.1.2. Selection of Neighborhood and Image Division

Apart from parameters k, the size of the neighborhood $n \times n$ that constitutes the structure of the GSLGS operator and the number of image division blocks M are also critical factors for finger trimodal recognition. Considering that n and M have a great influence on the recognition performance of the proposed algorithm, therefore, it is important to select suitable parameters. Here, we select different neighborhoods and image block sizes to perform the experiments. Some EERs of different parameters are listed in Table 2, with some ROCs shown in Figures 12 and 13.

Table 2. Comparisons on EER(%) for different parameters.

Blocks \ Neighborhood	6×6	7×7	8×8	9×9
3×3	0.22	0.16	0.086	0.37
5×5	0.08	0.029	0.022	0.024
7×7	0.14	0.075	0.056	0.082

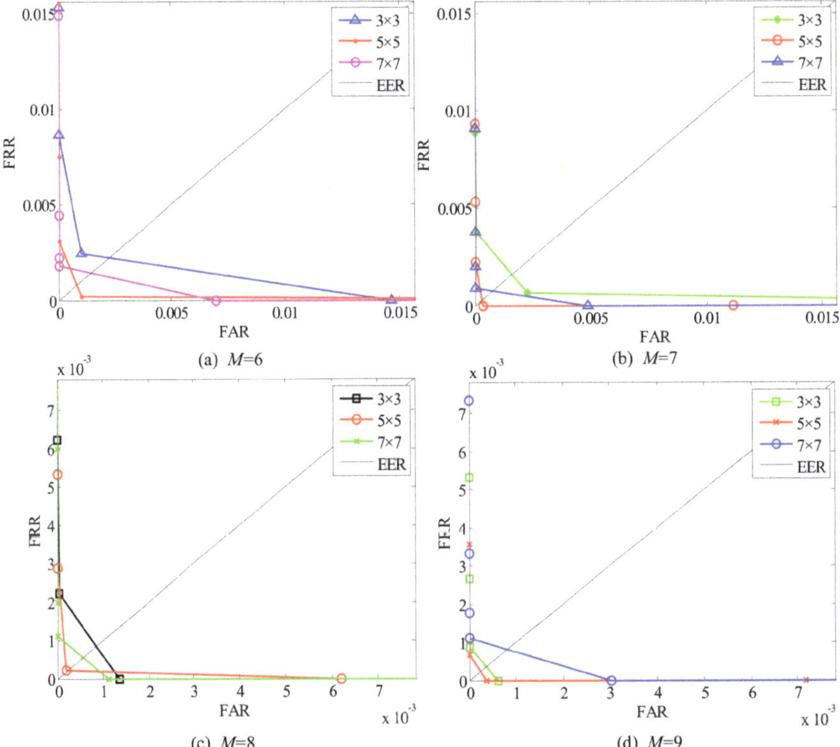

Figure 12. ROC of different neighborhoods in $M = 6, 7, 8, 9$.

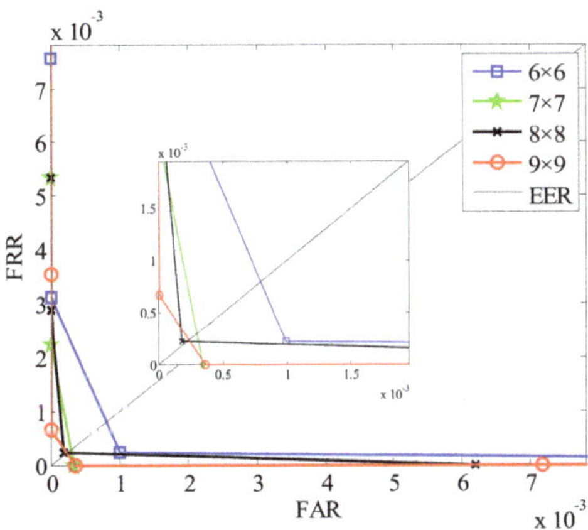

Figure 13. ROC of different division blocks M in a 5×5 neighborhood.

From Figure 12, it can be clearly seen that the ROC curves vary by changing n (n = 3, 5 or 7, respectively). This shows that different neighborhoods have different effects on the performance of finger trimodal recognition. By observing these obtained curves in the condition of the same image division block, such as M = 6, we find that the EER is lowest when the size of the neighborhood is selected as 5×5 (n = 5). Similarly, when M = 7, 8 or 9, respectively, the pixels selected in a 5×5 neighborhood for constructing the GSLGS operator also have optimal accuracy. The reason is that a 3×3 neighborhood is more sensitive to noise, while the 7×7 neighborhood is relatively weak in the capability of feature expression. However, the 5×5 neighborhood is preferred for feature expression among surrounding pixels and is insensitive to noise. Hence, n = 5 is the optimal parameter for constructing the proposed GSLGS operator.

The experimental results of different division blocks by using GSLGS with a 5×5 neighborhood are shown in Figure 13. From Figure 13, we can find that the proposed local coding algorithm obtains the best accuracy when the number of division blocks is 8×8 (M = 8). This shows that an appropriate image division scheme is beneficial for improving recognition accuracy rate. Hence, the image division blocks M = 8 is an optimal choice for the proposed Gabor-GSLGS approach in finger trimodal recognition.

6.2. Comparison of Unimodal and Multimodal

The proposed local coding algorithm of finger trimodal can also be applied for finger single modal recognition. Here, the experiments of finger unimodal and multimodal recognition are performed when n = 5 and M = 8. The experimental results of different modal combinations are listed in Table 3.

Table 3. Comparisons on EER (%) and time cost (single individual).

Modal	FP	FV	FKP	FV + FKP	FV + FP	FKP + FP	FP + FV + FKP
EER (%)	4.26	0.19	0.40	0.20	0.16	0.46	0.022
Time cost (s)	0.015	0.010	0.015	0.021	0.019	0.018	0.029

From Figure 14, we can see that the EER rate of different modal combinations are different. It is noted that the bimodal combination (FV + FKP and FV + FP) can achieve a better accuracy than single modal, especially for the FP trait and FKP trait.

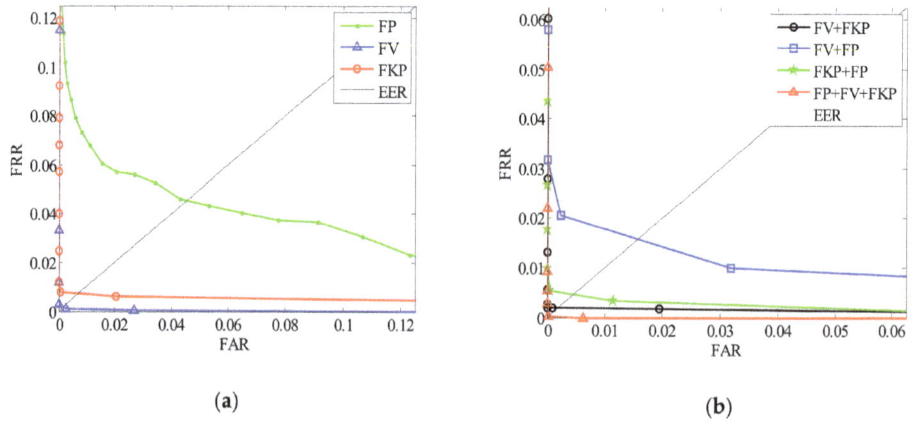

Figure 14. Comparison results of different modal combinations. (**a**) ROC of unimodal recognition; (**b**) ROC of multimodal recognition.

From Table 3, we can find that three modal combination have the best recognition accuracy, while the time cost increases with the increase of the modality number. It is noteworthy that in single modal recognition, FV trait performs better than FP trait and FKP trait. This shows that the FV trait is the most dominant trait in the three modalities.

In total, these results show that multimodal fusion recognition performs better than single modal. The reason is that the multimodal combination can make full use of the discrimination of different modalities and different modalities can complement each other in multimodal fusion recognition. However, the computational efficiency of multimodal recognition can still be improved.

6.3. Comparison of Different Methods

In order to further evaluate the proposed local coding method, here we compare it with some common feature extraction methods (LBP [16], GLBP [20], LLBP [17], SLGS [23], and MOW-LGS [24]). The ROCs are plotted in Figure 15, and the simulation results are listed in Table 4.

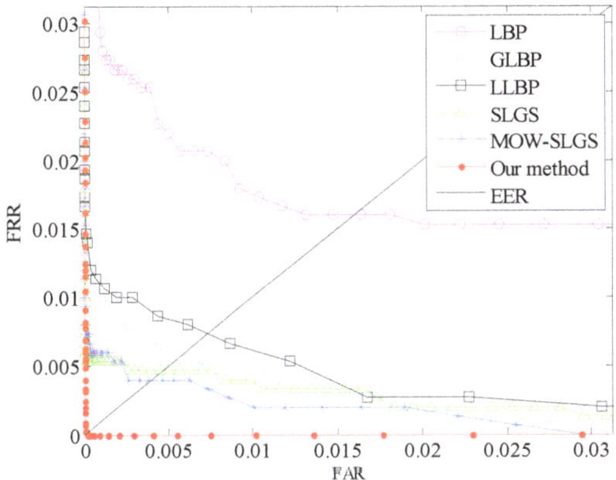

Figure 15. Comparisons of different methods.

Table 4. Comparisons on EER (%) and time cost (single individual).

Methods	LBP	GLBP	LLBP	SLGS	MOW-LGS	Our Method
EER (%)	1.60	0.58	0.74	0.46	0.42	0.022
Time cost (s)	0.129	0.186	0.261	0.110	0.192	0.029

From Figure 15, we observe that the EER of the proposed approach is lowest among these feature extraction methods. Hence, feature expression based on the proposed coding approach can effectively address the problem of illumination and finger posture variation in finger trimodal fusion recognition.

From Table 4, we can clearly see that the proposed local coding algorithm not only produces the best recognition accuracy, and also that the computation cost of feature extraction is also lowest as compared with other methods. This shows that our method is more robust to finger feature representation.

7. Conclusions

In this paper, a posture correction approach was first designed for reducing the finger pose variation. To solve the problem that the feature expression method was sensitive to illumination variation and posture rotation, a novel local coding algorithm was proposed for finger trimodal fusion recognition. On the one hand, the Gabor filter, to some extent, can effectively reduce the influence of illumination and noise in an image. On the other hand, the posture correction method and the local coding method were used to address the problem of finger posture variation. The proposed Gabor-GSLGS algorithm made full use of the texture features in multiple orientations between surrounding pixels. Furthermore, the proposed method assigned the same weights in symmetrical pixels, which improved the equilibrium of the feature representation of the finger images. The experimental results showed that our method could improve the accuracy and computational efficiency of finger trimodal fusion recognition.

As part of our future work, we will apply the proposed local coding algorithm to other public biometric databases. Moreover, we will focus on reducing the dimensions of the feature vector and improving the efficiency of finger multimodal fusion recognition. At the same time, we will aim to exploit a more robust and effective fusion method which can integrate multiple modal features for personal identification.

Author Contributions: S.L. and H.Z. conceived and designed the experiments; S.L. performed the experiments and analyzed the data; S.L., H.Z., Y.S. and J.Y. wrote the paper.

Funding: This work is supported by the National Natural Science Foundation of China (No. 61806208, No. 61502498) and the Fundamental Research Funds for the Central Universities (NO. 3122017001).

Conflicts of Interest: The authors declare no conflict of interest.

References

1. Jain, A.K.; Ross, A.; Prabhakar, S. An introduction to biometric recognition. *IEEE Trans. Circuits Syst. Video Technol.* **2004**, *14*, 4–820. [CrossRef]
2. Yang, J.F.; Zhang, X. Feature-level fusion of fingerprint and finger-vein for personal identification. *Pattern Recognit. Lett.* **2012**, *33*, 623–628. [CrossRef]
3. Yang, J.F.; Zhong, Z.; Jia, G.M.; Li, Y.N. Spatial circular granulation method based on multimodal finger feature. *Int. J. Electr. Comput. Eng.* **2016**, *2016*, 1–7. [CrossRef]
4. Zhang, L.; Zhang, L.; Zhang, D. Finger-knuckle-print: A new biometric identifier. In Proceedings of the International Conference on Image Processing (ICIP2009), Cairo, Egypt, 7–11 November 2009; pp. 1981–1984.
5. Evangelin, L.N.; Fred, A.L. Feature level fusion approach for personal authentication in multimodal biometrics. In Proceedings of the IEEE 2017 3th International Conference on Science Technology Engineering & Management (ICONSTEM), Chennai, India, 23–24 March 2017; pp. 148–151.
6. Peng, J.J.; Li, Y.N.; Li, R.R.; Jia, G.M.; Yang, J.F. Multimodal finger feature fusion and recognition based on delaunay triangular granulation. *Comput. Inf. Sci. Eng.* **2014**, *484*, 303–310.
7. Yang, J.F.; Shi, Y.H. Finger-vein ROI localization and vein ridge enhancement. *Pattern Recognit. Lett.* **2012**, *33*, 1569–1579. [CrossRef]
8. Yang, J.F.; Wei, J.Z.; Shi, Y.H. Accurate ROI localization and hierarchical hyper-sphere model for finger-vein recognition. *Neurocomputing* **2018**, *328*, 171–181. [CrossRef]
9. Xin, Y.; Kong, L.; Liu, Z.; Wang, C.; Xu, X. Multimodal feature-level fusion for biometrics identification System on IoMT platform. *IEEE Access* **2018**, *6*, 21418–21426. [CrossRef]
10. Yang, W.; Song, W.; Hu, J.; Zhang, G.; Valli, C. A Fingerprint and Finger-vein Based Cancelable Multi-biometric System. *Pattern Recognit.* **2017**, *78*, 242–251. [CrossRef]
11. Li, S.Y.; Zhang, H.G.; Jia, G.M.; Yang, J.F. Finger Vein Recognition Based on Weighted Graph Structural Feature Encoding. In Proceedings of the 13th Chinese Conference on Biometric Recognition, Xinjiang, China, 11–12 August 2018; pp. 29–37.
12. Yang, J.F.; Zhang, B.; Shi, Y. Scattering Removal for Finger-Vein Image Restoration. *Sensors* **2012**, *12*, 3627–3640. [CrossRef] [PubMed]
13. Yang, J.F.; Shi, Y.H. Finger-vein Network Enhancement and Segmentation. *Pattern Anal. App.* **2014**, *17*, 783–797. [CrossRef]
14. Liu, H.; Ji, R.; Wu, Y.; Huang, F.; Zhang, B. Cross-Modality Binary Code Learning via Fusion Similarity Hashing. In Proceedings of the 2017 IEEE Conference on Computer Vision and Pattern Recognition (CVPR), Honolulu, HA, USA, 21–26 July 2017; pp. 6345–6353.
15. Kong, W.K.; Zhang, D. Competitive coding scheme for palmprint verification. In Proceedings of the 17th International Conference on Pattern Recognition (ICPR2004), Cambridge, UK, 23–26 August 2004; pp. 520–523.
16. Ojala, T.; Pietikäinen, M.; Mäenpää, T. Gray Scale and Rotation Invariant Texture Classification with Local Binary Patterns. In Proceedings of the 2000 6th European Conference on Computer Vision (ECCV 2000), Dublin, Ireland, 26 June–1 July 2000.

17. Rosdi, B.A.; Shing, C.W.; Suandi, S.A. Finger Vein Recognition Using Local Line Binary Pattern. *Sensors* **2011**, *11*, 11357–11371. [CrossRef]
18. Lu, Y.; Yoon, S.; Xie, S.; Yang, J.C.; Wang, Z.H.; Park, D.S. Finger Vein Recognition Using Generalized Local Line Binary Pattern. *KSII Trans. Internet Inf.* **2014**, *8*, 1766–1784.
19. Meng, X.; Yang, G.; Yin, Y.; Xiao, R. Finger Vein Recognition Based on Local Directional Code. *Sensors* **2012**, *12*, 14937–14952. [CrossRef]
20. Peng, J.; Li, Q.; Abd El-Latif, A.A.; Wang, N.; Niu, X.M. Finger Vein Recognition with Gabor Wavelets and Local Binary Patterns. *IEICE Trans. Inf. Syst.* **2013**, *E96.D*, 1886–1889. [CrossRef]
21. Han, W.Y.; Lee, J.C. Palm vein recognition using adaptive Gabor filter. *Expert Syst. Appl.* **2012**, *39*, 13225–13234. [CrossRef]
22. Abusham, E.E.A.; Bashir, H.K. Face Recognition Using Local Graph Structure (LGS). In Proceedings of the 14th International Conference on Human-Computer Interaction. Interaction Techniques & Environments (HCI 2011), Orlando, FL, USA, 9–14 July 2011; pp. 169–175.
23. Abdullah, M.F.A.; Sayeed, M.S.; Muthu, K.S. Face recognition with Symmetric Local Graph Structure (SLGS). *Expert Syst. Appl.* **2014**, *41*, 6131–6137. [CrossRef]
24. Dong, S.; Yang, J.C.; Chen, Y.; Wang, C.; Zhang, X.Y.; Park, D.S. Finger Vein Recognition Based on Multi-Orientation Weighted Symmetric Local Graph Structure. *KSII Trans. Internet Inf. Sys.* **2015**, *9*, 4126–4142.
25. Yang, J.; Zhang, L.; Wang, Y.; Sun, W.H.; Park, D.S. Face Recognition based on Weber Symmetrical Local Graph Structure. *KSII Trans. Internet Inf. Sys.* **2018**, *12*, 1748–1759.
26. Yang, J.C.; Zhang, L.C.; Li, M.; Zhao, T.T.; Chen, Y.R.; Liu, J.Z.; Liu, N. Face Recognition with Facial Occlusion Based on Local Cycle Graph Structure Operator. *IntechOpen* **2018**, *4*, 597–609.
27. Zhang, H.G.; Li, S.Y.; Shi, Y.H.; Yang, J.F. Graph Fusion for Finger Multimodal Biometrics. *IEEE Access* **2019**, *7*, 28607–28615. [CrossRef]
28. Jia, W.; Hu, R.X.; Lei, Y.K.; Zhao, Y.; Gui, J. Histogram of Oriented Lines for Palmprint Recognition. *IEEE Trans. Syst. Man Cybern.* **2014**, *44*, 385–395. [CrossRef]
29. Luo, Y.T.; Zhao, L.Y.; Zhang, D.D.; Jia, W.; Xue, F.; Lu, J.T.; Zhu, Y.H.; Xu, B.Q. Local line directional pattern for palmprint recognition. *Pattern Recognit.* **2016**, *50*, 26–44. [CrossRef]
30. Kekre, H.B.; Bharadi, V.A. Fingerprint's core point detection using orientation field. In Proceedings of the International Conference on Advances in Computing, Control and Telecommunication Technologies (ACT 2009), Kerala, India, 28–29 December 2009; pp. 150–152.
31. Zhang, L.; Zhang, L.; Zhang, D.; Zhu, H.L. Online finger-knuckle-print verification for personal authentication. *Pattern Recognit.* **2010**, *43*, 2560–2571. [CrossRef]
32. Yang, J.F.; Shi, Y.H.; Jia, G.M. Finger-vein Image Matching based on Adaptive Curve Transformation. *Pattern Recognit.* **2017**, *66*, 34–43. [CrossRef]
33. Yang, J.F.; Yang, J.L. Multi-Channel Gabor Filter Design for Finger-Vein Image Enhancement. In Proceedings of the Fifth International Conference on Image and Graphics, Xi'an, China, 20–23 September 2009; pp. 87–91.
34. Nguyen, B.P.; Tay, W.L.; Chui, C.K. Robust Biometric Recognition from Palm Depth Images for Gloved Hands. *IEEE T. Hum. Mach. Syst.* **2015**, *45*, 1–6. [CrossRef]

Article

Combined Fully Contactless Finger and Hand Vein Capturing Device with a Corresponding Dataset

Christof Kauba *, Bernhard Prommegger and Andreas Uhl

Department of Computer Sciences, University of Salzburg, Jakob-Haringer-Str. 2, 5020 Salzburg, Austria; bprommeg@cs.sbg.ac.at (B.P.); uhl@cs.sbg.ac.at (A.U.)
* Correspondence: ckauba@cs.sbg.ac.at; Tel.: +43-662-8044-6334

Received: 15 October 2019; Accepted: 13 November 2019; Published: 17 November 2019

Abstract: Vascular pattern based biometric recognition is gaining more and more attention, with a trend towards contactless acquisition. An important requirement for conducting research in vascular pattern recognition are available datasets. These datasets can be established using a suitable biometric capturing device. A sophisticated capturing device design is important for good image quality and, furthermore, at a decent recognition rate. We propose a novel contactless capturing device design, including technical details of its individual parts. Our capturing device is suitable for finger and hand vein image acquisition and is able to acquire palmar finger vein images using light transmission as well as palmar hand vein images using reflected light. An experimental evaluation using several well-established vein recognition schemes on a dataset acquired with the proposed capturing device confirms its good image quality and competitive recognition performance. This challenging dataset, which is one of the first publicly available contactless finger and hand vein datasets, is published as well.

Keywords: finger vein recognition; hand vein recognition; contactless acquisition device; public vascular pattern dataset; biometric recognition performance evaluation

---

1. Introduction

Biometric authentication is gaining more and more attention and replaces traditional authentication methods like passwords, signatures and tokens. It offers higher security and increased user convenience compared to traditional methods. Biometric authentication techniques are based on so-called biometric traits, which are behavioural or physiological characteristics of a person. These biometric traits are unique to every person. The most commonly used biometric traits include fingerprints, face and iris. Recently, vascular pattern based biometrics, especially hand and finger based vascular patterns (usually denoted as hand and finger vein recognition) have become more popular as well. Since the first commercial contactless palm vein acquisition device from Fujitsu [1] became available in 2003, vascular pattern based biometrics have been employed in several application areas, especially in the banking area [2,3]. Vascular pattern based biometrics have several advantages over, for example, fingerprints [4]. This biometric trait is based on the patterns formed by the blood vessels, located underneath the skin, that is, it is an internal biometric trait. While fingerprints are susceptible to dirt and moisture on the skin, skin damage and abrasion, the vascular patterns are assumed to be insensitive to these skin conditions. Furthermore, vascular pattern based biometrics are more resistant to presentation attacks and forgery than are fingerprints and face [4] as the blood vessels are located beneath the skin and are only visible in near-infrared light.

1.1. Acquisition Principle and Capturing Devices

To render the patterns formed by the blood vessels visible, special acquisition devices are necessary. These devices are usually denoted as biometric scanners or sensors. The haemoglobin contained in the blood, which is flowing through the veins and arteries, has a higher light absorption coefficient in the near-infrared (NIR) wavelength spectrum (between 700 and 950 nm) than the surrounding tissue. Hence, the vascular pattern can be rendered visible by applying an NIR light source and capturing images using an NIR sensitive camera, which resembles the main parts of a finger or hand vein scanner. There are two distinct configurations depending on the relative positioning of the light source and the camera—light transmission and reflected light (see Figure 1 for an illustration). In the reflected light set-up, the camera and the light source are positioned on the same side of the finger/hand, whereas in the light transmission set-up, both are positioned on opposite sides of the finger/hand. A further distinction can be made regarding the side of the finger/hand which is captured—palmar or dorsal. Palmar refers to the bottom side of the finger/hand, while dorsal images are captured from the top side.

In both finger and hand vein recognition, usually the palmar side is utilised. While reflected light is the preferred set-up in hand vein recognition, finger vein scanners mainly capture the images using light transmission or light dispersion [5,6]. These days there are several commercial off-the-shelf (COTS) solutions for hand as well as finger vein recognition available for a wide range of application scenarios, from securing a personal computer over additional authentication at an automated teller machine (ATM) to high security access control systems at industry buildings. However, most of the COTS solutions have one major drawback for academia and research—the COTS scanners do not output the raw vein images. Instead, they only provide a template, encoded in a proprietary format which is defined by the manufacturer. These templates can only be used with the software provided by the manufacturer, hence limiting the use of those devices in research. Thus, research institutions began to construct their own, custom capturing devices for finger and hand vein images.

The main contribution of this work is the design of such a capturing device. We propose a fully contactless, combined finger and hand vein capturing device and the publication of a vascular pattern dataset, acquired with this device. Contactless acquisition devices have several advantages over touch based ones. The main advantage is that contactless devices achieve a higher user acceptability, mainly due to hygienic reasons and easier handling of the devices. Moreover, contactless acquisition preserves the vascular patterns from distortions [4]. On the other hand, contactless acquisition introduces some challenges as well—due to the higher degree of freedom in terms of finger/hand movement, the physical device design as well as the processing of the vascular pattern images has to account for different types of distortions/artefacts resulting from the image acquisition, including longitudinal finger rotation [7], finger bending and tilts as well as all kinds of translations and rotations of the fingers/hand. Besides these types of misplacements, one of the main challenges is to provide a uniform illumination within the whole range of the allowed relative position of the finger/hand to the capturing device. In the following we give an overview of related work on finger as well as hand vein capturing devices.

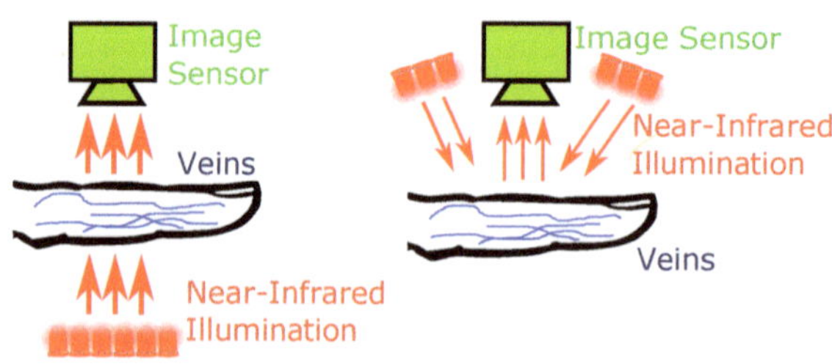

Figure 1. Light source and image sensor positioning, left: light transmission, right: reflected light.

1.2. Related Work

As the proposed capturing device design is a contactless one, we focus on contactless finger and hand vein capturing devices. While contactless acquisition has become common practice in hand vein recognition, the majority of the capturing devices in finger vein recognition are still contact based.

Almost all of the widely employed COTS finger vein capturing devices are contact based ones, capturing the finger from the palmar view using light transmission or light dispersion. The two major companies providing finger vein authentication solutions are Hitachi Ltd. (Tokyo, Japan) and Mofiria Ltd (Tokio, Japan). The most commonly used devices include the Hitachi H-1 USB finger vein scanner [8] and the Mofiria FVA-U3SX [5] as well as the Mofiria FVA-U4ST [6]. As those are commercial products, not many details about their design have been disclosed, except the recognition performance according to the manufacturers' data sheets. Due to the challenges and problems with contactless acquisition, there are only a few contactless finger vein capturing devices proposed in research. One of these devices is a mobile finger vein scanner for Android proposed by Sierro et al. [9]. Their prototype device captures contactless palmar finger vein images using reflected light. The illumination source consists of 12,850 nm LEDs, organised in 3 groups of 4 LEDs each (wide angle VSMG3700 and SFH4059 LEDs), providing global as well as optimised homogeneous illumination compensation. The power of each LED group can be adjusted using the Android software. The camera is a low-cost OV7670 colour one, using a CMOS sensor and a wide angle 2.1 mm lens with a maximum resolution of 640×480 pixel. They used an additional NIR pass-through filter with a cut-off wavelength of 740 nm. Another contactless finger vein capturing device was proposed by Kim et al. [10]. This device is based on NIR lasers and uses light transmission. The NIR lasers are manufactured by Lasiris Laser in StokerYale, Canada. A laser line generator lens (E43-475 from Edmund optics in Singapore) with a fixed pan angle is added in order to generate a line laser from the spot laser and should enable a uniform illumination along the finger. The image sensor is a GF 038B NIR CCD (charge coupled device) Camera from ALLIED Vision Technologies, Germany, which is equipped with an additional IR-pass filter. No further details about this device are available, the authors do not even include an image showing their capturing device in the paper. Another contactless device is proposed by Raghavendra et al. [11]. Their low-cost capturing device is able to acquire palmar finger vein images using light transmission as well as fingerprint images in a contactless manner and consists of a NIR light source, a physical structure to achieve a sufficient light intensity, a visible light source and a camera including a lens. The NIR light source is composed of 40 TSDD5210 NIR LEDs with a peak wavelength of 870 nm. The physical structure to achieve a sufficient illumination is wrapped with aluminium foil. The camera is a DMK 22BUC03 monochrome CMOS camera equipped with a T3Z0312CS 8 mm lens. The maximum resolution is 744×480 pixel. Even though the device is a contactless one, the images of the capturing device in the paper reveal that the range of motion for the finger is quite limited in every direction (x, y and z) due to the small opening of the device where the finger has to be placed in. None of the above mentioned capturing devices uses a special NIR enhanced camera. Thus, the resulting image quality in combination with an NIR light source is limited. A more recent device was proposed by Matsuda et al. [12]. It is a contactless walk-through style device which allows to capture multiple fingers at once in real time. It consists of an NIR camera and a depth camera, arranged below the finger placing part and an adaptive, multi-light source arranged vertically on the side of the finger placing part. No further technical details about this device are available but there is an official website from Hitachi [13] showing some images of the sensor prototype.

In the early stage of hand vein recognition, most capturing devices used almost closed box devices having a glass plate and some kind of pegs to force the hand to be placed in a defined position [14,15]. The users found this way of providing their biometric inconvenient and thus, the capturing devices developed from semi contactless ones (e.g., only using some hand attachment or guide [16,17] or a glass plate only [18]) to fully contactless ones. The following review of contactless hand vein capturing devices is not exhaustive but shall provide an overview over the major types of different device designs. The most well-known COTS hand vein authentication system is Fujitsu's PalmSecure™ [19]

one. Their capturing device [20] is contactless and small sized: $35 \times 35 \times 27$ mm. There are many non-commercial devices which have been proposed in several research papers as well, for example, the capturing device originally used to acquire the CASIA Multi-Spectral Palmprint Image Database V1.0 [21]. This device captures palmar hand images using six different wavelengths. It is a box with an opening in the front where the data subject has to put the hand inside. The CCD camera is located at the bottom of the device and the LEDs in different wavelength spectra are located around the camera. Sierro et al. [9] also proposed two contactless palm vein capturing device prototypes. Both are using the reflected light set-up and are equipped with ultrasonic sensors to measure the distance between the camera and the hand. The first prototype uses 20,940 nm LEDs (TSAL6400) as a light source and a Sony ICX618 CCD camera in combination with a 920 nm long-pass filter. The second prototype is able to capture multi-spectral images and uses an additional PTFE (Teflon) sheet to achieve a more uniform illumination. Michael et al. [22] proposed a low-cost contactless capturing device. It has one NIR and one visible light camera to capture both, palm vein and palm print images. The NIR camera has a NIR pass-through filter with a cut-off wavelength of 900 nm. The light source consists of 3 rows of 8 NIR LEDs and 3 yellowish light bulbs to capture the palm prints. The light source is covered by a diffusor paper. Zhang et al. [23] presented an approach to match hand veins using 3D point cloud matching. They use a binocular stereoscopic vision device as contactless capturing device. The hand is place above an NIR light source, consisting of 850 nm LEDs. Dorsal hand vein and knuckle shape images are captured by two NIR sensitive CCD cameras in a stereoscopic set-up, both having an additional NIR pass-through filter. Fletcher et al. [24] developed a mobile hand vein biometric system for health patient identification. They proposed two capturing devices; the first one uses an android smart phone in combination with a rechargeable 850 nm LED light source. The second one employs a low-cost webcam (Gearhead WC1100BLU USB) with integrated 940 nm LEDs and an optical filter, which is powered and controlled by an Android tablet. Both acquire contactless palmar hand vein images. Debiasi et al. [25] presented an illumination add-on for mobile hand vein image acquisition. This device can be used in combination with a modified smart phone (NIR blocking filter removed) to acquire contactless hand vein images from the palmar as well as the dorsal side. They also published a dataset containing palmar and dorsal hand vein images in the scope of the PROTECT Multimodal Biometric Database [26].

While most of the above mentioned capturing device designs are based on low-cost modified cameras, our design is based on a special NIR-enhanced industrial camera in combination with an optimal lens and an additional NIR pass-through filter to reduce image distortions and achieve the best possible image quality. Furthermore, in contrast to other existing designs we employ NIR laser modules instead of LEDs which enable a higher range of finger movement without impacting the image quality. Our capturing device is the first of its kind, able to use light transmission as well as reflected light. Moreover, it is the first combined capturing device, able to acquire finger as well as hand vein images. Finally, we do not only present a new capturing device design including all its technical details, but we also publish a corresponding dataset together with image quality and baseline recognition performance evaluation results on that dataset, which makes this work particularly valuable in the field of finger and hand vein recognition.

1.3. Main Contributions

The main contributions of this paper are:

- Design of a novel fully contactless combined finger and hand vein capturing device featuring laser modules instead of NIR LEDs, a special NIR enhanced industrial camera with an additional NIR pass-through filter to achieve the best possible image quality, an optimal lens and distance between the finger/hand and the camera to allow for minimal image distortions as well as an automated illumination control to provide a uniform illumination throughout the finger/hand surface and to arrive at the best possible contrast and image quality.

- Publication of all major technical details of the capturing device design—in this work we describe all the major components of the proposed capturing device design. Further technical details are available on request, which makes it easy to reproduce our design.
- Public finger and hand vein image database established with the proposed capturing device—together with this paper we publish the finger and hand vein datasets acquired with the proposed capturing device. These datasets are publicly available free of charge for research purposes and the finger vein one is the first publicly available contactless finger vein recognition dataset. Due to the nature of contactless acquisition, these datasets are challenging in terms of the different types of the finger/hand misplacements they include.
- Evaluation of the acquired database in terms of image quality and biometric recognition performance—the images acquired with our sensor are evaluated using several image quality assessment schemes. Furthermore, some well-established vein recognition methods implemented in our already open source vein recognition framework are utilised to evaluate the finger and hand vein datasets. This ensures full reproducibility of our published results. The achieved recognition performance during our evaluation is competitive with other state-of-the-art finger and hand vein acquisition devices, validate the advantages of our proposed capturing device design and prove the good image quality and recognition performance of our capturing device.

The remainder of this paper is organised as follows: Section 2 explains our proposed contactless finger and hand vein capturing device design, introduces the dataset acquired with the help of the proposed capturing device and it explains the experimental set-up, including the utilised recognition tool-chain, the evaluation protocol and the processing of the captured vein images. Section 3 lists the evaluation results of both, the acquired finger and hand vein dataset in terms of image quality and recognition accuracy, as well as the recognition accuracy of the considered fusion combinations. A discussion of the evaluation results, including a comparison with recognition performance results achieved by other capturing devices is provided in Section 4. Section 5 concludes this paper and gives an outlook on future work.

2. Materials and Methods

As mentioned in the introduction, a typical finger or hand vein capturing device consists of an NIR sensitive camera and some kind of NIR light source. In the following, the general design of our proposed contactless finger and hand vein capturing device as well as all the individual parts, including technical details and the design decisions are given. Afterwards, the acquired dataset and the utilised biometric recognition tool-chain are described.

2.1. Contactless Finger and Hand Vein Acquisition Device

Figure 2 shows our contactless finger and hand vein capturing device with all its individual parts annotated. It consists of an NIR enhanced camera together with a suitable lens and an additional NIR pass-through filter, two NIR illuminators, one laser module based for light transmission as well as one NIR LED based for reflected light, an illumination control board, a touchscreen display to assist the user during the acquisition process and its metal frame together with the wooden housing parts. All its parts are either standard parts which can be easily bought at a local hardware store or custom designed parts which are either laser cut plywood or 3D printed plastic parts and can easily be reproduced as well. The 3D models and technical drawings of those parts are provided on request. The following list summarises the main advantages and differences of our proposed design over the existing ones presented in Section 1.2.

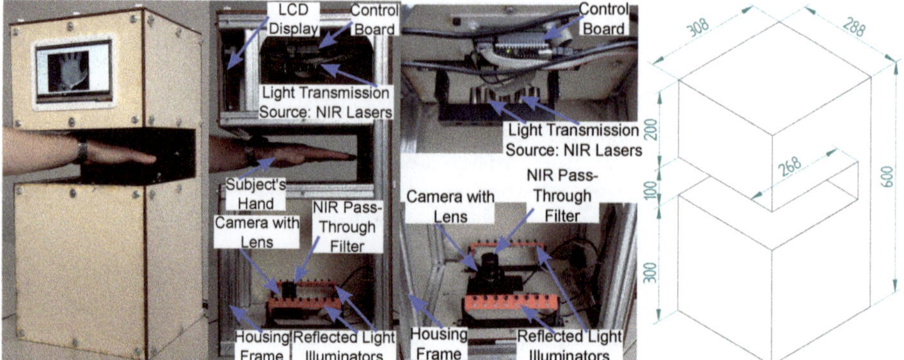

Figure 2. Contactless finger and hand vein capturing device, from left to right: device in use during acquisition, side view with annotated parts, top side view and bottom side detail view, housing including dimensions.

- Reflected light as well as light transmission—it is the first acquisition device of its kind, able to acquire reflected light as well as light transmission images. This extends the range of possible uses of this capturing device and speeds up the acquisition process if both types of illumination set-ups shall be investigated.

- Suitable for finger as well as hand vein images—it is possible to acquire palmar finger as well as hand vein images with the same device. Again, this is the first capturing device able to acquire both using the same device. In the default configuration, finger vein images are captured using light transmission while hand vein images are captured using reflected light but this can be changed in the set-up so there is a high flexibility in terms of possible acquisition configurations.

- NIR laser modules for light transmission illumination—the application of NIR laser modules has not been that common in finger vein recognition so far. In a contactless acquisition set-up, laser modules exhibit several advantages over LEDs, especially if it comes to increased range of finger/hand movement as well as an optimal illumination and image contrast [27]. Hence, we decided to equip our capturing device with NIR laser modules.

- Illumination control board and automated brightness control algorithm—the integrated brightness control board handles the illumination intensity of both, the light transmission and the reflected light illuminators. Each of the laser modules in the light transmission illuminator can be brightness controlled separately and independent from the others. This illumination control in combination with our automated brightness control algorithm enables an optimal image contrast without having the operator do any manual settings.

- Special NIR enhanced industrial camera—our capturing device uses a special NIR enhanced industrial camera. In contrast to modified (NIR blocking filter removed) visible light cameras, those NIR enhanced camera have an increased quantum efficiency in the NIR spectrum. This leads to a higher image contrast and quality compared to cheap, modified visible light cameras.

- Optimal lens set-up and distance between camera and finger/hand—in contrast to many other, mainly smaller devices (in terms of physical size of the device), we decided to use a lens with a focal length of 9 mm. This allows for minimal image distortions all over the image area, especially at the image borders at the cost of an increased distance between the camera and the finger/hand. Hence, our capturing device is rather big compared to others.

- Easy to reproduce design—in contrast to most other proposed capturing devices, for which only very few details are available, we provide references to the data sheets and technical details of all of the capturing device's parts. Furthermore, we provide the 3D models and technical drawings for the frame parts and the 3D printed parts on request. Hence, it is easy to reproduce our proposed capturing device design.

- Fast data acquisition—due to the automated brightness control and the automated acquisition process, sample data acquisition is fast. Capturing a hand vein image only takes less than a second and capturing a finger vein images takes between 2–4 s once the data subject placed their finger/hand.

- Ease of use during data acquisition—in contrast to other available vein capturing devices, for our proposed device the data subjects do not need to align their fingers/hands with some contact surface or pegs. This is one of the main advantages of our contactless design, making the data acquisition easier for the data subjects as well as for the operators. The automated illumination control algorithm and the intuitive graphical capturing software further contribute to a smooth and easy data acquisition process. Moreover, the integrated touchscreen display assists the data subjects by indicating which finger/hand to place at the sensor, how to place it and indicates potential misplacements.

- Biometric fusion can be employed to increase the recognition performance—our proposed capturing device acquires finger vein images as well as hand vein images using two different wavelengths of illumination. Hence, it is easily possible to increase the recognition performance by applying biometric fusion at sensor level with different fusion combinations. An evaluation of selected combinations is done in Section 3.3.

After this general overview of our capturing device we now describe its individual parts.

2.1.1. Camera, Lens and Filter

The camera is an IDS Imaging UI-ML3240-NIR [28] with a maximum resolution of 1280×1024 pixels and a maximum frame rate of 60 fps. It is based on the EV76C661ABT CMOS monochrome image sensor, having a colour depth of 8 bit, a maximum resolution of 1.31 Megapixels, with a pixel size of 5.3 µm and a sensor diagonal of 1/1.84 inches. The main advantage of this camera compared to modified webcams and other visible light cameras is that it is an NIR enhanced industrial camera, which is specifically designed to achieve a high quantum efficiency within the NIR spectrum. Due to its increased NIR sensitivity, an NIR enhanced camera achieves a higher image contrast in the NIR spectrum than a visible wavelength one, which is shown in Figure 3 left, depicting its quantum efficiency chart. The peak wavelengths of our NIR LEDs (850 nm + 950 nm) and NIR laser modules (808 nm) are within the increased sensitivity range of the image sensor.

The camera is equipped with a Fujifilm HF9HA-1B 9 mm fixed focal lens [29]. A lens with an increased focal length has less image distortions but requires a larger distance from the finger, thus increasing the overall size of the scanner. A shorter focal length reduces the minimum distance to the finger but increases the image distortions, especially at the image boundaries. Thus, we decided to use a 9 mm focal length as the best trade-off between the distance to the finger, that is, the overall scanner dimensions and the image distortions introduced due to the lens. A MIDOPT FIL LP780/27 [30] NIR pass-through filter is mounted on top of the lens to further suppress the negative influence of ambient light. The filter transmission chart is depicted in Figure 3 on the right.

2.1.2. Light Sources—Reflected Light and Light Transmission

The capturing device uses two different light sources—a light transmission and a reflected light one. The light transmission illuminator consists of 5 laser diodes [31] including an adjustable constant-current laser diode driver printed circuit board (PCB) [32] and a TO-18 housing with a focus adjustable lens [33] for each of the laser modules (the combination of laser diode + control PCB + housing is denoted as laser module or laser). The laser diodes have a peak wavelength of 808 nm and an optical output power of 300 mW. Each laser module can be brightness controlled separately. The main advantages of the laser modules over LEDs is their higher optical output power and their narrow radiation half angle. This enables a higher degree of vertical finger movement without degrading the image quality [27].

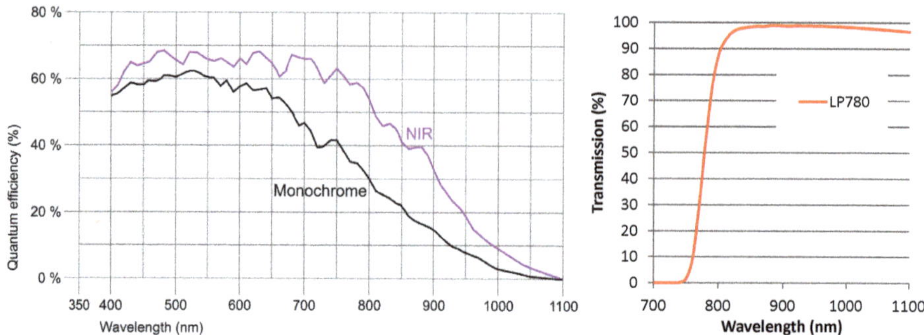

Figure 3. (**Left**) IDS UI-ML-3240-NIR quantum efficiency chart, (**right**) LP780 transmission chart.

The reflected light illuminator consists of 2 individual illuminators, one at each side of the camera (left and right). Each illuminator is composed of two rows of 8 LEDs each. The first row consists of 850 nm LEDs (Osram SFH 4550 [34] with a radiation half angle of ±3° and a max. radiant intensity of 700 mW/sr). The second row consists of 950 nm LEDs (Vishay Semiconductors CQY 99 [35] with a radiation half angle of ±22° and a maximum radiant intensity of 35 mW/sr). These two types of LEDs have peak wavelengths that are within the recommended spectrum for vascular pattern recognition. Each row can be brightness controlled as well, however only the whole row instead of each individual LED can be set to a certain brightness level. The emission spectra of the 850nm LEDs and the NIR laser modules can be seen in Figure 4, left and right, respectively.

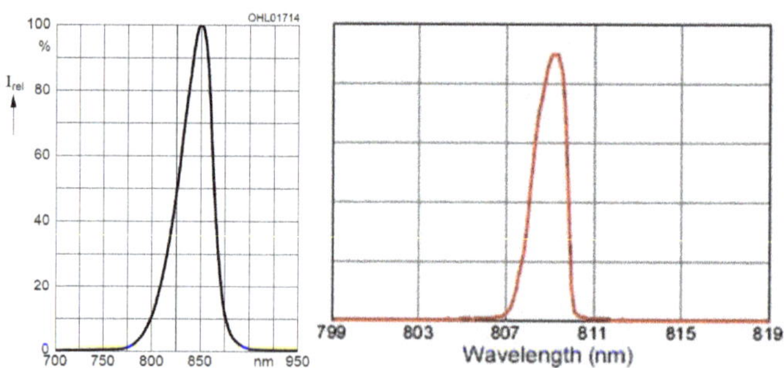

Figure 4. Emission spectrum of the 850nm near infrared (NIR) LEDs (**left**) and the NIR laser modules (**right**), taken from the data sheet [34].

2.1.3. Illumination Control Board and Brightness Control Algorithm

The schematic structure of the control board is depicted in Figure 5. The two main components of the illumination control board are an Arduino Nano board [36] and a Texas Instruments TLC59401 [37]. The Arduino Nano is a complete, breadboard-friendly microcontroller development board based on the Microchip ATmega328P microcontroller [38,39]. The Texas Instruments TLC59401 is an integrated 16-channel LED driver with dot correction and greyscale pulse width modulation (PWM) control enabling a convenient brightness control of LEDs without the need for external components like dropping resistors. Each output can be controlled separately (4096 steps) and has a drive capability of 120 mA. It operates as a constant-current sink and the desired current can be set using one external resistor only. In addition there are external PNP transistors (BC808-25 [40]) to drive the laser modules as their operating current exceeds the maximum current of the TLC59401. The reflected light illuminators are connected to one of the PWM outputs on the Arduino Nano using some external n-channel MOSFET transistors (AO3418 [41]) to drive them. The whole control board is interfaced using a simple, fixed-length, text-based serial protocol to control each of the individual LEDs/laser modules as well as the reflected light illuminators, to set a whole stripe at once and to turn off all illuminators again. On the PC side there is a graphical user interface based capturing control software which facilitates an easy and straight forward data acquisition. At the moment, the capturing process is initiated manually once the data subject placed their hand/finger in the sensor. This process will be automated in the future as well.

The brightness control algorithm controls each of the single light transmission illuminator's laser modules as well as the reflected light illuminators as a whole. We decided to implement a simple, iterative algorithm based on a comparison against a target grey level, which works as follows—at first the laser centres have to be configured, including the determination of the area of influence for each laser, which is the area in the image a single laser illuminates. Then all lasers are set to an initial intensity level/brightness value which is half of their maximum intensity (I_{max}). The live image of the camera is analysed and the current grey level in the circle of influence of each laser is determined ($GL_{current}$) and compared against the set target grey level (GL_{target}). The new brightness value is then set according to: $I_{n+1} = I_n + I_{corr}$, where I_{n+1} is the new intensity level, I_n is the current intensity level and $I_{corr} = \frac{GL_{target} - GL_{current}}{GL_{max}} \cdot \frac{I_{max}}{2 \cdot n}$, where GL_{max} is the maximum grey value and n is the current iteration. The iteration stops if either the target grey level GL_{target} has been reached or if no more intensity changes are possible. The algorithm finishes in at most $log_2(I_{max})$ iterations. Both, the Arduino Nano firmware as well as the capturing software, including our brightness control algorithm are available on request as well.

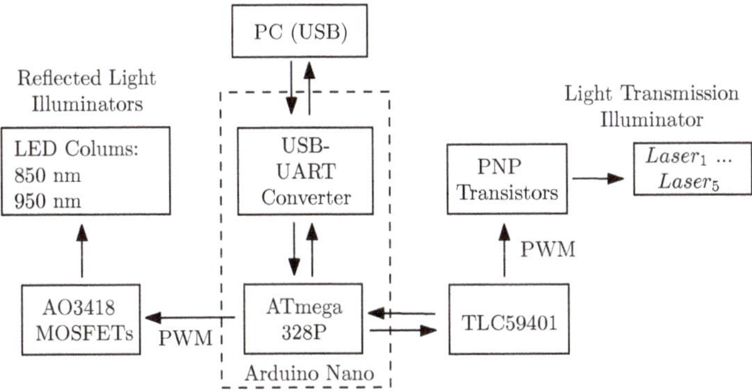

Figure 5. Schematic structure of the illumination control board.

2.1.4. Frame, Housing and Touchscreen

The outer frame is assembled using Coaxis® [42] aluminium profiles. The Coaxis® system is easy to use with several different profiles and connectors, which can be put together in many different ways. Another advantage is that this system provides a good stability and durability. On top of the aluminium frame there are laser cut plywood (4 mm beech wood) boards as side walls/cover. Figure 2 right shows the outside of the housing including its dimensions. A Waveshare 7inch HDMI LCD (C) touchscreen [43] is located the top front part of the capturing device. This touchscreen is connected to the acquisition PC and displays the live image stream from the camera, including an overlay of the optimal finger/hand position in order to help the data subjects in positioning their fingers/hand and also displays other information about the data acquisition, for example, which finger/hand to place next. The next revision of the capturing device will be a fully embedded one, that is, there is no need for an external PC and the whole data acquisition can be controlled using the device itself with the help of the integrated touchscreen display.

2.2. PLUSVein-Contactless Finger and Hand Vein Data Set

To validate our proposed capturing device design and to show the good recognition performance that can be achieved, we established a data set with the help of this device. Due to the contactless acquisition, these datasets are challenging in terms of finger/hand normalisation to compensate for the different types of finger/hand misplacements contained in the data (tilts, bending, in-planar and non-planar rotations). The dataset will be publicly available for research purposes together with the publication of this paper (http://www.wavelab.at/sources/PLUSVein-Contactless/). It contains two subsets—a palmar finger and a palmar hand vein one, including 42 subjects, 6 fingers/2 hands per subject and 5 images per finger/hand in one session. Hence, the finger vein subset contains 1260 images and the hand vein one contains 840 images (2 illumination configurations, 850 and 950 nm, 420 images each) in total. The raw images have a resolution of 1280×1024 pixels and are stored in 8 bit greyscale png format. The visible area of the finger in the images is 600×180 pixels and for the hand 750×750 pixels on average. Some example images are shown in Figure 6. The age and information about the handedness of the data subjects was recorded as well. Besides this information, no other sensitive private information about the subjects was collected. All subjects gave their informed consent for inclusion before they participated in the study. The study was conducted in accordance with the Declaration of Helsinki, and the protocol was approved by the Ethics Committee of the University of Salzburg (PLUSVein Contactless Data Acquisition 2019).

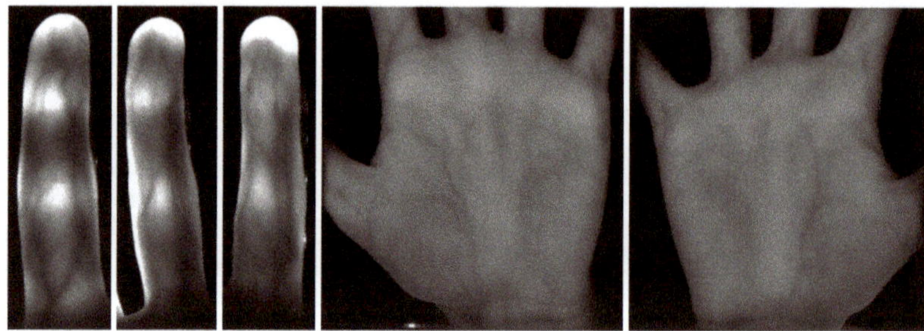

Figure 6. Example images of the PLUSVein-Contactless finger (**left**) and hand (**right**) vein dataset.

2.3. Finger and Hand Vein Recognition Tool-Chain

The recognition tool-chain includes all steps of a biometric recognition system starting with the extraction of the region of interest (ROI) to pre-processing, feature extraction and comparison, which are depicted in Figure 7 and described in the following. In addition, the utilised image quality assessment methods and biometric fusion, especially score level fusion, are explained as well. All of the utilised methods are implemented within our open source vein recognition framework PLUS OpenVein Toolkit (http://www.wavelab.at/sources/OpenVein-Toolkit/).

ROI Extraction

The key aim of the region of interest (ROI) extraction is to select the best suitable image part for the subsequent feature extraction and to automatically normalise the used finger/hand region in a way to avoid shifts, rotations and to account for scale changes. The ROI extraction and finger/hand normalisation is a crucial step, especially in contactless acquisition, to account for the higher degree of freedom and to compensate the different types of finger/hand misplacements. Different ROI extraction methods have been utilised for finger and hand vein images.

For the finger vein images, the finger is aligned and normalised according to a modified version of the method proposed by Lu et al. [44]. This alignment places the finger in the same position in every image, having the same finger width (different scales due to different finger positions). At first the finger outlines (edge between finger and the background of the image) are detected and the centre line (in the middle of the two finger lines) is determined. Afterwards, the centre line of the finger is rotated and translated in a way that it is placed in the middle of the image and the image region outside of the finger is masked out by setting the pixels to black. Then the finger outline is normalised to a pre-defined width. The final step is the actual extraction of a rectangular ROI of a fixed size (450 × 150 pixels) with its top border located at the fingertip. These steps are visualised in Figure 8.

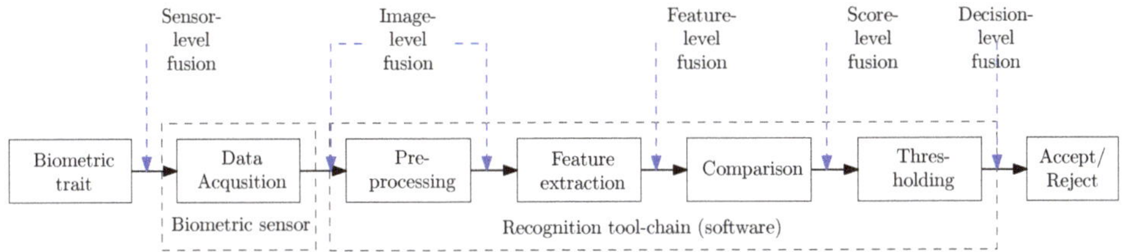

Figure 7. Biometric recognition tool-chain and different levels of biometric fusion.

The ROI method for hand vein images is a modified and extended version of the approach proposed by Zhou and Kumar [45]. At first the hand region is segmented by binarising the image using a local adaptive thresholding technique. Then the local minima and maxima points in the image are found. The local maxima correspond to the finger tips while the local minima correspond to the finger valleys. For the palmar view and the left hand, the second and fourth minima corresponds to the valley between the index and middle finger and the ring and the pinky finger, respectively. A line is fitted between those two valley points and then the image is rotated such that this line becomes horizontal. Afterwards, a square ROI is fitted inside the hand area, with its centre at the centre of mass of the hand (foreground in the segmented image). The size of the square ROI is adjusted such that its size is the maximum square without including any background pixels. The hand ROI extraction steps are shown in Figure 9. As a last step, the ROI image is scaled to a size of 384 × 384 pixels.

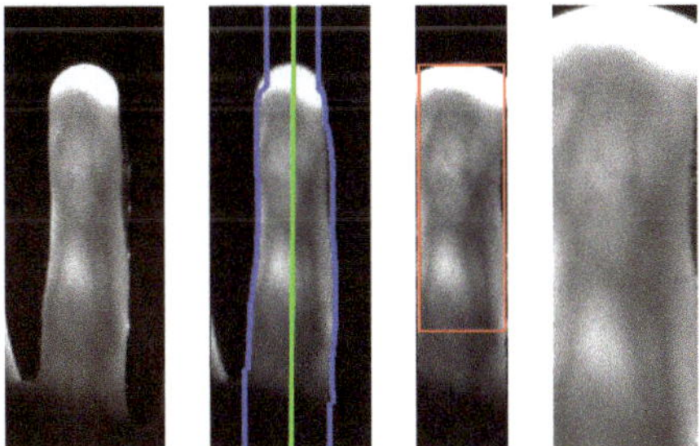

Figure 8. Finger vein region of interest (ROI) extraction process, from left to right: input image, finger outline and centre line detection, finger aligned, masked and normalised ROI boundary, final ROI.

Pre-Processing

Pre-processing approaches try to enhance the low contrast and improve the image quality. Simple **Contrast Limited Adaptive Histogram Equalisation (CLAHE)** [46] or other local histogram equalisation techniques are most prevalent for this purpose. Global contrast equalisation techniques tend to over-amplify bright areas in the image while some other dark areas are not sufficiently enhanced. A localised contrast enhancement technique like CLAHE is a suitable baseline tool to enhance the vein images as they exhibit unevenly distributed contrast. CLAHE has an integrated contrast limitation (clip limit) which should avoid the amplification of noise.

High Frequency Emphasis Filtering (HFEF) [47] tries to enhance the vein images in the frequency domain. At first the discrete Fourier transform of the image is computed, followed by the application of a Butterworth high-pass filter in the frequency domain. Afterwards the inverse Furier transform is computed to give prominence to the vein texture. In order to improve the image contrast the authors also apply a global histogram equalisation as a final step. We applied CLAHE instead of the global histogram equalisation.

Circular Gabor Filter (CGF) as proposed by Zhang and Yang [48] is another finger vein image enhancement technique which is rotation invariant and achieves an optimal joint localisation in both the spatial and the frequency domain. The authors originally suggested using grey level grouping for contrast enhancement and to reduce illumination fluctuations. Afterwards an even symmetric circular Gabor filter is applied to further attenuate the vein ridges in the image. Gabor filters are widely used to enhance images containing a high amount of texture and to analyse image texture information. In contrast to usual Gabor filters, a CGF does not have a direction, thus it amplifies the vein ridges in each direction. The bandwidth and the sigma of the CGF has to be tuned according to the visible vein information in the images (vein width in pixels).

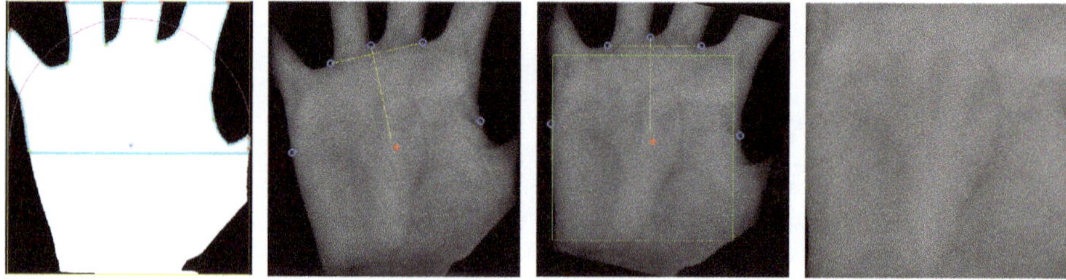

Figure 9. Hand vein ROI extraction process, from left to right: Segmented hand including outline and minima/maxima points, appropriate finger valleys and centre of mass selected, rotationally aligned hand image with maximum possible ROI fitted, final extracted ROI.

Furthermore, the images were resized to half of their original size, which not only speeded up the comparison process but also improved the results. For more details on the preprocessing methods the interested reader is referred to the authors' original publications. Each of the above mentioned pre-processing techniques is at least used for one of the feature extraction methods, but not necessarily with the same parameters for each method. The actual methods and parameters used for each feature extraction method are stated in the settings files (cf. Section 2.4).

Feature Extraction

Three vein pattern based techniques, which aim to extract the vein pattern from the background resulting in a binary image (vein pattern based methods) followed by a comparison of these binary images using a correlation measure and a general purpose key-point based technique were used, which are all algorithms well-established finger vein recognition algorithms.

Maximum Curvature (MC [49]) is a curvature based approach which is insensitive to varying vein widths as it aims to emphasise only the centre lines of the veins. At first the centre positions of the veins are extracted by determining the local maximum curvature in cross-sectional profiles obtained by calculating the first and second derivatives in four directions—horizontal, vertical and the two oblique directions. Each profile is classified as either being concave or convex. Vein lines are indicated by local maxima in concave profiles, hence only the concave ones are used. A score is assigned to each centre position which corresponds to the width and curvature of the maxima region. Afterwards, the centre positions of the veins are connected using a filtering operation in all four directions taking the

8-neighbourhood of pixels into account to account for misclassifications at the first step due to noise and other imperfections in the images. The output feature vector is essentially a binary image which is obtained by thresholding the recorded score values using the median of all scores as a threshold.

Principal Curvature (PC [50]) is another curvature based approach, which is not based on the derivates but on the gradient field of the image. Hence, the first step is the calculation of the gradient field. Hard thresholding to filter out small gradients by setting their values to zero is performed to prevent amplification of small noise components. Afterwards the normalised gradient field is obtained by normalising the magnitude to 1 at each pixel, which is then smoothed by applying a Gaussian filter. The actual principal curvature calculation is then done based on this smoothed normalised gradient field by computing the Eigenvalues of the Hessian matrix at each pixel. The two Eigenvalues are the principal curvatures and the corresponding Eigenvectors of the Hessian matrix represent the directions of the maximum and minimum curvature. The bigger Eigenvalue corresponds to the maximum curvature among all directions and is recorded and further used. The final step is again a threshold-based binarisation of the principal curvature values to obtain the output feature vector which is essentially a binary vein image.

Gabor Filter (GF [4]) is a Gaussian kernel function modulated by a sinusoidal plane wave. Gabor filters are inspired by the human visual system's multichannel processing of visual information. Several 2D even symmetric Gabor filters with different orientations (in $\frac{\pi}{k}$ steps where k is the number of orientations) form a filter bank. The image is filtered using this filter band to extract k different feature vectors. The single feature vectors from the previous step are fused and thresholded to get a resulting feature vector. To remove small noise components, this vector is further post-processed using morphological operations, resulting in the final output feature vector, which is again a binary image.

Scale Invariant Feature Transform (SIFT [51]) is a key-point based technique. In contrast to the three vein pattern based ones, key-point based techniques use information from the most discriminative points as well as consider the neighbourhood and context information around these points. This is achieved by extracting key-point locations at stable and distinct points in the image and then assigning a descriptor to each detected key-point location. The approach we used is based on the general purpose SIFT descriptor in combination with additional key-point filtering along the finger boundaries. This filtering is done to suppress information originating from the finger shape (outside boundary) instead of the vascular pattern. We originally presented this additional key-point filtering in Reference [52].

Comparison

The three vein pattern based features (MC, PC and GF) are compared using an extended version of the approach proposed by Miura et al. [49]. The input features (binary vein images) are not registered to each and only coarsely aligned (by the preceding ROI extraction). To account for small shifts and rotations, the correlations between the input feature vector and in x- and y-direction shifted as well as rotated versions of the reference feature vector are calculated. The final output score is the maximum among those individual correlation values, representing the best possible overlay/match between the two feature vectors. For the SIFT feature vectors a typical approach for key-point based features is utilised. At first the nearest neighbour for each key-point is found by simply calculating the distance between this key-point and all key-points in the reference feature vector. The nearest neighbours/best correspondences is the one with the highest similarity score. If this score is below a set threshold, the key-point does not have a matching one in the reference feature vector. The final comparison score is the ratio of the matched points and the maximum number of detected key-points in both images (which is the maximum number of possible matches).

Vein Specific Image Quality Assessment

In contrast to fingerprint recognition where there are standardised quality metrics like the NIST Fingerprint Image Quality (NFIQ) [53] and the newer version NFIQ 2.0, there are no standardised metrics in finger- and hand-vein recognition yet. Thus, the finger- and hand-vein images were analysed using GCF [54] as it is a general image contrast metric and hence, independent of the image content. With the help of GCF the image contrast can be quantified exclusively disregarding the actual image content. As we aim to quantify the image quality of vein images, of course two vein specific NIR image quality metrics, namely the approach proposed by Wang et al. [55] (Wang17) and the approach proposed by Ma et al. [56] (HSNR) were included as well. The first approach evaluates the vein image quality fusing a brightness uniformity and a clarity criterion, which is obtained by analysing the local pixel neighbourhoods. The HSNR approach, which is especially tailored for non-contact finger vein recognition, simulates the human visual system by calculating an HSNR index and integrates an effective area index, a finger shifting index and a contrast index to arrive at the final image quality value.

Score Level Fusion

According to the ISO/IEC TR 24722:2015 standard [57], biometric fusion can be regarded as a combination of information from multiple sources, that is, sensors, characteristic types, algorithms, instances or presentations in order to improve the overall system's performance and to increase the systems robustness. Biometric fusion can be categorised according to the level of fusion and the origin of input data. The different levels of fusion correspond to the components of a biometric recognition system—sensor-level, image-level, feature-level, score-level and decision-level fusion, which are indicated in Figure 7. Sensor-level fusion is also called multisensorial fusion and describes using multiple sensors for capturing samples of one biometric instance [57]. This can either be done by the sensor itself or during the biometric processing chain. Hence, we perform sensor-level fusion as our capturing device acquires finger as well as two different kinds of hand vein images. The actual fusion is done during the biometric processing chain at score level (fusing the output scores of the individual modalities—finger veins, hand veins 850nm and hand veins 950nm).

The following combinations of different acquired modalities are evaluated:

1. Hand veins 850 nm + hand veins 950 nm
2. Hand veins 850 nm + finger veins
3. Hand veins 950 nm + finger veins
4. Hand veins 850 nm + hand veins 950 nm + finger veins

Note that, for the combinations including finger veins, only one finger is included in the fusion. We evaluated the combinations including a finger for all fingers of the respective hand and used the best performing finger, which turned out to be the middle finger for both hands. Acquiring images of several, distinct fingers takes more time as only one finger is captured at a time, the same applies to acquiring both hands. Thus, we restricted to the evaluated combinations to the above listed ones which do not considerably increase the acquisition time. The actual score level fusion is performed using the BOSARIS tool-kit [58], which provides a MATLAB based framework for calibrating, fusing and evaluating scores from binary classifiers and has originally been developed for automatic speaker recognition. A 5 fold random split of training and test data with 20 runs was used to train and fuse the scores using BOSARIS. The reported performance results are the average values of the 20 individual runs.

2.4. Experimental Setup and Evaluation Protocol

The evaluation is split into three parts—image quality assessment, baseline recognition performance evaluation for the individual subsets and recognition performance evaluation of the fusion combinations. The image quality assessment and the baseline recognition performance evaluation is done separately for the finger dataset and the two hand vein datasets (850 nm and 950 nm illuminator). The three image quality assessment schemes are evaluated for each individual image per dataset. The results are the average values over the whole dataset, that is, there is a single value for the finger vein and the hand vein 850 nm as well as the hand vein 950 nm dataset for each image quality metric. For the recognition performance DET plots as well as the EER (the point where the FMR equals the

FNMR), the FMR1000 (the lowest FNMR for FMR = 0.1%) and the ZeroFMR (the lowest FNMR for FMR = 0%) are provided. At first the parameters for the pre-processing and feature extraction are optimised on a training dataset. Each dataset is divided into two roughly equal sized subsets, based on the contained subjects, that is, all fingers/hands of the same person are in one subset. The best parameters are determined on each subset and then applied to the other subset for determining the comparison scores. This ensures a full separation of the training and test set. The final results are based on the combined scores of both test runs. The FVC2004 [59] test protocol is applied for calculating the comparison scores in order to determine the FMR/FNMR: for the genuine scores, all possible genuine comparisons are evaluated, resulting in $n_{gen} = \frac{5 \cdot (5-1)}{2} \cdot (42 \cdot 6) = 2520$ and $n_{gen} = \frac{5 \cdot (5-1)}{2} \cdot (42 \cdot 2) = 840$ genuine scores for the finger and hand vein subset, respectively. For the impostor scores only the first template of a finger/hand is compared against the first template of all other fingers/hands, resulting in $n_{imp} = \frac{(42 \cdot 6) \cdot (42 \cdot 6 - 1)}{2} = 31{,}626$ impostor comparisons for the finger vein subset as well as $n_{imp} = \frac{(42 \cdot 2) \cdot (42 \cdot 2 - 1)}{2} = 3486$ impostor comparisons for the hand vein ones. The EER/FMR1000/ZeroFMR values are given in percentage terms, for example, 0.47 means 0.47%. The full results including the image quality values for each single image, the comparison scores and plots as well as the settings and script files to reproduce the experiments can be downloaded here: http://www.wavelab.at/sources/Kauba19c/.

3. Results

This section presents the results of the image quality assessment as well as the recognition performance evaluation on the acquired datasets and the score level fusion combination of the data sets.

3.1. Image Quality Assessment

Table 1 lists the image quality assessment results for the three tested metrics, namely GCF, Wang17 and HSNR. The GCF values range from 0 to 8, the Wang17 values from 0 to 1 and the HSNR values from 0 to 100. Higher values correspond to higher image quality. Note that a cross-modality comparison (finger vs. hand veins) using those metrics does not lead to meaningful results as the underlying input data (images) are fundamentally different. To enable a meaningful quality assessment and a comparison with other, available finger and hand vein dataset, we evaluated several other finger and hand vein datasets by using the same quality metrics. The evaluated finger vein datasets include SDUMLA-HMT [60], HKPU-FID [4], UTFVP [61], MMCBNU_6000 [44], FV-USM [62] and PLUSVein-FV3 [27]. The image quality was evaluated for the following hand vein datasets—Bosphorus Hand Vein [63], Tecnocampus Hand Image [64], Vera Palm Vein [65] and PROTECT HandVein [66]. The discussion of the image quality assessment results is done in Section 4.

Table 1. Image quality assessment results for the proposed datasets (bold face) and several available finger- and hand vein datasets. Best results per quality metric and modality are highlighted **bold face**.

	Dataset	GCF	Wang17	HSNR
Finger Vein	**Finger Vein**	**1.72**	0.256	**92.16**
	SDUMLA-HMT [60]	0.986	0.165	80.32
	HKPU-FID [4]	1.46	0.166	88.13
	UTFVP [61]	1.47	**0.356**	87.15
	MMCBNU_6000 [44]	1.52	0.121	87.39
	FV-USM [62]	0.69	0.136	83.35
	PLUSVein-FV3 [27]	1.48	0.306	89.78
Hand Vein	**Hand Vein 850 nm**	1.42	**0.682**	90.43
	Hand Vein 950 nm	1.87	0.656	**91.76**
	Bosphorus Hand Vein [63]	2.69	0.329	86.12
	Tecnocampus Hand Image [64]	2.31	0.373	54.33
	Vera Palm Vein [65]	1.31	0.43	85.09
	PROTECT HandVein [66]	**2.8**	0.563	82.43

3.2. Recognition Performance

The recognition performance results should serve as a baseline for further experiments/research conducted on these contactless finger and hand vein datasets. Table 2 lists the performance results in terms of EER, FMR1000 and ZeroFMR where the best results per subset (finger vein, hand vein 850 nm and hand vein 950 nm) are highlighted **bold face**. The corresponding DET plots are shown in Figure 10.

It is evident that MC performed best on all subsets in terms of EER, FMR1000 as well as ZeroFMR except for the finger vein one where it performed second best in terms of EER (but still best in terms of FMR1000 and ZeroFMR). The overall best performance was achieved on the hand vein 850 nm subset using MC and resulting in an EER of 0.35%. In terms of EER, on the finger vein subset SIFT performed best, followed by MC and GF while PC performed worst. On the hand vein 850 nm subset, PC performed second best, followed by SIFT and GF performed worst, while on the 950 nm subset SIFT performed second best, followed by PC and again, GF performed worst.

Table 2. Single modality recognition performance results.

Modality		MC	PC	GF	SIFT
	EER [%]	5.61	8.22	6.63	**3.66**
Finger Vein	FMR1000 [%]	**13.12**	23.99	18.39	16.61
	ZeroFMR [%]	**18.75**	42.19	28.76	36.11
	EER [%]	**0.35**	0.95	1.55	0.95
Hand Vein 850 nm	FMR1000 [%]	**0.95**	1.67	2.26	1.9
	ZeroFMR [%]	**1.67**	2.26	2.74	2.74
	EER [%]	**0.38**	0.83	0.72	0.82
Hand Vein 950 nm	FMR1000 [%]	**0.59**	1.43	1.19	1.67
	ZeroFMR [%]	**1.07**	1.67	1.67	2.02

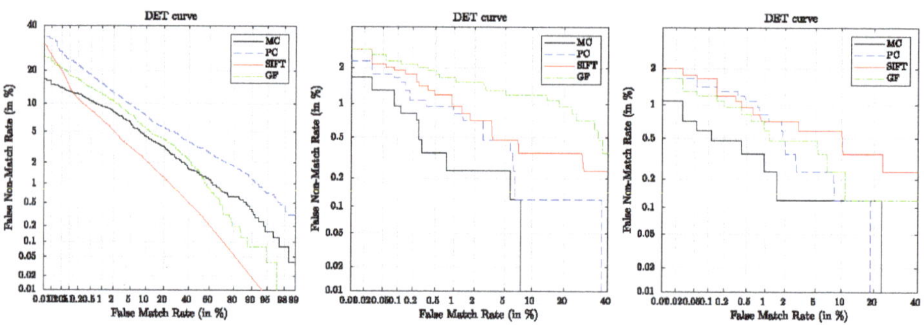

Figure 10. DET plots for finger vein (**left**), hand vein 850 nm (**middle**) and hand vein 950 nm (**right**).

3.3. Biometric Fusion Results

Table 3 shows the results for the tested fusion combination together with the relative performance increase of the combination. The relative performance increase (RPI) refers to the best performing single modality included in the fusion combination (usually the hand vein 850 or hand vein 950 nm result). Each fusion combination improved the results over the respective baseline ones. The overall best results of the tested fusion combinations was the combination of hand vein 850 nm + middle finger achieving an EER of 0.03% which corresponds to a relative performance increase of 1183%. The average improvement in terms of EER (over all feature types) compared to the best baseline (hand veins 850 nm) result for combination 1 is 148%, for combination 2 it is 373%, for combination 3 the average improvement is 140% and for combination 4 it is 365%.

Table 3. Score level fusion recognition performance results and improvement over baseline results. Best EER result (combination 2 for MC, combination 1 for PC, 3 for GF and 4 for SIFT) per feature type is highlighted **bold face**.

	Combination		MC	RPI	PC	RPI	GF	RPI	SIFT	RPI
1	Hand 850 Hand 950	EER [%]	0.24	44%	**0.16**	405%	0.60	19%	0.37	123%
		FMR1000 [%]	0.36	162%	0.21	586%	0.77	54%	0.65	155%
		ZeroFMR [%]	0.70	139%	4.90	−66%	0.92	82%	1.49	35%
2	Hand 850 Middle Finger	EER [%]	**0.03**	1183%	0.57	65%	0.64	144%	0.48	98%
		FMR1000 [%]	0.01	7862%	0.97	71%	1.19	90%	0.52	268%
		ZeroFMR [%]	0.14	1112%	1.32	72%	1.70	61%	0.79	246%
3	Hand 950 Middle Finger	EER [%]	0.14	171%	0.37	122%	**0.48**	50%	0.26	218%
		FMR1000 [%]	0.14	333%	0.71	102%	0.74	61%	0.37	352%
		ZeroFMR [%]	0.28	289%	1.28	30%	1.11	51%	2.35	−14%
4	Hand 850 Hand 950 Middle Finger	EER [%]	0.04	849%	0.22	272%	0.57	26%	**0.20**	311%
		FMR1000 [%]	0.00	-	0.36	298%	0.62	91%	0.30	449%
		ZeroFMR [%]	0.17	861%	11.47	−85%	0.74	127%	2.08	−3%

4. Discussion

At first, we discuss the image quality of our datasets in comparison with other publicly available finger- and hand-vein datasets. The evaluation results are listed in Table 1. Considering finger veins, our dataset achieved the best results for GCF and HSNR while it is ranked third for Wang17. These results confirm decent image quality in terms of contrast and also a good image quality in terms of vein specific properties. Considering hand veins our 850 nm data set achieved the best results for Wang17 and the second best ones for HSNR, while the 950 nm dataset achieved the best results for HSNR and second best for Wang17. In terms of GCF, both hand vein data sets are only ranked second and third to last, indicating that the general image contrast is lower than for other datasets. However, the vein specific quality metrics still indicate good image quality, despite the inferior image contrast compared to the other datasets. There are several works in the literature [67–70] that confirm that quality metric results do not necessarily have to correlate with recognition accuracy. The recognition accuracy is the most important aspect of a sensor design and dataset, thus we decided to focus on the recognition accuracy evaluation instead of evaluating the image quality only. Also note that other data set and sensor papers do not report the image quality, thus a direct comparison is not possible based on the image quality.

In the following, we compare our recognition performance evaluation results to other results reported in the literature. Matsuda et al. [12] reported an EER of 0.19% for their walk through style finger vein recognition system. Raghavendra et al. [11] et al. reported an EER of 1.74% for their systems in case of finger veins only. Kim et al. [10] arrived at an EER of 3.6% for their NIR laser based contactless acquisition set-up. Sierro et al. [9] did not present any performance evaluation of their dataset. None of the mentioned authors disclosed their dataset, so their results are not reproducible. As we aim for reproducibility, all the results listed in Tables 4 and 5 are evaluated on publicly available finger and hand vein datasets and we base our discussion on those results only.

Table 4 lists performance results (in terms of the EER unless indicated otherwise) achieved on various finger vein datasets, ordered by the year of publication, where the last row corresponds to the results presented in this work. The listed results are the best reported ones from the original dataset authors given that they were available and indicated by "-" if they were not available. Note that all of the listed datasets, except the one presented in this work, have been acquired in a non contactless way. Especially compared to the PLUSVein-FV3 dataset, which has been acquired using the same type of NIR enhanced camera and NIR laser modules, the results on our proposed dataset are clearly inferior (5.61% EER vs. 0.06% EER). However, given the increased level of difficulty and challenges of this new, contactless finger vein dataset, the results are still in an acceptable range. The proposed acquisition device design and thus, the acquired dataset, allows for more degrees of freedom in terms of finger/hand movement and thus, more unrestricted finger/hand positioning. This introduces different kinds of finger/hand misplacements, including tilts, bending, in-planar and non-planar rotations. Prommegger et al. showed, that especially those kinds of misplacements cause severe performance degradations for other publicly available datasets [7,71], especially for the SDUMLA-HMT finger vein dataset [60]. Also note that our results should only serve as a baseline and allow room for further improvements. We did not apply any special kind of finger misplacement corrections. If advanced correction schemes are applied, the results can of course be improved.

Table 4. Performance results of other publicly available finger vein datasets, ordered by publication year, "-" means that this information is not available. The "cla" column indicates contactless acquisition.

Name and Reference	Images/Subjects	cla	Feature Type	Performance (EER)	Year
PKU [72]	50,700/5208	no	WLD [72]	0.87%	2008
THU-FVFDT [73]	6540/610	no	MLD [73]	98.3% ident. rate	2009
SDUMLA-HMT [60]	3816/106	no	Minutia [74]	98.5% recogn. rate	2010
HKPU-FID [4]	6264/156	no	Gabor Filter [4]	0.43% (veins only)	2011
UTFVP [61]	1440/60	no	MC [49]	0.4%	2013
MMCBNU_6000 [44]	6000/100	no	-	-	2013
CFVD [75]	1345/13	-	-	-	2013
FV-USM [62]	5940/123	no	POC and CD [62]	3.05%	2013
VERA FV-Spoof [76,77]	440/110	no	MC [49]	6.2%	2014
PMMDB-FV [26]	240/20	no	MC [49]	9.75%	2017
PLUSVein-FV3 [27]	3600/60	no	MC [49]	0.06%	2018
Contactless FingerVein	840/42	**yes**	MC [49]	3.66%	2019

Regarding contactless hand vein recognition, Michael et al. [22] report an EER of 0.71% using palm veins only. Zhang et al. [23] only evaluate the KC value as measure of the registration between two feature vectors and thus, as an indicator for the recognition performance but they did not evaluate the actual recognition performance. The highest KC value they achieved was 1.1039. Fletcher et al. [24] reported an EER of 6.3% for their fully contactless hand vein based system for health patient identification. Table 5 summarises the achieved recognition performance for several publicly available hand vein recognition datasets, where the last row corresponds to the best results we achieved on our proposed contactless one so far. Note that, except for our proposed dataset and the PROTECT Mobile HandVein [25] dataset, all datasets have been acquired in a non contactless way. In the light of that and taking into account that we used only simple but well-established vein recognition schemes, the achieved recognition performance on our dataset is clearly competitive with other results reported in the literature. It is more than ten times better than the results published for the PROTECT Mobile HandVein [25] dataset and the results published by Fletcher et al. [24], and still twice as good as the results reported by Michael et al. [22], even though the only kind of hand normalisation we applied was the adaptive ROI extraction, correcting different scales, that is, different distances between the camera and the hand. No further tilt or non-planar rotation correction was applied. Again, note that our performance results should only serve as a baseline and can of course be improved.

Table 5. Related performance results of publicly available hand vein recognition datasets, ordered by publication year. The "cla" column indicates contactless acquisition.

Name and Reference	Images/Subjects	cla	Feature Type	Performance (EER)	Year
CIE [18]	2400/50	no	Thresholding [78]	1.1%	2011
Bosphorus Hand Vein [63]	1575/100	no	Geometry [79]	2.25%	2011
CASIA Multispectral [21]	7200/100	no	LBP/LDP [80]	0.09%	2011
Tecnocampus Hand Image [64]	6000/100	no	BDM [64]	98% recogn. rate	2013
Vera Palm Vein [65]	2200/110	no	LBP [81]	3.75%	2015
PROTECT HandVein [66]	2400/40	no	SIFT [51]	0.093%	2018
PROTECT Mobile HandVein [25]	920/31	**yes**	MC [49]	4.13%	2018
Contactless HandVein	420/42	**yes**	MC [49]	0.35%	2019

While the achieved baseline results for the contactless hand vein dataset are quite competitive compared to the contactbased hand vein datasets, the results for the contactless finger vein datasets are clearly inferior to the ones that can be achieved for contactbased finger vein recognition. Contactless finger vein recognition is more challenging than contactless hand vein one for several reason. One reason is that the finger has a much smaller area than the palm of the hand. Thus, small misplacements can lead to a reduced visibility of the vein patterns and a reduced image quality in general, making the recognition more difficult. Moreover, the vascular pattern structure within the finger is more fine-grained than within the palm of the hand. Hence, tilts, rotation and bending of the finger have a more severe effect on the acquired images in terms of the resulting distortions present in the image, again leading to complications during the recognition process. These challenges have to be tackled by suitable normalisation and correction schemes in order to improve the recognition performance for the contactless finger vein data.

The sensor level fusion results clearly indicate that by combining different acquisition modes (finger vein, hand vein 850nm and hand vein 950 nm) the recognition performance can be considerably improved. By combining the hand vein images in the two different wavelengths, the average performance improvement over the best baseline one is 148%. By combining one finger and one hand sample, the best results (MC) are improved by 1183% and 171% over the baseline results for 850 nm and 950 nm hand veins, respectively. By combining all three modes the results were improved as well and are more than 8 times better than the best baseline one (MC). Hence, applying sensor level fusion is an easy way to further enhance the recognition performance of our capturing device.

5. Conclusions

We proposed a new capturing device, which is able to acquire finger as well as hand vein images in a fully contactless way. Contactless acquisition has many advantages in terms of hygiene and user acceptance. In addition to the design and technical details of the acquisition device, we also provide a novel, contactless finger and hand vein dataset available for research purposes (can be downloaded here: http://www.wavelab.at/sources/PLUSVein-Contactless/). This dataset is the first available contactless finger vein dataset and one of the first available contactless hand vein datasets. It is a challenging dataset due to the contactless acquisition allowing for more unrestricted finger/hand movement and the resulting finger/hand misplacements. An image quality assessment using three vein tailored metrics has been conducted and confirmed the decent image quality which can be achieved using our proposed capturing device. Moreover, a recognition performance evaluation using several well-established vein recognition schemes has been carried out on this dataset in order to provide baseline results for further research. Those baseline results are competitive for the hand vein data (EER of 0.35%) and within range of other biometric technologies for the finger vein data (EER of 3.66%). Furthermore, biometric sensor level fusion experiments have been conducted to show the additional improvement in the recognition performance which can be achieved by combining finger vein and hand vein data (resulting in an overall best EER of 0.03%).

Our future work includes some improvements on the capturing device itself. The next version of the device should be an embedded device, eliminating the need for an additional PC to control the acquisition process. The capturing device has a built-in touch screen display already which can be used to control it via the graphical user interface. The only thing which is currently missing is the porting of the capturing software to an embedded platform like the Raspberry Pi and the automated start of the capturing process once the data subjects placed their finger/hand. Furthermore, we will extend our contactless finger and hand vein dataset. We are currently acquiring additional subjects and plan to enlarge the dataset to include a total of at least 100 subjects by the end of 2019. Moreover, we aim to do a thorough analysis on which types of finger/hand misplacements are present in the dataset, similar to the work has been done for other finger vein datasets [71]. Based on this analysis we will be able to apply different correction and normalisation schemes in order to improve the recognition performance.

Author Contributions: Conceptualization, C.K. and B.P.; methodology, C.K.; software, C.K. and B.P.; validation, C.K. and B.P.; formal analysis, C.K.; investigation, C.K. and B.P.; resources, C.K., B.P. and A.U.; data curation, C.K. and B.P.; writing—original draft preparation, C.K.; writing—review and editing, C.K.; visualization, C.K.; supervision, A.U.; project administration, C.K. and A.U.; funding acquisition, A.U.

Funding: This research was funded by European Union's Horizon 2020 research and innovation program under grant agreement number 700259, project PROTECT—Pervasive and UseR Focused BiomeTrics BordEr ProjeCT. This research received further funding by the Austrian Science Fund (FWF) and funding by the Salzburg state government, project no. P32201—Advanced Methods and Applications for Fingervein Recognition. Open Access Funding provided by the Austrian Science Fund (FWF).

Acknowledgments: First of all we really appreciate the spent time and effort of all the participants during our data collection and want to express our gratitude for their voluntary participation. Furthermore, we want to thank our colleagues Michael Linortner and Simon Kirchgasser who helped during the data acquisition. The Open Access Funding was kindly provided by the Austrian Science Fund (FWF).

Conflicts of Interest: The authors declare no conflict of interest. The funders had no role in the design of the study; in the collection, analyses, or interpretation of data; in the writing of the manuscript, or in the decision to publish the results.

Abbreviations

The following abbreviations are used in this manuscript:

ATM	Automated Teller Machine
CCD	Charge Coupled Device
CGF	Circular Gabor Filter
CLAHE	Contrast Limited Adaptive Histogram Equalisation
CMOS	Complimentary Metal Oxid Semiconductor
COTS	Commercial Off The Shelf
DET	Detection Error Tradeoff
EER	Equal Error Rate
FMR	False Match Rate
FNMR	False Non Match Rate
HFEF	High Frequency Emphasis Filtering
LED	Light Emitting Diode
NFIQ	NIST Fingerprint Image Quality
NIR	Near Infrared
NIST	National Institutes of Standards and Technology
PTFE	Polytetrafluoroethylene
PWM	Pulse Width Modulation
ROI	Region Of Interest

References

1. Fujitsu Laboratories Ltd. Fujitsu Develops Technology for World's First Contactless Palm Vein Pattern Biometric Authentication System. Available online: https://www.fujitsu.com/global/about/resources/news/press-releases/2003/0331-05.html (accessed on 6 October 2019).
2. Hitachi-Omron Terminal Solutions, Corp. Taiwan's CTBC Bank Adopts Finger Vein Authentication Solution for ATMs—Hitachi News. Available online: http://www.hitachi-omron-ts.com/news/pdf/201607-001.pdf (accessed on 20 June 2018).
3. Hitachi Group, Corp. Finger Vein Technology for Bank BPH (Poland)—Hitachi Europe News. Available online: http://www.hitachi.eu/en-gb/case-studies/finger-vein-technology-bank-bph-poland (accessed on 20 June 2018).
4. Kumar, A.; Zhou, Y. Human identification using finger images. *Image Process. IEEE Trans.* **2012**, *21*, 2228–2244. [CrossRef] [PubMed]
5. Mofiria Corp. Mofiria FVA-U3SXE Finger Vein Reader Data Sheet. Available online: https://www.mofiria.com/wp/wp-content/uploads/2017/08/FVA-U3SXE.pdf (accessed on 20 June 2018).
6. Mofiria Corp. Mofiria FVA-U4BT Finger Vein Reader Data Sheet (FVA-U4ST Is the Same Device Except for the USB Instead of Bluetooth Connection). Available online: https://www.mofiria.com/wp/wp-content/uploads/2017/08/FVA-U4BT_E.pdf (accessed on 20 June 2018).
7. Prommegger, B.; Kauba, C.; Linortner, M.; Uhl, A. Longitudinal Finger Rotation—Deformation Detection and Correction. *IEEE Trans. Biom. Behav. Identity Sci.* **2019**, 1–17. [CrossRef]
8. Hitachi Group, Corp. Hitachi H-1 Finger-Vein Scanner Product Page. Available online: http://www.hitachi.co.jp/products/it/veinid/global/products/embedded_devices_u.html (accessed on 20 June 2018).
9. Sierro, A.; Ferrez, P.; Roduit, P. Contact-less palm/finger vein biometrics. In Proceedings of the 2015 International Conference of the Biometrics Special Interest Group (BIOSIG), Darmstadt, Germany, 9–11 September 2015; pp. 1–12.
10. Kim, J.; Kong, H.J.; Park, S.; Noh, S.; Lee, S.R.; Kim, T.; Kim, H.C. Non-contact finger vein acquisition system using NIR laser. *Proc. SPIE* **2009**, *7249*, 72490Y-1–72490Y-8. [CrossRef]
11. Raghavendra, R.; Raja, K.B.; Surbiryala, J.; Busch, C. A low-cost multimodal biometric sensor to capture finger vein and fingerprint. In Proceedings of the 2014 IEEE International Joint Conference on Biometrics (IJCB), Clearwater, FL, USA, 29 September–2 October 2014; pp. 1–7.
12. Matsuda, Y.; Miura, N.; Nonomura, Y.; Nagasaka, A.; Miyatake, T. Walkthrough-style multi-finger vein authentication. In Proceedings of the 2017 IEEE International Conference on Consumer Electronics (ICCE), Las Vegas, NV, USA, 8–10 January 2017; pp. 438–441.
13. Hitachi Group, Corp. Making Society Safe and Convenient with High-Precision Walkthrough Finger Vein Authentication. Available online: https://www.hitachi.com/rd/portal/contents/story/fingervein/index.html (accessed on 6 October 2019).
14. Zhang, D.; Kong, W.K.; You, J.; Wong, M. Online palmprint identification. *IEEE Trans. Pattern Anal. Mach. Intell.* **2003**, *25*, 1041–1050. [CrossRef]
15. Zhang, Y.B.; Li, Q.; You, J.; Bhattacharya, P. Palm vein extraction and matching for personal authentication. In Proceedings of the International Conference on Advances in Visual Information Systems, Shanghai, China, 28–29 June 2007; pp. 154–164.
16. Badawi, A.M. Hand Vein Biometric Verification Prototype: A Testing Performance and Patterns Similarity. *IPCV* **2006**, *14*, 3–9.
17. Distler, M.; Jensen, S.; Myrtue, N.G.; Petitimbert, C.; Nasrollahi, K.; Moeslund, T.B. Low-cost hand vein pattern recognition. In Proceedings of the IEEE International Conference on Signal and Information Processing (CSIP), Shanghai, China, 28 October 2011; pp. 1–4.
18. Kabacinski, R.; Kowalski, M. Vein pattern database and benchmark results. *Electron. Lett.* **2011**, *47*, 1127–1128. [CrossRef]
19. Fujitsu Limited. Fujitsu Identity Management and PalmSecure Whitepaper. 2015. Available online: https://www.fujitsu.com/nz/Images/PalmSecure_white_paper-eu-en.pdf (accessed on 6 October 2019).
20. Fujitsu Limited. Fujitsu PalmSecure Datasheet. Available online: https://www.fujitsu.com/global/Images/PalmSecure_Datasheet.pdf (accessed on 6 October 2019).

21. Chinese Academy of Sciences' Institute of Automation (CASIA). CASIA Multispectral Palmprint V1.0. Available online: http://biometrics.idealtest.org/dbDetailForUser.do?id=6 (accessed on 6 October 2019).

22. Michael, G.K.O.; Connie, T.; Teoh, A.B.J. A contactless biometric system using palm print and palm vein features. In *Advanced Biometric Technologies*; IntechOpen: London, UK, 2011.

23. Zhang, Q.; Zhou, Y.; Wang, D.; Hu, X. Personal authentication using hand vein and knuckle shape point cloud matching. In Proceedings of the 2013 IEEE Sixth International Conference on Biometrics: Theory, Applications and Systems (BTAS), Arlington, VA, USA, 29 September–2 October 2013; pp. 1–6. [CrossRef]

24. Fletcher, R.R.; Raghavan, V.; Zha, R.; Haverkamp, M.; Hibberd, P.L. Development of mobile-based hand vein biometrics for global health patient identification. In Proceedings of the IEEE Global Humanitarian Technology Conference (GHTC 2014), San Jose, CA, USA, 10–13 October 2014; pp. 541–547.

25. Debiasi, L.; Kauba, C.; Prommegger, B.; Uhl, A. Near-Infrared Illumination Add-On for Mobile Hand-Vein Acquisition. In Proceedings of the IEEE 9th International Conference on Biometrics: Theory, Applications, and Systems (BTAS2018), Redondo Beach, CA, USA, 22–25 October 2018; pp. 1–9. [CrossRef]

26. University of Reading. PROTECT Multimodal DB Dataset. 2017. Available online: http://projectprotect.eu/dataset/ (accessed on 6 October 2019)

27. Kauba, C.; Prommegger, B.; Uhl, A. Focussing the Beam—A New Laser Illumination Based dataset Providing Insights to Finger-Vein Recognition. In Proceedings of the IEEE 9th International Conference on Biometrics: Theory, Applications, and Systems (BTAS2018), Los Angeles, CA, USA, 22–25 October 2018; pp. 1–9. [CrossRef]

28. IDS Imaging Development Systems GmbH. UI-ML3240-NIR NIR-Enhanced Industrial Camera Data Sheet. Available online: https://en.ids-imaging.com/IDS/datasheet_pdf.php?sku=AB00442 (accessed on 6 October 2019).

29. Fujifilm Corp. Fujifilm HF9HA-1B Product Page. Available online: http://www.fujifilmusa.com/products/optical_devices/machine-vision/2-3-15/hf9ha-1b/index.html (accessed on 20 June 2018).

30. MIDOPT Corp. MIDOPT LP780 NIR Pass-Through Filter Product Page. Available online: http://midopt.com/filters/lp780/ (accessed on 20 June 2018).

31. Aliexpress. TO-18 300 mW 808 nm NIR Laser Diode Product Page. Available online: https://www.aliexpress.com/item/5Pcs-lot-High-Quality-808nm-300mW-High-Power-Burning-Infrared-Laser-Diode-Lab/32272128336.html?spm=a2g0s.9042311.0.0.27424c4drx8E2d (accessed on 20 June 2018).

32. Aliexpress. Double IC Two Road ACC Circuit Laser Dode Driver Board 650nm 2.8-5v Adjustable Constant Current 0-390mA 780nm 808nm 980nm Laser Product Page. Available online: https://www.aliexpress.com/item/Double-IC-Two-Road-ACC-Circuit-laser-Dode-Driver-Board-650nm-2-8-5v-Adjustable-Constant/32818824875.html?spm=a2g0s.9042311.0.0.27424c4drx8E2d (accessed on 20 June 2018).

33. Aliexpress. 10x Focusable 1230 Metal Housing w Lens for TO-18 5.6mm Laser Diode LD Product Page. Available online: https://www.aliexpress.com/item/10x-Focusable-1230-Metal-Housing-w-Lens-for-TO-18-5-6mm-Laser-Diode-LD/32665828682.html?spm=a2g0s.9042311.0.0.27424c4drx8E2d (accessed on 20 June 2018).

34. Osram Opto Semiconductors AG. Osram SFH-4550 850 nm High Power Infrared LED Data Sheet. Available online: https://dammedia.osram.info/media/resource/hires/osram-dam-5580407/SFH%204550_EN.pdf (accessed on 20 June 2018).

35. Vishay Semiconductors. TSUS540 Series Infrared Emitting Diode, 950 nm, GaAs Data Sheet. Available online: https://www.vishay.com/docs/81056/tsus5400.pdf (accessed on 20 June 2018).

36. Arduino LLC. Arduino Nano Manual. Available online: https://www.arduino.cc/en/uploads/Main/ArduinoNanoManual23.pdf (accessed on 20 June 2018).

37. Texas Instruments Corporation. Texas Instruments TLC59401 16-Channel LED Driver with Dot Correction and Greyscale PWM Control Data Sheet. Available online: http://www.ti.com/lit/ds/sbvs137/sbvs137.pdf (accessed on 20 June 2018).

38. Microchip Corp. Microchip AVR ATmega328P 8-Bit Microcontroller Product Page. Available online: https://www.microchip.com/wwwproducts/en/ATmega328P (accessed on 20 June 2018).

39. Microchip Corp. Microchip AVR ATmega328P 8-Bit Microcontroller Full Data Sheet. Available online: http://ww1.microchip.com/downloads/en/DeviceDoc/Atmel-7810-Automotive-Microcontrollers-ATmega328P_Datasheet.pdf (accessed on 20 June 2018).

40. ON Semiconductor. BC808 PNP SMD General Purpose Transistor Data Sheet. Available online: http://www.onsemi.com/pub/Collateral/BC808-25LT1-D.PDF (accessed on 20 June 2018).

41. Alpha&Omega Semiconductor. AO3418 30V N-Channel MOSFET SMD Data Sheet. Available online: http://aosmd.com/pdfs/datasheet/AO3418.pdf (accessed on 20 June 2018).

42. alfer aluminium GmbH. Combitech Coaxis Online Product Catalog. Available online: https://products.alfer.com/out/media/97010.pdf (accessed on 6 October 2019).

43. Waveshare. Waveshare 7inch HDMI LCD (C) Wiki Page. Available online: http://www.waveshare.net/wiki/7inch_HDMI_LCD_(C) (accessed on 6 October 2019).

44. Lu, Y.; Xie, S.J.; Yoon, S.; Wang, Z.; Park, D.S. An available database for the research of finger vein recognition. In Proceedings of the 2013 6th International Congress on Image and Signal Processing (CISP), Hangzhou, China, 16–18 December 2013; Volume 1, pp. 410–415.

45. Zhou, Y.; Kumar, A. Human identification using palm-vein images. *IEEE Trans. Inf. Forensics Secur.* **2011**, *6*, 1259–1274. [CrossRef]

46. Zuiderveld, K. Contrast Limited Adaptive Histogram Equalization. In *Graphics Gems IV*; Heckbert, P.S., Ed.; Morgan Kaufmann: Burlington, MA, USA, 1994; pp. 474–485.

47. Zhao, J.; Tian, H.; Xu, W.; Li, X. A New Approach to Hand Vein Image Enhancement. In Proceedings of the Second International Conference on Intelligent Computation Technology and Automation, ICICTA'09, Zhangjiajie, China, 10–11 October 2009; Volume 1, pp. 499–501.

48. Zhang, J.; Yang, J. Finger-vein image enhancement based on combination of gray-level grouping and circular Gabor filter. In Proceedings of the International Conference on Information Engineering and Computer Science, Wuhan, China, 19–20 December 2009; pp. 1–4.

49. Miura, N.; Nagasaka, A.; Miyatake, T. Extraction of finger-vein patterns using maximum curvature points in image profiles. *IEICE Trans. Inf. Syst.* **2007**, *90*, 1185–1194. [CrossRef]

50. Choi, J.H.; Song, W.; Kim, T.; Lee, S.R.; Kim, H.C. Finger vein extraction using gradient normalization and principal curvature. *Proc. SPIE* **2009**, *7251*, 9. [CrossRef]

51. Lowe, D.G. Object recognition from local scale-invariant features. In Proceedings of the Seventh IEEE International Conference on Computer Vision (CVPR'99), Kerkyra, Greece, 20–27 September 1999; Volume 2, pp. 1150–1157.

52. Kauba, C.; Reissig, J.; Uhl, A. Pre-processing cascades and fusion in finger vein recognition. In Proceedings of the International Conference of the Biometrics Special Interest Group (BIOSIG'14), Darmstadt, Germany, 10–12 September 2014, pp. 1–6.

53. Tabassi, E.; Wilson, C.; Watson, C. Nist fingerprint image quality. NIST Res. Rep. NISTIR7151 2004, 5. Available online: https://www.nist.gov/sites/default/files/documents/2016/12/12/tabassi-image-quality.pdf (accessed on 16 November 2019)

54. Matkovic, K.; Neumann, L.; Neumann, A.; Psik, T.; Purgathofer, W. Global Contrast Factor-a New Approach to Image Contrast. *Comput. Aesthet.* **2005**, *2005*, 159–168.

55. Wang, C.; Zeng, X.; Sun, X.; Dong, W.; Zhu, Z. Quality assessment on near infrared palm vein image. In Proceedings of the 2017 32nd Youth Academic Annual Conference of Chinese Association of Automation (YAC), Hefei, China, 19–21 May 2017; pp. 1127–1130.

56. Ma, H.; Cui, F.P.; Oluwatoyin, P. A Non-Contact Finger Vein Image Quality Assessment Method. *Appl. Mech. Mater.* **2013**, *239*, 986–989. [CrossRef]

57. ISO/IEC JTC 1/SC 37 . Information Technology – Biometrics – Multimodal and Other Multibiometric Fusion. ISO/IEC TR 24722:2015, 2015. Available online: https://www.iso.org/standard/64061.html (accessed on 16 November 2019).

58. Brümmer, N.; de Villiers, E. The BOSARIS toolkit. *arXiv* **2013**, arXiv:1304.2865.

59. Maio, D.; Maltoni, D.; Cappelli, R.; Wayman, J.L.; Jain, A.K. *FVC2004: Third Fingerprint Verification Competition*; Springer: Berlin/Heidelberg, Germany, 2004; Volume 3072, pp. 1–7.

60. Yin, Y.; Liu, L.; Sun, X. SDUMLA-HMT: A multimodal biometric database. In *Biometric Recognition*; Springer: Berlin/Heidelberg, Germany, 2011; pp. 260–268.

61. Ton, B.; Veldhuis, R. A high quality finger vascular pattern dataset collected using a custom designed capturing device. In Proceedings of the International Conference on Biometrics, ICB 2013, Madrid, Spain, 4–7 June 2013, pp. 1–5.

62. Asaari, M.S.M.; Suandi, S.A.; Rosdi, B.A. Fusion of band limited phase only correlation and width centroid contour distance for finger based biometrics. *Expert Syst. Appl.* **2014**, *41*, 3367–3382. [CrossRef]
63. Bogazici University. Bosphorus Hand Database. Available online: http://bosphorus.ee.boun.edu.tr/hand/Home.aspx (accessed on 6 October 2019).
64. Faundez-Zanuy, M.; Mekyska, J.; Font-Aragonès, X. A new hand image database simultaneously acquired in visible, near-infrared and thermal spectrums. *Cogn. Comput.* **2014**, *6*, 230–240. [CrossRef]
65. Tome, P.; Marcel, S. On the Vulnerability of Palm Vein Recognition to Spoofing Attacks. In Proceedings of the 8th IAPR International Conference on Biometrics (ICB), New Delhi, India, 29 March–1 April 2015; pp. 319–325.
66. Kauba, C.; Uhl, A. Shedding Light on the Veins—Reflected Light or Transillumination in Hand-Vein Recognition. In Proceedings of the 11th IAPR/IEEE International Conference on Biometrics (ICB'18), Gold Coast, Australia, 20–23 February 2018; pp.1– 8. [CrossRef]
67. Li, G.; Yang, B.; Busch, C. Autocorrelation and dct based quality metrics for fingerprint samples generated by smartphones. In Proceedings of the 2013 18th International Conference on Digital Signal Processing (DSP), Fira, Greece, 1–3 July 2013; pp. 1–5.
68. Yang, B.; Li, G.; Busch, C. Qualifying fingerprint samples captured by smartphone cameras. In Proceedings of the 2013 IEEE International Conference on Image Processing, Melbourne, Australia, 15–18 September 2013; pp. 4161–4165.
69. Hämmerle-Uhl, J.; Pober, M.; Uhl, A. Systematic evaluation methodology for fingerprint-image quality assessment techniques. In Proceedings of the 2014 37th International Convention on Information and Communication Technology, Electronics and Microelectronics (MIPRO), Opatija, Croatia, 26–30 May 2014; pp. 1315–1319.
70. Hämmerle-Uhl, J.; Pober, M.; Uhl, A. General purpose bivariate quality-metrics for fingerprint-image assessment revisited. In Proceedings of the 2014 IEEE International Conference on Image Processing (ICIP), Paris, France, 27–30 October 2014; pp. 4957–4961.
71. Prommegger, B.; Kauba, C.; Uhl, A. On the Extent of Longitudinal Finger Rotation in Publicly Available Finger Vein datasets. In Proceedings of the 12th IAPR/IEEE International Conference on Biometrics (ICB'19), Crete, Greece, 4–7 June 2019; pp. 1–8.
72. Huang, B.; Dai, Y.; Li, R.; Tang, D.; Li, W. Finger-vein authentication based on wide line detector and pattern normalization. In Proceedings of the 2010 20th International Conference on Pattern Recognition (ICPR), Istanbul, Turkey, 23–26 August 2010; pp. 1269–1272.
73. Yang, W.; Yu, X.; Liao, Q. Personal authentication using finger vein pattern and finger-dorsa texture fusion. In Proceedings of the 17th ACM international conference on Multimedia, Beijing, China, 19–24 October 2009; pp. 905–908.
74. Ong, T.S.; Teng, J.H.; Muthu, K.S.; Teoh, A.B.J. Multi-instance finger vein recognition using minutiae matching. In Proceedings of the 2013 6th International Congress on Image and Signal Processing (CISP), Hangzhou, China, 16–18 December 2013; Volume 3, pp. 1730–1735. [CrossRef]
75. Zhang, C.; Li, X.; Liu, Z.; Zhao, Q.; Xu, H.; Su, F. The CFVD reflection-type finger-vein image database with evaluation baseline. In *Biometric Recognition*; Springer: Berlin/Heidelberg, Germany, 2013; pp. 282–287.
76. Tome, P.; Vanoni, M.; Marcel, S. On the Vulnerability of Finger Vein Recognition to Spoofing. In Proceedings of the IEEE International Conference of the Biometrics Special Interest Group (BIOSIG), Darmstadt, Germany, 10–12 September 2014, pp. 1–10.
77. Vanoni, M.; Tome, P.; El Shafey, L.; Marcel, S. Cross-database evaluation using an open finger vein sensor. In Proceedings of the 2014 IEEE Workshop on Biometric Measurements and Systems for Security and Medical Applications (BIOMS) Proceedings, Rome, Italy, 17 October 2014; pp. 30–35.
78. Shahin, M.; Badawi, A.; Kamel, M. Biometric authentication using fast correlation of near infrared hand vein patterns. *Int. J. Biol. Med Sci.* **2007**, *2*, 141–148.
79. Yuksel, A.; Akarun, L.; Sankur, B. Hand vein biometry based on geometry and appearance methods. *IET Comput. Vis.* **2011**, *5*, 398–406. [CrossRef]

80. Mirmohamadsadeghi, L.; Drygajlo, A. Palm vein recognition with local binary patterns and local derivative patterns. In Proceedings of the 2011 International Joint Conference on Biometrics (IJCB), Washington, DC, USA, 11–13 October 2011; pp. 1–6.
81. Mirmohamadsadeghi, L.; Drygajlo, A. Palm vein recognition with local texture patterns. *IET Biom.* **2014**, *3*, 198–206. [CrossRef]

Article

Recognition of Dorsal Hand Vein Based Bit Planes and Block Mutual Information

Yiding Wang [1], Heng Cao [1,*], Xiaochen Jiang [1] and Yuanyan Tang [2]

[1] Department of Communication Engineering, School of Information, North China University of Technology, No. 5, Jinyuanzhuang Road, Shijingshan District, Beijing 100043, China
[2] Department of Computer and Information Science, School of Technology, University of Macau, Macau 999078, China
* Correspondence: kevin_andrew@163.com; Tel.: +86-136-9127-0341

Received: 20 May 2019; Accepted: 22 August 2019; Published: 28 August 2019

Abstract: The dorsal hand vein images captured by cross-device may have great differences in brightness, displacement, rotation angle and size. These deviations must influence greatly the results of dorsal hand vein recognition. To solve these problems, the method of dorsal hand vein recognition was put forward based on bit plane and block mutual information in this paper. Firstly, the input gray image of dorsal hand vein was converted to eight-bit planes to overcome the interference of brightness inside the higher bit planes and the interference of noise inside the lower bit planes. Secondly, the texture of each bit plane of dorsal hand vein was described by a block method and the mutual information between blocks was calculated as texture features by three kinds of modes to solve the problem of rotation and size. Finally, the experiments cross-device were carried out. One device was used to be registered, the other was used to recognize. Compared with the SIFT (Scale-invariant feature transform, SIFT) algorithm, the new algorithm can increase the recognition rate of dorsal hand vein from 86.60% to 93.33%.

Keywords: bit planes; block; mutual information; cross-device; dorsal hand vein recognition

1. Introduction

Biometric is a technique that uses inherent and unique biometric feature to recognize people identification [1]. Biometric authentication systems are well established today as they exhibit many advantages over traditional password and token-based ones [2]. Dorsal hand vein recognition mainly uses the subcutaneous vein tissue structure of the dorsal hand for personal identification, the vein structure of the back of hands is highlighted because of the different infrared light absorption rates [3]. Anatomical works [4] have proved that the structure of dorsal hand vein is unique in the process of growth and development. Therefore, research on the recognition of the dorsal hand vein is becoming more and more important in terms of value.

In recent years, more and more researchers have begun to pay attention to the algorithm of hand vein recognition. These algorithms for feature extraction are roughly divided into global and local texture features. The global texture feature, such as PCA(Principal components analysis, PCA) [5], it utilizes the geometric texture of the hand vein and the texture mapping of the ROI (region of interest), but to a certain extent, it ignores the local information which is separable. Its performance is easily affected by the change of viewing angle, illumination intensity, distortion and occlusion. Local texture feature, such as LBP(Local binary pattern, LBP) [6] and SIFT [7], pay attention to the relationship between key pixels and surrounding pixels, so the matching with local key features is more robust to the above-mentioned interference factors, Because the texture details of the hand vein is rather few, combining both a global and a local method was proposed and the performance has been improved. Zhang et al. proposed a Gaussian distribution based random key-point generation (GDRKG) [8] which can obtain a reasonable number of key points with good coverage, so it could improve recognition performance. Wang et al. proposed cross-device hand vein recognition based on improved SIFT [9] which is based on the traditional SIFT, but optimized for the scale factor σ, using an extreme searching neighborhood structure and matching threshold R. It not only has had a significant improvement in the recognition rate in single-device experiments, but also a higher recognition rate than the traditional SIFT in cross-device experiments. Li et al. proposed hand dorsal vein recognition by matching using a width skeleton model, which uses the width skeleton model (WSM) [10] containing width and structural information. It makes full use of the global shape information, making the ability to characterize vein features stronger.

Although the above methods have achieved a high recognition rate, research on the dorsal hand vein are mostly based on a database acquired by a single device. Considering the diversity of imaging acquisition devices, as well as changes in environment and growth, hand vein recognition is very limited. At present, most published research papers are carried out under the strong constraints of controlled environment and user cooperation to achieve higher recognition accuracy. How to improve the cross-device hand vein recognition rate in the condition of seldom cooperation for users is the main problem solved in this paper. We propose a feature extraction method based on bit planes and block mutual information. The optimal bit plane was selected to overcome the influence of brightness and noise. The texture features of dorsal hand vein were described by a block method, and the optimal number of blocks was determined by the average entropy matrix of different blocks. Then the mutual information among different blocks was calculated as texture features by three kinds of mutual information calculation modes. Finally, the Euclidean distance classifier was used for classification recognition. The recognition rate of dorsal hand vein images under a cross-device increased to 93.33%.

2. Bit Plane Generation

2.1. Image Acquisition of Dorsal Hand Vein

The dorsal hand vein exists under subcutaneous tissues. Using a general camera is difficult to capture clear images of dorsal hand vein in the condition of visible light source, so an infrared light source was adopted in our devices [11]. The appearance of acquisition equipment is shown in Figure 1a,b is the internal structure. Figure 2 is the captured images, Figure 2a,b are both dorsal hand vein images for the same person but captured by two different devices.

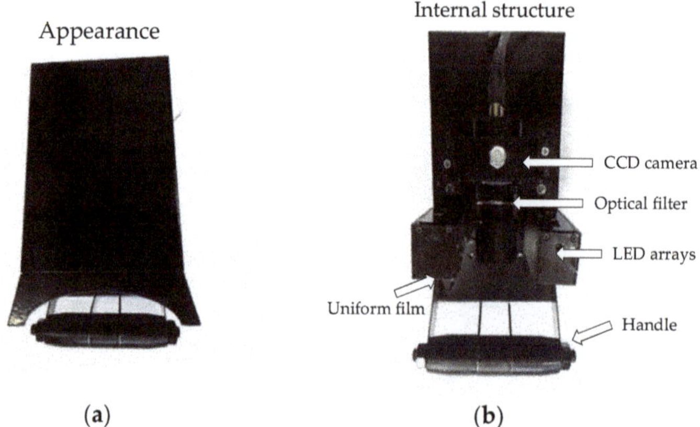

Figure 1. Image acquisition device. (**a**) Appearance; (**b**) internal structure.

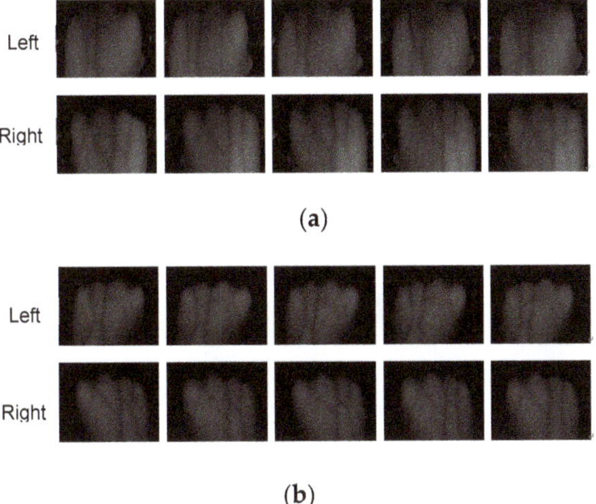

Figure 2. Dorsal hand vein images captured by two different devices. (**a**) Captured by No. 1 device; (**b**) captured by No. 2 device.

As we can see in Figure 2, the vein texture is clear and the details are rich. In order to research the image recognition of the dorsal hand vein under weak constraints, we need to create a database with diversity. Dorsal hand vein images of the same person under different devices are quite different, including changes in rotation angle, size, brightness and noise. This is mainly due to differences in parameters such as contrast, brightness, focal length and lens optical performance of different collection devices, as well as in the state of the collector's hand. Therefore, it is more difficult for cross-device dorsal hand vein recognition.

2.2. Image Preprocessing

As mentioned above, dorsal hand vein images captured by the same person on different devices also have great differences in brightness, noise, size and rotation angle [12]. These factors will have a great impact on the recognition results, and simple scale normalization is not conducive to extract texture features of samples. In this paper, a centroid adaptive method was used to determine the ROI region of dorsal hand vein images. Find the centroid $C(x_0, y_0)$ of the ROI hand vein image through the length and width of the vein area, and the centroid was taken as the center of the maximum inscribed circle of dorsal hand vein area which is shown in Figure 3a. The diameter (d) of the maximum inscribed circle, was taken as the standard of size normalization. After scale normalization, the ROI region with a size of 400 × 400 is intercepted, as shown in Figure 3b.

Figure 3. Captures the region of interest (ROI) area. (**a**) Maximum inscribed circle, (**b**) ROI area.

In addition, because of the difference about lighting condition and each thickness of hand, distribution of gray level in images can't be equal. Thus, we need to normalize gray level from 0 to 255 by Formula (1).

$$N(x, y) = ((R(x, y) - min) \times 255) / (max - min) \tag{1}$$

where $R(x, y)$ represents image gray level of the ROI region, max and min represent respectively the maximum and minimum gray value of images, $N(x, y)$ represents the normalized gray level. The result after gray normalization is shown in Figure 4.

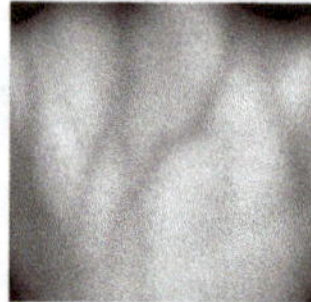

Figure 4. Gray-normalized ROI dorsal hand vein image.

In order to obtain the texture contour of dorsal hand vein, gradient based image segmentation method [13] was adopted in this paper. The segmented binary image is shown in Figure 5.

Figure 5. The segmented binary image.

However, the binary image may lose lots of gray information, therefore, multiplying inverted binary image and normalized gray image to obtain the gray image that only retains the contour of dorsal hand vein, as shown in Figure 6.

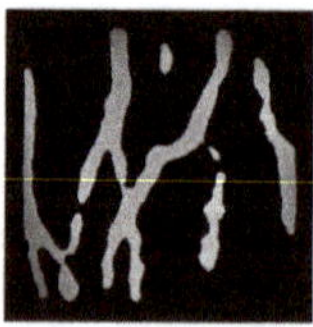

Figure 6. The gray image that only retains the contour of dorsal hand vein.

2.3. Selection of Bit Plane

In order to obtain more abundant gray information and overcome the interference of brightness and noise caused by the collection environment, we studied the bit planes generated by gray image that only retains the contour of dorsal hand vein. The concept of bit planes is now illustrated by a 256-level gray image. If per pixel value of the input gray image is within [0, 255], then each pixel can be denoted by a binary number of eight bits, i.e., b_7, b_6, b_5, b_4, b_3, b_2, b_1, b_0, as shown in Formula (2). From b_7 to b_0 are the highest to the lowest bit plane respectively as shown in Figure 7.

$$I = b_7 \times 2^7 + b_6 \times 2^6 + b_5 \times 2^5 + b_4 \times 2^4 + b_3 \times 2^3 + b_2 \times 2^2 + b_1 \times 2^1 + b_0 \times 2^0 \tag{2}$$

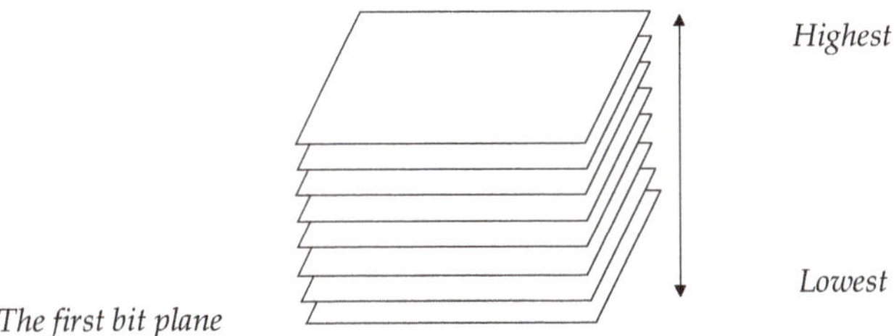

Figure 7. Bit plane stratification.

Each item in Formula (2) denotes a bit plane of a pixel, and eight bit planes are shown in Figure 8.

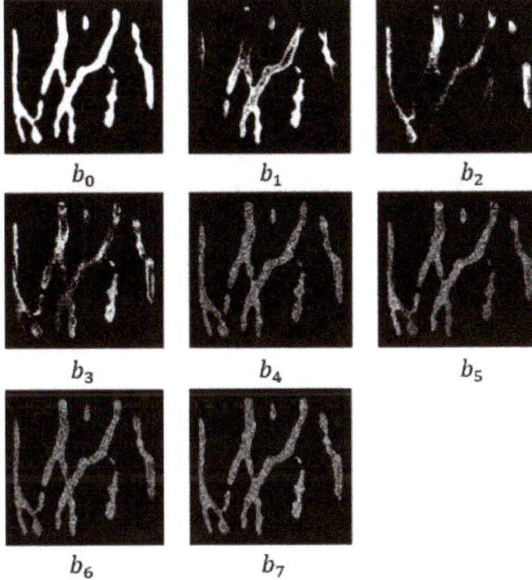

Figure 8. Eight-bit planes.

As we can see in Figure 8, the lower bit planes are close to binary images, which is easily interfered with by the noise from the collection environment and equipment, and the higher bit planes contain more gray information, which is close to the gray image that only retains the contour of the dorsal hand vein. It is susceptible to illumination and brightness during acquisition. Therefore, we chose the intermediate optimal bit plane to solve these problems effectively. In the following, the eight-bit planes are respectively divided into blocks to calculate mutual information, and the statistical recognition rate will be used to obtain the optimal bit plane to improve the accuracy and robustness of the hand vein recognition.

3. Hand Vein Recognition Based on Block Mutual Information

3.1. Mutual Information Calculation

Calculating the correlation of different bit planes and finding the best match is an important issue in this research. The correlation between different bit planes indicates the similarity of their contents, and their correlation can be characterized by mutual information [14].

For discrete random variables, let X be a random variable, $p(x)$ is the probability that this variable X takes the value x, then the entropy H(X) describing its uncertainty is expressed as:

$$H(X) = -\sum_{x \in X} p(x) \log p(x) \tag{3}$$

The introduction of mutual information is to measure the amount of information that contains another random variable in a random variable, which denotes closeness between two random variables. With two random variables X and Y, the probability distributions are $p(x)$ and $p(y)$, respectively, and the mutual information between them is expressed as:

$$I(X;Y) = -\sum_{x \in X} \sum_{y \in Y} p(x,y) \log \frac{p(x,y)}{p(x)p(y)} \tag{4}$$

Mutual information of images denotes the correlation between images [15], and it can be expressed as:

$$I(A;B) = \sum_{a \in K_a} \sum_{b \in K_b} p(a,b) \log \frac{p(a,b)}{p(a)p(b)} \tag{5}$$

In Equation (5), $p(a)$ and $p(b)$ are respectively the probability distributions of image A and image B, $p(a, b)$ is the joint distribution probability, K_a and K_b are gray levels. The larger $I(A;B)$, the higher correlation between two images.

3.2. Optimal Number of Blocks

As mentioned above, the mutual information can indicate the correlation between images, however calculating that between each bit plane not only is a large amount of calculation, but also the information entropy obtained cannot distinguish different categories well. Therefore, we used a block method to describe the texture of dorsal hand vein, which not only solves the above problems, but in addition; the texture relationship between blocks can eliminate the effects of image rotation and scale changes. The image is divided into m × n blocks as shown in Figure 9.

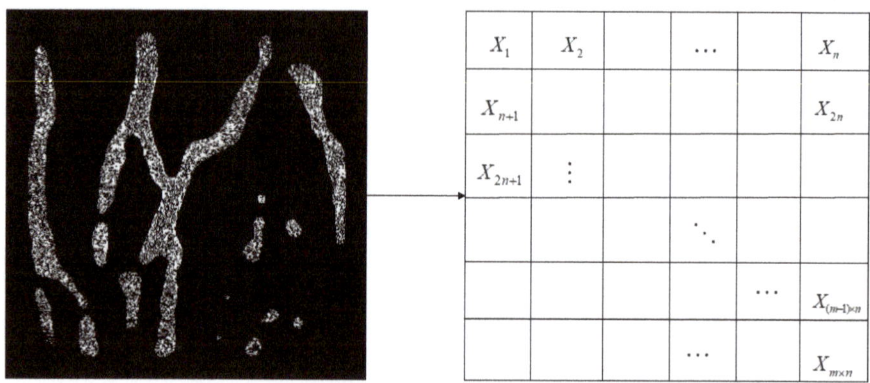

Figure 9. Divide image into blocks.

The number of blocks will affect the extraction of texture features, the appropriate number of blocks can not only minimize dimension of the image, but also largely retain the texture information of the dorsal hand vein, so it is necessary to find the most appropriate number of blocks. According to the principle of pattern recognition, the optimal number of blocks should meet the requirement that the variance of the average entropy matrix based the average threshold as large as possible [16], so as to maximize the difference in average entropy between different blocks. In other words, the difference in texture information is obviously reflected and has good separability [17].

The image is divided from 1 × 1 to 25 × 25 blocks, and the grayscale symbiosis matrix of each sub-block is calculated to obtain the average entropy matrix of each image [18]. We used the Otsu method [19] to obtain the global threshold of each average entropy matrix, and then calculated the average threshold of all images under the same number of blocks, the result is shown in Figure 10.

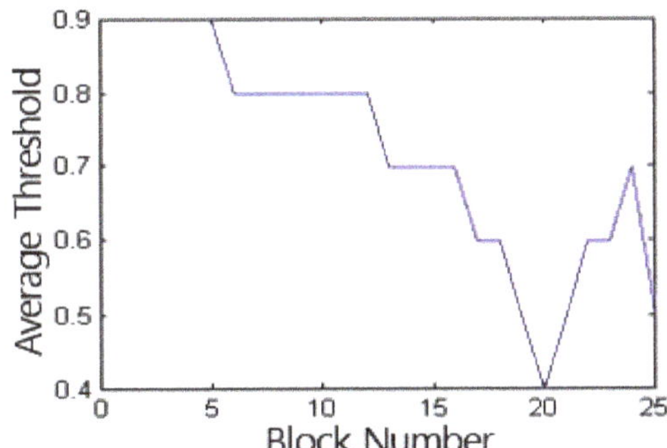

Figure 10. Average threshold distribution of average entropy matrix.

As the number of blocks increases, the average threshold gradually decreases. This is because the sub-image becomes smaller as the number of blocks increases, so the energy of the grayscale symbiosis matrix is reduced. Calculate the corresponding variance according to the average threshold distribution of the average entropy matrix, the formula is:

$$V = \sum \left(f_{ij} - t\right)^2 \tag{6}$$

In Formula (6), t is the average threshold corresponding to average entropy matrix, f_{ij} is the global threshold corresponding to average entropy matrix of each dorsal hand vein image, i is the category to which image belongs in this experiment, and j is the order in which image are arranged in this category, the result is shown in Figure 11.

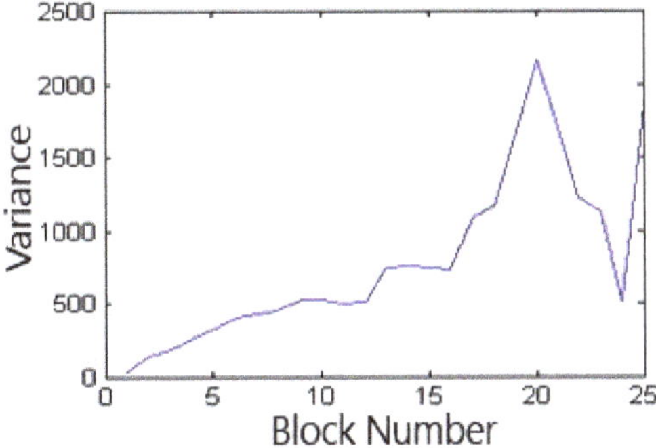

Figure 11. The variance of the average entropy matrix.

It can be seen from Figure 11, that when the number of blocks is 20 × 20, the variance is the largest, that is, its threshold value is the best for the classification of the average entropy matrix.

3.3. Block-based Mutual Information Feature Vector Calculation Mode

In the previous section, the number of blocks with the best classification effect has been obtained. Next, it is the main problem of this paper to quantify the texture relationship between blocks by means of mutual information. For the calculation of mutual information, we proposed three calculation modes, namely horizontal traversal, vertical traversal and eight-neighborhood traversal. Calculating mutual information of adjacent blocks by the horizontal traversal as shown in Figure 12.

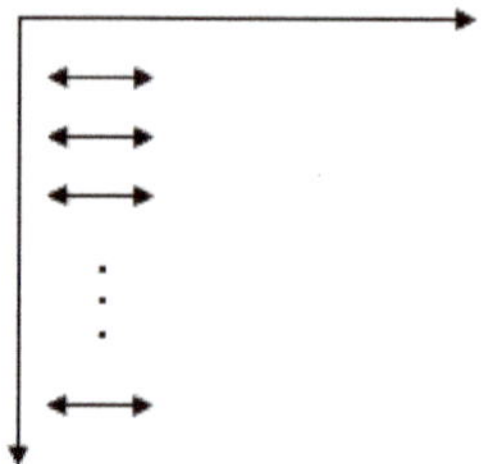

Figure 12. The horizontal traversal.

According to Formula (5), the mutual information between adjacent blocks x_1 and x_2, x_2 and $x_3, \cdots,$ $x_{m \times n-1}$ and $x_{m \times n}$ is calculated by horizontal traversing from the first row to the last. They are $I_1{}^r$, $I_2{}^r, \cdots,$ $I_{m \times (n-1)}{}^r$. In order to facilitate the next classification and recognition research, the $m \times (n-1)$ mutual information obtained above is stacked, and then a feature vector R_r is obtained, as in Formula (7).

$$
R_r = \begin{pmatrix} I_1^r \\ I_2^r \\ \vdots \\ I_{m \times (n-1)}^r \end{pmatrix}
\tag{7}
$$

The vertical traversal mode as shown in Figure 13.

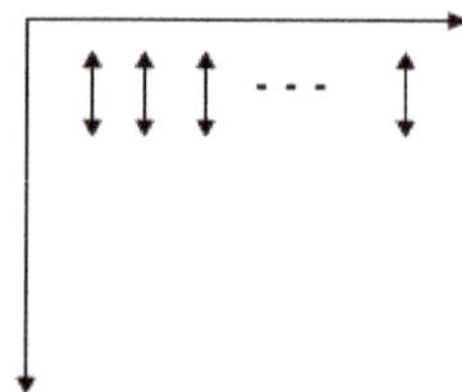

Figure 13. The vertical traversal.

Similarly, the mutual information between the adjacent blocks x_1 and x_{n+1}, x_{n+1} and $x_{2n+1}, \cdots$ $, x_{m \times (n-1)}$ and $x_{m \times n}$ is calculated by vertical traversal from the first column to the last. The $n \times (m-1)$ mutual information obtained above is stacked, and then a feature vector R_c is obtained, as in Formula (8).

$$
R_c = \begin{pmatrix} I_1^c \\ I_2^c \\ \vdots \\ I_{n \times (m-1)}^c \end{pmatrix}
\tag{8}
$$

The eight-neighborhood traversal mode as shown in Figure 14.

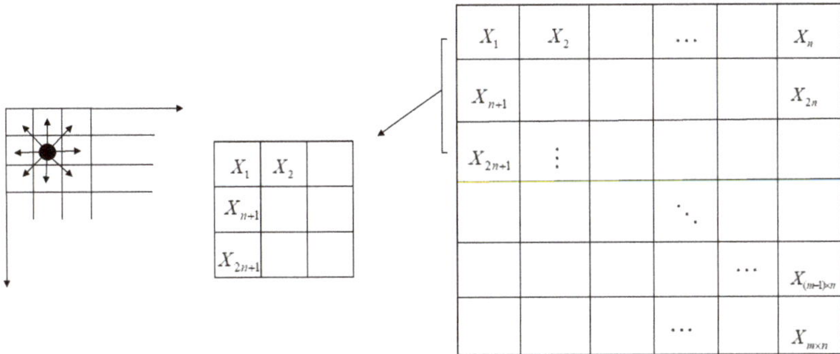

Figure 14. The eight-neighborhood traversal.

First, calculate the mutual information of block x_{n+2} and its surrounding eight neighbors x_1, x_2, x_3, x_{n+1}, x_{n+3}, x_{2n+1}, x_{2n+2}, x_{2n+3}, then calculate the eight neighborhood mutual information of x_{n+3}, x_{n+4}, $\cdots$, $x_{(m-1)\times n-1}$, which are respectively I_1^e, I_2^e, $\cdots$, $I_{(m-2)\times(n-2)\times 8}^e$. Performing a stacking operation on $(m-2)\times(n-2)\times 8$ mutual information to obtain a feature vector R_e, as in Equation (9).

$$R_e = \begin{pmatrix} I_1^e \\ I_2^e \\ \vdots \\ I_{(m-2)\times(n-2)\times 8}^e \end{pmatrix} \tag{9}$$

In this paper, the training set and test set are processed separately in the above three calculation modes, and then the optimal mutual information calculation mode is determined by the experimental results.

3.4. Classification Identification

The above mentioned that the bit plane is processed by 20×20 blocks, and then the mutual information between the blocks is calculated in three modes of horizontal, vertical and eight-neighborhood. Furthermore, the feature vectors R_r, R_c and R_e are obtained as the feature extraction of dorsal hand vein. The training samples R_t' from the device 1, and the test samples R_t from the device 2. The feature vector of the training samples in the three calculation modes is defined as $R_t' = \left[I_{t1}'\ I_{t2}'\ ...\ I_{tk}'\right]$, $t = 1, 2, ..., n$, and the feature vector of test samples is defined as $R_t = \left[I_{t1}\ I_{t2}\ ...\ I_{tk}\right]$, $t = 1, 2, ..., n$, where t represents the category of samples, and k is the number of mutual information.

We have carried out experiments in the cross-device and single-device scenario. In the single-device scenario, there are 10 images of each hand, we only took five samples to match. In the cross-device scenario, we also took five samples in the test sample of device two to match. An n-dimension distance vector matrix $dis_t(t = 1, 2, 3, 4 \cdots, n)$ is obtained by calculating the Euclidean distance [20] between the test sample (from device 2) feature R_t and training sample (from device 1) features R_t', as in Formula (10).

$$dis_t = \left\|R_t - R_t'\right\| = \sqrt{\sum_{i=1}^{k}\left(I_{ti} - I_{ti}'\right)^2}, t = 1, 2, \cdots, n \tag{10}$$

Get the minimum value d of the feature distance.

$$d = \min_{t=1}^{n}\{dis_t\} \tag{11}$$

Then the test sample R_t is identified as the training sample R_t' through the minimum value d.

4. Experiment Analysis

In order to fully prove the result of cross-device hand vein image recognition based on the bit-plane mutual information, this experiment used two different parameters of the device, labeled as first device and second device to collect and classify 50 peoples' hand vein images. Their right and left hand were collected by 10 images, respectively; a total of 2000 dorsal hand vein images with a size of 400×400 were taken. Due to the disparity between the vein networks, right and left hands are considered as different subjects, which makes the number of classes double. In addition, there are differences in parameters such as contrast, brightness, focal length, and lens optical performance of two different devices. The data were collected twice by different devices with a time span of 12 months.

The experiment uses one device for registration and the other for recognition. Data acquisition uses two generations of different acquisition systems. The two devices are two generations of different acquisition systems, their illumination module adopts reflectance illumination scheme of infrared LED array with different wavelength and bandwidth. Device1 uses the 700 nm ~ 1000 nm near-infrared diode source (wideband source) as the active incident source. Device2 uses the near-infrared diode light source with a central band of 850 nm and a radius bandwidth of 50 nm (narrow-band light source) and increases the number of LED array. In the image acquisition module, device1 uses a common camera, the main parameters are as follows: Resolution: 420 lines, output pixels: 640×480, signal to noise ratio: 40 dB, device2 uses an industrial grade camera, the main parameters: Resolution: 570 lines, output pixels: 768×494, signal to noise ratio: 46 dB. In the interface module, the two devices also use different acquisition cards.

In order to ensure the distribution of cross-device dorsal hand vein images, it used automatic collection and didn't limit the volunteers' posture. In addition, the parameter difference between different devices makes it more difficult for recognition based heterogeneous images. We used different types of images (gray-normalized image, binary image, the gray image that only retains the contour of dorsal hand vein and the bit plane image) to experiment separately. This experiment chose optimal number of blocks, bit plane and mutual information calculation mode to compare the result of our algorithm with other algorithms for cross-device images, and then the robustness of the algorithm was verified by the recognition rate.

Due to changes of the collection environment, the images collected by two devices are significantly different, which are mainly reflect in the changes of brightness, displacement and rotation.

The images have a distinct brightness difference in the brightest and darkest areas as shown in Figure 15. it affects the recognition rate to a large extent.

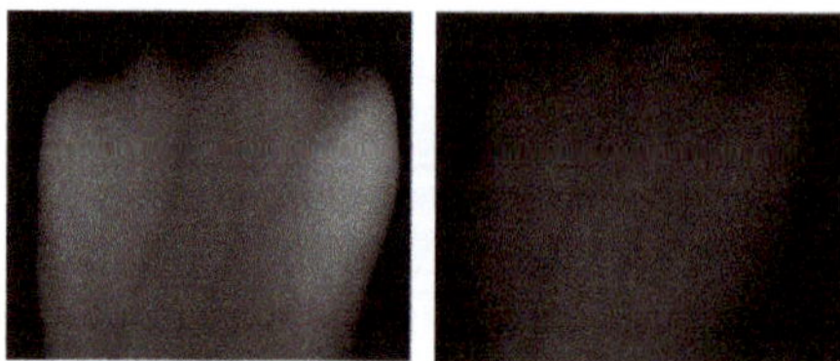

Figure 15. Difference in brightness.

The difference in the posture of the person and the handle width of different devices, the back of hand produces a certain displacement, as shown in Figure 16. When the displacement is large, some information on the back of hand will be covered, therefore, it affects the recognition rate to a certain extent.

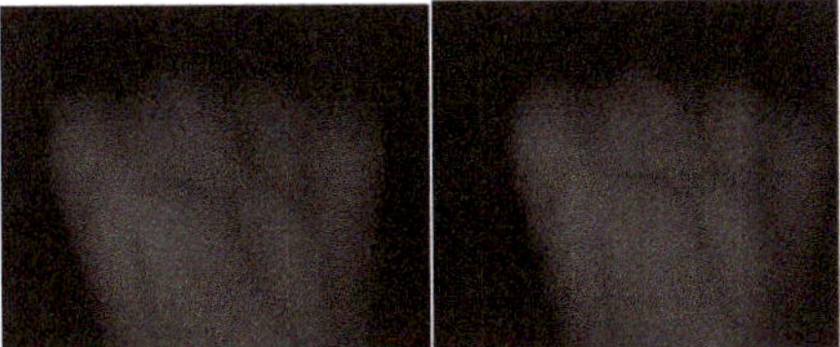

Figure 16. Difference in displacement.

Since the different angles of collector's hands, dorsal hand vein images are deformed, as shown in Figure 17. It can also affect the recognition rate of dorsal hand vein.

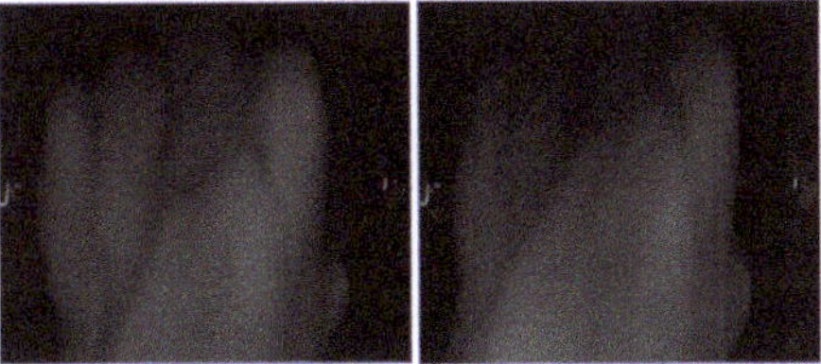

Figure 17. Difference in rotation angle.

These differences can lead to a significant increase in the difficulty and complexity of recognition of cross-device dorsal hand vein images. Experimental comparison is conducted below to verify that method of this paper has a better effect on overcoming the effects of brightness, displacement and rotation.

First, the gray-normalized image (Figure 4), the binary image (Figure 5) and the gray image that only retains the contour of dorsal hand vein (Figure 6) were divided into 20 × 20 blocks, respectively. Then, the mutual information feature vector between the blocks was obtained by using the calculation modes of horizontal, vertical and eight-neighborhood respectively. Finally, the classification result was output by the Euclidean distance classifier. The recognition rates of three different types of dorsal hand vein images in three modes are shown in Table 1.

Table 1. Recognition rates of three different types of dorsal hand vein images in three modes.

Modes	Gray-Normalized	Binary	The Gray Image that Only Retains the Contour
Horizontal	48.30%	43.30%	46.33%
Vertical	86.60%	83.30%	83.30%
Eight-neighborhood	88.00%	86.70%	89.67%

Through experiments, it can be found that the recognition rate of the gray-normalized image is less than 50%, and the binary image reaches 86.60%, while the gray image that only retains the contour of dorsal hand vein reaches 89.67%. The gray-normalized image has the effect of the background such as skin, and the binary image completely loses the grayscale information, so the recognition rate is not as good as the gray image that only retains the contour of dorsal hand vein. In the three modes, the recognition rate of the eight-neighborhood mode is higher than the other two modes, which indicates that it is more accurate to calculate the mutual information of adjacent blocks by eight-neighborhood traversal as the texture feature of dorsal hand vein.

In order to make full use of the gray information of the dorsal hand vein and overcome the effects of illumination, brightness, rotation and scale changes in the acquisition environment, the eight bit planes generated by the gray image that only retains the contour of dorsal hand vein was tested separately, and the statistical recognition rate is shown in Figure 18.

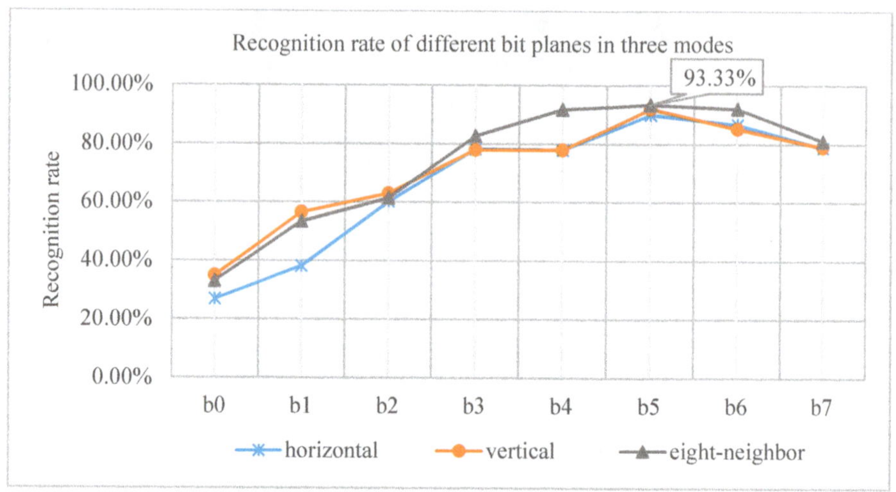

Figure 18. Recognition rate of different bit planes in three modes.

It can be seen that when the number of blocks is 20 × 20 and the mutual information calculation mode is eight-neighborhood traversal, the recognition rate of the sixth bit plane (b5) reaches the best in this paper, which is 93.33%. The sixth bit plane not only contains the original contour of dorsal hand vein, but also overcomes the influence of brightness and noise to a certain extent, and better reflects the texture features. At the same time, the experiment is compared with other methods on dorsal hand vein recognition. In the long-term research of the dorsal hand vein recognition, the Intelligent Recognition and Image Processing Laboratory of North China University of Technology (NCUT) reproduced some mainstream algorithms on the NCUT hand vein dataset. The results of the comparative experiment are shown in Table 2.

Table 2. Comparison of recognition rates about different algorithms under cross-device.

Algorithms	Recognition Rate (Single-Device)	Recognition Rate (Cross-Device)
Ours	>98%	93.33%
LBP	93.50%	73.42%
PCA	90.40%	54.83%
SIFT	97.50%	86.60%
Gaussian distribution based random key-point generation (GDRKG)	92.30%	71.38%
Improved SIFT	98.63%	90.8%

The LBP algorithm is used to research the local grayscale texture features, and it requires a high degree of registration about the position of dorsal hand vein, so the recognition rate is not high. The PCA algorithm treats the sample as a whole, and therefore ignores the local attribute, but the neglected part is likely to contain important separability information, so the effect of cross-device dorsal hand vein recognition is very poor. Although the SIFT algorithm has the characteristics of scale transformation, rotation and illumination invariance, there are fewer feature points taken by different devices, therefore, the recognition rate is also not very high. The position of the feature points generated by the Gaussian random distribution based on the GDRKG random feature point algorithm is not determined, so the probability of matching errors is greatly increased, and the recognition rate is not ideal. The improved SIFT algorithm has achieved a good recognition rate in cross-device experiments, but it relies too much on parameter settings and template selection, and the calculation speed is very slow. Our method is to calculate the mutual information between adjacent blocks of the bit planes to quantify the texture features of dorsal hand vein, and the Euclidean distance is used for classification. The high recognition rate achieved by the experiment fully demonstrates the effectiveness and feasibility of the proposed method.

5. Conclusions

Aiming at the problem that the recognition rate of the dorsal hand vein image collected by different devices is not high, this paper proposes a research method-based bit plane and block mutual information. The optimal block is determined by the variance corresponding to the average entropy matrix, the gray-normalized image, the binary image, the gray image that only retains the contour of dorsal hand vein, and the bit planes are tested respectively under various mutual information calculation modes. By comparing other algorithms used on cross-device hand vein recognition, the method proposed in this paper has been significantly improved. However, at present, only the one-bit plane is processed separately, therefore, the fusion and optimization of multiple bit planes will be the focus of further research in the later stage.

Author Contributions: Y.W. provided the ideas and methods of the whole article. H.C. designed the experiment and conducted experimental analysis on the proposed algorithm. Partial preparatory work was done with the help of X.J., Y.T. was finally responsible for the review of the thesis.

Funding: National Natural Science Fund Committee of China (NSFC No. 61673021).

Acknowledgments: This work was supported by the National Natural Science Fund Committee of China (NSFC No. 61673021).

Conflicts of Interest: The authors declare no conflicts of interest.

References

1. Wang, J.; Wang, G. Quality-specific hand vein recognition system. *IEEE Trans. Inf. Forensics Secur.* **2017**, *12*, 2599–2610. [CrossRef]
2. Kauba, C.; Uhl, A. Shedding Light on the Veins-Reflected Light or Transillumination in Hand-Vein Recognition. In Proceedings of the IEEE 2018 International Conference on Biometrics (ICB), Gold Coast, QLD, Australia, 20–23 February 2018.
3. Wang, L.; Leedham, G.; Cho, D.S.-Y. Minutiae feature analysis for infrared hand vein pattern biometrics. *Pattern Recognit.* **2008**, *41*, 920–929. [CrossRef]
4. S. Standring. Gray's anatomy, 39th ed. Edinburgh: Elsevier Churchill Livingston. 2006. (Monograph). Available online: http://xueshu.baidu.com/usercenter/paper/show?paperid=283b26b16aa526eeb86bfd2a6bdcf553&site=xueshu_se (accessed on 20 May 2019).
5. Khan, M.H.-M.; Subramanian, R.; Khan, N.A.M. Representation of hand dorsal vein features using a low dimensional representation integrating cholesky decomposition. In Proceedings of the International Congress on Image and Signal Processing, Tianjin, China, 17–19 October 2009.
6. Wang, Y.; Li, K.; Cui, J.; Shark, L.K.; Varley, M. Study of hand-dorsa vein recognition. In Proceedings of the International Conference on Intelligent Computing, Changsha, China, 18–21 August 2010; pp. 490–498.
7. Huang, D.; Tang, Y.; Wang, Y.; Chen, L.; Wang, Y. Hand-Dorsa Vein Recognition by Matching Local Features of Multisource Keypoints. *IEEE Trans. Cybern.* **2015**, *45*, 1. [CrossRef] [PubMed]
8. Zhang, R.; Huang, D.; Wang, Y.; Wang, Y. Improving Feature based Dorsal Hand Vein Recognition through Random Keypoint Generation and Fine-Grained Matching. In Proceedings of the International Conference on Biometrics (ICB), Phuket, Thailand, 19–22 May 2015.
9. Wang, Y.; Zheng, X. Cross-device hand vein recognition based on improved SIFT. *Int. J. Wavelets Multiresolution Inf. Process.* **2018**, *16*, 18400. [CrossRef]
10. Li, X.; Huang, D.; Zhang, R.; Wang, Y.; Xie, X. Hand dorsal vein recognition by matching Width Skeleton Models. In Proceedings of the 2016 IEEE International Conference on Image Processing (ICIP), Phoenix, AZ, USA, 25–28 September 2016.

11. Im, S.K.; Park, H.M.; Kim, Y.W.; Han, S.C.; Kim, S.W.; Kang, C.H.; Chung, C.K. A biometric identification system by extracting hand vein patterns. *J. Korean Phys. Soc.* **2001**, *38*, 268–272.

12. Pan, X.-p.; Wang, T.-f. Research on ROI Extraction Algorithm for Hand Dorsal Image. *Inf. Commun.* **2013**, *5*, 1–3.

13. Wang, Y.; Wang, H. Gradient based image segmentation for vein pattern. In Proceedings of the Fourth International Conference on Computer Sciences and Convergence Information Technology, Seoul, South Korea, 24–26 November 2009.

14. Zhi, L.; Zhang, S.; Zhao, D.; Zhao, H.; Lin, D. Similarity-combined image retrieval algorithm inspired by mutual information. *J. Image Graph.* **2018**, *16*, 1850–1857.

15. Guo, J.; Li, H.; Wang, C.; Li, S. An Image Registration Algorithm Based on Mutual Information. *J. Transduct. Technol.* **2013**, *26*, 958–960.

16. Wang, Y. *Identification Technique of Dorsal Vein in the Hand*; Science Press: Beijing, China, 2015.

17. Huang, D.; Zhu, X.; Wang, Y.; Zhang, D. Dorsal hand vein recognition via hierarchical combination of texture and shape clues. *Neurocomputing* **2016**, *214*, 815–828. [CrossRef]

18. Pan, Y.; Xie, B. Image Restoration Algorithm Combining Gray Level Co-occurrence Matrix and Entropy. *Microcomput. Appl.* **2012**, *31*, 44–46.

19. Otsu, N. A Threshold Selection Method from Gray-Level Histograms. *IEEE Trans. Syst. Man Cybern.* **1979**, *9*, 62–66. [CrossRef]

20. Schouten, T.E.; Broek, E.L.V.D. Fast Exact Euclidean Distance (FEED): A New Class of Adaptable Distance Transforms. *IEEE Trans. Pattern Anal. Mach. Intell.* **2014**, *36*, 2159–2172. [CrossRef] [PubMed]

Article

Wrist Vascular Biometric Recognition Using a Portable Contactless System

Raul Garcia-Martin * and Raul Sanchez-Reillo

University Group for ID Technologies (GUTI), University Carlos III of Madrid (UC3M), Av. de la Universidad 30, 28911 Leganés, Madrid, Spain; rsreillo@ing.uc3m.es
* Correspondence: raulgarc@ing.uc3m.es; Tel.: +34-695-891-871

Received: 6 February 2020; Accepted: 2 March 2020; Published: 7 March 2020

Abstract: Human wrist vein biometric recognition is one of the least used vascular biometric modalities. Nevertheless, it has similar usability and is as safe as the two most common vascular variants in the commercial and research worlds: hand palm vein and finger vein modalities. Besides, the wrist vein variant, with wider veins, provides a clearer and better visualization and definition of the unique vein patterns. In this paper, a novel vein wrist non-contact system has been designed, implemented, and tested. For this purpose, a new contactless database has been collected with the software algorithm TGS-CVBR®. The database, called UC3M-CV1, consists of 1200 near-infrared contactless images of 100 different users, collected in two separate sessions, from the wrists of 50 subjects (25 females and 25 males). Environmental light conditions for the different subjects and sessions have been not controlled: different daytimes and different places (outdoor/indoor). The software algorithm created for the recognition task is PIS-CVBR®. The results obtained by combining these three elements, TGS-CVBR®, PIS-CVBR®, and UC3M-CV1 dataset, are compared using two other different wrist contact databases, PUT and UC3M (best value of Equal Error Rate (EER) = 0.08%), taken into account and measured the computing time, demonstrating the viability of obtaining a contactless real-time-processing wrist system.

Keywords: vascular biometric recognition; wrist vein recognition; contactless dataset; identification; pattern recognition; infrared camera; non-contact devices; Scale-Invariant Feature Transform (SIFT®); Speeded Up Robust Features (SURF®); Oriented FAST and Rotated BRIEF (ORB)

---

1. Introduction

Nowadays, biometric recognition is a trendy technology that affects everyone's safety and privacy to a greater or lesser extent. In this sense, and according to Vascular Biometric Recognition (VBR), the lack of non-contact commercial and research systems observed in the state-of-the-art has been the motivation behind this work in order to contribute to the reduced social and market integration of this technology. As is known, a contactless vascular biometric system, as facial, iris, or voice recognition systems, provides essential improvements to the user in hygiene and usability but also increases the difficulty for researchers of preprocessing, feature extraction, and feature matching in the verification/identification process. In previous research [1], a portable contactless image capture device for VBR was implemented. In the current study, this capture device is integrated (processing and storing a novel contactless algorithm and database) in order to obtain and analyze a complete contactless VBR system.

The use of the wrist area or wrist vein modality avoids the palm vein modality pattern (Fujitsu©) [2] and the finger vein modality pattern (Hitachi©) [3]. In addition, the use of this area could be considered, for future researches, in combination with other biometric research systems or techniques like Electrocardiogram (ECG) [4] or even biomedicine solutions like [5].

1.1. Related Work

It is important to note that, as far as it is known, there are no well-integrated and well-known commercial systems on the market based on the wrist vein modality. However, there are several studies in the research stage, as is exposed in the state-of-the-art of wrist Vascular Biometric Recognition (VBR) summarized in Table 1. It is divided into three units: dataset, capture device, and software algorithms. As it can be extracted from Table 1 and as far as it is known, there are only limited recent works, and there is only one public database for wrist vein modality: PUT [6] (50 subjects × 2 wrists × 4 samples × 3 sessions = 1200 images, 1100 genuine intraclass or mated comparisons and 108,900 impostor interclass or non-mated comparisons). This database is used in several works, e.g., [7], that also presents a complete and updated state-of-the-art of wrist VBR.

The rest of the works presented in Table 1 use two privately-distributed databases: UC3M [8] and Singapore [9]. Other less extensive works, e.g., [10], which are not presented in the table, use private databases collected with their own-designed system, as it is the case of the present study.

As far as is known, the cameras used mounted CCD sensors and LED type illumination with a wavelength of approximately 850 nm (considered the best near-infrared value for VBR).

It is essential to point out that all these databases require physical contact between the subject and the hardware part of the system, which reinforces the motivation discussed previously.

According to the recognition algorithms, all the studies follow the traditional recognition process against the trendy deep learning methods: preprocessing, feature extraction algorithm, and feature matching algorithm based on distances or machine learning techniques.

The process always begins with the preprocessing and enhancement of the near-infrared (NIR) images. The starting point is usually monochromatic images whose vein patterns are enhanced, for better definition and visualization, in the following order: contrast increase (e.g., histogram equalization), noise reduction (filters), binarization and skeletonization (e.g., Zhang and Suen [11]). Then, the task could continue (only in [9] in Table 1), with the extraction of the Region of Interest (ROI). For feature extraction, several techniques are applied: minutiae extraction, as the own algorithm discussed in [8]; feature extraction base on Local Binary Pattern (Dense Local Binary Pattern), [12]; Hessian matrix, [13]; and convolution approach, [14]. The matching algorithms are based on distance (Hausdorff distance [9,14] and own minutiae algorithm [8]) and cross-correlation comparison [13].

Traditional Machine Learning methods for matching are only employed in [12] (Support Vector Machines).

Computing time for the entire software algorithm is given in the latest works, [12,13], revealing the evolution of biometric systems nowadays. As the results of the proposed system indicate (Section 3, Section 3.2.2), computing time is a critical variable in the integration of real-time biometric systems.

The performance for all works indicated in Table 1, based on the Equal Error Rate (EER), varies between 0.14% and 2.27%. These values should be considered reduced enough, but it is important to remark, as it has been mentioned, all the devices required physical contact with the users, fixing the wrist position and easing the recognition task. The images obtained are extremely invariant in scale and orientation, which translate into really high biometric performance, as it is demonstrated in the current work, with entirely similar features extracted. In this sense, a new scale-orientation-invariant algorithm is presented in the current study.

Another important factor noticed in the capture devices, due to the contact feature, is the immunity to the environmental or external light. This light does not reach the capture device due to the closed space between the camera and the wrist. Again, as a result, the similarity between the images is improved and, of course, the recognition performance. A non-contact system, as it is demonstrated in this paper, is affected by the external light conditions despite the extra capture light illumination. These two factors are the goals to overcome in order to improve wrist VBR recognition and obtain contactless devices.

Table 1. Summary of the state-of-the-art for wrist Vascular Biometric Recognition (VBR).

		Infrared Imaging of Hand Vein Patterns for Biometric Purposes [9] (2007)	Vascular Biometrics Based on a Minutiae Extraction Approach [8] (2011)	Spectral Minutiae for Vein Pattern Recognition [14] (2011)	A New Wrist Vein Biometric System [12] (2014)	Fast Cross-Correlation Based Wrist Vein Recognition Algorithm with Rotation and Translation Compensation [13] (2018)
Title						
Dataset	Name	Singapore (NIR, Own) [9]	UC3M (Own) [8] and Singapore [5]	UC3M [8] and Singapore [9]	PUT (Public) [6]	PUT [6]
	Subject	150	121	Same as UC3M and Singapore.	50	Same as PUT.
	Wrists	2	1 (right)	Same as UC3M and Singapore.	2	Same as PUT.
	Samples	3	5	Same as UC3M and Singapore.	12 (4 per session)	Same as PUT.
	Sessions	N/A	1	Same as UC3M and Singapore.	3	Same as PUT.
	Total Images	900	605	Same as UC3M and Singapore.	1200	Same as PUT.
Capture device	Images Acquisition	CCD KP-F2A Hitachi Denshi NIR camera, Hoya RM80 optical NIR 800 nm high pass filter	CCD Imaging Source DM 21BU054 camera, EYSEO TV8570 1/3″ objective and B + W 52 092 and B + W 52 093 optical NIR high pass filters LEDs (880 nm): DOM 1410 (DCM system)	Same as UC3M [b] and Singapore [a].	USB camera	Same as PUT.
	IR Light	LEDs (850 nm)		Same as UC3M and Singapore.	LEDs (850 nm)	Same as PUT.
	Type	Reflection	Reflection	Same as UC3M and Singapore.	Reflection	Same as PUT.
	Contactless	No	No	Same as UC3M and Singapore.	No	Same as PUT.
Software algorithms	Preprocessing	Monochromatic images, Noise reduction (Median filter + 2D Gaussian low pass filter), Normalization ([15]), Binarization (own thresholding) and Skeletonization (Zhang and Suen [15]).	Monochromatic images, Contrast increase (own histogram equalization), Noise reduction (Median filter + 2D Gaussian low pass filter), Normalization ([15]), Binarization (own thresholding) and Skeletonization (Zhang and Suen [15]).	Monochromatic images, Enhanced (Adaptive non-local means [16]), Noise reduction and edge enhancing [17]), Inversion, Binarization ([18]) and Skeletonization (fast marching algorithm [19]).	Monochromatic images, Adaptive histogram equalization [20] and Discrete Meyer Wavelet [21]	Monochromatic images, Gaussian filter and k-means++ algorithm [22]
	ROI	Sobel filter	No	No	No	No
	Feature Extraction	No	Minutiae extraction (own algorithm)	Convolution approach ([23]) and Location-Based Spectral Minutiae Representation (SML, [24]).	Dense Local Binary Pattern (D-LBP, own algorithm)	Hessian matrix
	Feature Matching	Hausdorff distance	Vector minutiae comparison (own algorithm)	Hausdorff distance, Modified Hausdorff (MHD) [25,26], Similarity-based Mix-matching (SMM) [27], SML correlation (SMLC, own) and SML fast rotate (SMLFR, own).	Support Vector Machines (SVMs) [28]	Cross-correlation based comparison
	Computing time (s)	N/A	N/A	N/A	0.771 (Windows, Matlab, i5 CPU)	0.92 (Linux, Python, i7-5930K CPU)
	Performance	No	EER_{UC3M} = 2.27% $EER_{SINGA.}$ = 1.63%	$EER_{UC3M\ (SMM)}$ = 1.18% $EER_{SINGA.\ (SMM)}$ = 0.14 % (SMM)	EER_{PUT} = 0.79%	$FNMR_{PUT}$ = 3.75% for $FMR_{PUT} \approx 0.1\%$

1.2. Contributions

The main goal of the work presented in this paper is to obtain and test a complete, low-cost, real-time, contactless vascular biometric system based on wrist vein recognition. For this purpose, the capture algorithm, TGS-CVBR®, and the capture device exposed in [1] are integrated and used in the current study to collect a contactless database (UC3M-CV1). Then, a new scale-orientation-invariant software algorithm, based on Scale-Invariant Feature Transform (SIFT®), Speeded Up Robust Features (SURF®), and Oriented FAST and Rotated BRIEF (ORB), is proposed and tested on the database: Preprocessing and Identification Software for Contactless Vascular Biometric Recognition (PIS-CVBR®). The present work is summarized in the supplementary video material.

2. Materials and Methods

The experimental procedure, material, and methods are summarized in Figure 1 (expired on ISO/IEC 19795-1:2019 [29]). For the research and implementation of a complete wrist VBR contactless system, the following elements have been defined:

(1) Hardware: present in all the subsystems of Figure 1 (capture, storage, signal processing, comparison, and decision).
(2) Software algorithms: divide into two software algorithms TGS-CVBR® and PIS-CVBR®. The first one is in charge of the data capture (yellow and left side of Figure 1) and the second one takes care of the and storage, preprocessing, comparison and decision tasks (green and right side of Figure 1). Combining these two algorithms a final system is obtained.

The following units of this section detail the procedure for obtaining these two elements: hardware (capture, processing, and storage devices) and software algorithms (TGS-CVBR®: capture algorithm and PIS-CVBR®: storage, signal processing, comparison, and decision).

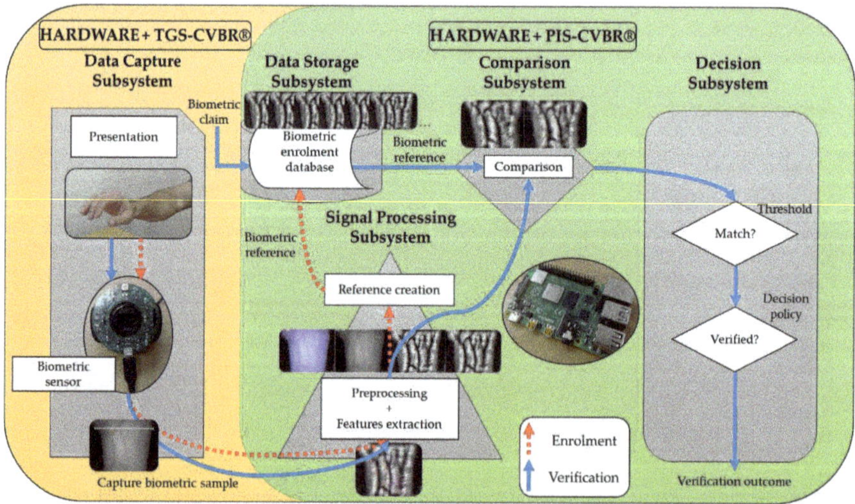

Figure 1. Components of the experimental wrist VBR system.

2.1. Hardware: Capture, Storage and Processing Devices

The hardware only implemented as a capture device in [1] was integrated and used in the current study as a capture, processing, and storage system. It consists of three parts: near-infrared camera (sensor for the capture) near-infrared Printed Circuit Board (PCB, LED illumination) and small computer (processor and storage). The camera selected and modified was the commercial USB webcam

Logitech® HD Webcam C525 [30]. For the infrared lighting, a PCB with eight infrared LEDs (OSRAM© SFH 4715 A [31], 850 nm) was designed and manufactured. In the current work, the small computer Raspberry® Pi 4 Model B [32] was used, instead of the Raspberry® Pi 3 Model B [33] of [1], for VBR processing and database storage.

2.2. Software Algorithms

As was mentioned, the software algorithm is divided into two fragments. The first one, Three-Guideline Software for Contactless Vascular Biometric Recognition (TGS-CVBR®, presented in [1] only as a capture algorithm), is used to guide users on how to position the wrist in the database collection (image capture and visualization). The second one, Preprocessing and Identification Software for Contactless Vascular Biometric Recognition (PIS-CVBR®), is the recognition algorithm.

2.2.1. TGS-CVBR®

The real-time video of the camera capture (640 × 480 resolution) was displayed on a monitor together with the three fixed guidelines, as is shown in Figure 2 (step 2, right side). This algorithm provided feedback to the user on how he/she was positioning the wrist and was used for database collection (UC3M-CV1, in this case) and user recognition (combined with PIS-CVBR®). The guidelines were useful because they fixed the user's wrist, obtaining scale-orientation-invariant images in order to improve the recognition algorithm task: the largest horizontal guideline sets the wrist orientation, and the two smaller guidelines establish the distance between the wrist and the camera.

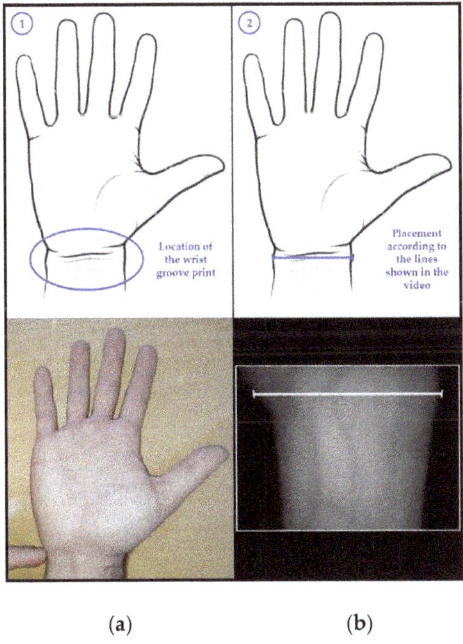

Figure 2. Three-Guideline Software for Contactless Vascular Biometric Recognition (TGS-CVBR®), wrist positioning steps (based on [1]). (**a**) Step 1: the location of the wrist groove line. (**b**) Step 2: match of the wrist groove print and the guideline.

This software was developed using Python™ 3.4.2 due to the quick and easy way to access to the USB camera and the well-integration of the language with deep learning libraries, in order to be used in future works.

The user should follow the steps shown in Figure 2:

1. Locate the wrist groove print or mark.
2. Align/match it with the guide trace displayed.

2.2.2. PIS-CVBR®

After the database collection (experimental process explained in the later section), the next step was to recognize the user: authentication/verification (1:1 user comparison) or identification (1:N user comparison). For this purpose, Preprocessing and Identification Software for Contactless Vascular Biometric Recognition (PIS-CVBR®) is proposed in this paper. It was divided into three parts or steps: preprocessing, feature extraction, and feature matching.

This software has been also developed using Python™ 3.4.2.

Preprocessing

The main goal of preprocessing was to enhance, normalize, and define the vein patterns in other to extract the features later on. This process is summarized in Figure 3. The infrared RGB images were captured in 640 × 480 resolution and ".jpg" compressed format (Figure 3a) with TGS-CVBR® and the modified camera (RGB camera). The first step, RGB to greyscale (monochromatic image with values from 0, black, to 255, white) transformation, is shown in Figure 3b.

In order to obtain a higher contrast between veins and the rest living tissue, the adaptive histogram equalization technique Contrast Limited Adaptive Histogram Equalization (CLAHE) [20] was used (Figure 3c). To reduce the high-frequency noise (salt-and-pepper and Gaussian noise in this case) generated by this algorithm and the camera sensor, several low-pass software filters were employed (Figure 3d) in the following order: Gaussian filter, Median filter, and Averaging filter. The kernel size of all of them was 11 × 11. This was the last step of the preprocessing task.

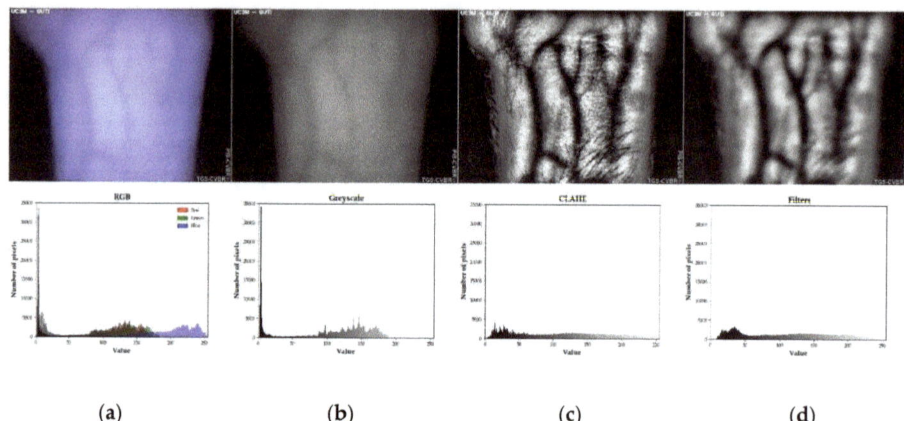

(a) (b) (c) (d)

Figure 3. Preprocessing and Identification Software for Contactless Vascular Biometric Recognition (PIS-CVBR®): Preprocessing steps for User 0. Example images (above) and their histograms (below): (**a**) RGB image. (**b**) Image after greyscale conversion. (**c**) Image after greyscale conversion and Contrast Limited Adaptive Histogram Equalization (CLAHE) algorithm. (**d**) Image after greyscale conversion, CLAHE algorithm, and filtered by Gaussian filter, Median filter, and Averaging (11 × 11 kernel).

Finally, it is important to remark that in this paper, and at this moment, the ROI extraction was not considered required for this software. However, it would probably improve system performance and is a step to contemplate in the future.

Feature Extraction

For the extraction of unique features from the wrist vein patterns, three scale-orientation-invariant algorithms for homography have been used and tested: Scale-Invariant Feature Transform (SIFT®) [34], Speeded Up Robust Features (SURF®) [35] and Oriented FAST and Rotated BRIEF (ORB) [36]. They have been selected and used, along with the TGS-CVBR® algorithm, in order to avoid the variability of the size and orientation of the wrist area, caused by the non-contact feature.

The first algorithm, SIFT®, patented in 2004 [34], was based on the Harris Corner Detector, whose variant-scale features were the motivation for improvement. SIFT® is a well-known algorithm due to its excellent performance but also to its high computing time. In order to reduce this time, SURF® was patent in 2006 [35]. Finally, the ORB algorithm, a fusion of the modified FAST and BRIEF algorithms, was published in 2011 as an open-use and a faster alternative.

In VBR, only SIFT® has been used in the wrist variant, in [37]. However, in the current study, these three algorithms have been compared, and also, with a contactless dataset. After the preprocessing, the performance of the feature extraction (100 key points) for each algorithm, with scale and orientation, is shown in Figure 4a–c, respectively.

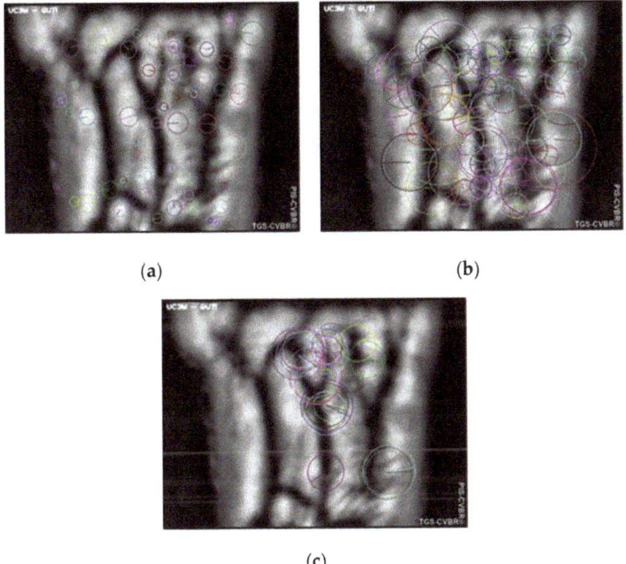

(a)   (b)

(c)

Figure 4. PIS-CVBR®: Feature extraction for User 0. Scale and orientation of the 100 key points extracted with the three algorithms used: (**a**) Scale-Invariant Feature Transform (SIFT®). (**b**) Speeded Up Robust Features (SURF®). (**c**) Oriented FAST and Rotated BRIEF (ORB).

Feature Matching

For the feature or key points matching, two algorithms were used:

1. Brute Force Matcher (BFM): for the descriptors of the features extracted with ORB.
2. Fast Library for Approximate Nearest Neighbors (FLANN) [38]: for the descriptors of the features extracted with SURF® and SIFT®.

The matching between the wrist pattern image of one user (User 0) and real-time video capture was taken and shown, for the two matching algorithms, in Figure 5.

The BFM and the FLANN algorithms provided distances between the features matched. These distances are similarity values between matched features or key points. For the BFM, the Hamming distance was selected. A higher value of distance means that the points were more separated, i.e., they were less similar. To decide if these matched points are suitable, the Lowe's ratio test [34] was used for FLANN (SIFT® and SURF®), and a simple distance score value was set for BFM (ORB). The result of performance per each algorithm is discussed in Section 3.2.2.

So as to obtain a real-time authentication and identification system, analyzing the computational performance of the proposal software algorithms, TGS-CVBR® and PIS-CVBR® are combined. Figure 6 shows and summarized the authentication and verification process made in this work.

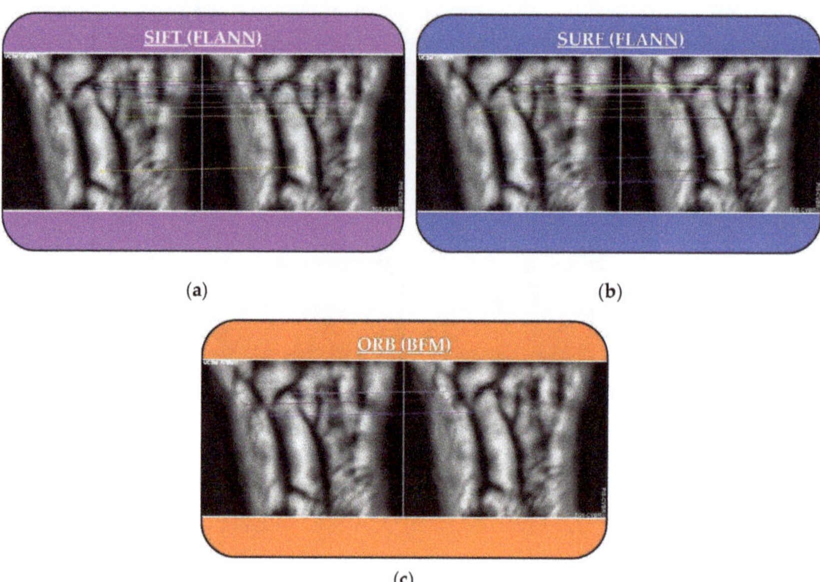

(a)

(b)

(c)

Figure 5. PIS-CVBR®: feature matching for User 0. Correct matching points for two samples of User 0 with the three feature extraction algorithms: (**a**) SIFT® (with Fast Library for Approximate Nearest Neighbors (FLANN)). (**b**) SURF® (with FLANN). (**c**) ORB (with Brute Force Matcher (BFM)).

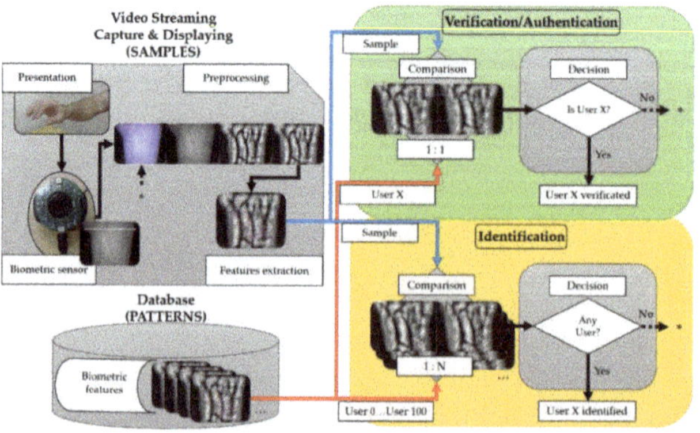

Figure 6. TGS-CVBR® and PIS-CVBR® union: Real-time authentication and identification process.

For the authentication or verification task (green block in Figure 6, 1:1 user comparison), the unique user image pattern (User X extracted from the database) was compared with the real-time video capture (samples), i.e., the features extracted from the image (Figure 7, left) were matched with the features extracted from the streaming video (Figure 7, right). Please, check the supplementary video material for better comprehension.

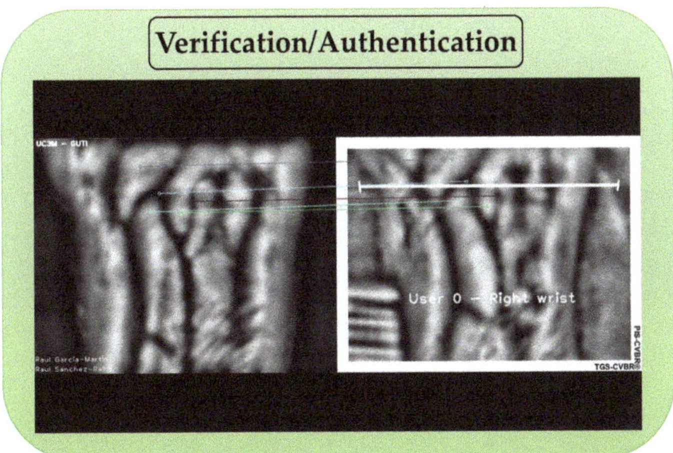

Figure 7. TGS-CVBR® and PIS-CVBR® union: User 0 real-time authentication (screenshot). Unique user image pattern (**left** side) comparison with video (**right** side) using (SIFT algorithm, 7–8 FPS). In the video, the word "User" refers to the subject, and the wrist is predefined (User 0 = Subject 0 and Right wrist, User 1 = Subject 0 and Left wrist).

For the identification task (yellow block in Figure 6, 1:N user comparison), once the unique features had been extracted from each user (User 0 to User 100) at the initialization of the program, they were compared with real-time video capture. Figure 8 shows two identification examples of two users. It is important to notice that according to the normative ISO/IEC 19795-1:2019 [29], this software does not identify because does not provide a rank index, R, of the number of users considered as potential candidates selected with a threshold T.

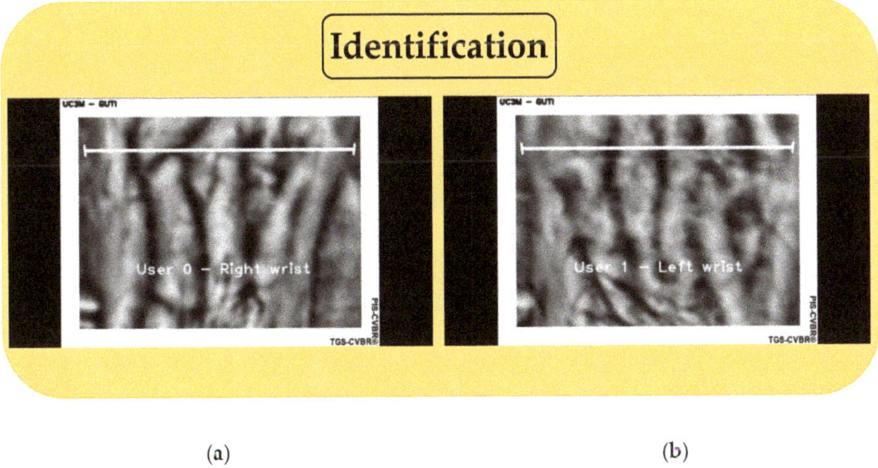

(a)   (b)

Figure 8. Final system software: User 0 and User 1 real-time identification using TGS-CVBR® and PIS-CVBR® (SIFT algorithm). (**a**) User 0 capture. (**b**) User 1 capture.

The computing performance for these two tasks is detailed in the results section, Section 3.

2.3. Dataset Collection: Experimental and Evaluation Procedure

The database acquired in this work was named UC3M-Contactless Version 1 (UC3M-CV1) database. It was collected with the proposed TGS-CVBR®, and the hardware described previously. The two other databases detailed in Table 1 and acquired with physical contact, UC3M [8] and PUT [6], were employed in this study in order to compare, with contact and non-contact dataset, the results obtained with the software algorithms proposed: TGS-CVBR® and PIS-CVBR®.

2.3.1. Parameters

Subject Conditions

The UC3M-CV1 database was made of 1200 infrared greyscale 640 × 480 images captured from 100 users: both wrist of 50 subjects (25 females and 25 males) from Europe (43), America (4), Africa (1) and Asia (2) aged between 21 and 75 years (39.92 years on average, 17.74 standard deviation).

The age and the skin color distribution, according to the Fitzpatrick phototypes scale [39] and the von Luschan chromatic scale, is shown in Figure 9. As is pretended in recent works as [40], studying different environmental (light, temperature, and humidity) and subject conditions, the idea was to introduce new concepts that may affect the vein visualization. In this case, one subject condition is reflected: skin color. It is claimed that skin damage [41] and skin pigmentation [42] do not affect the visualization of the veins in the palm and the finger vein modalities. In these areas, the melanin concentration is lower due to the thickness of the skin. However, the wrist region had slightly higher levels of melanin. These levels increased in dark-skinned subjects, and, as detected in this work, without conclusive results, they could affect the process of vein visualization. For this reason, in Figure 9, the subject phototype distribution is reflected. The chromatic scale distribution was clearly displaced to values under 21 (phototypes I to IV), and the age was mainly distributed between 20–30 and 60–70 years.

The influence of the continent-region was also another issue that is not addressed in this work but is also a factor to take into account in future researches. The origin distribution should be higher.

Six samples per session were captured for each subject wrist: 50 subjects × 2 wrists × 6 samples × 2 sessions = 1200 images. These monochromatic images have been stored in ".jpg" compressed format. More than two weeks and less than four weeks was the distance between sessions.

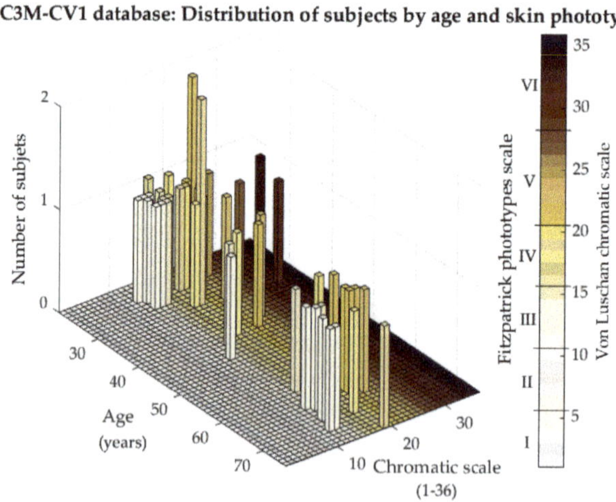

Figure 9. UC3M-CV1: Distribution of subject by age and skin phototype, according to the Fitzpatrick phototypes scale and the von Luschan chromatic scale.

The size of this dataset could be increased in future works, but it is essential to remark that it is larger in the number of sessions and samples than the UC3M dataset but smaller in the number of sessions than the PUT dataset.

Environmental Conditions

The samples have been taken under uncontrol environmental conditions:

1. <u>Temperature</u>: Approximately 20–23 °C.
2. <u>Humidity</u>: Dry ambient.
3. <u>External light</u>: Different daytimes, places (outdoor/indoor), and external artificial lights (usually without direct sunlight).

2.3.2. Collection Method

For the generation of the database, the next steps have been followed:

- The volunteers were informed of the experiment they will be part of and their rights according to the last General Data Protection Regulation (GDPR, applied since May 25th, 2018) [43]. Then, they signed the explicit consent.
- Registration of the personal data of the subject.
- Brief demonstration for the subject, following and showing it in Figure 2, on how to position the wrist correctly according to TGS-CVBR®.
- One operator took one capture when the user's wrist was placed correctly. The operator helped the user (voice indications) if it detected that the subject was placing the wrist in an extremely wrong way: too far/near from the camera (not following the two small guidelines) or with an incorrect orientation (not placing the wrist grove print aligned with the largest guideline).
- The capture process was repeated, obtaining 12 samples per each subject (six samples per wrist): one session per subject. The external light conditions between the different subjects were not the same: different days at a different time in different places (outdoor/indoor).
- Two weeks after the first session, steps 4 and 5 were repeated in the second session obtaining 24 samples per each subject (12 samples per wrist) in total.

3. Results

In order to evaluate the different parts exposed in this paper, this section follows the structure of the previous one. The experimental evaluation procedures have been detailed.

3.1. Hardware: Capture, Storage and Processing Devices

As was presented in the previous work [1], the NIR camera and the NIR PCB illumination provided homogeneous light distribution and good quality images, avoiding excessively bright or dark areas. In this paper, in order to evaluate the response of the capture device in different ambient light conditions, several images were taken. The main goal of this evaluation was to introduce the environmental light influence concept as a critical issue to use this type of system outdoors. Figure 10 shows the comparison of one image of the right wrist of two subjects, User 0 and User 82, in three different outdoor ambient light conditions: darkness, sunny daylight, and cloudy daylight, respectively Figure 10a–c. As a first approach, these conditions were heuristics because the luminous intensity had been not measured.

As can be seen, veins patterns were also visible and recognizable in sunny and cloudy day scenarios, considered as unfavorable light ambient conditions. However, the influence of the conditions of the scenarios is also remarkable.

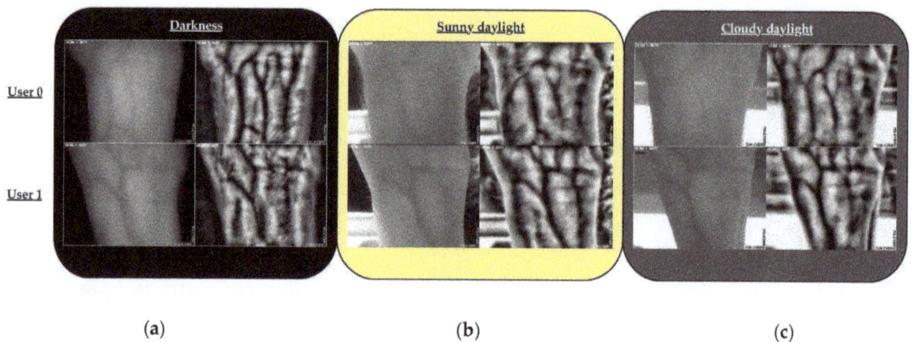

(a)   (b)   (c)

Figure 10. Results: outdoor external light conditions for User 0 (top row) and User 82 (bottom row). (**a**) Darkness. (**b**) Sunny daylight. (**c**) Cloudy daylight.

In the darkness, the resulting images are quite similar, with homogenous light diffusion, to the ones obtained with the contact device used at [8]. The quality of the images was slightly lower in this work, but it is worth taking into account the reduction in size and cost of the camera and illumination. Otherwise, as has been mentioned, it is important to point out that most of the images collected for the UC3M-CV1 were taken indoor with artificial light conditions but without direct sunlight.

The processing time performance was a hardware-software relation requirement that is analyzed and discussed in the next section (Section 3.2.2, Processing-time performance).

3.2. Software Algorithms

3.2.1. TGS-CVBR®

This software component is also evaluated in [1], demonstrating a reduced variation in size and orientation (illumination in consequence) of the wrists. The recognition process was improved, although the algorithms selected and used in this paper (PIS-CVBR®), SIFT®, SURF®, and ORB were already scale-orientation-invariant algorithms.

Figure 11 shows the results of the use of TGS-CVBR®. The repetitiveness in the samples is evident.

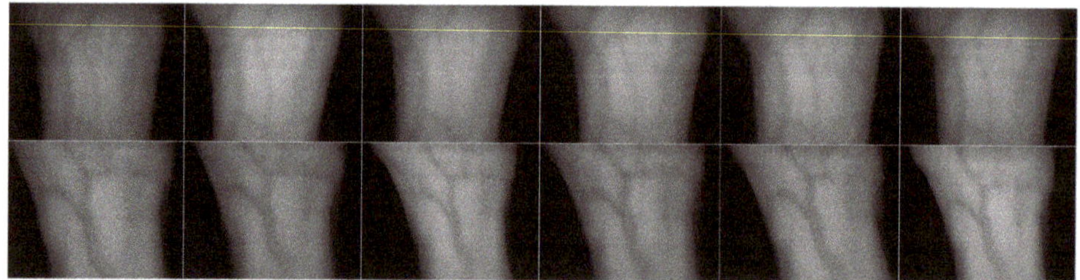

Figure 11. Three-Guideline Software for Contactless Vascular Biometric Recognition (TGS-CVBR®) images comparison for six samples of User 0 (**top row**) and User 82 (bottom row) [1].

In addition, although a usability test for the subjects was not realized for the collection of the UC3M-CV1 dataset, they have indicated that they felt comfortable as the sessions and sample capture have been going on. In the future, a usability test should be passed for a complete evaluation.

3.2.2. PIS-CVBR®

The PIS-CVBR®, software algorithm, was analyzed, according to the normative ISO/IEC 19795-1:2019 [29], in two different ways for the three algorithms used: biometric system performance and processing-time performance. In this way, the three algorithms selected have been tested: SIFT®, SURF®, and ORB.

Biometric System Performance

For this purpose, the database generated, UC3M-CV1, was used. As is mandatory by the normative [29], the False Match Rate (FMR) and False Non-Match Rate (FNMR) were provided in a Detection Error Trade-Off (DET) plot (recommended). Failure-To-Enrol Rate (FTER) and Failure-To-Acquire Rate (FTAR) were unknown.

The first approach of the biometric performance obtained from the UC3M-CV1 database, Figure 12a–c shows, respectively, for each algorithm the error rate in % versus the threshold (number of coincident key points that determine the acceptance or rejection of the user). These graphics discussed the FMR and the FNMR according to the threshold for the 1100 intraclass or mated comparisons (50 subjects × 2 wrist patterns × 11 samples), and the 108,900 interclass or non-mated comparisons (100 wrist patterns × 99 wrist patterns × 11 samples) made.

The threshold, shown in the decision subsystem in Figure 1, was the number of coincident points to choose and accept a user. The green curve represents as a percentage the False Match Rate (FMR) or value of samples compared that should be rejected but are accepted by the algorithm. The red curve represents as a percentage the False Non-Match Rate (FNMR) or value of samples compared that should be accepted but are rejected by the algorithm.

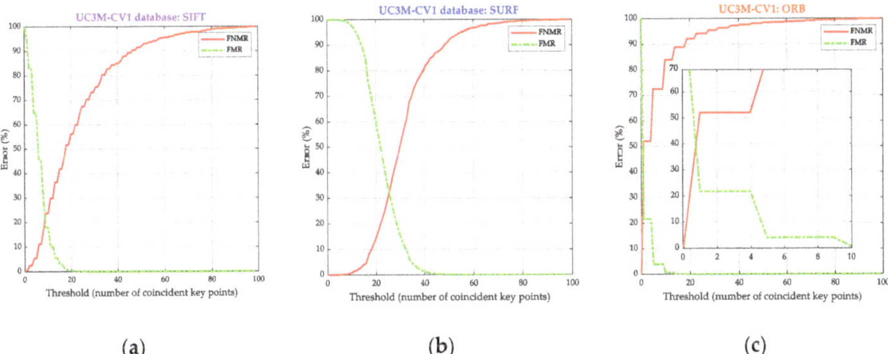

(a)   (b)   (c)

Figure 12. Results: biometric system performance. It is represented the percentage of error, False Match Rate (FMR) (line-dot green curve), and False Non-Match Rate (FNMR) (continuous red curve) versus the threshold of the number of coincident key points. (**a**) SIFT®. (**b**) SURF®. (**c**) ORB.

These graphics anticipate the best biometric performance for SIFT® and the worst for ORB, just analyzing the high value for the crossing point, EER: 21.76%, 32.29%, and 39.94% for respectively, SIFT®, SURF®, and ORB, and the integer thresholds of 9, 25, and 1.

In order to verify this prediction, the Detection Error Trade-Off (DET) curves were obtained (Figure 13) for the collected UC3M-CV1 database and according to the three algorithms (SIFT®, SURF®, and ORB, respectively in Figure 13a–c) and the two capture sessions (Session 1 with green curve, Session 2 with cyan line-dot curve and the entire dataset with yellow gap-line curve).

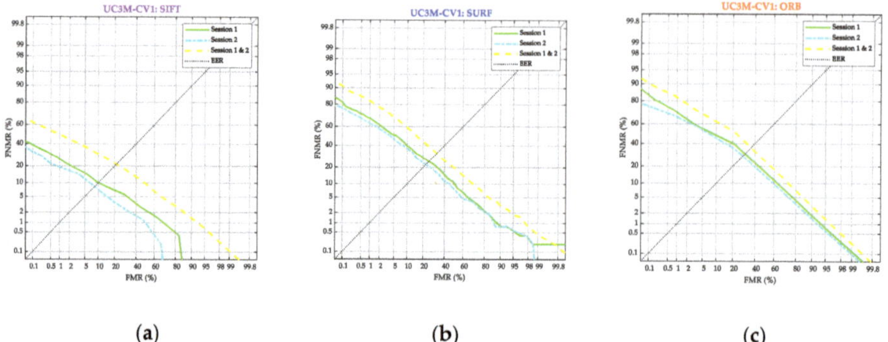

(a)   **(b)**   **(c)**

Figure 13. Results: Biometric system performance. Detection Error Trade-Off curve. The False Non-Match Rate is represented versus the False Match Rate. The green (continuous), cyan (line-dot), and yellow (line-line) curves are respectively for Session 1, Session 2, and the full database. (**a**) SIFT®. (**b**) SURF®. (**c**) ORB.

The results for SIFT® were confirmed as significantly better for all the sessions. The curves were clearly closer to small values for the SIFT® case. It is interesting to point out that, for the three algorithms, the biometric performance improved in the second session, comparing it with the first, showing a more homogeneous placement of the wrist by the subjects in the capture. Otherwise, the curve performance for the entire UC3M-CV1 dataset gets worse. These results have been clearly treated in the discussion section.

In order to compare them with the results obtained with a physical contact database, the UC3M [8] and PUT [6] dataset have been processed with the same software algorithms. The compared results are shown below.

Although the EER value is deprecated according to the normative [29], Figure 14 shows this value obtained for SIFT®, SURF®, and ORB.

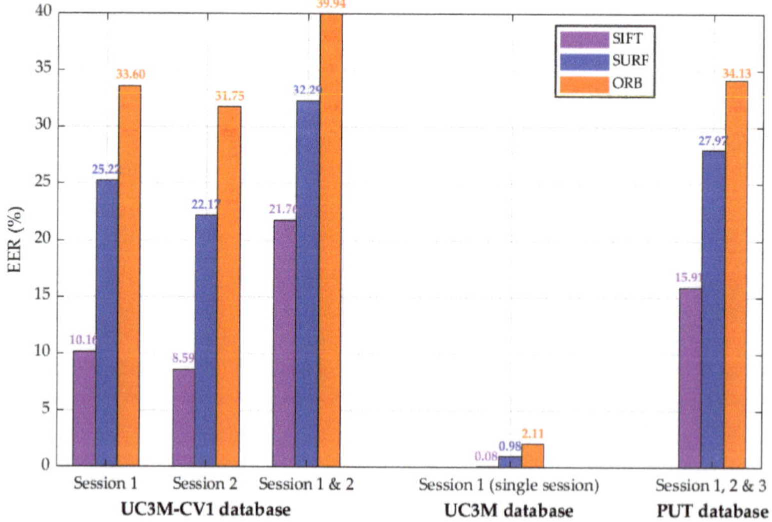

Figure 14. Results: Biometric system performance. Equal Error Rate (EER) obtained for each database using TGS-CVBR® and PIS-CVBR® with SIFT®, SURF®, and ORB algorithms.

The better performance for this working point for the UC3M-CV1 and the two other databases was obtained using, respectively, SIFT®, SURF®, and ORB algorithms.

For the UC3M-CV1 database and the three algorithms, the EER value is reduced in the second session (from 10.16 in Session 1 to 8.59 in Session 2, in SIFT® case), most probably, as it was mentioned, due to the practice of the subject using the system. However, obtaining the results for both sessions, the full database, the EER reached inadmissible values: 21.76%, 32.29%, and 39.94%. These EERs should not be compared with the results obtained and presented in the current state-of-the-art with physical contact devices.

For the UC3M database, the results were much better, probably due to three factors:

1. Single session: the UC3M database was collected in one session. This fact avoids subject usability variability.
2. Contact system: The system used in [8] fixed extremely the position (scale and orientation) of each user with the contact capture device (non-portable big size system). In addition, the device isolated the wrist from the external illumination conditions.
3. Images quality: For the generation of this database, probably, the quality of the images obtained was better with the usage of a dedicated industrial camera [8] (larger size and higher price).

For the PUT database, the EER values were higher than in the UC3M database but lower than in the UC3M-CV1. The capture contact system [6] obtained worse quality images but homogeneous in illumination without external light influence.

In order to ratify these results, the DET graphics for the two other databases used are shown in Figure 15a (UC3M) and Figure 15b (PUT).

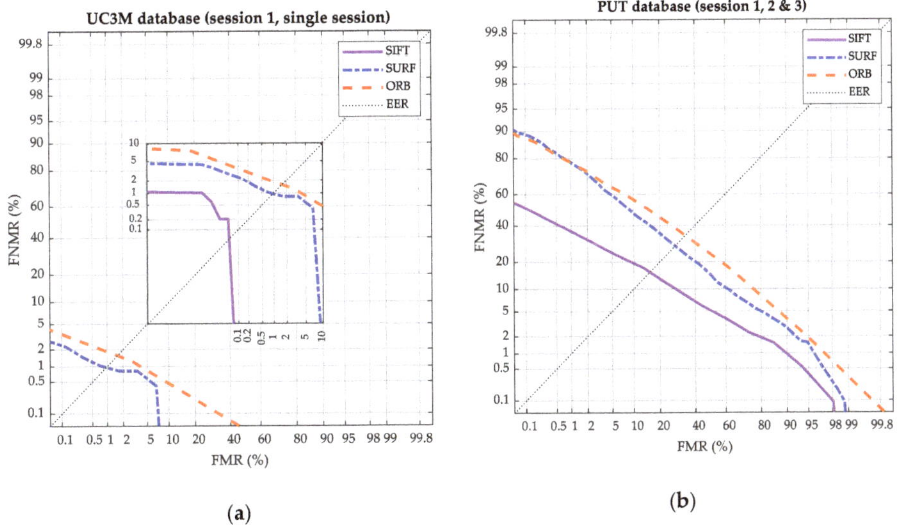

Figure 15. Results: Biometric system performance. DET curve. The FNMR percentage is represented versus the FMR percentage. The purple (continuous), blue (line-dot), and orange (line-line) curves are respectively for the SIFT®, SURF®, and ORB. (**a**) UC3M. (**b**) PUT.

The EER obtained for the UC3M dataset with SIFT®, shown in the extended purple DET curve of Figure 15a, reached 0.08%. This value was significantly lower than the obtained in all the studies of Table 1 (state-of-the-art summary). In the case of the PUT dataset, the results are clearly improvable, comparing the current values with ones of the state-of-the-art. This comparison denotes the high correlation that exists between the design of the software algorithms and the datasets in which they are tested, in agreement with the way the datasets have been collected (mainly the capture device).

Finally, it is essential to remark that in all studied cases, SIFT® and SURF® algorithms obtained better results, but also, the computational cost was significantly higher for the key points extraction and matching, as it is analyzed in processing-time performance unit.

The results were obtained using Python™ 3.4.2. and Matlab® R2019b.

Processing-Time Performance

Figure 16, for the three algorithms, shows the computing time spent in the completed UC3M-CV1, UC3M [8] and PUT [6] databases for the preprocessing, the feature extraction (generation of key points and its descriptors), the intraclass (mated) and interclass (non-mated) comparison (feature matching) and the total time. The hardware used for processing, Raspberry® Pi 4 Model B [32], provided with 64-bit ARM-Cortex A72 (1.5 GHz, quad-core), 4 GB RAM and 128 GB external flash memory, is compared with the Asus® K55V-SX441H [44] 64-bit laptop provided with Intel® Core™ i7 3630QM (2.4 GHz, quad-core), 8 GB RAM and 1 TB of SSD memory. As can be seen, the processing time was considerably reduced in the laptop.

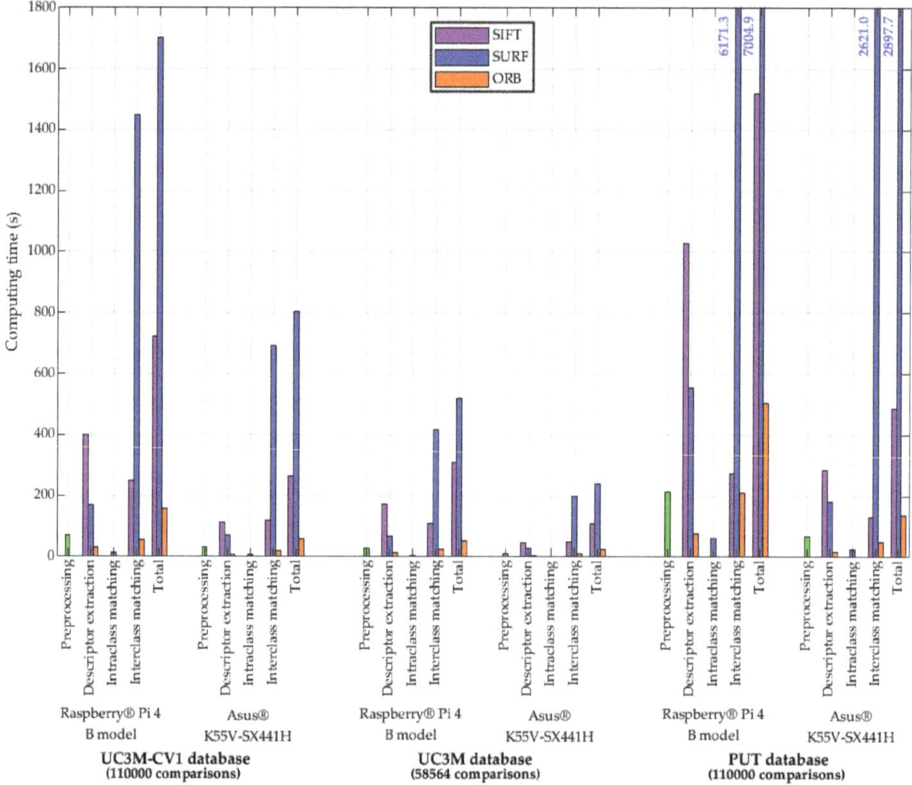

Figure 16. Results: Processing-time performance in seconds for each database and its computing hardware using TGS-CVBR® and PIS-CVBR® with SIFT®, SURF®, and ORB algorithms. Preprocessing time is in green color due to its independence with feature extraction algorithms.

The ORB algorithm was faster in all aspects. SURF® was always slower matching features than SIFT®, that is slower extracting descriptors. The processing time of the PUT database was higher than in the UC3M-CV1 database (same number of images, subjects, and comparisons) due to the ".bmp" image non-compressed format. It has been verified that there was no loss of information with the ".jpg" format for the algorithms used. The results were obtained using Python™ 3.4.2. and Matlab® R2019b.

The processing time performance of the real-time authentication and identification system is summarized in Table 2 for the three algorithms used and the UC3M-CV1 database.

Table 2. Results: final system software. Processing-time performance in Frames Per Second (FPS) for the UC3M-CV1 database and its computing hardware using TGS-CVBR® and PIS-CVBR® with SIFT®, SURF®, and ORB algorithms.

Hardware	Authentication (FPS, 1 User)			Identification (FPS, 100 Users)		
	SIFT®	**SURF®**	**ORB**	**SIFT®**	**SURF®**	**ORB**
Raspberry® Pi 4 Model B	2 (*)	4–5	8	2 (*)	2–3 (*)	4
Asus® K55V-SX441H	7–8	9–10	15	6	9	15

* Framerate too low to obtain a real-time processing system.

As has been indicated previously, in the processing time performance of PIS-CVBR®, SIFT® and SURF® were slow algorithms with a high computational cost. This was reflected in obtaining reduced values of Frames Per Second. As the footer of Table 2 indicates, values below 2–3 FPS were considered too low rates to obtain a real-time processing system, producing an unacceptable lag effect.

The results have been obtained using Python™ 3.4.2.

4. Discussion

In this work, a wrist vein non-contact capture system (hardware and software) for Vascular Biometric Recognition (VBR) has been designed, implemented, and tested. For this purpose, 1200 near-infrared images have been taken and analyzed with a novel contactless capture algorithm. According to the current state-of-the-art, this system tries to contribute to the VBR research world obtaining a system with the following remarkable main features and goals:

1. Contactless.
2. Real-time processing.
3. Portable: small size (85.60 mm × 56.5 mm × 17 mm) and weight (0.2 kg).
4. Reduced price (less than 200 $).
5. Invariant to environmental light conditions.

All these aspects have been demonstrated and fulfilled, except for the invariance to external light conditions. In order to obtain them, in the hardware part, a homogeneous NIR PCB illumination has been integrated. Two new software algorithms have been registered: TGS-CVBR® and PIS-CVBR®. The first one fixes, in a contactless way, the orientation and the scale of the wrist, in order to avoid differences in the illumination and to ease the feature extraction process. The second one, PIS-CVBR®, is in charge of preprocessing (enhancing and increasing vein patterns visualization despite suboptimal environmental light conditions) and of the identification process. For the identification process, the texture based-on homography algorithms, SIFT® [34], SURF® [35], ORB [36] are used. These well-known algorithms are invariant to scale and orientation, a property a priori advantageous for the purpose.

In order to test the biometric and the processing time performance, a new contactless database has been generated, UC3M-CV1, with 100 users (both wrists of 50 subjects) in two sessions.

Finally, the results reflect the following conclusions:

1. The portable and cheap hardware allows obtaining homogenous illumination, avoiding dark and bright areas, although it is not completely immune to the environmental (sunlight an artificial light) conditions (Figure 10). As a required improvement, precise control of the sensor's

near-infrared wavelength sensitivity and the pass-band near-infrared filter would be essential, but probably not definitive, for the achievement of this goal. In addition, the quality of the sensor could be improved. In this sense, the results obtained with PIS-CVBR® reflect that the biometric performance for the two sessions is clearly better in a separate way than for the full UC3M-CV1 database. According to the processing or computing time, it is thought that the small computer used, Raspberry® Pi 4 Model B [32], mounts an enough powerful computing hardware for real-time processing these types of recognition software algorithms, despite the issues evinced by the slowest, SIFT®.

2. The TGS-CVBR® fulfills its goal providing scale-orientation-invariant images, i.e., wrists with the same orientation or positioning and the same size for each user. Nevertheless, user interaction with this type of guiding-feedback algorithms, related to the biometric performance, presents an unexplored field that should be researched in the future.

3. The PIS-CVBR® evinces, as it was stated in point 1, that the biometric performance is completely linked to the environmental light conditions. As far as it is known, this issue had not been addressed in other works that usually employ devices statically in a laboratory and with the sensor isolated from the external light influence. It is thought that the preprocessing step is correct due to the high and clear visualization of the vein patterns. However, the recognition results are entirely not acceptable. In the future, in order to obtain a contactless real-time-processing VBR system, all efforts will be focused on the improvement of the algorithm (biometric performance) and its execution speed (processing time performance), according to the hardware selected.

5. Conclusions

In this paper, a novel vein wrist non-contact VBR system has been designed, implemented, and tested. For this purpose, a contactless device has been integrated with a guiding algorithm, TGS-CVBR®. A novel preprocessing registered method for pattern vein definition has been created. A new non-contact database with 100 different wrists and 1200 infrared images, UC3M-CV1, has been collected. Three scale-orientation-invariant algorithms, SIFT® [34], SURF® [35], and ORB [36], have been tested on it and two other databases (physical contact datasets). Selecting the SIFT® algorithm as the one with the best biometric performance (but worst processing time performance), the results denote the need to continue researching on wrist VBR contactless algorithms, although the improvement against the state-of-the-art results (EER = 0.08% for the UC3M database).

In the future, the lines of research will continue, firstly, with the enhancement of the system invariance against the environmental light and the integration of these devices, introduced in this work. Secondly, the biometric performance will be improved taking into account the scale and orientation of the wrist in the image strongly related to the external light influence. For this purpose, new embedded devices and, against the traditional recognition process, deep learning algorithms are being researched.

6. Patents

From the work reported in this paper, no patents have resulted nevertheless, two software algorithms have been registered: TGS-CVBR® and PIS-CVBR®.

Supplementary Materials: The following are available online at https://zenodo.org/record/3696767#.Xl-o6aj0na8, Video S1: UC3M-Wrist_Vascular_Biometric_Recognition_Using_a_Portable_Contactless_System.mp4.

Author Contributions: Conceptualization, R.S.-R. and R.G.-M.; methodology, R.S.-R. and R.G.-M.; software, R.G.-M.; validation, R.S.-R. and R.G.-M.; formal analysis, R.G.-M.; investigation, R.G.-M.; resources, R.S.-R. and R.G.-M.; data curation, R.G.-M.; writing—original draft preparation, R.G.-M.; writing—review and editing, R.S.-R. and R.G.-M.; visualization, R.G.-M.; supervision, R.S.-R.; project administration, R.S.-R.; funding acquisition, R.S.-R. All authors have read and agreed to the published version of the manuscript.

Funding: This research received no external funding.

Conflicts of Interest: The authors declare no conflict of interest.

References

1. Garcia-Martin, R.; Sanchez-Reillo, R.; Suarez-Pascual, J.E. Wrist Vascular Biometric Capture Using a Portable Contactless System. In Proceedings of the 2019 International Carnahan Conference on Security Technology (ICCST), Chennai, India, 1–3 October 2019; pp. 1–6.
2. Endoh, T.; Aoki, T.; Goto, M.; Watanabe, M. Individual Identification Device. US2005/0148876A1, 29 May 2012.
3. Kitane, K. Fingervein Authentication Unit. US2011/0222740A1, 15 September 2011.
4. Kim, H.; Chun, S.Y. Cancelable ECG Biometrics Using Compressive Sensing-Generalized Likelihood Ratio Test. *IEEE Access.* **2019**, *7*, 9232–9242. [CrossRef]
5. Chui, K.T.; Lytras, M.D. A Novel MOGA-SVM Multinomial Classification for Organ Inflammation Detection. *Appl. Sci.* **2019**, *9*, 2284. [CrossRef]
6. Kabaciński, R.; Kowalski, M. Vein pattern database and benchmark results. *Electron. Lett.* **2011**, *47*, 1127–1128. [CrossRef]
7. Mohamed, C.; Akhtar, Z.; Eddine, B.N.; Falk, T.H. Combining left and right wrist vein images for personal verification. In Proceedings of the 2017 Seventh International Conference on Image Processing Theory, Tools and Applications (IPTA), Montreal, QC, Canada, 28 November–1 December 2017; pp. 1–6.
8. Uriarte-Antonio, J.; Hartung, D.; Pascual, J.E.S.; Sanchez-Reillo, R. Vascular biometrics based on a minutiae extraction approach. In Proceedings of the 2011 Carnahan Conference on Security Technology, Barcelona, Spain, 18–21 October 2011; pp. 1–7.
9. Wang, L.; Leedham, G.; Cho, S. Infrared imaging of hand vein patterns for biometric purposes. *IET Comput. Vis.* **2007**, *1*, 113–122. [CrossRef]
10. Raghavendra, R.; Busch, C. A low cost wrist vein sensor for biometric authentication. In Proceedings of the 2016 IEEE International Conference on Imaging Systems and Techniques (IST), Chania, Crete Island, Greece, 4–6 October 2016; pp. 201–205.
11. Suen, C.Y.; Zhang, T.Y. A fast parallel algorithm for thinning digital patterns. *Commun. ACM* **1984**, *27*, 236–239.
12. Das, A.; Pal, U.; Ballester, M.A.F.; Blumenstein, M. A new wrist vein biometric system. In Proceedings of the 2014 IEEE Symposium on Computational Intelligence in Biometrics and Identity Management (CIBIM), Orlando, FL, USA, 9–12 December 2014; pp. 68–75.
13. Nikisins, O.; Eglitis, T.; Anjos, A.; Marcel, S. Fast cross-Correlation based wrist vein recognition algorithm with rotation and translation compensation. In Proceedings of the 2018 International Workshop on Biometrics and Forensics (IWBF), Sassari, Italy, 7–8 June 2018; pp. 1–7.
14. Hartung, D.; Olsen, M.A.; Xu, H.; Busch, C. Spectral minutiae for vein pattern recognition. In Proceedings of the 2011 International Joint Conference on Biometrics (IJCB), Washington, DC, USA, 11–13 October 2011; pp. 1–7.
15. Hong, L.; Wan, Y.; Jain, A. Fingerprint image enhancement: Algorithm and performance evaluation. *IEEE Trans. Pattern Anal. Mach. Intell.* **1998**, *20*, 777–789. [CrossRef]
16. Struc, V.; Pavesic, N. Illumination Invariant Face Recognition by Non-Local Smoothing. In *Biometric ID Management and Multimodal Communication*; Springer: Berlin/Heidelberg, Germany, 2009.
17. Weickert, J. Applications of Nonlinear Diffusion in Image Processing and Computer Vision. *Acta Math. Univ. Comen.* **2001**, *70*, 33–50.
18. Frangi, R.F.; Niessen, W.J.; Vincken, K.L.; Viergever, M.A. Multiscale Vessel Enhancement Filtering. In *Medical Image Computing and Computer-Assisted Intervention*; Springer: Berlin/Heidelberg, Germany, 1995; pp. 130–137.
19. Telea, A.; van Wijk, J.J. An Augmented Fast Marching Method for Computing Skeletons and Centerlines. In Proceedings of the 2002 Joint Eurographics and IEEE TCVG Symposium on Visualization, VisSym 2002, Barcelona, Spain, 27–29 May 2002; pp. 251–260.
20. Pizer, S.M.; Amburn, E.P.; Austin, J.D. Adaptive Histogram Equalization and Its Variations. *Comput. Vis. Graph. Image Process.* **1987**, *39*, 355–368. [CrossRef]
21. Daubechies, I. Ten lectures on wavelets. In *CBMS-NSF Conference Series in Applied Mathematics*; SIAM: Philadelphia, PA, USA, 1992; pp. 117–119.

22. Arthur, D.; Vassilvitskii, S. K-means++: The advantages of careful seeding. In Proceedings of the Eighteenth Annual ACM-SIAM Symposium on Discrete Algorithms, Philadelphia, PA, USA, 7–9 January 2007; pp. 1027–1035.

23. Olsen, M.A.; Hartung, D.; Busch, C.; Larsen, R. Convolution approach for feature detection in topological skeletons obtained from vascular patterns. *IEEE Symp. Ser. Comput. Intell.* **2011**, 163–167.

24. Xu, H.; Veldhuis, R.N.J.; Bazen, A.M.; Kevenaar, T.A.M.; Akkermans, T.A.H.M.; Gokberk, B. Fingerprint Verification Using Spectral Minutiae Representations. *IEEE Trans. Inf. Forensics Secur.* **2009**, *4*, 397–409.

25. Dubuisson, M.; Jain, A.K. A modified Hausdorff distance for object matching. In Proceedings of the 12th International Conference on Pattern Recognition, Jerusalem, Israel, 9–13 October 1994; Volume 1, pp. 566–568.

26. Wang, L.; Leedham, G.; Cho, D.S.-Y. Minutiae feature analysis for infrared hand vein pattern biometrics. *Pattern Recognit.* **2008**, *41*, 920–929. [CrossRef]

27. Chen, H.; Lu, G.; Wang, R. A New Palm Vein Method Based on ICP Algorithm. *Int. Conf. Inf. Syst.* **2009**, 1207–1211.

28. William, H.; Saul, A.; William, T.; Flannery, B.P. Support Vector Machines. In *Numerical Recipes: The Art of Scientific Computing*, 3rd ed.; Cambridge University Press: New York, NY, USA, 2007; ISBN 978-0-521-88068-8.

29. American National Standards Institute. SO/IEC 19795-1:2019. In *Information Technology—Biometric Performance Testing and Reporting—Part 1: Principles and Framework*; ANSI: Washington, DC, USA, 2019.

30. Logitech®HD Webcam C525 Specifications. Logitech. Available online: https://www.logitech.com/en-us/product/hd-webcam-c525/ (accessed on 5 March 2020).

31. OSLON Black, SFH 4715A. OSRAM. Available online: https://www.osram.com/ecat/OSLON%C2%AE%20Black%20SFH%204715A/com/en/class_pim_web_catalog_103489/global/prd_pim_device_2219803/ (accessed on 5 March 2020).

32. Raspberry®Pi 4 Model, B. Raspberry. Available online: https://www.raspberrypi.org/products/raspberry-pi-4-model-b/ (accessed on 5 March 2020).

33. Raspberry®Pi 3 Model, B. Raspberry. Available online: https://www.raspberrypi.org/products/raspberry-pi-3-model-b/ (accessed on 5 March 2020).

34. Lowe, D.G. *Distinctive Image Features from Scale-Invariant Keypoints*; University of British Columbia: Vancouver, BC, Canada, 2004.

35. Bay, H.; Tuytelaars, T.; Gool, L. SURF: Speeded up robust features. In Proceedings of the 9th European Conference on Computer Vision, Graz, Austria, 7–13 May 2006; pp. 404–417.

36. Rublee, E.; Rabaud, V.; Konolige, K.; Bradski, G. ORB: An efficient alternative to SIFT or SURF. In Proceedings of the 2011 International Conference on Computer Vision, Barcelona, Spain, 6–13 November 2011; pp. 2564–2571.

37. Clotet, P.F.; Findling, R.D. Mobile Wrist Vein Authentication using SIFT Features. In Proceedings of the 16th International Conference, Las Palmas de Gran Canaria, Spain, 19–24 February 2017.

38. Muja, M.; Lowe, D.G. *FLANN-Fast Library for Approximate Nearest Neighbor User Manual*; INSTICC Press: Setubal, Portugal, 2013.

39. Fitzpatrick, T.B. Soleil et peau. *J. de Médecine Esthétique* **1975**, *2*, 33–34.

40. Uhl, A.; Busch, C.; Marcel, S.; Veldhuis, R. *Handbook of Vascular Biometrics*; Springer International Publishing: Cham, Switzerland, 2020; pp. 179–199.

41. Kauba, C.; Prommegger, B.; Uhl, A. Combined Fully Contactless Finger and Hand Vein Capturing Device with a Corresponding Dataset. *Sensors* **2019**, *19*, 5014. [CrossRef] [PubMed]

42. Kisku, R.D.; Gupta, P.; Sing, J.K. *Design and Implementation of Healthcare Biometric Systems*; IGI Global: Pennsylvania, PA, USA, 2019; pp. 14–15.

43. Regulation (EU) 02016R0679 of the European Parliament and of the Council of 27 April 2016. The General Data Protection Regulation (GDPR). April, 2016. Available online: https://eur-lex.europa.eu/legal-content/EN/TXT/PDF/?uri=CELEX:02016R0679-20160504&from=EN (accessed on 5 March 2020).

44. Asus K55VD-SX441H i7-3630/8GB/1TB/GT 610/15.6. PC Components, 2013. Available online: https://www.pccomponentes.com/asus-k55vd-sx441h-i7-3630-8gb-1tb-gt-610-15-6- (accessed on 5 March 2020).

Article

Ear Detection Using Convolutional Neural Network on Graphs with Filter Rotation

Arkadiusz Tomczyk * and Piotr S. Szczepaniak

Institute of Information Technology, Lodz University of Technology, ul. Wolczanska 215, 90-924 Lodz, Poland; Piotr.Szczepaniak@p.lodz.pl
* Correspondence: arkadiusz.tomczyk@p.lodz.pl; Tel.: +48-42-632-97-57

Received: 12 November 2019; Accepted: 6 December 2019; Published: 13 December 2019

Abstract: Geometric deep learning (GDL) generalizes convolutional neural networks (CNNs) to non-Euclidean domains. In this work, a GDL technique, allowing the application of CNN on graphs, is examined. It defines convolutional filters with the use of the Gaussian mixture model (GMM). As those filters are defined in continuous space, they can be easily rotated without the need for some additional interpolation. This, in turn, allows constructing systems having rotation equivariance property. The characteristic of the proposed approach is illustrated with the problem of ear detection, which is of great importance in biometric systems enabling image based, discrete human identification. The analyzed graphs were constructed taking into account superpixels representing image content. This kind of representation has several advantages. On the one hand, it significantly reduces the amount of processed data, allowing building simpler and more effective models. On the other hand, it seems to be closer to the conscious process of human image understanding as it does not operate on millions of pixels. The contributions of the paper lie both in GDL application area extension (semantic segmentation of the images) and in the novel concept of trained filter transformations. We show that even significantly reduced information about image content and a relatively simple, in comparison with classic CNN, model (smaller number of parameters and significantly faster processing) allows obtaining detection results on the quality level similar to those reported in the literature on the UBEAR dataset. Moreover, we show experimentally that the proposed approach possesses in fact the rotation equivariance property allowing detecting rotated structures without the need for labor consuming training on all rotated and non-rotated images.

Keywords: geometric deep learning; ear detection; structured prediction; semantic segmentation; rotation equivariance; Gaussian mixture model; superpixels

---

1. Introduction

A biometric can be defined as a measurable, physical characteristic, which can be used to identify individuals. There are various types of biometrics used in practical applications: voice recordings, fingerprints, signatures, DNA, hand geometry, iris and face images, or even keystroke dynamics, to mention only a few. A good biometric should have several properties [1]. It should be universal (everyone should possess this characteristic), distinctive (it should allow discriminating between people) and permanent (ideally, it should not change in time). Moreover, the process of acquisition should be inexpensive, generally acceptable, and not troublesome (in some applications, it should be even discreet). Finally, the identification system using such a biometric should be hard to circumvent. The biometrics mentioned above meet those expectations to varying degrees. It is relatively easy to forge a signature, whereas a DNA test usually is hard to falsify. Similarly, collecting face images, as a rule, is treated as a violation of privacy, while taking fingerprints seems to be natural.

In this work, ear images are considered [2,3]. There are several factors that make the research and application of ear recognition important and attractive: people can be identified on that basis;

this biometric does not change in time; and its gathering does not create a great deal of controversy. Furthermore, technology enables acquisition of ear images from a distance, which may be of great importance for police investigations and in security systems. It has, however, at least two consequences. Firstly, before the individual can be identified [4,5], the localization of the ear must be precisely detected. Secondly, since the acquisition process is not controlled, the orientation of the head can significantly vary. As a result, the need for detection methods arises, which will be able to cope with ear transformations.

Convolutional neural networks (CNNs) became a state-of-the-art solution of many image analysis problems [6,7]. Their main component is convolutional layers designed to apply the same, trainable filters (represented by a rectangular mask) locally to every part of the image. It allows extracting the spatial distribution of image characteristic features (so-called feature maps). These kinds of layers together with downsampling/upsampling mechanisms and classic fully connected layers allow solving many typical tasks. These are, among others: image classification, object localization and detection, semantic and instance segmentation, etc. However, despite unquestionable advantages, convolutional neural networks are not free of vices. First of all, they operate on pixels. Bearing in mind the dimensions of currently processed images, to solve practical problems, the structure of such networks must be quite complex (deep architectures). This results in a large number of parameters that need to be trained and, consequently, huge training datasets that need to be prepared. The second problem is their sensitivity to object rotations. If rotation invariance/equivariance is required, the complexity of the trained model must be further increased. Both of those problems can be overcome by the geometric deep learning (GDL) technique presented in this paper.

1.1. Ear Detection

In [8], the authors emphasized the extremely challenging character of ear detection as this part of the human head can be presented on images in various sizes, rotations, shapes, and colors. Moreover, the images can be of diverse quality, and fragments of ears can be partly hidden. To overcome and manage those problems, and to offer solutions that are applicable in practice, in the last few years, machine learning methods have become more and more popular in use. Motivated by the current direction of research development, in the following, we continue the surveys [2,3] and present a short, but comprehensive view on progress made since 2016 in machine learning application to the problem under consideration.

The approach published in [9] was based on geometric morphometrics and deep learning. It was proposed for automatic ear detection and feature extraction in the form of landmarks. A convolutional neural network (CNN) was trained and results compared with a set of manually landmarked examples.

A two step approach was described in [10]. In the first step, the detection of three regions of different scales allowed obtaining information about the ear location context within the image. In the second step, a filtering was performed with the aim of extracting the correct ear region and eliminating the false positives ones. This technique used convolutional neural networks (called here multiple scale Faster R-CNN) to detect ears from profile images.

In [8], the authors applied a convolutional encoder-decoder network to perform the binary classification of image pixels as belonging to either the ear or to the non-ear class. Temphhe result was improved by a post-processing procedure based on anthropometric knowledge and deletion of spurious image regions. The paper involved comparative results with a state-of-the-art known from the literature.

A detection technique applying an ensemble of convolutional neural networks (CNNs) was presented in [11]. The weighted average of the outputs of three trained CNNs was considered as result of detection of the ear regions. A better performance observed for the ensemble of networks compared to the use of single CNN models was reported. A similar approach was described in [12], where also an ensemble of three networks was used. This time, however, members of the ensemble did not differ in network architecture, but they were trained with regions of the image taken at different cropping scales.

1.2. Rotation Equivariance

Transformation invariance and equivariance are terms sometimes mistakenly used interchangeably. The first means that the system will respond in the same way regardless of the transformation, which is applied at its input, while the latter indicates that transformation of the input will result in the same transformation of the output. In the context of image analysis, if the classification task is considered, transformation invariance can be expected. In other problems like semantic segmentation discussed in this work, the desired property of the system is its transformation equivariance.

CNNs manifest natural translation equivariance since the same filters (feature detectors) are applied at different locations of the image. Thanks to the additional pooling operations, they also possess approximate invariance to translation. Input rotation, however, is still a problem for those networks.

Three main groups of techniques used to overcome it can be distinguished. The first one uses data augmentation, generating rotated versions of the training images, forcing the network to learn all the possible orientations of the objects. That approach, of course, requires very complex network architectures having sufficient flexibility. Two alternative methods use either an input image [13,14] or filter [15,16] rotations with some kind of result aggregation. In addition, to avoid excessive increase of the parameter number, trained weights are shared between different processing paths. The rotation of images and filters in classic CNNs in general is problematic as both are represented by a regular, rectangular grid of values. Consequently, some interpolation algorithms must be used, and the object of interest should be located in the image center to avoid artifacts at the border.

In [13], the authors combined augmentation with input image rotation. In the latter case, only $0°$ and $45°$ angles were considered together with additional image cropping and flipping. As a result, 16 variants of the same image were processed by the network. The output feature maps were concatenated and passed to dense layers serving as a classifier. A slightly different approach was presented in [14]. Here, also input image rotations were used, but this time as an aggregation method, transformation invariant pooling was used where the element-wise maximum was taken from resulting feature maps.

The authors of [15] rotated filters instead of input images. In this case, bicubic interpolation was applied to generate a group of rotated filters. As a response for each group, max-pooling was used. While training, gradients were passed through the element in a filter group with the largest activation. Such an orientation pooling could also be found in [16]. This time, however, to avoid interpolation, a set of dedicated atomic, circular filters were prepared, and the actual network filters were sought for as a linear combination of these.

At the end of this short survey, one more approach should be mentioned, as it does not fit any of the above-mentioned categories. In [17], the authors decided to embed inside CNN additional processing blocks, transforming, in particular rotating, feature maps. After transformation, the outputs were concatenated and processed by the further part of the network.

1.3. Geometric Deep Learning

GDL is a dynamically developing area of research in recent years [18]. It, inter alia, tries to generalize and apply the concept of CNN for structures less regular than images (graphs) and for continuous domains (manifolds). This adaptation requires the proper definition of the convolution operation, which should be able to compute features of given elements based on their local neighborhood. GDL was successfully applied to various practical problems. Two of the most popular fields of application are: prediction of chemical molecules' properties [19,20] and document classification taking into account citation links [21,22]. In the first case, final prediction is assigned to a graph as a whole, while in the second, every graph node is considered separately. Surprisingly, very few of the approaches were tested on images. In most of the cases, the problem was the initial definition of graph convolution itself, which required a fixed structure of the graph (in the case of images, graphs are different depending on the content). The existing approaches were used only for handwritten digit classification (MNIST dataset) [23–25], where either a grid of pixels was treated as a graph or image content was represented by an irregular graph of superpixels [26].

The latter approach, although not popular in the GDL community, has undeniable advantages. The change of image representation, where its content is described with a significantly smaller set of spatially distributed elements, leads to a reduction of model complexity, which is required for its processing. Moreover, such a representation is more human friendly. Conscious understanding of image content operates rather on regions and borders separating them than on thousands of millions of pixels. This, in turn, enables simpler interpretability of the results and simpler acquisition of additional expert knowledge there, where the number of training samples is limited (e.g., medicine, biometry, etc.).

1.4. Contribution

The main contribution of this paper lies in the application of GDL to semantic segmentation of the images and in the introduction of trained filters' rotations. Both of those features are illustrated with the ear detection problem, but can also be applied in other object detection, semantic segmentation, or image classification tasks.

The proposed approach to object detection is a novel area of application of GDL in image the analysis domain since, so far, in [24] and in similar works, these kinds of networks were used only to assign labels to image as a whole. When graph nodes were interpreted separately, as was done in this work (Figure 1), only the other application areas were explored in [21,22]. What is more, the originality of the method lies also in a specific approach to semantic segmentation itself. Typical applications of classic CNNs in this field use some downsampling/upsampling layers to reduce the number of parameters [27–29]. Only in a few specific applications described in [30,31], such mechanisms were not required. In this work, such techniques are not used as well, because reduced image representation allows using relatively simple models.

The proposed method is, of course, new in the context of ear detection as well. Until now, CNNs were applied only at the pixel level both for semantic segmentation [10] and direct object detection [8]. Here, alternative superpixel representation is used, showing its usefulness in these kinds of applications and allowing significantly simplifying the architecture of the trained network.

The additional novelty of this work lies in proving that training of the proposed model with a limited number of samples and rotation of trained filters allowed detecting the rotated structures as well (Figure 2). Consequently, after a simple training process, we could obtain the rotation equivariance property of the considered network. This approach is different from the filter rotations described in [15] or [16], as there these filters are also rotated when the network is trained, which complicates the whole procedure. It should be also emphasized that, since filters are defined by GMM, there are no interpolation problems, typical in classic CNNs, when filters are rotated.

The above described property can be useful in ear detection problems where profile images are acquired. If a limited number of training samples, in particular in only one orientation, can be gathered, we can prepare a rotation equivariant detection model, which should able to locate ears when the head is rotated in the image plane.

The content of this work is split into several sections. In Section 1, a short literature review in the areas of ear detection and rotation invariance/equivariance is presented. It contains also a description of the paper's novelty and contribution. Section 2 contains the details of the proposed approach, as well as the results of an experiment verifying its properties. In Section 3, results and a detailed discussion of the main experiments are described. A summary of the conducted research concludes the paper.

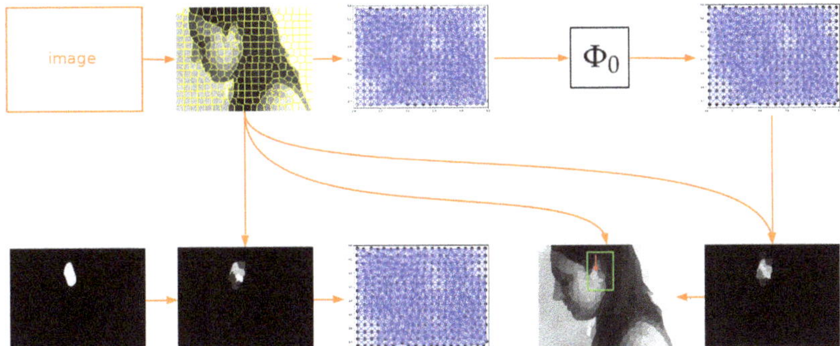

Figure 1. Image processing flow of a convolutional network operating on graphs. As an input, the image from the UBEAR dataset [32] (it is not shown here to protect the identity of the depicted person) and corresponding binary mask, indicating precise position of the ear, are given. In the first step, the SLIC algorithm [26] is used to transform image content into its superpixel representation. Next, taking this representation and original mask into account, the expected values are assigned to superpixels. They indicate the correspondence between superpixels and the ear region and constitute the goal of training (second column). Further, graphs for both the transformed image and expected mask are constructed (third column). The edges connecting nodes are created based on the spatial adjacency of superpixels. It is worth noticing that the expected values are assigned to graph nodes and not to the graph as a whole. Those graphs are used to train network Φ_0, which learns to localize ears for basic head orientation. When the network is trained, it can process any other input graph. Knowing the output graph and the distribution of superpixels, it is possible to construct the output mask where the maximum should reflect the position of the ear (last column).

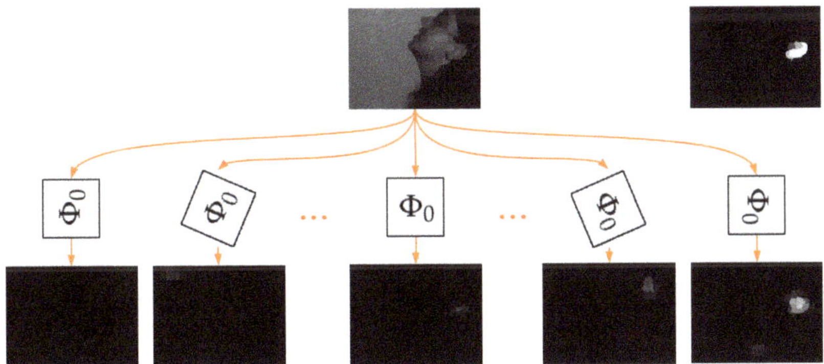

Figure 2. System with the rotation equivariance property. Network Φ_0 is trained using only samples with basic head orientation (Figure 1). After rotation of GMM filters, it is able to detect the corresponding rotated structures as well. Consequently, selecting the output mask with the maximum value as the output of the whole system, we can obtain the desired rotation equivariance property. Thanks to that, we are able to localize ears even for non-standard head orientations. The top right image presents the expected output mask.

2. Materials and Method

This section presents the characteristics of the dataset used, as well as the GDL technique discussed in this work. Naturally, to apply a convolutional neural network on graphs, additional pre- and post-processing had to be performed. The whole processing flow is shown in Figure 1; however, the majority of the details of the methods used will be presented in sections devoted to the specific experiments. Here, only the functioning of the neural network will be explained. Moreover, while describing the details of the GDL approach, the hypothesis about its properties will be formulated. That is why, at the end of this section, the results of the experiment, verifying the correctness of the proposed assumptions, are described as well.

2.1. Dataset

There are several publicly available benchmark datasets dedicated to ear biometric tasks. They differ, however, significantly. The CP dataset [33] and the IITD dataset [34,35] are very similar. Both contain grayscale, tightly cropped, and aligned images. The AMI dataset [36] and the WPUT dataset [37] are comprised of color images with ears and surrounding head fragments. In all those sets, images were acquired in controlled, laboratory conditions and can be used for training purposes and evaluation of ear identification systems. The last two sets described below were prepared in a different way. The UBEAR dataset [32] is composed of profile, grayscale images taken from video sequences (selected frames). Images in the AWE dataset [38] were collected from the web. They are better suited for ear detection problems as they contain also original, not cropped, pictures. It should be also emphasized that, except the first two, in all those datasets, to a different extent, some additional difficulties are present. To mention only a few of them: images were acquired in different illumination conditions and with different background; ears were occluded; head was rotated, leading to ear transformations, etc.

The UBEAR and AWE datasets together with sets dedicated to other problems (e.g., face recognition) are exploited in the ear detection literature. In this work, in all the experiments, only the UBEAR dataset was used. There are several reasons for that choice. First of all, we did not have access to the original, not cropped, AWE images. Secondly, it corresponded to our initial idea of an ear detection system where people could be identified discreetly from video sequences. Thirdly, this set is relatively large. It contains 4429 images taken from 126 persons. Fourthly, in this dataset, binary masks, indicating precise ear positions, are available, which is a rare case in these kinds of datasets (usually only bounding boxes are annotated). What is more, it contains information about head poses, which allows identifying images with a specific head orientation. There were 5 poses identified. Every pose had a unique letter assigned: M means that person was stepping ahead (normal head orientation), whereas U, D, O, and T indicate that the head was rotated upwards, downwards, outwards, and towards, respectively. Finally, it is a quite challenging dataset for analysis. Not only all the above mentioned difficulties were present, but also motion artifacts could be observed. Sample images from the UBEAR dataset are presented in Figure 3.

2.2. Method

In this work, the existing GDL method, presented in [24], was further developed. That approach defined convolutional filters using Gaussian mixture model (GMM) in a pseudo-coordinate space. Assuming that nodes of input and output graphs are described with vectors in N and M dimensions (channels), respectively, the operation of a single convolutional layer ϕ can be expressed in the following way (Figure 4):

$$h^m(s) = \Psi\left(b^m + \sum_{n=1}^{N} \sum_{t \in \mathcal{N}(s)} \varphi^{n,m}(\mathbf{u}(s,t))f^n(t)\right) \tag{1}$$

where:

$$\varphi^{n,m}(\mathbf{u}) = \sum_{j=1}^{J} g_j^{n,m} \exp\left(-\frac{1}{2}(\mathbf{u} - \mu_j^{n,m})^T (\mathbf{K}_j^{n,m})^{-1}(\mathbf{u} - \mu_j^{n,m})\right) \tag{2}$$

and $m = 1, \ldots, M$. In the above equations, s and t are node indices, f and h represent vectors describing features of the input and output graphs' nodes, respectively, J denotes the number of Gaussians, and $\mathcal{N}$

is the neighborhood function identifying adjacent nodes. Mapping $\mathbf{u}$ calculates pseudo-coordinates of node t relative to given node s. These coordinates are d-dimensional vectors. Finally, Ψ is an activation function applied element-wise for every graph node, and b represents additional, optional bias. Every convolutional layer defined in this way contains M groups of N filters φ. The trainable parameters of those filters are: real numbers g, vectors μ of size d, and diagonal $d \times d$ matrices $\mathbf{K}$ (only d non-zero elements). This gives a total number of parameters per filter equal to $J(2d + 1)$ and in the whole layer equal to $MN(J(2d + 1) + 1)$.

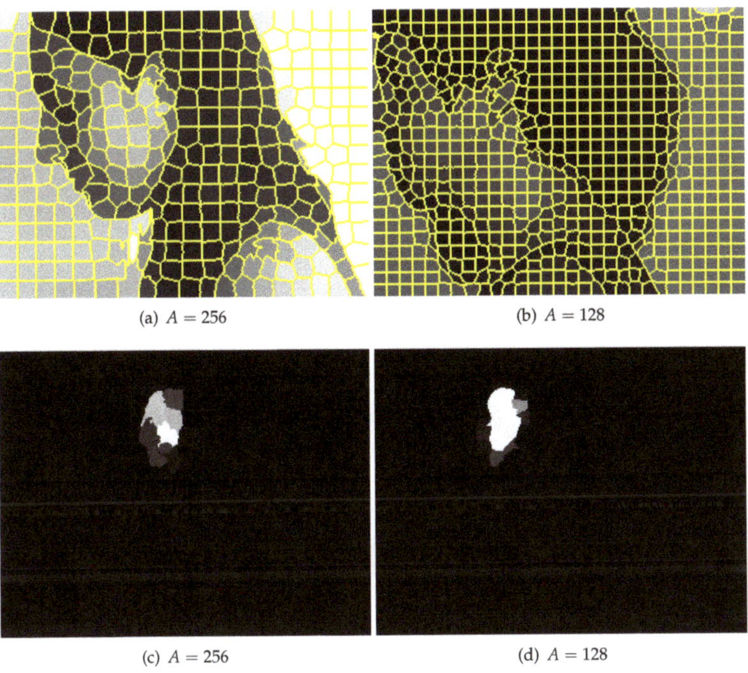

(a) $A = 256$ (b) $A = 128$

(c) $A = 256$ (d) $A = 128$

Figure 3. Sample images from the UBEAR dataset (pose M). Their content was described using superpixels with different average areas A of a single superpixel (here and in the whole work, the original images are not used on purpose to protect the identity of the depicted people): (**a**) original image, (**b**) binary mask with ear localization. The color assigned to every superpixel is the average color of covered pixels.

The above formulation differs slightly from the original one presented in [24]. In that paper it was not clear whether every pair of input and output channels had its own fully trainable filter. In our experiments we have used PyTorch Geometric library [39] where, in its earlier versions, some of the filter parameters were shared. The presented extension was added to the library by the authors of this work and is available in PyTorch Geometric starting from version 1.3.1.

To construct a network operating on graphs and useful for semantic segmentation tasks, we must ensure that input and output graphs have the same size and structure. In classic CNNs, to achieve that goal without excessive growth of the number of parameters, additional downsampling/upsampling blocks are used. Here, thanks to the reduced representation of the image content, it was sufficient to consider only a sequential composition of the above layers:

$$h = \Phi(f) = (\phi^L \circ \ldots \circ \phi^1)(f) \tag{3}$$

where L denotes the number of layers. Naturally, the number of input N and output dimensions M (as long as they are consistent in successive layers), as well as activation functions Ψ may vary between layers.

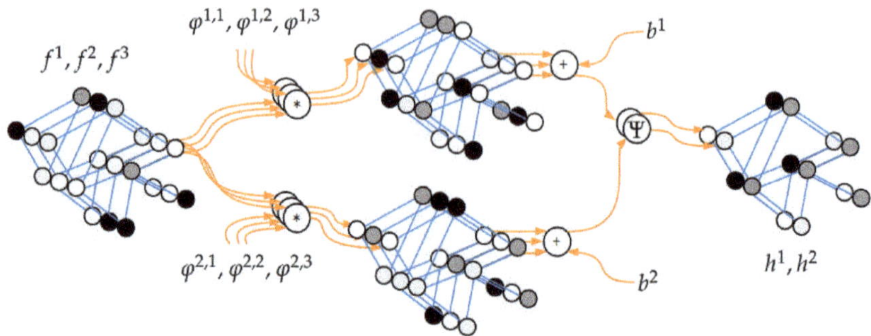

Figure 4. The processing scheme of a single layer ϕ. Here, the input and output graphs are described with vectors in $N = 3$ and $M = 2$ dimensions, respectively. For illustration purposes, every dimension (channel) is shown separately, but of course, the graph structure is always the same. It can be observed that every pair of input f^n and output h^m channels has its own fully trainable GMM filter $\varphi^{n,m}$.

Every filter φ is described by the corresponding GMM in pseudo-coordinate space defined by mapping $\mathbf{u}$. When Cartesian coordinates are used in the image plane ($d = 2$), GMM can be rotated around the origin of the coordinate system $(0,0)$ by an angle θ, resulting in a new filter φ_θ. If the original filter φ detects some specific node configurations in a graph, the filter φ_θ should have a high response for those nodes, which exhibit the same, but rotated characteristic of their neighborhood. Consequently, if all the filters in convolutional layers ϕ_θ are rotated, the whole network Φ_θ should possess the same property. To confirm this hypothesis, a verification experiment, described in the next subsection, was proposed. This confirmation is crucial to facilitate the construction of the system possessing the rotation equivariance property shown in Figure 2. We can train then a network Φ_0 capable of recognizing ears for basic head pose (a smaller training set is required) and, after successive rotations of filters, use it to detect rotated ears as well.

2.3. Verification

To verify the hypothesis about the capability of the trained model to detect rotated structures, the experiment presented below was conducted.

First, from UBEAR dataset [32], fragments of binary masks with precise ear localization (100 samples) were extracted. Next, their graph representations were created using the SLIC algorithm [26]. Nodes of those graphs corresponded to generated superpixels and were described by their average intensity normalized to the $[0, 1]$ interval (Figure 5a). Directed edges connected superpixels' centroids (Figure 5b) and had Cartesian pseudo-coordinates assigned.

Those graphs allowed generating a family of training $\mathcal{D}_\theta^{TR}$, validation $\mathcal{D}_\theta^{VA}$ and test $\mathcal{D}_\theta^{TE}$ sets with 60, 20, and 20 samples, respectively, where for every input graph, the expected output graph was prepared. The values assigned to nodes of output graphs indicated whether the node approximately corresponded to an outer edge at θ angle. They were calculated using the following formula:

$$h_\theta(s) = \max\left(\frac{1}{\Omega_\theta(s)} \sum_{t \in \mathcal{N}(s):\omega_\theta(s,t)>0} \omega_\theta^2(s,t)(f(t) - f(s)), 0\right) \tag{4}$$

where:

$$\Omega_\theta(s) = \sum_{t \in \mathcal{N}(s):\omega_\theta(s,t)>0} \omega_\theta^2(s,t) \tag{5}$$

and:

$$\omega_\theta(s) = \sin(\alpha(s,t) - \theta) \tag{6}$$

As before, f and h represent feature vectors of input and output graphs' nodes and $\alpha(s,t)$ is the angle between the horizontal axis and edge connecting nodes s and t. The sample, horizontal edge found in this way for $\theta = 0$ is depicted in Figure 5c.

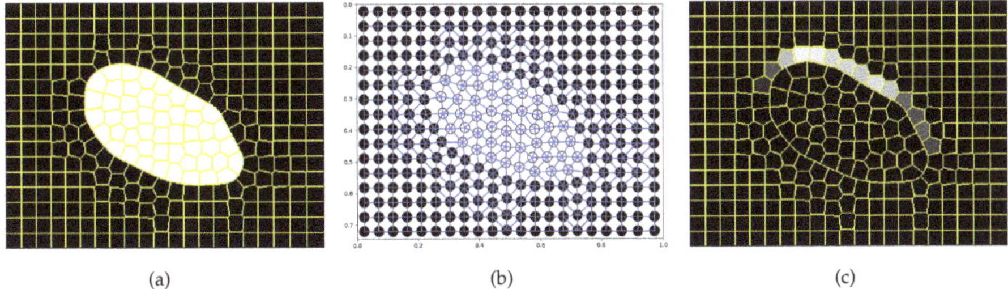

(a)   (b)   (c)

Figure 5. Images and graph used for initial verification of the proposed approach: (**a**) original image with superpixels detected; (**b**) its graph representation; (**c**) the expected output. In this experiment, the expected average size of the superpixel was equal to $A = 400$, and graph nodes were connected only if the corresponding superpixels were adjacent. It is worth noticing that the irregularities of superpixels and consequently irregularities of the graph structure are present only if image colors are not uniform. This is a typical situation when the SLIC algorithm is used for superpixel generation.

Having these data prepared, the CNN with GMM filters was trained using only $\mathcal{D}_0^{TR}$. The MSE loss and Adam optimizer with the learning rate equal to 10^{-3} were applied. The validating samples were used to check if the model was not overfitted and to select the optimal one. The network contained $L = 2$ convolutional layers. There were 10 groups with 1 filter in the first layer ϕ^1 and 1 group with 10 filters in the second one ϕ^2. Every filter contained $J = 4$ Gaussian functions. In the first layer, ReLU and, in the second, identity activation functions were used. In Figure 6, the sample trained GMM filter φ from the first layer together with its rotated version φ_θ are presented. The results depicted in Figure 7 prove that the proposed concepts behaved correctly not only for training samples, but they were also able to generalize and give reasonable results for unseen graphs. In Table 1, a systematic evaluation is presented. It is evident that if filters in the original network Φ_0 were rotated by an angle θ, the resulting network Φ_θ was able to detect structures rotated by the same angle (significantly smaller MSE error) effectively.

3. Results and Evaluation

This section contains the results of the experiments conducted on the UBEAR dataset [32]. First, the convolutional network was trained to detect ears in images with a normal head orientation. Next, it is applied for the detection of ears for other, selected head poses. At the end of this section, the discussion of the experiments' outcomes is presented.

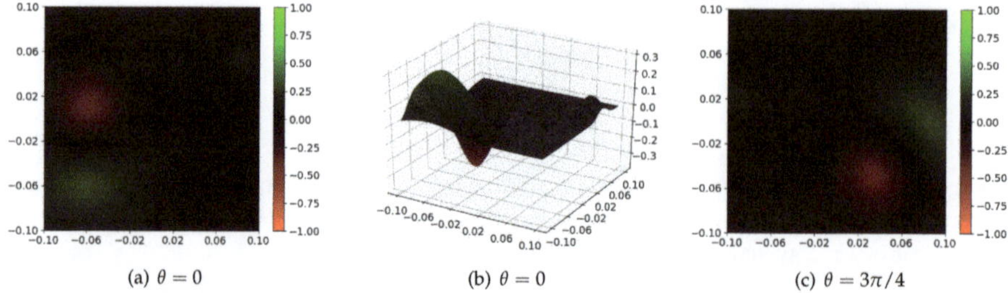

(a) $\theta = 0$ (b) $\theta = 0$ (c) $\theta = 3\pi/4$

Figure 6. Sample GMM filter in layer ϕ^1 of the CNN trained in the described experiment: (**a,b**) 2D and 3D filter visualization, respectively; (**c**) GMM filter rotated by an angle θ. In all cases, red color represents a negative value, and green color represents a positive one. Black color corresponds to values close to 0.

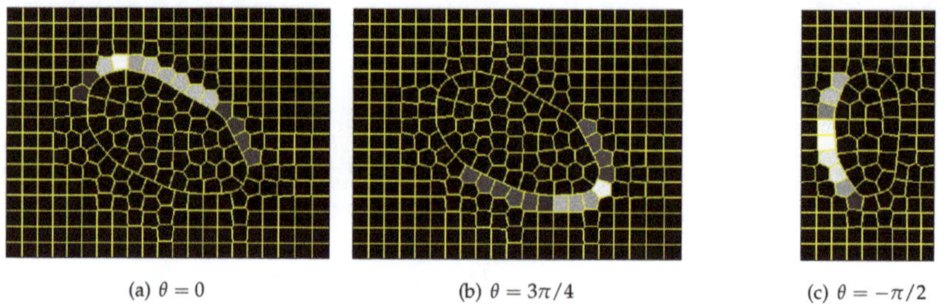

(a) $\theta = 0$ (b) $\theta = 3\pi/4$ (c) $\theta = -\pi/2$

Figure 7. Outputs of the trained CNN for different filter rotation angles θ: (**a,b**) results for training image shown in Figure 5; (**c**) result for the image from test set. It should be noticed that network Φ_0 trained to detect structures in their basic orientation is able to give a reasonable answer when its rotated version Φ_θ is used.

Table 1. MSE errors of network Φ_0 (trained using $\mathcal{D}_0^{TR}$ to detect horizontal edges) and its rotated versions Φ_θ calculated for datasets $\mathcal{D}_\theta$ with different expected orientation of edges. As was expected, when the edge rotation corresponds to network (filter) rotation, the MSE errors are significantly smaller than errors obtained for the original network Φ_0. The presented errors in fact involve only a small number of superpixels as both in network outputs and in expected graphs, most of the values are equal to 0.

	$\theta =$	0	$3\pi/4$	$-\pi/2$	
	$\mathcal{D}_\theta^{TR}$	**4**	32.7	32.9	
Φ_0	$\mathcal{D}_\theta^{VA}$	**4.1**	33	31.7	$\times\,10^{-4}$
	$\mathcal{D}_\theta^{TE}$	**9.3**	36	30.5	
	$\mathcal{D}_\theta^{TR}$	**4**	6.1	7.2	
Φ_θ	$\mathcal{D}_\theta^{VA}$	**4.1**	5.2	7.8	$\times\,10^{-4}$
	$\mathcal{D}_\theta^{TE}$	**9.3**	11.4	11.4	

3.1. Assumptions

To apply the proposed approach, the content of every image in the UBEAR dataset needed to be represented as a graph. For that purpose, first, images and binary masks with precise ear localization were scaled down to have only 0.25 of their original size. Next, superpixel detection was performed with the use of the SLIC algorithm [26]. The expected number of superpixels, which is a parameter of SLIC algorithm, was determined by their expected average area A. Two possible configurations were considered with $A = 256$ and $A = 128$. They led to around 300 and 600 superpixels per image respectively.

Having superpixels generated, two graphs were created: input graph and expected, output graph. Nodes of both graphs corresponded to superpixels. In the input graph, the feature vector assigned to node contained the average intensity of image pixels covered by a given superpixel (Figure 3a,b). In the expected, output graph, it was the average intensity of pixels taken from binary masks multiplied by scaling constant $W > 0$. It should be noted that in the latter case, values assigned to nodes need not be equal either to zero or W as the borders of superpixels need not coincide with the borders of the ear region (Figure 3c,d). Two possible values of scaling constant were considered. These were $W = 1$ and $W = 100$. This constant was introduced based on our earlier experience with classic CNN applications. Such a procedure allowed avoiding, during network training, local minima with all responses equal to zero.

Nodes in the considered graphs must be connected with directed edges. To determine which airs of nodes should be connected, first, the adjacency of all superpixels was examined. Two nodes were connected with an edge if there existed a path (path in this context means a sequence of superpixels) of length shorter than or equal to a given number D, connecting corresponding superpixels. Here, also two configurations were analyzed, where $D = 1$ and $D = 2$ (Figure 8). Selecting a higher value of D, we should be able to increase the size of the visual field, i.e., increase the number of input nodes, which influences the single output node. Loops connecting nodes were allowed because thanks to that the node could express its influence on the output assigned to this node.

After a series of initial trials, a network architecture with $L = 4$ layers ϕ was found to be the optimal one. The number of filter groups in those layers, and hence the number of output graphs, was equal to 20, 10, 5, and 1, respectively. The number of filters in the given group corresponded to the number of layer inputs. In the first layer, it was one, and in the subsequent layers, it depended on the output of the previous layer. In all layers except the last one, the ReLU activation function was used. The last layer had an identity activation function assigned. The number of GMM components in all filters φ was equal to $J = 4$. As before, while training, MSE loss, as well as the Adam optimizer were used. This time, however, a smaller learning rate, equal to 10^{-4}, was considered. For weight initialization, the Glorot scheme was used [40].

To detect ears based on the output of the network, simply the node (superpixel) with the highest response was sought (Figure 9a,b). In order to evaluate if this detection was correct, it was checked whether the superpixel (its centroid) lied inside the bounding box surrounding the ear region. That rectangle was found using the original binary masks provided in the UBEAR dataset and was slightly enlarged to take into account the size of the superpixels (Figure 9c,d).

In all the experiments, images from the UBEAR dataset were split into three sets: training $\mathcal{D}^{TR}$, validation $\mathcal{D}^{VA}$, and test set $\mathcal{D}^{TE}$. The split was made made based on person identifiers, i.e., images of the same person were assigned always to the same set. This should allow checking if the trained models were able to generalize the acquired knowledge and respond correctly for new people. The validation set was used to prevent overfitting by the selection of the optimal model from among models created in the training phase. The number of people in the discussed sets was equal to 75, 25, and 26, respectively. Only left ears were considered since the proposed approach should work for right ears in the same way. What is more, mirror transformation of GMM filters should also allow applying the network trained with left ears to work for right ears, as well.

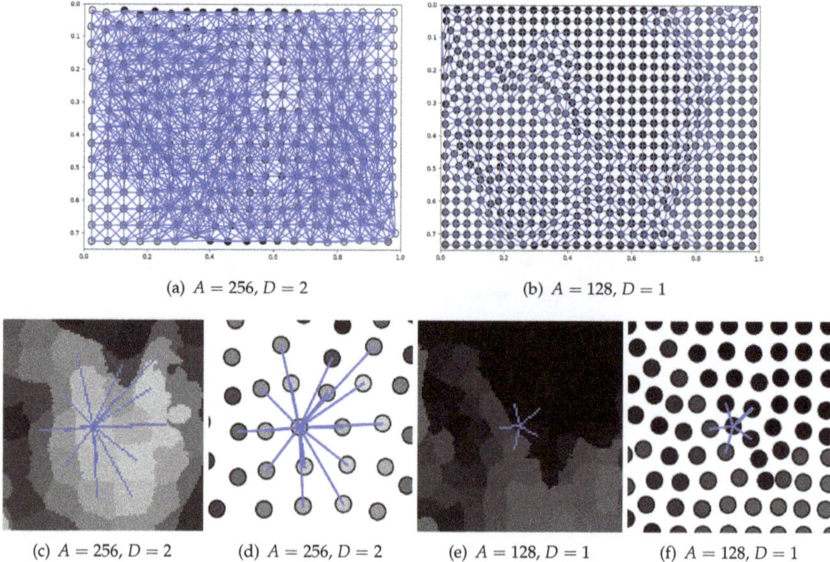

(a) $A = 256, D = 2$ (b) $A = 128, D = 1$

(c) $A = 256, D = 2$ (d) $A = 256, D = 2$ (e) $A = 128, D = 1$ (f) $A = 128, D = 1$

Figure 8. Graphs generated for images shown in Figure 3 for different superpixels' number and different node neighborhoods (parameters A and D, respectively): (**a,b**) full graph with all edges; (**c,e**) superpixels with the local neighborhood of the selected node; (**d,f**) selected graph node with its neighborhood. As the image scale was preserved, it can be observed that when A was smaller, the smaller image region was processed by CNN. Consequently, not only D but also A influenced the size of the effective visual field.

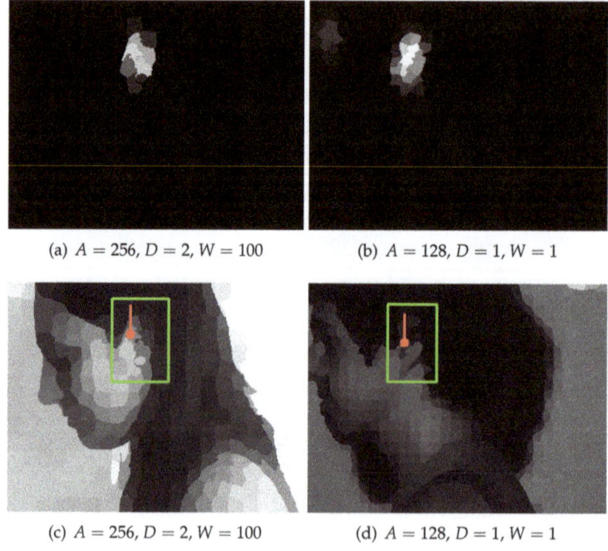

(a) $A = 256, D = 2, W = 100$ (b) $A = 128, D = 1, W = 1$

(c) $A = 256, D = 2, W = 100$ (d) $A = 128, D = 1, W = 1$

Figure 9. Detection results for images shown in Figure 3: (**a,b**) network output (it was scaled back using W and cut to $[0, 255]$ interval); (**c,d**) detection visualization (the green rectangle represents the expected bounding box, and the red dot indicates selected superpixel; the red line shows the orientation of filters in the network, and the vertical line corresponds to basic orientation).

3.2. Experiment I

In the first experiment, the CNN network was trained on a UBEAR subset $\mathcal{D}_M^{TR}$ containing only heads in their standard orientation (pose M in the UBEAR dataset). Every combination of parameters A, D, and W was tested to select the optimal one. The obtained results and cardinalities of the considered training, validation, and test sets are gathered in Table 2. In all cases, results were satisfactory with the correct detection rate bigger than 90%. It is worth noting that in three cases, the detection accuracy for the training set was equal to 100%. This seemed, however, to be slightly overfitted, and as the best configuration, the one with $A = 256$, $D = 2$, and $W = 100$ was indicated.

The closer analysis of the results revealed that, surprisingly, representation with $A = 256$ was not worse than representation with a bigger number of superpixels obtained when $A = 128$. It could be expected that in the latter case, when more details are given, the accuracy would increase. The explanation can be the fact that in both cases, the same network architecture was used, and consequently, for smaller A, the effective visual field was also smaller. What is more, in both cases, the same number 2000 of training iterations was used, and perhaps, more details required longer training. Nevertheless, since for $A = 256$, the results were satisfactory and, thanks to the simpler representation, graph processing was faster, this seemed to be a reasonable choice for the discussed problem. In the case of other parameters, the configurations with $D = 2$ and $W = 100$ seemed to lead to models with better generalization abilities. For them, the detection accuracy was higher when validation and test sets were considered. Those observations were also confirmed by the training characteristic depicted in Figure 10. For optimal values of the parameters (Figure 10a), the best model, the one with the smallest validation error, could be easily selected in the early stage of training. In other cases (Figure 10b), both errors seemed to decrease slowly, and maybe, further training could provide a better solution. Not without significance is also the random initialization of network weights.

Table 2. The ear detection accuracy of the networks trained using the $\mathcal{D}_M^{TR}$ set for different combinations of parameters A, D, and W. Additionally, the last column contains the cardinality of every considered set. It is worth noticing that networks trained using only samples for basic head orientation (pose M) can successfully detect ear not only for other people, but also for different orientations (poses U and D). Naturally, in the latter case, detection accuracy was significantly smaller.

$A =$	256				128				
$D =$	1		2		1		2		#$\mathcal{D}$
$W =$	1	100	1	100	1	100	1	100	
$\mathcal{D}_M^{TR}$	99.52%	97.83%	100%	**99.76%**	98.07%	98.07%	100%	100%	415
$\mathcal{D}_M^{VA}$	91.85%	97.04%	94.7%	**97.04%**	92.59%	93.33%	93.33%	95.56%	135
$\mathcal{D}_M^{TE}$	92.86%	96.1%	90.91%	**98.05%**	94.16%	93.51%	99.35%	96.75%	154
$\mathcal{D}_{U,D}^{TR}$	73.73%	75.36%	70.88%	**80.45%**	74.95%	77.19%	80.45%	79.23%	491
$\mathcal{D}_{U,D}^{VA}$	79.19%	81.88%	73.15%	**87.92%**	78.52%	77.18%	85.23%	81.88%	149
$\mathcal{D}_{U,D}^{TE}$	74.73%	83.33%	77.42%	**81.72%**	77.42%	74.73%	82.26%	81.18%	186

In Figure 11, additional samples of good and wrong detections of the optimal network are presented. It can be observed that correct detections were possible in different illumination conditions and with different background, as well as in situations when there were additional objects located in the ear neighborhood. The typical reasons for the detection mistakes were: questionable annotations (head orientation slightly different than expected), position of the ear close at the image border, and highly occluded ears.

To have a better insight into the process of single graph processing, also the selected outputs of convolutional layers ϕ are presented in Figure 12. In classic CNNs the first layers are usually responsible for the detection of some local image characteristics. Although here, the interpretation was not that obvious, that kind of behavior to a certain extent can also be observed. In Figure 12a, for example, the detection of vertical edges seemed to take place.

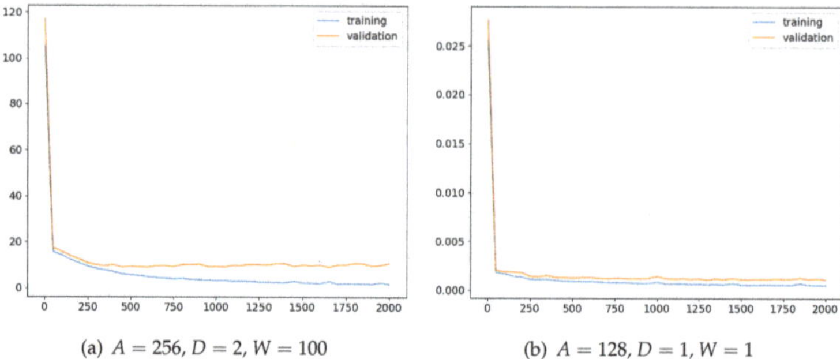

(a) $A = 256, D = 2, W = 100$

(b) $A = 128, D = 1, W = 1$

Figure 10. Training characteristics of two models. The plots present errors for training $\mathcal{D}_M^{TR}$ and validation $\mathcal{D}_M^{VA}$ sets that were calculated every 50 epochs. On the left, the training run for the best parameter combination is depicted. It can be observed that to select the optimal network, the model generated after 600 epochs should be chosen. On the right, another combination of parameters was used. This time, however, no epoch can be indicated where model overfitting seemed to take place. Probably, the training should be continued further.

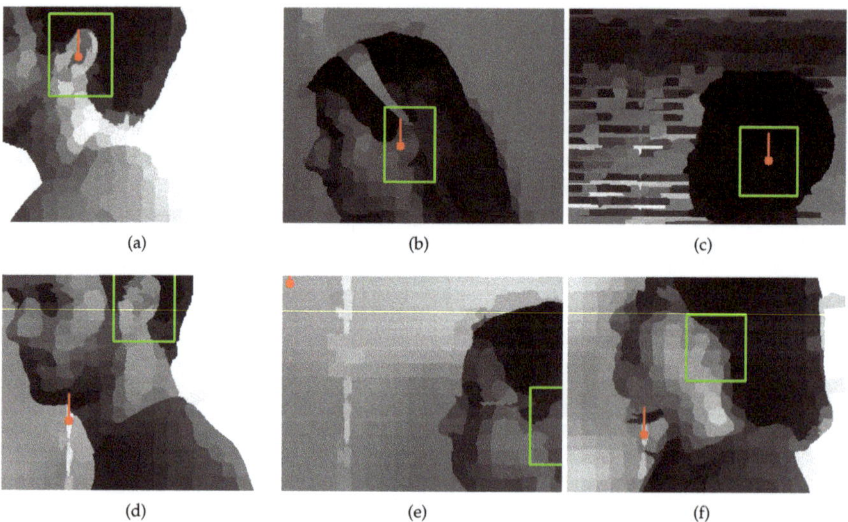

(a)

(b)

(c)

(d)

(e)

(f)

Figure 11. Examples of detections for the best, trained network for images in $\mathcal{D}_M^{VA}$ and $\mathcal{D}_M^{TE}$: (**a**–**c**) correct detections; (**d**–**f**) wrong detections. The network is able to respond correctly in different illumination conditions and for different backgrounds. Problems can be observed when ears are not fully visible (image border or occlusion) and when the head orientation is different than expected (wrong annotation). The convention of the result presentation was described in Figure 9's caption.

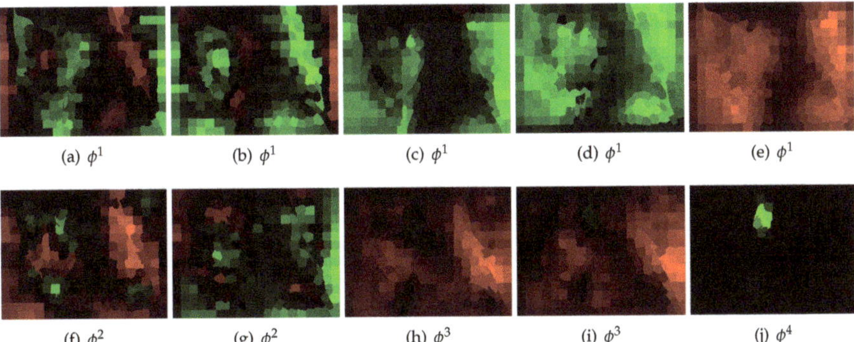

(a) ϕ^1 (b) ϕ^1 (c) ϕ^1 (d) ϕ^1 (e) ϕ^1

(f) ϕ^2 (g) ϕ^2 (h) ϕ^3 (i) ϕ^3 (j) ϕ^4

Figure 12. Selected, raw outputs of convolutional layers for the image shown in Figure 3a and the optimal network. In the first two layers (ϕ^1 and ϕ^2), the person outlines can be still observed, so probably some local image characteristics are extracted here. This behavior is typical also for classic CNNs. In the final layer ϕ^4, the output allows finding ear position. For visualization purposes, every output was scaled separately (they cannot be compared). Red color denotes negative and green positive values.

3.3. Experiment II

In the second experiment, the best network, trained to detect ears in their basic orientation (pose M), was applied to detect ears when head was rotated in the image plane (poses U and D). This network, trained only using samples from $\mathcal{D}_M^{TR}$, will be further denoted as Φ_0 to indicate that it was not rotated. In Table 2, initial detection results for subsets of $\mathcal{D}_{U,D}$ obtained with this network are presented. These results, between 80% and 90%, were surprisingly good. Three explanations seem to be possible. Firstly, apparently, in the UBEAR dataset, most of the cases with U and D pose were similar to pose M (the head rotation was not large). Secondly, ear orientation relative to the head is an individual feature. Slightly rotated ears in $\mathcal{D}_M^{TR}$ could allow Φ_0 to learn how to recognize them in $\mathcal{D}_{U,D}$. Finally, which in general is an interesting hypothesis, to detect ears, it may be sufficient to observe only the configuration of image regions containing head. Humans need not see ear details to make correct detections. Perhaps, our CNN, working on reduced graph representations, did the same thing.

To check if network rotations can help in the detection of rotated ears, we prepared a set containing 11 networks Φ_θ. They were created based on network Φ_0 where filters φ were rotated by angles θ equally distributed in interval $[-\pi, \pi]$. Of course, the original network Φ_0 was in this set. Next, every image was processed by all those networks, and that output was considered to be the final result, for which the maximum node value was observed (Figure 13). It was expected that such system would possess the rotation equivariance property, i.e., the network Φ_θ with correct angle θ correlated with ear orientation should give the rotated response of the network Φ_0 for basic ear orientation.

The results obtained in this way are presented in Table 3. To our surprise, they were worse than the result of separate network Φ_0. This means that the network rotations introduce additional maxima in the wrong regions of the image. They can be observed in Figures 13 and 14. Two typical reasons for incorrect detections are presented there. Firstly, areas, which locally, at a certain angle, can be considered similar to the ear, were indicated (Figure 14f). This behavior could be expected and cannot be avoided at this level of image representation. Secondly, there are maxima in completely unexpected locations (Figure 14b). We have a suspicion that those artifacts are caused by the characteristic of superpixels generated by the SLIC algorithm in regions of uniform color. They were very regular and, since the network Φ_0 was trained only for one orientation, it was not able to give correct responses in situations when the graph was rotated.

After further analysis of the results, we also noticed that even if the output with maximum node value did not allow improving the detection results, the ear localization was frequently indicated correctly by one of the networks Φ_θ (Figure 14). To check if this was a general rule, we conducted an additional experiment where the results were accepted (correct detection) if any network Φ_θ was able to solve the task. Those results are also shown in Table 3. Accuracy calculated in this way, for all poses, was above 96%. It proved that the required information was not lost and rotations of the trained filters allowed constructing a satisfactory solution.

3.4. Discussion

In the literature, several results can be found concerning the UBEAR dataset. In [10], authors reported the accuracy of their method to be equal to 98.22%. This is a significantly better result than other, mentioned in that paper, techniques. The traditional Faster R-CNN was able to reach 65.56%, whereas AdaBoost gained 51.74%. An ensemble of classic CNNs, described in [12], achieved an accuracy equal to 75.08%. Our results cannot be directly compared with these values because of several reasons. Firstly, in the mentioned papers, models were trained using data coming from datasets other than UBEAR. Secondly, other methodologies were used to indicate detections (bounding boxes), and consequently, other evaluation measures had to be used to verify their correctness. Thirdly, we took into account only left ears assuming that for the right ones, our approach would behave in a similar way. Finally, our network was trained using only one pose of the head.

That is why, to show convincingly and objectively the quality of our method, we decided to train our model using set $\mathcal{D}_{ALL}^{TR}$ with all the poses. The obtained results are included in Table 3. The network Φ_{ALL}, trained in this way, was able to recognize images in the training, validation, and test sets with accuracies of 98.75%, 92.91%, and 94.15%, respectively, which were very satisfactory. Nevertheless, since the training data were more complex, to further improve them, we extended our basic architecture by an additional layer on the input of the network. This layer consisted of 20 filter groups. The results of that network, denoted as Φ_{ALL}^+, are also shown in Table 3. One can observe an improvement, in particular for images in validation set $\mathcal{D}_{ALL}^{VA}$.

Table 3. Comparison of the detection accuracy for different subsets of the UBEAR dataset and different detection models. The reference solution is network Φ_0, which was the best network trained using only only samples from $\mathcal{D}_M^{TR}$ (basic head orientation). The second column represents the approach where the maximum of networks Φ_θ outputs indicates the ear position. In the third column, detection was considered successful if at least one Φ_θ output showed correct ear position. It presents the maximum possible accuracy if the correct θ can be found. The fifth and sixth column contain results of models trained using images with all head poses gathered in set $\mathcal{D}_{ALL}^{TR}$. Network Φ_{ALL} has the same architecture as Φ_0 (4 layers), whereas Φ_{ALL}^+ has an additional layer at the beginning (5 layers). The last column contains the cardinality of every considered set.

	max Φ_θ	any Φ_θ	Φ_0	Φ_{ALL}	Φ_{ALL}^+	#$\mathcal{D}$
$\mathcal{D}_M^{TR}$	94.22%	**100**%	99.76%	99.28%	**100**%	415
$\mathcal{D}_M^{VA}$	83.7%	**100**%	97.04%	94.81%	**100**%	135
$\mathcal{D}_M^{TE}$	84.42%	**99.35**%	98.05%	95.45%	95.45%	154
$\mathcal{D}_{U,D}^{TR}$	62.93%	**96.54**%	80.45%	99.39%	99.8%	491
$\mathcal{D}_{U,D}^{VA}$	69.8%	**98.66**%	87.92%	97.99%	97.32%	149
$\mathcal{D}_{U,D}^{TE}$	71.51%	**98.82**%	81.72%	94.62%	97.85%	186
$\mathcal{D}_{ALL}^{TR}$	67.67%	93.09%	83.25%	98.75%	**99.56**%	1361
$\mathcal{D}_{ALL}^{VA}$	68.79%	97.87%	88.89%	92.91%	**96.45**%	423
$\mathcal{D}_{ALL}^{TE}$	67.45%	93.37%	81.87%	94.15%	**94.15**%	513

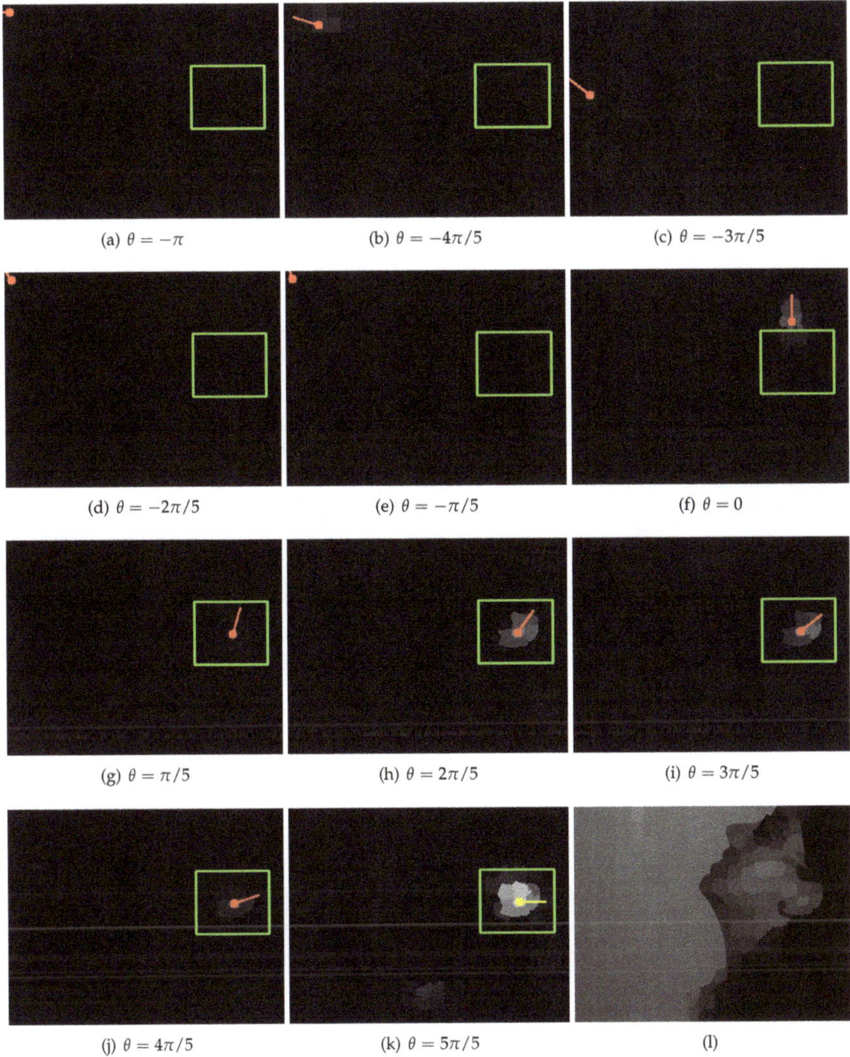

Figure 13. Pictures (**a**–**k**) present the outputs of successive networks Φ_θ. The green rectangle shows the expected ear location. Yellow and red dots with a line indicate superpixels with maximum value (yellow identifies the maximum among all the outputs; the line shows the orientation of filters in the network). The last picture (**l**) shows the input image. Starting from angle $\theta = \pi/5$, networks are able to detect ear correctly. In (**b,k**), unexpected artifacts can be noticed. Picture (**f**) demonstrates an alternative detection region, which, when observed locally, can be indeed mistakenly recognized as an ear.

The presented results revealed also indirectly that there was redundant information in image pixels. In order to detect cars, it was not required to operate on millions of pixels, which usually leads to very complex models. Classic CNNs, usually used for semantic segmentation (e.g., FCN [27], DeepLab [28], or SegNet [29]), require a big computational effort because they have tens of millions trainable parameters. Our network had only several thousands of parameters. This, in turn, sped up both the training and processing of single images. On the same CPU, the forward pass of one graph took on average 0.14 s, while FCN processed the one scaled-down image about 15.6 times slower, i.e., in 2.19 s. Even taking into account the time required to transform the image into a graph (superpixel generation with the use of SLIC algorithm lasted 1.65 s), our approach allowed getting very good results significantly faster.

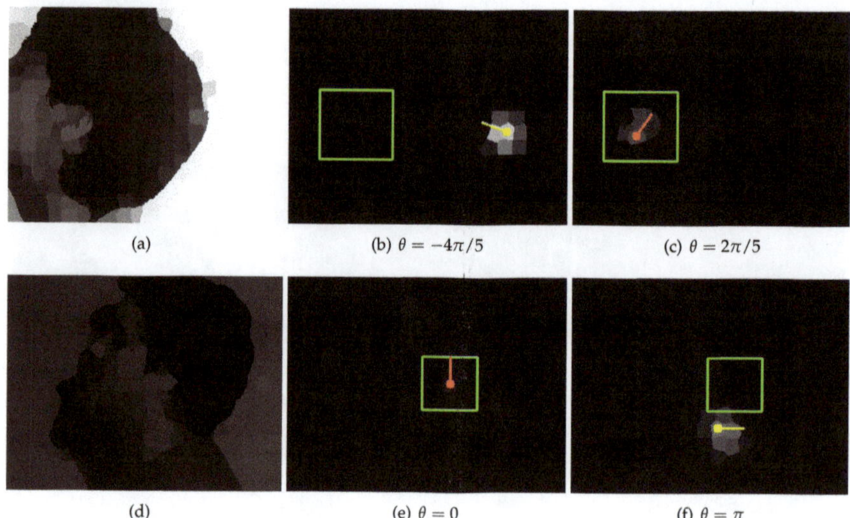

Figure 14. Selected outputs of networks Φ_θ for images where the maximum node value does not allow detecting ear correctly: (**a,d**) input image; (**c,e**) output with correct detection; (**b,f**) output with the node having the maximum value, wrong detection. The unexpected local maxima can be observed in areas of uniform color. The convention of the result presentation was described in Figure 13 caption.

4. Summary and Future Work

In this work, it was proven that the application of CNN operating on graphs for semantic segmentation allowed effectively solving a biometric task of ear detection. The best trained model was able to achieve more than 94% (depending on the considered subset) accuracy on the UBEAR dataset, which contained images with different head orientations, different illumination conditions, and where occlusions and motion artifacts were present as well. This result (Table 3) was comparable with the best results reported in the literature on the same dataset. What is more, the reduced, superpixel based image representation (hundreds of superpixels instead of millions of pixels) allowed constructing a relatively simple model (fewer parameters), which processed data faster than classic CNNs.

We also showed that the specific, used in this paper, network with GMM filters could be used to construct a system having the rotation equivariance property (Figure 2). It can be trained with a limited amount of data, where only structures in one orientation are available and no augmentation is used. The experiments revealed that such a model potentially allows achieving results even better than the model trained using structures in all their possible orientations (Table 3). Additional investigation is, however, required to understand and filter out the artifacts that appear.

The superpixel based image representation used in this work is not the only option. We are currently exploring other alternatives focusing not only on the character of the elements describing image content, but also on their faster generation. This and the short processing time of our networks should allow constructing systems able to detect ears efficiently on devices with only a CPU available.

Another interesting research direction is further theoretical analysis of the presented approach. It was shown that even a very coarse representation (relatively small number of superpixels) of image content allowed getting satisfactory results. This is not fully surprising since humans can do that with 100% accuracy for images in the UBEAR dataset. Apparently, ear details are not required to detect them correctly. It is suspected that for humans, it is enough to identify only the region containing head. Further research should show if such a mechanism also takes place in our networks. All the more, such an analysis, thanks to the reduced image content representation, should be easier than the analysis of classic CNNs. Firstly, this is because, we were not operating on a huge set of pixels, and consequently, the analyzed network was simpler (only a few layers are enough to cover a large visual field). Secondly, this is because humans are not operating consciously directly on pixels, and the explanation of the algorithm behavior in terms of easily understandable, small, and homogeneous regions will be more natural and convincing. That potential of explainability can be considered as an additional advantage of the presented technique.

Author Contributions: Conceptualization, A.T. and P.S.S.; methodology, A.T. and P.S.S.; software, A.T.; validation, A.T. and P.S.S.; formal analysis, A.T.; investigation, A.T.; resources, A.T.; data curation, A.T.; writing, original draft preparation, A.T. and P.S.S.; writing, review and editing, P.S.S.; visualization, A.T.; supervision, P.S.S.; project administration, P.S.S.; funding acquisition, A.T.

Funding: This project has been partly funded with support from the National Science Centre, Republic of Poland, Decision Number DEC-2012/05/D/ST6/03091.

Conflicts of Interest: The authors declare no conflict of interest.

References

1. Gutiérrez, L.; Melin, P.; López, M. Modular Neural Network for Human Recognition from Ear Images Using Wavelets. In *Soft Computing for Recognition Based on Biometrics*; Melin, P.; Kacprzyk, J.; Pedrycz, W., Eds.; Springer: Berlin/Heidelberg, Germany, 2010; pp. 121–135.
2. Žiga Emeršič.; Štruc, V.; Peer, P. Ear recognition: More than a survey. *Neurocomputing* **2017**, *255*, 26–39. [CrossRef]
3. Pflug, A.; Busch, C. Ear biometrics: a survey of detection, feature extraction and recognition methods. *IET Biometr.* **2012**, *1*, 114–129. [CrossRef]
4. Nanni, L.; Lumini, A. A multi-matcher for ear authentication. *Pattern Recognit. Lett.* **2007**, *28*, 2219–2226. [CrossRef]
5. Tian, L.; Mu, Z. Ear recognition based on deep convolutional network. In Proceedings of the 2016 9th International Congress on Image and Signal Processing, BioMedical Engineering and Informatics (CISP-BMEI), Datong, China, 15–17 October 2016; pp. 437–441.
6. Cireşan, D.C.; Meier, U.; Masci, J.; Gambardella, L.M.; Schmidhuber, J. Flexible, High Performance Convolutional Neural Networks for Image Classification. In Proceedings of the Twenty-Second International Joint Conference on Artificial Intelligence—Volume Two (IJCAI'11), Barcelona, Spain, 19–22 July 2011.
7. LeCun, Y.; Bengio, Y. Convolutional Networks for Images, Speech, and Time-Series. In *The Handbook of Brain Theory and Neural Networks*; Arbib, M.A., Ed.; MIT Press: Cambridge, MA, USA, 1995.
8. Emeršič, Ž.; Gabriel, L.L.; Štruc, V.; Peer, P. Pixel-wise Ear Detection with Convolutional Encoder-Decoder Networks. *arXiv* **2017**, arXiv:1702.00307.
9. Cintas, C.; Quinto-Sánchez, M.; Acuña, V.; Paschetta, C.; de Azevedo, S.; Cesar Silva de Cerqueira, C.; Ramallo, V.; Gallo, C.; Poletti, G.; Bortolini, M.C.; et al. Automatic ear detection and feature extraction using Geometric Morphometrics and convolutional neural networks. *IET Biometr.* **2017**, *6*, 211–223. [CrossRef]
10. Zhang, Y.; Mu, Z. Ear Detection under Uncontrolled Conditions with Multiple Scale Faster Region-Based Convolutional Neural Networks. *Symmetry* **2017**, *9*, 53. [CrossRef]
11. Ganapathi, I.I.; Prakash, S.; Dave, I.R.; Bakshi, S. Unconstrained ear detection using ensemble-based convolutional neural network model. *Concurr. Comput. Pract. Exp.* **2019**, e5197. [CrossRef]
12. Raveane, W.; Galdámez, P.L.; González Arrieta, M.A. Ear Detection and Localization with Convolutional Neural Networks in Natural Images and Videos. *Processes* **2019**, *7*, 457. [CrossRef]
13. Dieleman, S.; Willett, K.W.; Dambre, J. Rotation-invariant convolutional neural networks for galaxy morphology prediction. *Mon. Not. R. Astron. Soc.* **2015**, *450*, 1441–1459. [CrossRef]
14. Laptev, D.; Savinov, N.; Buhmann, J.M.; Pollefeys, M. TI-POOLING: Transformation-Invariant Pooling for Feature Learning in Convolutional Neural Networks. *arXiv* **2016**, arXiv:1604.06318.

15. Marcos, D.; Volpi, M.; Tuia, D. Learning rotation invariant convolutional filters for texture classification. In Proceedings of the 2016 23rd International Conference on Pattern Recognition (ICPR), Cancun, Mexico, 4–8 December 2016.

16. Weiler, M.; Hamprecht, F.A.; Storath, M. Learning Steerable Filters for Rotation Equivariant CNNs. *arXiv* **2017**, arXiv:cs.LG/1711.07289.

17. Tarasiuk, P.; Pryczek, M. Geometric Transformations Embedded into Convolutional Neural Networks. *J. Appl. Comput. Sci.* **2016**, *24*, 33–48.

18. Bronstein, M.M.; Bruna, J.; LeCun, Y.; Szlam, A.; Vandergheynst, P. Geometric Deep Learning: Going beyond Euclidean data. *IEEE Signal Process. Mag.* **2017**, *34*, 18–42. [CrossRef]

19. Duvenaud, D.K.; Maclaurin, D.; Iparraguirre, J.; Bombarell, R.; Hirzel, T.; Aspuru-Guzik, A.; Adams, R.P. Convolutional Networks on Graphs for Learning Molecular Fingerprints. In *Advances in Neural Information Processing Systems 28*; Cortes, C., Lawrence, N.D., Lee, D.D., Sugiyama, M., Garnett, R., Eds.; Curran Associates, Inc.: New York, NY, USA, 2015; pp. 2224–2232.

20. Niepert, M.; Ahmed, M.; Kutzkov, K. Learning Convolutional Neural Networks for Graphs. In Proceedings of the 33rd International Conference on Machine Learning, New York, NY, USA, 19–24 June 2016; Balcan, M.F., Weinberger, K.Q., Eds.; PMLR: New York, NY, USA, 2016; Volume 48, pp. 2014–2023.

21. Atwood, J.; Towsley, D. Diffusion-Convolutional Neural Networks. *arXiv* **2015**, arXiv:cs.LG/1511.02136.

22. Hamilton, W.L.; Ying, R.; Leskovec, J. Inductive Representation Learning on Large Graphs. In Proceedings of the 31st International Conference on Neural Information Processing Systems (NIPS'17), Long Beach, CA, USA, 4–9 December 2017; Curran Associates Inc.: New York, NY, USA, 2017; pp. 1025–1035.

23. Fey, M.; Lenssen, J.E.; Weichert, F.; Müller, H. SplineCNN: Fast Geometric Deep Learning with Continuous B-Spline Kernels. *arXiv* **2017**, arXiv:1711.08920.

24. Monti, F.; Boscaini, D.; Masci, J.; Rodolà, E.; Svoboda, J.; Bronstein, M.M. Geometric Deep Learning on Graphs and Manifolds Using Mixture Model CNNs. In Proceedings of the 2017 IEEE Conference on Computer Vision and Pattern Recognition, CVPR 2017, Honolulu, HI, USA, 21–26 July 2017; IEEE Computer Society: Washington, DC, USA, 2017; pp. 5425–5434.

25. Defferrard, M.; Bresson, X.; Vandergheynst, P. Convolutional Neural Networks on Graphs with Fast Localized Spectral Filtering. In *Advances in Neural Information Processing Systems 29*; Lee, D.D., Sugiyama, M., Luxburg, U.V., Guyon, I., Garnett, R., Eds.; Curran Associates, Inc.: New York, NY, USA, 2016; pp. 3844–3852.

26. Achanta, R.; Shaji, A.; Smith, K.; Lucchi, A.; Fua, P.; Susstrunk, S. SLIC Superpixels Compared to State-of-the-Art Superpixel Methods. *IEEE Trans. Pattern Anal. Mach. Intell.* **2012**, *34*, 2274–2282. [CrossRef] [PubMed]

27. Long, J.; Shelhamer, E.; Darrell, T. Fully convolutional networks for semantic segmentation. In Proceedings of the 2015 IEEE Conference on Computer Vision and Pattern Recognition (CVPR), Boston, MA, USA, 7–12 June 2015; pp. 3431–3440.

28. Chen, L.C.; Papandreou, G.; Schroff, F.; Adam, H. Rethinking Atrous Convolution for Semantic Image Segmentation. *arXiv* **2017**, arXiv:cs.CV/1706.05587.

29. Badrinarayanan, V.; Kendall, A.; Cipolla, R. SegNet: A Deep Convolutional Encoder-Decoder Architecture for Image Segmentation. *IEEE Trans. Pattern Anal. Mach. Intell.* **2017**, *39*, 2481–2495. [CrossRef] [PubMed]

30. Stasiak, B.; Tarasiuk, P.; Michalska, I.; Tomczyk, A. Application of convolutional neural networks with anatomical knowledge for brain MRI analysis in MS patients. *Bull. Polish Acad. Sci. Tech. Sci.* **2018**, *66*, 857–868.

31. Tomczyk, A.; Stasiak, B.; Tarasiuk, P.; Gorzkiewicz, A.; Walczewska, A.; Szczepaniak, P. Localization of Neuron Nucleuses in Microscopy Images with Convolutional Neural Networks. In Proceedings of the 11th International Joint Conference on Biomedical Engineering Systems and Technologies (BIOSTEC 2018), Funchal, Portugal, 19–21 January 2018.

32. Raposo, R.; Hoyle, E.; Peixinho, A.; Proença, H. UBEAR: A Dataset of Ear Images Captured On-the-move in Uncontrolled Conditions. In Proceedings of the 2011 IEEE Workshop on Computational Intelligence inBiometrics and Identity Management (SSCI 2011 CIBIM), Paris, France, 11–15 April 2011; pp. 84–90.

33. Carreira-Perpinan, M.A. Compression Neural Networks for Feature Extraction: Application to Human Recognition from Ear Images. Master's Thesis, Technical University of Madrid, Madrid, Spain, 1995. (In Spanish)

34. Kumar, A.; Wu, C. Automated human identification using ear imaging. *Pattern Recognit.* **2012**, *45*, 956–968. [CrossRef]
35. IIT Delhi Ear Database. Available online: http://www.comp.polyu.edu.hk/~csajaykr/IITD/Database_Ear.htm (accessed on 20 May 2019).
36. AMI Ear Database. Available online: http://ctim.ulpgc.es/research_works/ami_ear_database (accessed on 20 May 2019).
37. Frejlichowski, D.; Tyszkiewicz, N. The West Pomeranian University of Technology Ear Database—A Tool for Testing Biometric Algorithms. In Proceedings of the 7th International Conference on Image Analysis and Recognition, ICIAR 2010, Póvoa de Varzin, Portugal, 21–23 June 2010; Proceedings, Part II; Springer: Berlin/Heidelberg, Germany, 2010; pp. 227–234.
38. Annotated Web Ears (AWE). Available online: http://awe.fri.uni-lj.si (accessed on 20 May 2019).
39. Fey, M.; Lenssen, J.E. Fast Graph Representation Learning with PyTorch Geometric. In Proceedings of the ICLR Workshop on Representation Learning on Graphs and Manifolds, New Orleans, LA, UAS, 6 May 2019.
40. Glorot, X.; Bengio, Y. Understanding the difficulty of training deep feedforward neural networks. In Proceedings of the Thirteenth International Conference on Artificial Intelligence and Statistics, Sardinia, Italy, 13–15 May 2010; Teh, Y.W., Titterington, M., Eds.; PMLR: Sardinia, Italy, 2010; Volume 9, pp. 249–256.

Article

Face Detection Ensemble with Methods Using Depth Information to Filter False Positives

Loris Nanni [1], Sheryl Brahnam [2,*] and Alessandra Lumini [3]

[1] Department of Information Engineering, University of Padova, Via Gradenigo, 6, 35131 Padova, Italy; nanni@dei.unipd.it
[2] Department of Information Technology and Cybersecurity, Missouri State University, 901 S. National Street, Springfield, MO 65804, USA
[3] Dipartimento di Informatica—Scienza e Ingegneria, Università di Bologna, Via Sacchi 3, 47521 Cesena, Italy; alessandra.lumini@unibo.it
* Correspondence: sbrahnam@missouristate.edu

Received: 10 October 2019; Accepted: 25 November 2019; Published: 28 November 2019

Abstract: A fundamental problem in computer vision is face detection. In this paper, an experimentally derived ensemble made by a set of six face detectors is presented that maximizes the number of true positives while simultaneously reducing the number of false positives produced by the ensemble. False positives are removed using different filtering steps based primarily on the characteristics of the depth map related to the subwindows of the whole image that contain candidate faces. A new filtering approach based on processing the image with different wavelets is also proposed here. The experimental results show that the applied filtering steps used in our best ensemble reduce the number of false positives without decreasing the detection rate. This finding is validated on a combined dataset composed of four others for a total of 549 images, including 614 upright frontal faces acquired in unconstrained environments. The dataset provides both 2D and depth data. For further validation, the proposed ensemble is tested on the well-known BioID benchmark dataset, where it obtains a 100% detection rate with an acceptable number of false positives.

Keywords: face detection; depth map ensemble; filtering

---

1. Introduction

One of the most fundamental yet difficult problems in computer vision and human–computer interaction is face detection, the object of which is to detect and locate all faces within a given image or video clip. Face detection is fundamental in that it serves as the basis for many applications [1] that involve the human face, such as face alignment [2,3], face recognition/authentication [4–7], face tracking and tagging [8], etc. Face detection is a hard problem because unlike face localization, no assumptions can be made regarding whether any faces are located within an image [9,10]. Moreover, faces vary widely based on gender, age, facial expressions, and race, and can dramatically change in appearance depending on such environmental conditions as illumination, pose (out-of-plane rotation), orientation (in-plane rotation), scale, and degree of occlusion and background complexity. Not only must a capable and robust face detection system overcome these difficulties, but for many of today's applications, it must also be able to do so in real time.

These challenges have resulted in a large body of literature reporting different methods for tackling the problem of face detection [11]. Yang et al. [12], who published a survey of face detection algorithms developed in the last century, have divided these earlier algorithms into four categories: knowledge-based methods, feature invariant approaches, template-matching methods, and appearance-based methods, the latter demonstrating some superiority compared with the other algorithms thanks to the rise in computing power. In general, these methods formulate face detection as a two-class pattern recognition problem that divides a 2D image into subwindows that are then classified as either containing a face or not [13]. Moreover, these approaches take a monocular perspective in the sense that they forgo any additional sensor or contextual information that might be available.

Around the turn of the century, Viola and Jones [14] presented a 2D detection method that has since become a major source of inspiration for many subsequent face detectors. The famous Viola–Jones (VJ) algorithm achieved real-time object detection using three key techniques: an integral image stratagem for efficient Haar feature extraction, a boosting algorithm (AdaBoost) for an ensemble of weak classifiers, and an attentional cascade structure for fast negative rejection. However, there are some significant limitations to the VJ algorithm that are due to the suboptimal cascades, the considerable pool size of the Haar-like features, which makes training extremely slow, and the restricted representational capacity of Haar features to handle, for instance, variations in pose, illumination, facial expression, occlusions, makeup, and age-related factors [15]. These problems are widespread in unconstrained environments, such as those represented in the Face Detection Dataset and Benchmark (FDDB) [16] where the VJ method fails to detect most faces [17].

Some early Haar-like extensions and enhancements intended to overcome some of these shortcomings include rotated Haar-like features [18], sparse features [19], and polygon features [20]. Haar-like features have also been replaced by more powerful image descriptors, such as local binary patterns (LBP) [21], spatial histogram features [22], histograms of oriented gradients (HoG) [23], multidimensional local Speeded-Up Robust Features (SURF) patches [24], and, more recently, by normalized pixel difference (NPD) [17] and aggregate channel features [25], to name but a few.

Some older feature selection and filtering techniques for reducing the pool size, speeding up training, and improving the underlying boosting algorithm of the cascade paradigm include the works of Brubaker et al. [26] and Pham et al. [27]. In Küblbeck et al. [28], the illumination invariance and speed were improved with boosting combined with modified census transform (MCT); in Huang et al. [29], a method for detecting faces with arbitrary rotation in-plane and rotation off-plane angles in both still images and videos is proposed. For an excellent survey of face detection methods prior to 2010, see [11].

Some noteworthy 2D approaches produced in the last decade include the work of Li et al. [15] at Intel labs, who introduced a two-pronged strategy for the faster convergence speed of the SURF cascade, first by adopting, as with [24], multidimensional SURF features rather than single-dimensional Haar features to describe local patches, and second, by replacing decision trees with logistic regression. Two simple approaches that are also of note are those proposed in Mathias et al. [30], which obtained top performance compared with such commercial face detectors as Google Picasa, Face.com, Intel Olaworks, and Face++. One method is based on rigid templates, which is similar in structure to the VJ algorithm, and the other detector uses a simple deformable part model (DPM), which, in brief, is a generalizable object detection approach that combines the estimation of latent variables for alignment and clustering at the training time with multiple components and deformable parts to manage intra-class variance.

Four 2D models of interest in this study are the face detectors proposed by Nilsson et al. [31], Asthana et al. [32], Liao et al. [33], and Markuš et al. [34]. Nilsson et al. [31] used successive mean quantization transform (SMQT) features that they applied to a Split up sparse Network of Winnows (SN) classifier. Asthana et al. [32] employed face fitting, i.e., a method that models a face shape with a set of parameters for controlling a facial deformable model. Markuš et al. [34] combined a modified VJ method with an algorithm for localizing salient facial landmark points. Liao et al. [33], in addition to proposing the aforementioned scale-invariant NPD features, expanded the original VJ tree classifier with two leaves to a deeper quadratic tree structure.

Another powerful approach for handling the complexities of 2D face detection is deep learning [35–41]. For instance, Girshick et al. [36] were one of the first to use Convolutional Neural Networks (CNN) in combination with regions for object detection. Their model, appropriately named Region-CNN (R-CNN), consists of three modules. In the testing phase, R-CNN generates approximately

2000 category-independent region proposals (module 1), extracts a fixed-length deep feature vector from each proposal using a CNN (module 2), and then classifies them with Support Vector Machines (SVMs) (module 3). In contrast, the deep dense face detector (DDFD) proposed by Farfade et al. [37] requires no pose/landmark annotations and can detect faces in many orientations using a single deep learning model. Zhang et al. [39] proposed a deep learning method that is capable of extracting tiny faces, also using a single deep neural network.

Motivated by the development of affordable depth cameras, another way to enhance the accuracy of face detection is to go beyond the limitations imposed by the monocular 2D approach and include additional 3D information, such as that afforded by the Minolta Vivid 910 range scanner [42], the MU-2 stereo imaging system [43], the VicoVR sensor, the Orbbec Astra, and Microsoft's Kinect [44], the latter of which is arguably the most popular 3D consumer-grade device on the market. Kinect combines a 2D RGB image with a depth map (RGB-D) that initially (Kinect 1) was computed based on the structured light principle of projecting a pattern onto a scene to determine the depth of every object but which later (Kinect 2) exploited the time-of-flight principle to determine depth by measuring the changes that an emitted light signal encounters when it bounces back from objects.

Since depth information is insensitive to pose and changes in illumination [45], many researchers have explored depth maps and other kinds of 3D information [46]; furthermore, several benchmark datasets using Kinect have been developed for both face recognition [44] and face detection [47]. The classic VJ algorithm was adapted to consider depth and color information a few years after Viola and Jones published their groundbreaking work [48,49]. To improve detection rates, most 3D face detection methods combine depth images with 2D gray-scale images. For instance, in Shieh et al. [50], the VJ algorithm is applied to images to detect a face, and then its position is refined via structured light analysis.

Expanding on the work of Shotton et al. [51], who used pair-wise pixel comparisons in depth images to quickly and accurately classify body joints and parts from single depth images for pose recognition, Mattheij et al. [52] compared square regions in a pair-wise fashion for face detection. Taking cues from biology, Jiang et al. [53] integrated texture and stereo disparity information to filter out locations unlikely to contain a face. Anisetti et al. [54] located faces by applying a course detection method followed by a technique based on a 3D morphable face model that improves accuracy by reducing the number of false positives, and Taigman et al. [6] found that combining a 3D model-based alignment with DeepFace trained on the Labeled Faces in the Wild (LFW) dataset [55] generalized well in the detection of faces in an unconstrained environment. Nanni et al. [9] overcame the problem of increased false positives when combining different face detectors in an ensemble by applying different filtering steps based on information in the Kinetic depth map.

The face detection system proposed in this paper is composed of an ensemble of face detectors that utilizes information extracted from the 2D image and depth maps obtained by Microsoft's Kinect 1 and Kinect 2 3D devices. The goal of this paper, which improves the method presented in [9], is to test a set of filters, which includes a new wave-based filter proposed here, on a new collection of face detectors. The main objective of this study is to find those filters that preserve the ensemble's increased rate of true positives while simultaneously reducing the number of false positives. Creating an ensemble of classifiers is a feasible method for improving performance in face detection (see [9]), as well as in many other classification problems. The main reason that ensembles improve face detection performance is that the combination of different methods increases the number of candidate windows and thus the probability of including a previously lost true positive. However, the main drawback of using ensembles in face detection is the increased generation of false positives. The rationale behind the proposed approach is to use some filtering steps to reduce false positives. The present work extends [9] by adding to the proposed ensemble additional face detectors.

The best performing system developed experimentally in this work is validated on the challenging dataset presented in [9] that contains 549 samples with 614 upright frontal faces. This dataset includes depth images as well as 2D images. The results in the experimental section demonstrate that the filtering steps succeed in significantly decreasing the number of false positives without significantly affecting the detection rate of the best-performing ensemble of face detectors. To validate the strength of the proposed new even system further, we validate it on the widely used BioID dataset [56], where it obtains a 100% detection rate with a limited number of false positives. Our best ensemble/filter combination outperforms the method proposed by Markuš et al. [34], which has been shown to surpass the performance of these well-known state-of-the-art commercial face detection systems: Google Picasa, Face++, and Intel Olaworks.

The organization of this paper is as follows. In Section 2, the strategy taken in this work for face detection is described along with the face detectors tested in the ensembles and the different filtering steps. In Section 3, the experiments on the two above-mentioned datasets are presented, along with a description of the datasets, definition of the testing protocols, and a discussion of the experimental results. The paper concludes, in Section 4, by providing a summary with some notes regarding future directions. The MATLAB code developed for this paper, along with the dataset, is freely available at https://github.com/LorisNanni.

2. Materials and Methods

The basic strategy taken in this work is to develop experimentally a high-performing face detection ensemble composed of well-known face detectors. The goal is to obtain superior results without significantly increasing the number of false positives. The system proposed here, as illustrated in Figure 1, is a three-step process.

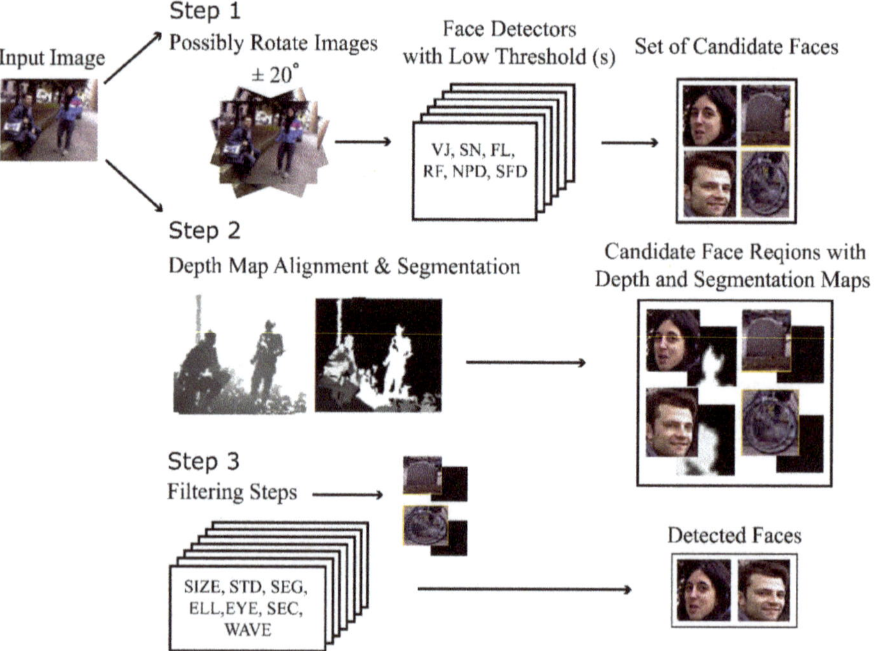

Figure 1. Schematic of the proposed face detection system.

In Step 1, high recall is facilitated by first performing face detection on the color images. A set of six face detectors (experimentally derived, as described in the experimental section) are applied to each image. The face detection algorithms tested in this paper are described in Section 2.2. Before detection, as also illustrated in Figure 1, color images are sometimes rotated {20°, −20°} to handle faces that are not upright. The addition of rotated images is noted in the experimental section whenever these are included in the dataset.

Since this first step is imprecise and therefore produces many false positives, the purpose of Step 2 is to align the depth maps to the color images so that false positives can be winnowed out in Step 3 by applying seven filtering approaches that take advantage of the depth maps. Alignment is accomplished by first calibrating the color and depth data using the calibration technique proposed in Herrera et al. [57]. The positions of the depth samples in 3D space are determined using the intrinsic parameters (focal length and principal point) of the depth camera. Then, these positions are reprojected in 2D space by considering both the color camera's intrinsic parameters and the extrinsic parameters of the camera pair system. Next, color and depth values are associated with each sample, as described in Section 2.1. This operation is applied only to regions containing a candidate face to reduce computation time. Finally, in Step 3, these regions are filtered, as detailed in Section 2.3, to remove false positives from the candidate faces.

2.1. Depth Map Alignment and Segmentation

The color images and depth maps are jointly segmented by a procedure similar to that described in Mutto et al. [58] that has two main stages. In Stage 1, each sample is transformed into a six-dimensional vector. In Stage 2, the point set is clustered using the mean shift algorithm [59].

Every sample in the Kinetic depth map corresponds to a 3D point, p_i, $i = 1, \ldots, N$, with N the number of points. The joint calibration of the depth and color cameras, as described in [57], allows a reprojection of the depth samples over the corresponding pixels in the color image so that each point is associated with the 3D spatial coordinates (x, y, and z) of p_i and its RGB color components. Since these two representations lie in entirely different spaces, they cannot be compared directly, and all components must be comparable to extract multidimensional vectors that are appropriate for the mean shift clustering algorithm. Thus, a conversion is performed so that the color values lie in the CIELAB uniform color space, which represents color in three dimensions expressed by values representing lightness (L) from black (0) to white (100), a value (a) from green (−) to red (+), and a value (b) from blue (−) to yellow (+). This introduces a perceptual significance to the Euclidean distance between the color vectors that can be used in the mean shift algorithm.

Formally, the color information of each scene point in the CIELAB color space, c, can be described with the 3D vector:

$$p_i^c = \begin{bmatrix} L(p_i) \\ a(p_i) \\ b(p_i) \end{bmatrix}, \quad i = 1, \ldots, N. \tag{1}$$

The geometry, g, can be represented simply by the 3D coordinates of each point, thus:

$$p_i^g = \begin{bmatrix} x(p_i) \\ y(p_i) \\ z(p_i) \end{bmatrix}, \quad i = 1, \ldots, N. \tag{2}$$

The scene segmentation algorithm needs to be insensitive to the relative scaling of the point-cloud geometry. Moreover, the geometry and color distances must be brought into a consistent framework. For this reason, all the components of p_i^g are normalized with respect to the average of the standard deviations of the point coordinates in the three dimensions $\sigma_g = (\sigma_x + \sigma_y + \sigma_z)/3$. Normalization produces the vector:

$$\begin{bmatrix} \bar{x}(p_i) \\ \bar{y}(p_i) \\ \bar{z}(p_i) \end{bmatrix} = \frac{3}{\sigma_x + \sigma_y + \sigma_z} \begin{bmatrix} x(p_i) \\ y(p_i) \\ z(p_i) \end{bmatrix} = \frac{1}{\sigma_g} \begin{bmatrix} x(p_i) \\ y(p_i) \\ z(p_i) \end{bmatrix}. \tag{3}$$

To balance the relevance of color and geometry in the merging process, the color information vectors are normalized as well. The average of the standard deviations of the L, a, and b color components are computed producing the final color representation:

$$
\begin{bmatrix} \overline{L}(p_i) \\ \overline{a}(p_i) \\ \overline{b}(p_i) \end{bmatrix} = \frac{3}{\sigma_L + \sigma_a + \sigma_b} \begin{bmatrix} L(p_i) \\ a(p_i) \\ b(p_i) \end{bmatrix} = \frac{1}{\sigma_c} \begin{bmatrix} L(p_i) \\ a(p_i) \\ b(p_i) \end{bmatrix}.
\tag{4}
$$

Once the geometry and color information vectors are normalized, they can be combined for a final representation f:

$$
p_i^f = \begin{bmatrix} \overline{L}(p_i) \\ \overline{a}(p_i) \\ \overline{b}(p_i) \\ \lambda_{\overline{x}} \\ \lambda_{\overline{y}} \\ \lambda_{\overline{z}} \end{bmatrix},
\tag{5}
$$

with the parameter λ adjusting the contribution to the final segmentation of color (low values of λ indicating high color relevance) and geometry (low values indicating high geometry relevance). By adjusting λ, the algorithm can be reduced to a color-based segmentation ($\lambda = 0$) or to a geometry (depth)-only segmentation ($\lambda \rightarrow \infty$) (see [58] for a discussion of the effects that this parameter produces and for automatically tuning λ to an optimal value).

Once the final vectors p_i^f are calculated, they can be clustered by the mean shift algorithm [59] to segment the acquired scene. This algorithm offers an excellent trade-off between segmentation accuracy and computational complexity. For final refinement, regions are removed that are smaller than a predefined threshold, since they are typically due to noise. In Figure 2, examples of a segmented image are shown.

Figure 2. Color image (**left**), depth map (**middle**), and segmentation map (**right**).

2.2. Face Detectors

We perform experiments on the fusion of six face detectors: the four detectors tested in [9] (the canonic VJ algorithm [14], a method using the Split up sparse Network of Winnows (SN) classifier [31], a modification of the VJ algorithm with fast localization (FL) [34], and a face detector based on Discriminative Response Map Fitting (DRMF) [32]), as well as two additional face detectors (the VJ modification using NPD features (NPD) [33] and a high-performance method implemented here: http://dlib.net/face_detector.py.html. In the following, this latter method is called Single Scale-invariant Face Detector (SFD). Each of these face detection algorithms is briefly described below.

2.2.1. VJ

The canonical VJ algorithm [14] is based on Haar wavelets extracted from the integral image. Classification is performed, as noted in the introduction, by combining an ensemble of AdaBoost classifiers that select a small number of relevant descriptors with a cascade combination of weak learners.

The disadvantage of this approach is that it requires considerable training time. However, it is relatively fast during the testing phase. The precision of VJ relies on the threshold s, which is used to classify a face within an input subwindow.

2.2.2. SN

SN [31], available in MATLAB (http://www.mathworks.com/matlabcentral/fileexchange/loadFile.do?objectId=13701&objectType=FILE), feeds SMQT features, as briefly discussed in the Introduction, to a Split up Sparse Network of Winnows (SN) classifier. SMQT enhances gray-level images. This enhancement reveals the structure of the data and additionally removes some negative properties such as gain and bias. This is how SMQT features overcome to some extent the illumination and noise problem.

SMQT features are extracted by moving a patch across the image while repeatedly downscaling and resizing it to detect faces of different sizes. The detection task is performed by the SN classifier, i.e., a sparse network of linear units over a feature space that can be used to create lookup tables.

2.2.3. FL

FL (Fast Localization) [34] is a method that combines a modification of the standard VJ algorithm with a component for localizing a salient facial landmark. An image is scanned with a cascade of binary classifiers that considers a set of reasonable positions and scales. Computing a data structure, such as integral images, an image pyramid, or HoG features, etc., is not required with this method. An image region is classified as having a face when all the classifiers are in agreement that the region contains one. At this stage, another ensemble calculates the position of each facial landmark point. Each binary classifier in the cascade is an ensemble of decision trees that have pixel intensity comparisons in their internal nodes as binary tests. Moreover, they are based on the same feature type, unlike the VJ algorithm that uses five types of Haar-like features. Learning takes place with a greedy regression tree construction procedure and a boosting algorithm.

2.2.4. RF

RF [32] is a face detector based on Discriminative Response Map Fitting (DRMF), which is a specific face fitting technique. DRMF is a discriminative regression method for the Constrained Local Models (CLMs) framework. Precision is adjusted in RF using the sensitivity parameter s that sets both a lower and a higher sensitivity value.

2.2.5. NPD

NPD [33] extracts the illumination and blur invariant NPD features mentioned in the Introduction. NPD is computed as the difference-to-sum ratio between two pixels and is extremely fast because it requires only one memory access using a lookup table. However, because NPD contains redundant information, AdaBoost is applied to select the most discriminative feature set and to construct strong classifiers. The Gentle AdaBoost algorithm [60] is adopted for the deep quadratic trees. The splitting strategy consists in quantizing the feature range into l discrete bins ($l = 256$ in the original paper and here), and an exhaustive search is performed to determine whether a feature lies within a given range $[\theta_1, \theta_2]$. The weighted mean square error is applied as the optimal splitting criterion.

2.3. Filtering Steps

As noted in Figure 1, some of the false positives generated by the ensemble of classifiers are extracted by applying several filtering approaches that take advantage of the depth maps. The filters tested in this work are the set of six tested in [9] (viz. SIZE, STD, SEG, ELL, EYE, and SEC) and a new filter proposed here (viz. WAV), which is based on processing the image with different wavelets.

Each of these filtering techniques is described below. Figure 3 illustrates images rejected by the seven types of filters.

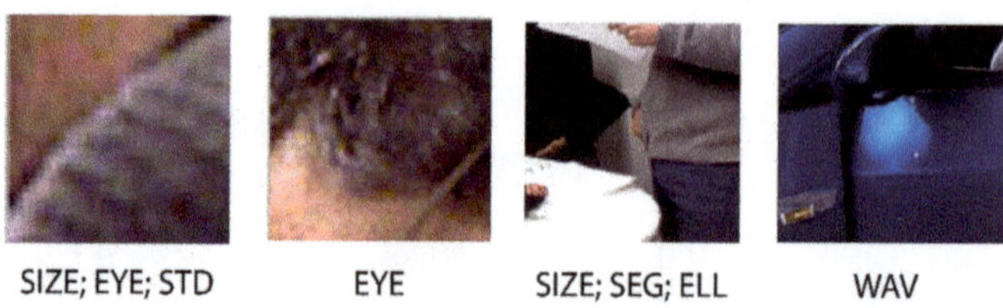

Figure 3. Examples of images rejected by the different filtering methods.

2.3.1. Image Size Filter (SIZE)

SIZE [10] rejects candidate faces based on the size of the face region extracted from the depth map. First, the 2D position and dimension (W_{2D}, h_{2D}) in pixels of a candidate face region are identified by the face detector. Second, this information is used to estimate the corresponding 3D physical dimension in mm (W_{3D}, h_{3D}) as follows:

$$W_{3D} = W_{2D}\frac{\bar{d}}{f_x} \text{ and } h_{3D} = h_{2D}\frac{\bar{d}}{f_x}, \tag{6}$$

where f_x and f_y are the Kinect camera focal lengths computed by the calibration algorithm in [57], and $\bar{d}$ is the average depth of the samples in the candidate bounding box. Face candidate regions are rejected when they lie outside the fixed range in cm [0.075, 0.35]. Note that $\bar{d}$ is defined as the median of the depth samples and is necessary for reducing the impact of noisy samples in the average computation.

2.3.2. Flatness/Unevenness Filter (STD)

STD, as proposed in [9], extracts information from the depth map that relates to the flatness and unevenness of candidate face regions. Flat and uneven faces detected by the classifiers are then removed using the depth map and a segmentation method based on the depth map.

The filtering method is a two-step process. In Step 1, a segmentation procedure using the depth map is applied; in Step 2, the standard deviation (STD) of the pixels of the depth map that belong to the larger segment (i.e., the region obtained by the segmentaion procedure) is calculated from each face candidate region. Those regions whose STD lies outside the range of [0.01, 2.00] are rejected.

2.3.3. Segmentation-Based Filtering (SEG and ELL)

SEG and ELL, proposed in [9], apply the segmented version of the depth image to compare its dimension to its bounding box in SEG or to its shape (which should approximate that of an ellipse) in ELL. From this information, two simple but useful evaluations can be made. In the case of SEG, the relative dimension of the larger area can be compared to the entire candidate image. The candidate regions where the area of the larger region is less than 40% of the entire area are rejected. In the case of ELL, the larger region is given a fitness score using the least-squares criterion to determine its closeness to an elliptical model. This score is calculated here using the MATLAB function fit_ellipse [61]. The candidate regions with a score higher than 100 are rejected.

2.3.4. Eye-Based Filtering (EYE)

EYE, as proposed in [9], uses the presence of eyes in a region to detect a face. In EYE, two robust eye detectors are applied to candidate face regions [62,63]. Regions with a low probability of containing two eyes are rejected.

One of the eye detectors [62] used in EYE is a variant of the Pictorial Structures (PS) model. PS is a computationally efficient framework that represents a face as an undirected graph $G = (V, E)$, where the vertices V correspond to facial features. The edges E describe the local pairwise spatial relationships between the feature set. PS is expanded in [62] so that it can deal with complications in appearance as well as with many of the structural changes that eyes undergo in different settings.

The second eye detector, presented in [63], makes use of color information to build an eye map that highlights the iris. A radial symmetry transform is applied to both the eye map and the original image once the area of the iris is identified. The cumulative results of this enhancement process provide the positions of the eye. Face candidates are rejected in those cases where detection of the eyes fall outside a threshold of 1 for the first approach [62] and of 750 for the second approach [63].

2.3.5. Filtering Based on the Analysis of the Depth Values (SEC)

SEC, as proposed in [9], takes advantage of the fact that most faces, except those where people are lying flat, are on top of the body, while the remaining surrounding volume is often empty. With SEC, candidate faces are rejected when the neighborhood manifests a different pattern from that which is expected.

The difference in the expected pattern is calculated as follows. First, the rectangular region defining a candidate face is enlarged so that the neighborhood of the face in the depth map can be analyzed.

Second, the enlarged region is then partitioned into radial sectors (eight in this work, see Figure 4), each emanating from the center of the candidate face. For each sector Sec_i, the number of pixels n_i are counted whose depth value d_p is close to the average depth value of the face $\bar{d}$, thus:

$$n_i = \left| \left\{ p : \left| d_p - \overline{d} \right| < t_d \wedge p \in Sec_1 \right\} \right| \tag{7}$$

where t_d is a measure of closeness ($t_d = 50$ cm here).

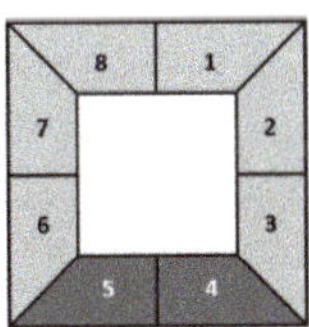

Figure 4. Examples of partitioning of a neighborhood of the candidate face region into 8 sectors (gray area). The lower sectors Sec_4 and Sec_5 that should contain the body are depicted in dark gray [9].

Finally, the number of pixels per sector is averaged on the two lower sectors (Sec_4 and Sec_5) and then again on the remaining sectors, from which two of the values, n_u and n_l respectively, are obtained. The ratio between n_u and n_l is then computed as:

$$\frac{n_l}{n_u} = \frac{\frac{1}{2}(n_4 + n_5)}{\frac{1}{6}(n_1 + n_2 + n_3 + n_6 + n_7 + n_8)}. \tag{8}$$

If the ratio drops below a certain threshold, t_r (where $t_r = 0.8$ here), then the candidate face is removed.

2.3.6. WAV

WAV is a filtering technique that processes an image with different wavelets. With WAV, statistical indicators are extracted (e.g., the mean and variance) and used for discarding candidate images with no faces. Rejection is based on five criteria.

The first criterion applies phase congruency [64] to the depth map of the largest cluster, and the average value is used to discriminate between face/non-face. The segmentation process divides the image into multiple clusters, and only the largest cluster (that is, the one that is most likely to contain the face) is considered. Phase congruency has higher values when there are edges. WAV keeps only those candidates with an acceptable value, i.e., those with a number of edges that is neither too high nor too low, and deletes all others since they most likely contain no faces.

WAV is used here in two ways, but in both cases, Haar-like waves are selected since they often give the best results, as demonstrated in [65]. The first method (second criterion) works on the same principle as the phase congruency test: the Haar wave is applied to each image, and the average value is calculated for each one. However, the second test (third criterion) follows the approach in [50], where edge maps are first extracted and then fitted to an ellipse (the typical shape of a face). If an ellipse is found, then the image is rotated by an angle given by the intersection between the origin and the major axis of the ellipse, and the filter is applied to the rotated image. If no elliptical shape is found, the filter is applied to the original unrotated image. To conclude, the WAV filter produces higher values when it encounters specific features, especially abrupt changes that are typically not present in many non-faces.

Two remaining tests (fourth and fifth criteria) are based on Gabor's logarithmic wavelet filter for finding the symmetry of the shape of the largest cluster. We calculate the phase symmetry of points in an image. This is a contrast invariant measure of symmetry [64]. High values indicate the presence of symmetry, which can mean the presence of a symmetrical shape, such as an ellipse, and therefore that have a good probability of containing a face. The first test discriminates based on the average of the scores, while the latter uses variance instead of the mean.

3. Results and Discussion

3.1. Datasets

Four datasets—Microsoft Hand Gesture (MHG) [66], Padua Hand Gesture (PHG) [67], Padua FaceDec (PFD) [10], and Padua FaceDec2 (PFD2) [9]—were used to experimentally develop the system proposed in this work. The faces in these datasets were captured in unconstrained environments. All four datasets contain colored images and their corresponding depth maps. All faces are upright and frontal with each possessing limited degrees of rotation. Originally, for two datasets, the faces were collected for gesture recognition rather than face detection. In addition, a separate set of images was collected for preliminary experiments and for parameter tunings. These faces were extracted from the Padua FaceDec dataset [10]. As in [9], these datasets were merged to form a challenging dataset for face detection.

In addition to the merged datasets, experiments are reported on the BioID dataset [56] so that comparisons with the system proposed here can be made with other face detection systems. Each of these five datasets is discussed below, with important information about each one summarized in Table 1.

MHG [66] was collected for the purpose of gesture recognition. This dataset contains images of 10 different people performing a set of gestures, which means that not only does each image in the dataset include a single face, but the images also exhibit a high degree of similarity. As in [9], a subset of 42 MHG images was selected, with each image manually labeled with the face position.

PHG [67] is a dataset for gesture recognition. It contains images of 10 different people displaying a set of hand gestures, and each image contains only one face. A subset of 59 PHG images were manually labeled.

PFD [10] was acquired specifically for face detection. PFD contains 132 labeled images that were collected outdoors and indoors with the Kinect 1 sensor. The images in this dataset contain zero, one, or more faces. Images containing people show them performing many different daily activities in the wild. Images were captured at different times of the day in vary lighting conditions. Some faces also exhibit various degrees of occlusion.

PFD2 [9] contains 316 images captured indoors and outdoors in different settings with the Kinect 2 sensor. For each scene, a 512×424 depth map and a 1920×1080 color image were obtained. Images contain zero, one, or more faces. Images of people show them in various positions with their heads tilted or next to objects. The outdoor depth data collected by Kinect 2 are highly noisy compared to the images collected with Kintect 1. This makes PFD2 an even more challenging dataset. The depth data was retroprojected over the color frame and interpolated to the same resolution to obtain two aligned depth and color fields.

Table 1. Characteristics of the six datasets. MHG: Microsoft Hand Gesture, PHG: Padua Hand Gesture, PFD: Padua FaceDec, and PFD2: Padua FaceDec2.

Dataset	Number Images	Color Resolution	Depth Resolution	Number Faces	Difficulty Level
MHG	42	640×480	640×480	42	Low
PHG	59	1280×1024	640×480	59	Low
PFD	132	1280×1024	640×480	150	High
PFD2	316	1920×1080	512×424	363	High
MERGED	549	—	—	614	High
BioID	1521	384×286	—	1521	High

The MHG, PHG, PFD, and PFD2 datasets were merged, as in [9], to form a larger, more challenging dataset, called MERGED, containing 549 images with 614 total faces. Only upright frontal faces with a maximum rotation of ±30° were included. Parameter optimization of the face detectors was manually performed and fixed for all images even though they came from four datasets with different characteristics.

As a final dataset for validating the approach proposed in this work, we chose one of the leading benchmark datasets for upright frontal face detection: the BioID dataset [56]. It contains 1521 images of 23 people collected during several identification sessions. The images in BioID are gray-scale and do not include depth map information. Moreover, the degree of rotation in the facial images is small. As a consequence, most of the filters applied to the ensembles were not transferable to the BioID dataset. Despite this shortcoming, this dataset is useful in demonstrating the effectiveness of the ensembles developed in this work.

3.2. Performance Indicators

The following two well-known performance indicators are reported here:

- Detection rate (DR): the ratio between the number of faces correctly detected and the total number of faces in the dataset. The faces were manually labeled. DR is evaluated at different precision levels considering different values of "eye distance". Let d_l, (d_r) be the Euclidean distance between the manually extracted C_l, (C_r) and the detected $C\prime_l$, $(C\prime_r)$ left (right) eye positions. The relative error of detection is defined as $ED = \max(d_l, d_r)/d_{lr}$, where the normalization factor d_{lr} is the Euclidean distance of the expected eye centers used to make the measurement independent of the scale of the face in the image and of the image size. There is a general agreement [56] that $ED \leq$ 0.25 is a good criterion for claiming eye detection, since this value roughly corresponds to an eye distance smaller than the eye width. Some face detectors (i.e., FL and RF) give the positions of the eye centers as the output, whereas for others (i.e., VJ and SN), the eye position is assumed to be a fixed position inside the face bounding box.
- False positives (FP): the number of candidate faces that do not include a face.

3.3. Experiments

The first experiment compares the detection rates of the six face detectors, along with some of their combinations, by adjusting (1) the sensitivity values of s, where applicable, and (2) the detection procedure which either does or does not involved the addition of poses constructed by rotating images 20°/−20°.

The value for the sensitivity threshold s is shown in parentheses in Table 1. To reduce the number of false positives (FP), all output images having a distance of their centroid ≤30 pixels are merged as in [9].

As evident in the results in Table 2, the addition of rotated poses is of little value for the RF face detector, since this detector was originally trained on images that contained rotated faces. Thus, the addition of rotated poses increased the number of false positives.

Table 2. Performance of the six face detectors and the best performing ensembles (see the last seven rows) on the MERGED dataset (* denotes the addition of the 20°/−20° rotated images/poses in the dataset). As in [9], a face is considered detected in an image if the eye distance $ED < 0.35$. DR: detection rate, FL: fast localization, FP: false positives, NPD: normalized pixel difference, SFD: Single Scale-invariant Face Detector, SN: Split up sparse Network of Winnows, VJ: Viola–Jones.

Face Detector(s)/Ensemble	+Poses	DR	FP
VJ(2)	No	55.37	2528
RF(−1)	No	47.39	4682
RF(−0.8)	No	47.07	3249
RF(−0.65)	No	46.42	1146
SN(1)	No	66.61	508
SN(10)	No	46.74	31
FL	No	78.18	344
NPD	No	55.70	1439
SFD	No	81.27	186
VJ(2) *	Yes	65.31	6287
RF(−1) *	Yes	49.67	19,475
RF(−0.8) *	Yes	49.67	14,121
RF(−0.65) *	Yes	49.02	5895
SN(1) *	Yes	74.59	1635
SN(10) *	Yes	50.16	48
FL *	Yes	83.39	891
NPD *	Yes	64.17	10,431
FL + RF(−0.65)	No	83.06	1490
FL + RF(−0.65) + SN(1)	No	86.16	1998
FL + RF(−0.65) + SN(1) *	Mixed	88.44	3125
FL * + SN(1) *	Yes	87.79	2526
FL * + RF(−0.65) + SN(1) *	Mixed	90.39	3672
FL * + RF(−0.65) + SN(1) * + SFD	Mixed	91.21	3858
FL * + RF(−0.65) + SN(1) * + NPD * + SFD	Mixed	**92.02**	16,325

Only the most interesting results are reported for the ensembles of classifiers. As can be seen in Table 2, high-performing approaches in an ensemble increase the detection rates while also generating more false negatives.

In Table 3, the performance of the face detectors presented in Table 2 are reported on the BioID dataset. As noted in [9], the addition of rotated poses is not needed when images are acquired in constrained environments. Although there is no significant difference in performance when adding the rotated poses, a difference is evident in the number of false positives that the rotated poses produce: they increase the false positives.

Table 3. Performance of the six face detectors and ensembles reported above on the BioID dataset (note: some values are taken from [9]).

Face Detector(s)/Ensemble	+Poses	DR (ED < 0.15)	DR (ED < 0.25)	DR (ED < 0.35)	(FP)
VJ(2)	No	13.08	86.46	99.15	517
RF(−1)	No	87.84	98.82	99.08	80
RF(−0.8)	No	87.84	98.82	99.08	32
RF(−0.65)	No	87.84	98.82	99.08	21
SN(1)	No	71.27	96.38	97.76	12
SN(10)	No	72.06	98.16	99.74	172
FL	No	92.57	94.61	94.67	67
SFD	No	99.21	99.34	99.34	1
VJ(2) *	Yes	13.08	86.46	99.15	1745
RF(−1) *	Yes	90.53	99.15	99.41	1316
RF(−0.8) *	Yes	90.53	99.15	99.41	589
RF(−0.65) *	Yes	90.53	99.15	99.41	331
SN(1) *	Yes	71.33	96.52	97.90	193
SN(10) *	Yes	72.12	98.36	99.87	1361
FL *	Yes	92.57	94.61	94.67	1210
FL + RF(−0.65)	No	98.42	99.74	99.74	88
FL + RF(−0.65) + SN(10)	No	99.15	99.93	99.93	100
FL + RF(−0.65) + SN(1) *	Mixed	99.15	100	100	281
FL * + SN(1) *	Yes	98.03	99.87	99.93	260
FL * + RF(−0.65) + SN(1) *	Mixed	99.15	100	100	1424
FL * + RF(−0.65) + SN(1) * + SFD	Mixed	**99.41**	**100**	100	1425

In Table 3, we also discover that each of the face detectors identifies a different set of faces. This diversity in the individual face detectors is what enables the ensemble to improve the best standalone approaches. It is also noteworthy that the same classifier can perform differently on the MERGED versus BioID dataset. For instance, RF works well on BioID but not so well on MERGED; perhaps this is because it contains low-quality faces.

In Table 4, an experiment is reported that evaluated the seven filtering steps, as detailed in Section 2.3, along with their combinations. The first experiments showed that the best ensemble (considering the trade-off between performance and false positives) is FL + RF(−0.65) + SN(1)* + SFD. For this reason, the filtering sets are tested only for this detector.

Table 4. Performance of FL + RF(−0.65) + SN(1)* + SFD obtained combining different filtering steps on MERGED.

Filter Combination	DR	FP
SIZE	91.21	1547
SIZE + STD	91.21	1514
SIZE + STD + SEG	91.21	1485
SIZE + STD + SEG + ELL	91.04	1440
SIZE + STD + SEG + ELL + EYE	90.55	1163
SIZE + STD + SEG + ELL + SEC + EYE	90.39	1132
SIZE + STD + SEG + ELL + SEC + EYE + WAV	90.07	1018

SIZE is clearly the best method for removing false positive candidates from a set of faces detected by FL + RF(−0.65) + SN(1)* + SFD. The next best filter is EYE. However, because EYE is computationally expensive, it cannot be used in all applications. Although the other filters, when considered individually, are of less value because of their low computational costs, they are useful for reducing the number of false positives when applied sequentially. If real-time detection is not required (which is typically the case when tagging faces), then EYE filtering can be used to reduce the number of false positives produced by an ensemble without decreasing the number of true positives.

The results presented in the previous tables shows that the proposed approach performs better than FL and SPD, both of which are considered two of the best face detectors in the literature. It is true that the results reported here have been obtained on two rather small datasets; nonetheless, MERGED is highly realistic. Thus, it is reasonable to predict that the best ensemble proposed in this work would perform comparatively well in real-world conditions. The images contained in MERGE include those containing a single frontal face as well as those containing multiple faces acquired "in the wild".

Finally, in order to evaluate the computational cost of our approach, the processing time per 640×480 image on a i7-7700HQ PC system is reported in Table 5 for each detection method of "FL* + RF(−0.65) + SN(1)* + SFD" and each additional filter (on a candidate region of size 78×78 pixels). All the tests are performed without parallelizing the code. However, it should be noted that the filters and face detectors can run in parallel, resulting in a significant reduction of computation time.

Table 5. Average processing time per image in ms.

Detection Method/Filter	ms
RF	12,571
SN	1371
FL	170
SPD	175
SIZE	0.33
STD	10.86
SEG	8.808
ELL	10.24
EYE	19,143
WAV	179.4

4. Conclusions

In this paper, an ensemble of state-of-the-art face detectors is combined with a set of filters calculated from both the depth map and the color image. The filters reduce the number of false positives produced by the ensemble while maximizing the detection rate. A set of seven filters based on the size, the flatness, or the unevenness of the candidate face regions, or on the size of the larger cluster of the depth map of the candidate face regions, or on eye detection or the degree of ellipse fitting are evaluated, including a new method proposed here that is based on processing the candidate region with different wavelets. The method proposed in this work for developing an ensemble of face detectors uses the depth map to obtain increased effectiveness even under many indoor and outdoor illumination settings.

The experimental results demonstrate that the filtering steps significantly reduce the number of false positives (from 16,325 to 1018) without significantly decreasing the detection rate (from 92.02 to 90.07) on a challenging dataset containing images with cluttered and complicated backgrounds. The performance of the proposed system is also reported on the challenging BioID benchmark to validate the approach presented here further and to compare the best performing ensemble with the state-of-the-art in face detection.

The face detector named SFD is shown to outperform all other standalone methods. However, an ensemble proposed here that combines SFD with other types of face detectors is shown to boost the standalone performance of SFD. Obviously, increasing the number of face detectors included in ensembles increases the number of false positives; however, as the experiments in this work demonstrate, the application of a new cascade of filters reduces this number to acceptable levels.

Author Contributions: Conceptualization, L.N. and A.L.; methodology, L.N.; software, L.N. and A.L.; validation, L.N., S.B. and A.L.; formal analysis, L.N.; investigation, A.L.; resources, S.B.; writing—original draft preparation, A.L. and S.B.; writing—review and editing, S.B.; visualization, S.B. and A.L.

Funding: This research received no external funding.

Conflicts of Interest: The authors declare no conflict of interest.

References

1. Zeng, Z.; Pantic, M.; Roisman, G.I.; Huang, T.S. A Survey of Affect Recognition Methods: Audio, Visual, and Spontaneous Expressions. *IEEE Trans. Pattern Anal. Mach. Intell.* **2009**, *31*, 39–58. [CrossRef] [PubMed]
2. Zhu, X.; Liu, X.; Lei, Z.; Li, S.Z. Face Alignment in Full Pose Range: A 3D Total Solution. *IEEE Trans. Pattern Anal. Mach. Intell.* **2019**, *41*, 78–92. [CrossRef] [PubMed]
3. Xiong, X.; Torre, F.D. Supervised Descent Method and Its Applications to Face Alignment. In Proceedings of the 2013 IEEE Conference on Computer Vision and Pattern Recognition, Washington, DC, USA, 23–28 June 2013; pp. 532–539.
4. Xie, X.; Jones, M.W.; Tam, G.K.L. Deep face recognition. In *British Machine Vision Conference (BMVC)*; Xie, X., Jones, M.W., Tam, G.K.L., Eds.; BMVA Press: Durham, UK, 2015; pp. 41.1–41.12.
5. Schroff, F.; Kalenichenko, D.; Philbin, J. FaceNet: A unified embedding for face recognition and clustering. In Proceedings of the 2015 IEEE Conference on Computer Vision and Pattern Recognition (CVPR), Boston, MA, USA, 7–12 June 2015; pp. 815–823.
6. Taigman, Y.; Yang, M.; Ranzato, M.A.; Yang, M.; Wolf, L. DeepFace: Closing the Gap to Human-Level Performance in Face Verification. In Proceedings of the 2014 IEEE Conference on Computer Vision and Pattern Recognition, Columbus, OH, USA, 23–28 June 2014; pp. 1701–1708.
7. Zhu, X.; Lei, Z.; Yan, J.; Yi, D.; Li, S.Z. High-fidelity Pose and Expression Normalization for face recognition in the wild. In Proceedings of the 2015 IEEE Conference on Computer Vision and Pattern Recognition (CVPR), Boston, MA, USA, 7–12 June 2015; pp. 787–796.
8. Kim, M.; Kumar, S.; Pavlovic, V.; Rowley, H.A. Face tracking and recognition with visual constraints in real-world videos. In Proceedings of the 2008 IEEE Conference on Computer Vision and Pattern Recognition, Anchorage, AK, USA, 24–26 June 2008; pp. 1–8.
9. Nanni, L.; Lumini, A.; Minto, L.; Zanuttigh, P. Face detection coupling texture, color and depth data. In *Advances in Face Detection and Facial Image Analysis*; Kawulok, M., Celebi, M., Smolka, B., Eds.; Springer: Cham, Switzerland, 2016; pp. 13–33.
10. Nanni, L.; Lumini, A.; Dominio, F.; Zanuttigh, P. Effective and precise face detection based on color and depth data. *Appl. Comput. Inform.* **2014**, *10*, 1–13. [CrossRef]
11. Zhang, C.; Zhang, Z. *A Survey of Recent Advances in Face Detection*; Microsoft: Redmond, WA, USA, 2010.
12. Yang, M.H.; Kriegman, D.J.; Ahuja, N. Detecting faces in images: A survey. *IEEE Trans. Pattern Anal. Mach. Intell.* **2002**, *24*, 34–58. [CrossRef]
13. Jin, H.; Liu, Q.; Lu, H. Face detection using one-class-based support vectors. In Proceedings of the Sixth IEEE International Conference on Automatic Face and Gesture Recognition, Seoul, Korea, 19 May 2004; pp. 457–462.
14. Viola, P.A.; Jones, M.P. Rapid object detection using a boosted cascade of simple features. In Proceedings of the 2001 IEEE Computer Society Conference on Computer Vision and Pattern Recognition (CVPR), Kauai, HI, USA, 8–14 December 2001; p. 3.
15. Li, J.; Zhang, Y. Learning SURF Cascade for Fast and Accurate Object Detection. In Proceedings of the 2013 IEEE Conference on Computer Vision and Pattern Recognition, Washington, DC, USA, 23–28 June 2013; pp. 3468–3475.
16. Jain, V.; Learned-Miller, E. *FDDB: A Benchmark for Face Detection in Unconstrained Setting*; University of Massachusetts: Amherst, MA, USA, 2010.
17. Cheney, J.; Klein, B.; Klein, A.K.; Klare, B.F. Unconstrained Face Detection: State of the Art Baseline and Challenges. In Proceedings of the 8th IAPR International Conference on Biometrics (ICB), Phuket, Thailand, 19–22 May 2015.
18. Lienhart, R.; Maydt, J. An extended set of Haar-like features for rapid object detection. In Proceedings of the International Conference on Image Processing, Rochester, NY, USA, 22–25 September 2002; pp. I-900–I-903.
19. Huang, C.; Ai, H.; Li, Y.; Lao, S. Learning sparse features in granular space for multi-view face detection. In Proceedings of the 7th International Conference on Automatic Face and Gesture Recognition (FGR06), Southampton, UK, 10–12 April 2006; pp. 401–406.

20. Pham, M.T.; Gao, Y.; Hoang, V.D.; Hoang, V.D.; Cham, T.J. Fast polygonal integration and its application in extending haar-like features to improve object detection. In Proceedings of the CVPR, IEEE Computer Society Conference on Computer Vision and Pattern Recognition, San Francisco, CA, USA, 13–18 June 2010.

21. Jin, H.; Liu, Q.; Lu, H.; Tong, X. Face detection using improved LBP under bayesian framework. In Proceedings of the International Conference on Image and Graphics, Hong Kong, China, 18–20 December 2004; pp. 306–309.

22. Zhang, H.; Gao, W.; Chen, X.; Zhao, D. Object detection using spatial histogram features. *Image Vis. Comput.* **2006**, *24*, 327–341. [CrossRef]

23. Zhu, Q.; Yeh, M.C.; Cheng, K.T.; Avidan, S. Fast Human Detection Using a Cascade of Histograms of Oriented Gradients. In Proceedings of the 2006 IEEE Computer Society Conference on Computer Vision and Pattern Recognition, New York, NY, USA, 17–22 June 2006; Volume 2, pp. 1491–1498.

24. Jianguo, L.; Tao, W.; Yimin, Z. Face detection using SURF cascade. In Proceedings of the 2011 IEEE International Conference on Computer Vision Workshops (ICCV Workshops), Barcelona, Spain, 6–13 November 2011; pp. 2183–2190.

25. Bin, Y.; Yan, J.; Lei, Z.; Li, S.Z. Aggregate channel features for multi-view face detection. In Proceedings of the IEEE International Joint Conference on Biometrics, Clearwater, FL, USA, 29 September–2 October 2014; pp. 1–8.

26. Brubaker, S.C.; Wu, J.; Sun, J.; Mullin, M.D.; Rehg, J.M. On the design of cascades of boosted ensembles for face detection. *Int. J. Comput. Vis.* **2008**, *77*, 65–86. [CrossRef]

27. Pham, M.T.; Cham, T.J. Fast training and selection of haar features during statistics in boosting-based face detection. In Proceedings of the CVPR, IEEE Computer Society Conference on Computer Vision and Pattern Recognition, Rio de Janeiro, Brazil, 14–21 October 2007.

28. Küblbeck, C.; Ernst, A. Face detection and tracking in video sequences using the modifiedcensus transformation. *Image Vis. Comput.* **2006**, *24*, 564–572. [CrossRef]

29. Huang, C.; Ai, H.; Li, Y.; Lao, S. High-Performance Rotation Invariant Multiview Face Detection. *IEEE Trans. Pattern Anal. Mach. Intell.* **2007**, *29*, 671–686. [CrossRef]

30. Mathias, M.; Benenson, R.; Pedersoli, M.; Gool, L.V. Face Detection without Bells and Whistles. In *ECCV*; Springer: Cham, Switzerland, 2014.

31. Nilsson, M.; Nordberg, J.; Claesson, I. Face Detection using Local SMQT Features and Split up Snow Classifier. In Proceedings of the 2007 IEEE International Conference on Acoustics, Speech and Signal Processing-ICASSP '07, Honolulu, HI, USA, 15–20 April 2007; pp. II-589–II-592.

32. Asthana, A.; Zafeiriou, S.; Cheng, S.; Pantic, M. *Robust Discriminative Response Map Fitting with Constrained Local Models, CVPR*; IEEE: Portland, OR, USA, 2013.

33. Liao, S.; Jain, A.K.; Li, S.Z. A Fast and Accurate Unconstrained Face Detector. *IEEE Trans. Pattern Anal. Mach. Intell.* **2016**, *38*, 211–223. [CrossRef]

34. Markuš, N.; Frljak, M.; Pandžić, I.S.; Ahlberg, J.; Forchheimer, R. Fast Localization of Facial Landmark Points. *arXiv* **2014**, arXiv:1403.6888.

35. Li, H.; Lin, Z.L.; Shen, X.; Brandt, J.; Hua, G. A convolutional neural network cascade for face detection. In Proceedings of the 2015 IEEE Conference on Computer Vision and Pattern Recognition (CVPR), Boston, MA, USA, 7–12 June 2015; pp. 5325–5334.

36. Girshick, R.; Donahue, J.; Darrell, T.; Malik, J. Rich Feature Hierarchies for Accurate Object Detection and Semantic Segmentation. In Proceedings of the 2014 IEEE Conference on Computer Vision and Pattern Recognition, Columbus, OH, USA, 23–28 June 2014; pp. 580–587.

37. Farfade, S.S.; Saberian, M.; Li, L.J. *Multi-View Face Detection Using Deep Convolutional Neural Networks*; Cornell University: Ithaca, NY, USA, 2015.

38. Yang, W.; Zhou, L.; Li, T.; Wang, H. A Face Detection Method Based on Cascade Convolutional Neural Network. *Multimed. Tools Appl.* **2018**, *78*, 1–18. [CrossRef]

39. Zhang, S.; Zhu, X.; Lei, Z.; Shi, H.; Wang, X.; Li, S.Z. S^3FD: Single Shot Scale-Invariant Face Detector. In Proceedings of the 2017 IEEE International Conference on Computer Vision (ICCV), Venice, Italy, 22–29 October 2017; pp. 192–201.

40. Yang, B.; Yan, J.; Lei, Z.; Li, S.Z. Convolutional Channel Features. In Proceedings of the 2015 IEEE International Conference on Computer Vision (ICCV), Boston, MA, USA, 7–13 December 2015; pp. 82–90.

41. Zhang, K.; Zhang, Z.; Li, Z.; Qiao, Y. Joint Face Detection and Alignment Using Multitask Cascaded Convolutional Networks. *IEEE Signal Process. Lett.* **2016**, *23*, 1499–1503. [CrossRef]

42. Faltemier, T.C.; Bowyer, K.W.; Flynn, P.J. Using a Multi-Instance Enrollment Representation to Improve 3D Face Recognition. In Proceedings of the 2007 First IEEE International Conference on Biometrics: Theory, Applications, and Systems, Crystal City, VA, USA, 27–29 September 2007; pp. 1–6.

43. Gupta, S.; Castleman, K.R.; Markey, M.K.; Bovik, A.C. Texas 3D Face Recognition Database. In Proceedings of the IEEE Southwest Symposium on Image Analysis and Interpretation, Austin, TX, USA, 23–25 May 2010; pp. 97–100.

44. Min, R.; Kose, N.; Dugelay, J. KinectFaceDB: A Kinect Database for Face Recognition. *IEEE Trans. Syst. Man Cybern. Syst.* **2014**, *44*, 1534–1548. [CrossRef]

45. Guo, Y.; Sohel, F.A.; Bennamoun, M.; Wan, J.; Lu, M. RoPS: A local feature descriptor for 3D rigid objects based on rotational projection statistics. In Proceedings of the 2013 1st International Conference on Communications, Signal Processing, and their Applications (ICCSPA), Sharjah, UAE, 12–14 February 2013; pp. 1–6.

46. Zhou, S.; Xiao, S. 3D face recognition: A survey. *Hum. Cent. Comput. Inf. Sci.* **2018**, *8*, 35. [CrossRef]

47. Hg, R.I.; Jasek, P.; Rofidal, C.; Nasrollahi, K.; Moeslund, T.B.; Tranchet, G. An RGB-D Database Using Microsoft's Kinect for Windows for Face Detection. In Proceedings of the 2012 Eighth International Conference on Signal Image Technology and Internet Based Systems, Naples, Italy, 25–29 November 2012; pp. 42–46.

48. Dixon, M.; Heckel, F.; Pless, R.; Smart, W.D. Faster and more accurate face detection on mobile robots using geometric constraints. In Proceedings of the 2007 IEEE/RSJ International Conference on Intelligent Robots and Systems, San Diego, CA, USA, 29 October–2 November 2007; pp. 1041–1046.

49. Burgin, W.; Pantofaru, C.; Smart, W.D. Using depth information to improve face detection. In Proceedings of the 6th International Conference on Human-Robot Interaction, Lausanne, Switzerland, 8–11 March 2011; pp. 119–120.

50. Shieh, M.Y.; Hsieh, T.M. Fast Facial Detection by Depth Map Analysis. *Math. Probl. Eng.* **2013**, *2013*, 1–10. [CrossRef]

51. Shotton, J.; Sharp, T.; Kipman, A.; Fitzgibbon, A.; Finocchio, M.; Blake, A.; Cook, M.; Moore, R. Real-time human pose recognition in parts from single depth images. *Commun. ACM* **2013**, *56*, 116–124. [CrossRef]

52. Mattheij, R.; Postma, E.; Van den Hurk, Y.; Spronck, P. Depth-based detection using Haarlike features. In Proceedings of the BNAIC 2012 Conference, Maastricht, The Netherlands, 25–26 October 2012, pp. 162–169.

53. Jiang, F.; Fischer, M.; Ekenel, H.K.; Shi, B.E. Combining texture and stereo disparity cues for real-time face detection. *Signal Process. Image Commun.* **2013**, *28*, 1100–1113. [CrossRef]

54. Anisetti, M.; Bellandi, V.; Damiani, E.; Arnone, L.; Rat, B. A3fd: Accurate 3d face detection. In *Signal Processing for Image Enhancement and Multimedia Processing*; Damiani, E., Dipanda, A., Yetongnon, K., Legrand, L., Schelkens, P., Chbeir, R., Eds.; Springer: Boston, MA, USA, 2008; pp. 155–165.

55. Huang, G.B.; Ramesh, M.; Berg, T.; Learned-Miller, E. *Labeled Faces in the Wild: A Database for Studying Face Recognition in Unconstrained Environments*; University of Massachusetts: Amherst, MA, USA, 2007.

56. Jesorsky, O.; Kirchberg, K.J.; Frischholz, R. Robust Face Detection Using the Hausdorff Distance. In *AVBPA*; Springer: Berlin/Heidelberg, Germany, 2001.

57. Herrera, D.C.; Kannala, J.; Heikkilä, J. Joint Depth and Color Camera Calibration with Distortion Correction. *IEEE Trans. Pattern Anal. Mach. Intell.* **2012**, *34*, 2058–2064. [CrossRef]

58. Mutto, C.D.; Zanuttigh, P.; Cortelazzo, G.M. Fusion of Geometry and Color Information for Scene Segmentation. *IEEE J. Sel. Top. Signal Process.* **2012**, *6*, 505–521. [CrossRef]

59. Comaniciu, D.; Meer, P. Mean shift: A robust approach toward feature space analysis. *IEEE Trans. Pattern Anal. Mach. Intell.* **2002**, *24*, 603–619. [CrossRef]

60. Friedman, J.; Hastie, T.; Tibshirani, R. Additive logistic regression: A statistical view of boosting. *Ann. Stat.* **2000**, *38*, 337–374. [CrossRef]

61. Gal, O. Fit_Ellipse. Available online: https://www.mathworks.com/matlabcentral/fileexchange/3215-fit_ellipse (accessed on 2 October 2003).

62. Tan, X.; Song, S.; Zhou, Z.H.; Chen, S. Enhanced pictorial structures for precise eye localization under uncontrolled conditions. In Proceedings of the IEEE Computer Society Conference on Computer Vision and Pattern Recognition (CVPR'09), Miami, FL, USA, 20–25 June 2009; pp. 1621–1628.

63. Skodras, E.; Fakotakis, N. Precise localization of eye centers in low resolution color images. *Image Vis. Comput.* **2015**, *36*, 51–60. [CrossRef]

64. Kovesi, P. Image features from Phase Congruency. *J. Comput. Vis. Res.* **1999**, *1*, 1–27.

65. Bobulski, J. Wavelet transform in face recognition. In *Biometrics, Computer Security Systems and Artificial Intelligence Applications*; Saeed, K., Pejaś, J., Mosdorf, R., Eds.; Springer Science + Business Media: New York, NY, USA, 2006; pp. 23–29.

66. Ren, Z.; Meng, J.; Yuan, J. Depth camera based hand gesture recognition and its applications in Human-Computer-Interaction. In Proceedings of the 2011 8th International Conference on Information, Communications & Signal Processing, Singapore, 13–16 December 2011; pp. 1–5.

67. Dominio, F.; Donadeo, M.; Zanuttigh, P. Combining multiple depth-based descriptors for hand gesture recognition. *Pattern Recognit. Lett.* **2014**, *50*, 101–111. [CrossRef]

Review

Face Recognition Systems: A Survey

Yassin Kortli [1,2,*]**, Maher Jridi** [1]**, Ayman Al Falou** [1] **and Mohamed Atri** [3]

1   AI-ED Department, Yncrea Ouest, 20 rue du Cuirassé de Bretagne, 29200 Brest, France; maher.jridi@isen-ouest.yncrea.fr (M.J.); ayman.alfalou@isen-ouest.yncrea.fr (A.A.F.)
2   Electronic and Micro-electronic Laboratory, Faculty of Sciences of Monastir, University of Monastir, Monastir 5000, Tunisia
3   College of Computer Science, King Khalid University, Abha 61421, Saudi Arabia; matri@kku.edu.sa
*   Correspondence: yassin.kortli@isen-ouest.yncrea.fr

Received: 15 October 2019; Accepted: 15 December 2019; Published: 7 January 2020

Abstract: Over the past few decades, interest in theories and algorithms for face recognition has been growing rapidly. Video surveillance, criminal identification, building access control, and unmanned and autonomous vehicles are just a few examples of concrete applications that are gaining attraction among industries. Various techniques are being developed including local, holistic, and hybrid approaches, which provide a face image description using only a few face image features or the whole facial features. The main contribution of this survey is to review some well-known techniques for each approach and to give the taxonomy of their categories. In the paper, a detailed comparison between these techniques is exposed by listing the advantages and the disadvantages of their schemes in terms of robustness, accuracy, complexity, and discrimination. One interesting feature mentioned in the paper is about the database used for face recognition. An overview of the most commonly used databases, including those of supervised and unsupervised learning, is given. Numerical results of the most interesting techniques are given along with the context of experiments and challenges handled by these techniques. Finally, a solid discussion is given in the paper about future directions in terms of techniques to be used for face recognition.

Keywords: face recognition systems; person identification; biometric systems; survey

---

1. Introduction

The objective of developing biometric applications, such as facial recognition, has recently become important in smart cities. In addition, many scientists and engineers around the world have focused on establishing increasingly robust and accurate algorithms and methods for these types of systems and their application in everyday life. All types of security systems must protect all personal data. The most commonly used type for recognition is the password. However, through the development of information technologies and security algorithms, many systems are beginning to use many biometric factors for recognition task [1–4]. These biometric factors make it possible to identify people's identity by their physiological or behavioral characteristics. They also provide several advantages, for example, the presence of a person in front of the sensor is sufficient, and there is no more need to remember several passwords or confidential codes anymore. In this context, many recognition systems based on different biometric factors such as iris, fingerprints [5], voice [6], and face have been deployed in recent years.

Systems that identify people based on their biological characteristics are very attractive because they are easy to use. The human face is composed of different structures and characteristics. For this reason, in recent years, it has become one of the most widely used biometric authentication systems, given its potential in many applications and fields (surveillance, home security, border control, and so on) [7–9]. Facial recognition system as an ID (identity) is already being offered to consumers outside of phones, including at airport check-ins, sports stadiums, and concerts. In addition, this system does not require the intervention of people to operate, which makes it possible to identify people only from images obtained from the camera. In addition, many biometric systems that are developed using different types of search provide good identification accuracy. However, it would be interesting to develop new biometric systems for face recognition in order to reach real-time constraints.

Owing to the huge volume of data generated and rapid advancement in artificial intelligence techniques, traditional computing models have become inadequate to process data, especially for complex applications like those related to feature extraction. Graphics processing units (GPUs) [4], central processing unit (CPU) [3], and programmable gate arrays (FPGAs) [10] are required to efficiently perform complex computing tasks. GPUs have computing cores that are several orders of magnitude larger than traditional CPU and allow greater capacity to perform parallel computing. Unlike GPUs, the FPGAs have a flexible hardware configuration and offer better performance than GPUs in terms of energy efficiency. However, FPGAs present a major drawback related to the programming time, which is higher than that of CPU and GPU.

There are many computer vision approaches proposed to address face detection or recognition tasks with high robustness and discrimination, such as local, subspace, and hybrid approaches [10–16]. However, several issues still need to be addressed owing to various challenges, such as head orientation, lighting conditions, and facial expression. The most interesting techniques are developed to face all these challenges, and thus develop reliable face recognition systems. Nevertheless, they require high processing time, high memory consumption, and are relatively complex.

Rapid advances in technologies such as digital cameras, portable devices, and increased demand for security make the face recognition system one of the primary biometric technologies.

To sum up, the contributions of this paper review are as follows:

1. We first introduced face recognition as a biometric technique.
2. We presented the state of the art of the existing face recognition techniques classified into three approaches: local, holistic, and hybrid.
3. The surveyed approaches were summarized and compared under different conditions.
4. We presented the most popular face databases used to test these approaches.
5. We highlighted some new promising research directions.

2. Face Recognition Systems Survey

2.1. Essential Steps of Face Recognition Systems

Before detailing the techniques used, it is necessary to make a brief description of the problems that must be faced and solved in order to perform the face recognition task correctly. For several security applications, as detailed in the works of [17–22], the characteristics that make a face recognition system useful are the following: its ability to work with both videos and images, to process in real time, to be robust in different lighting conditions, to be independent of the person (regardless of hair, ethnicity, or gender), and to be able to work with faces from different angles. Different types of sensors, including RGB, depth, EEG, thermal, and wearable inertial sensors, are used to obtain data. These sensors may provide extra information and help the face recognition systems to identify face images in both static images and video sequences. Moreover, three categories of sensors that may improve the reliability and the accuracy of a face recognition system by tackling the challenges include illumination variation, head pose, and facial expression in pure image/video processing. The first group is non-visual sensors, such as audio, depth, and EEG sensors, which provide extra information in addition to the visual dimension and improve the recognition reliability, for example, in illumination variation and position shift situation. The second is detailed-face sensors, which detect a small dynamic change of a face component, such as eye-trackers, which may help differentiate the background noise and the face images. The last is target-focused sensors, such as infrared thermal sensors, which can facilitate the face recognition systems to filter useless visual contents and may help resistance illumination variation.

Three basic steps are used to develop a robust face recognition system: (1) face detection, (2) feature extraction, and (3) face recognition (shown in Figure 1) [3,23]. The face detection step is used to detect and locate the human face image obtained by the system. The feature extraction step is employed to extract the feature vectors for any human face located in the first step. Finally, the face recognition step includes the features extracted from the human face in order to compare it with all template face databases to decide the human face identity.

- *Face Detection*: The face recognition system begins first with the localization of the human faces in a particular image. The purpose of this step is to determine if the input image contains human faces or not. The variations of illumination and facial expression can prevent proper face detection. In order to facilitate the design of a further face recognition system and make it more robust, pre-processing steps are performed. Many techniques are used to detect and locate the human face image, for example, Viola–Jones detector [24,25], histogram of oriented gradient (HOG) [13,26], and principal component analysis (PCA) [27,28]. Also, the face detection step can be used for video and image classification, object detection [29], region-of-interest detection [30], and so on.
- *Feature Extraction*: The main function of this step is to extract the features of the face images detected in the detection step. This step represents a face with a set of features vector called a "signature" that describes the prominent features of the face image such as mouth, nose, and eyes with their geometry distribution [31,32]. Each face is characterized by its structure, size, and shape, which allow it to be identified. Several techniques involve extracting the shape of the mouth, eyes, or nose to identify the face using the size and distance [3]. HOG [33], Eigenface [34], independent component analysis (ICA), linear discriminant analysis (LDA) [27,35], scale-invariant feature transform (SIFT) [23], gabor filter, local phase quantization (LPQ) [36], Haar wavelets, Fourier transforms [31], and local binary pattern (LBP) [3,10] techniques are widely used to extract the face features.
- *Face Recognition*: This step considers the features extracted from the background during the feature extraction step and compares it with known faces stored in a specific database. There are two general applications of face recognition, one is called identification and another one is called verification. During the identification step, a test face is compared with a set of faces aiming to find the most likely match. During the identification step, a test face is compared with a known face in the database in order to make the acceptance or rejection decision [7,19]. Correlation filters (CFs) [18,37,38], convolutional neural network (CNN) [39], and also k-nearest neighbor (K-NN) [40] are known to effectively address this task.

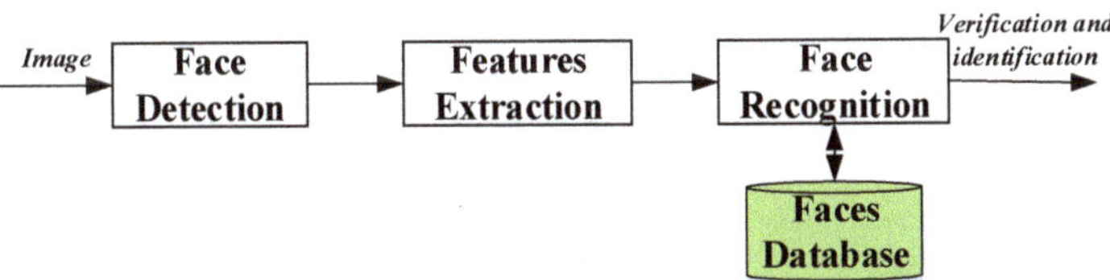

Figure 1. Face recognition structure [3,23].

2.2. Classification of Face Recognition Systems

Compared with other biometric systems such as the eye, iris, or fingerprint recognition systems, the face recognition system is not the most efficient and reliable [5]. Moreover, this biometric system has many constraints resulting from many challenges, despite all the above advantages. The recognition under the controlled environments has been saturated. Nevertheless, in uncontrolled environments, the problem remains open owing to large variations in lighting conditions, facial expressions, age, dynamic background, and so on. In this paper survey, we review the most advanced face recognition techniques proposed in controlled/uncontrolled environments using different databases.

Several systems are implemented to identify a human face in 2D or 3D images. In this review paper, we will classify these systems into three approaches based on their detection and recognition method (Figure 2): (1) local, (2) holistic (subspace), and (3) hybrid approaches. The first approach is classified according to certain facial features, not considering the whole face. The second approach employs the entire face as input data and then projects into a small subspace or in correlation plane. The third approach uses local and global features in order to improve face recognition accuracy.

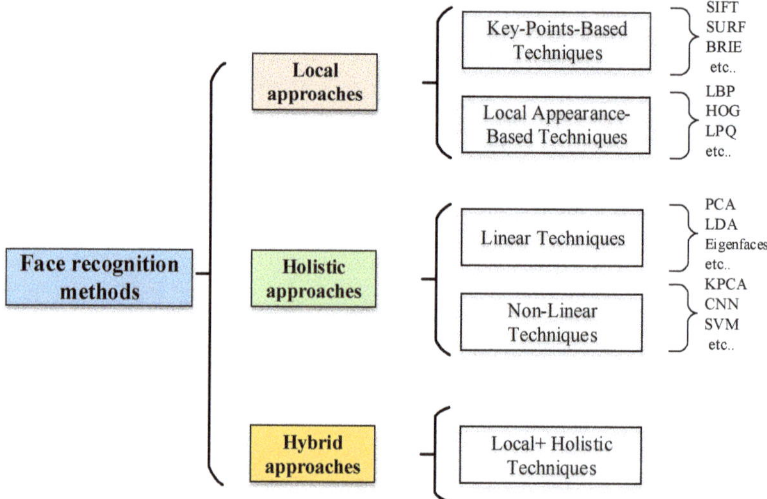

Figure 2. Face recognition methods. SIFT, scale-invariant feature transform; SURF, scale-invariant feature transform; BRIEF, binary robust independent elementary features; LBP, local binary pattern; HOG, histogram of oriented gradients; LPQ, local phase quantization; PCA, principal component analysis; LDA, linear discriminant analysis; KPCA, kernel PCA; CNN, convolutional neural network; SVM, support vector machine.

3. Local Approaches

In the context of face recognition, local approaches treat only some facial features. They are more sensitive to facial expressions, occlusions, and pose [1]. The main objective of these approaches is to discover distinctive features. Generally, these approaches can be divided into two categories: (1) local appearance-based techniques are used to extract local features, while the face image is divided into small regions (patches) [3,32]. (2) Key-points-based techniques are used to detect the points of interest in the face image, after which the features localized on these points are extracted.

3.1. Local Appearance-Based Techniques

It is a geometrical technique, also called feature or analytic technique. In this case, the face image is represented by a set of distinctive vectors with low dimensions or small regions (patches). Local appearance-based techniques focus on critical points of the face such as the nose, mouth, and eyes to generate more details. Also, it takes into account the particularity of the face as a natural form to identify and use a reduced number of parameters. In addition, these techniques describe the local features through pixel orientations, histograms [13,26], geometric properties, and correlation planes [3,33,41].

- Local binary pattern (LBP) and it's variant: LBP is a great general texture technique used to extract features from any object [16]. It has widely performed in many applications such as face recognition [3], facial expression recognition, texture segmentation, and texture classification.

The LBP technique first divides the facial image into spatial arrays. Next, within each array square, a 3×3 pixel matrix (p1......p8) is mapped across the square. The pixel of this matrix is a threshold with the value of the center pixel (p_0) (i.e., use the intensity value of the center pixel $i(p_0)$ as a reference for thresholding) to produce the binary code. If a neighbor pixel's value is lower than the center pixel value, it is given a zero; otherwise, it is given one. The binary code contains information about the local texture. Finally, for each array square, a histogram of these codes is built, and the histograms are concatenated to form the feature vector. The LBP is defined in a matrix of size 3×3, as shown in Equation (1).

$$LBP = \sum_{p=1}^{8} 2^p s\left(i_0 - i_p\right), \quad with\ s(x) = \begin{cases} 1 & x \geq 0 \\ 0 & x < 0 \end{cases}, \tag{1}$$

where i_0 and i_p are the intensity value of the center pixel and neighborhood pixels, respectively. Figure 3 illustrates the procedure of the LBP technique.

Khoi et al. [20] propose a fast face recognition system based on LBP, pyramid of local binary pattern (PLBP), and rotation invariant local binary pattern (RI-LBP). Xi et al. [15] have introduced a new unsupervised deep learning-based technique, called local binary pattern network (LBPNet), to extract hierarchical representations of data. The LBPNet maintains the same topology as the convolutional neural network (CNN). The experimental results obtained using the public benchmarks (i.e., LFW and FERET) have shown that LBPNet is comparable to other unsupervised techniques. Laure et al. [40] have implemented a method that helps to solve face recognition issues with large variations of parameters such as expression, illumination, and different poses. This method is based on two techniques: LBP and K-NN techniques. Owing to its invariance to the rotation of the target image, LBP become one of the important techniques used for face recognition. Bonnen et al. [42] proposed a variant of the LBP technique named "multiscale local binary pattern (MLBP)" for features' extraction. Another LBP extension is the local ternary pattern (LTP) technique [43], which is less sensitive to the noise than the original LBP technique. This technique uses three steps to compute the differences between the neighboring ones and the central pixel. Hussain et al. [36] develop a local quantized pattern (LQP) technique for face representation. LQP is a generalization of local pattern features and is intrinsically robust to illumination conditions. The LQP features use the disk layout to sample pixels from the local neighborhood and obtain a pair of binary codes using ternary split coding. These codes are quantized, with each one using a separately learned codebook.

- Histogram of oriented gradients (HOG) [44]: The HOG is one of the best descriptors used for shape and edge description. The HOG technique can describe the face shape using the distribution of edge direction or light intensity gradient. The process of this technique done by sharing the whole face image into cells (small region or area); a histogram of pixel edge direction or direction gradients is generated of each cell; and, finally, the histograms of the whole cells are combined to extract the feature of the face image. The feature vector computation by the HOG descriptor proceeds as follows [10,13,26,45]: firstly, divide the local image into regions called cells, and then calculate the amplitude of the first-order gradients of each cell in both the horizontal and vertical direction. The most common method is to apply a 1D mask, [–1 0 1].

$$G_x(x,\ y) = I(x+1,\ y) - I(x-1,\ y), \tag{2}$$

$$G_y(x,\ y) = I(x,\ y+1) - I(x,\ y-1), \tag{3}$$

where $I(x, y)$ is the pixel value of the point (x, y) and $G_x(x, y)$ and $G_y(x, y)$ denote the horizontal gradient amplitude and the vertical gradient amplitude, respectively. The magnitude of the gradient and the orientation of each pixel (x, y) are computed as follows:

$$G(x, y) = \sqrt{G_x(x, y)^2 + G_y(x, y)^2},$$
(4)

$$\theta(x, y) = \tan^{-1}\left(\frac{G_y(x, y)}{G_x(x, y)}\right).$$
(5)

The magnitude of the gradient and the orientation of each pixel in the cell are voted in nine bins with the tri-linear interpolation. The histograms of each cell are generated pixel based on direction gradients and, finally, the histograms of the whole cells are combined to extract the feature of the face image. Karaaba et al. [44] proposed a combination of different histograms of oriented gradients (HOG) to perform a robust face recognition system. This technique is named "multi-HOG".

The authors create a vector of distances between the target and the reference face images for identification. Arigbabu et al. [46] proposed a novel face recognition system based on the Laplacian filter and the pyramid histogram of gradient (PHOG) descriptor. In addition, to investigate the face recognition problem, support vector machine (SVM) is used with different kernel functions.

- Correlation filters: Face recognition systems based on the correlation filter (CF) have given good results in terms of robustness, location accuracy, efficiency, and discrimination. In the field of facial recognition, the correlation techniques have attracted great interest since the first use of an optical correlator [47]. These techniques provide the following advantages: high ability for discrimination, desired noise robustness, shift-invariance, and inherent parallelism. On the basis of these advantages, many optoelectronic hybrid solutions of correlation filters (CFs) have been introduced such as the joint transform correlator (JTC) [48] and VanderLugt correlator (VLC) [47] techniques. The purpose of these techniques is to calculate the degree of similarity between target and reference images. The decision is taken by the detection of a correlation peak. Both techniques (VLC and JTC) are based on the "$4f$" optical configuration [37]. This configuration is created by two convergent lenses (Figure 4). The face image F is processed by the fast Fourier transform (FFT) based on the first lens in the Fourier plane S_F. In this Fourier plane, a specific filter P is applied (for example, the phase-only filter (POF) filter [2]) using optoelectronic interfaces. Finally, to obtain the filtered face image F' (or the correlation plane), the inverse FFT (IFFT) is made with the second lens in the output plane.

For example, the VLC technique is done by two cascade Fourier transform structures realized by two lenses [4], as presented in Figure 5. The VLC technique is presented as follows: firstly, a 2D-FFT is applied to the target image to get a target spectrum S. After that, a multiplication between the target spectrum and the filter obtain with the 2D-FFT of a reference image is affected, and this result is placed in the Fourier plane. Next, it provides the correlation result recorded on the correlation plane, where this multiplication is affected by inverse FF.

The correlation result, described by the peak intensity, is used to determine the similarity degree between the target and reference images.

$$C = FFT^{-1}\{S^* \circ POF\},$$
(6)

where FFT^{-1} stands for the inverse fast FT (FFT) operation, * represents the conjugate operation, and $\circ$ denotes the element-wise array multiplication. To enhance the matching process, Horner and Gianino [49] proposed a phase-only filter (POF). The POF filter can produce correlation peaks marked with enhanced discrimination capability. The POF is an optimized filter defined as follows:

$$H_{POF}(u,v) = \frac{S^*(u,v)}{|S(u,v)|},$$

(7)

where $S^*(u,v)$ is the complex conjugate of the 2D-FFT of the reference image. To evaluate the decision, the peak to correlation energy (PCE) is defined as the energy in the correlation peaks' intensity normalized to the overall energy of the correlation plane.

$$PCE = \frac{\sum_{i,j}^{N} E_{peak}(i,j)}{\sum_{i,j}^{M} E_{correlation-plane}(i,j)},$$

(8)

where i, j are the coefficient coordinates; M and N are the size of the correlation plane and the size of the peak correlation spot, respectively; E_{peak} is the energy in the correlation peaks; and $E_{correlation-plane}$ is the overall energy of the correlation plane. Correlation techniques are widely applied in recognition and identification applications [4,37,50–53]. For example, in the work of [4], the authors presented the efficiency performances of the VLC technique based on the "4f" configuration for identification using GPU Nvidia Geforce 8400 GS. The POF filter is used for the decision. Another important work in this area of research is presented by Leonard et al. [50], which presented good performance and the simplicity of the correlation filters for the field of face recognition. In addition, many specific filters such as POF, BPOF, Ad, IF, and so on are used to select the best filter based on its sensitivity to the rotation, scale, and noise. Napoléon et al. [3] introduced a novel system for identification and verification fields based on an optimized 3D modeling under different illumination conditions, which allows reconstructing faces in different poses. In particular, to deform the synthetic model, an active shape model for detecting a set of key points on the face is proposed in Figure 6. The VanderLugt correlator is proposed to perform the identification and the LBP descriptor is used to optimize the performances of a correlation technique under different illumination conditions. The experiments are performed on the Pointing Head Pose Image Database (PHPID) database with an elevation ranging from −30° to +30°.

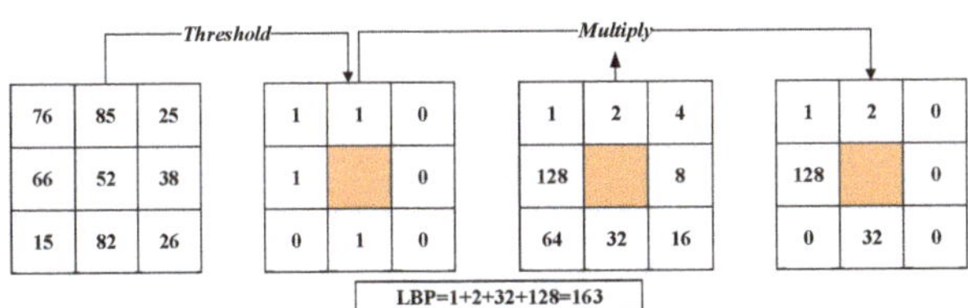

Figure 3. The local binary pattern (LBP) descriptor [19].

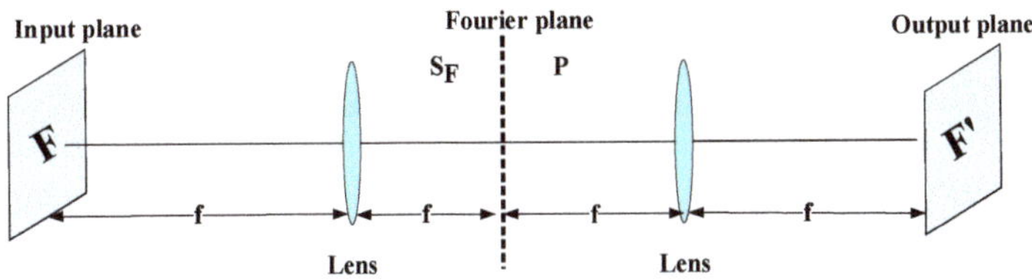

Figure 4. All "4f" optical configuration [37].

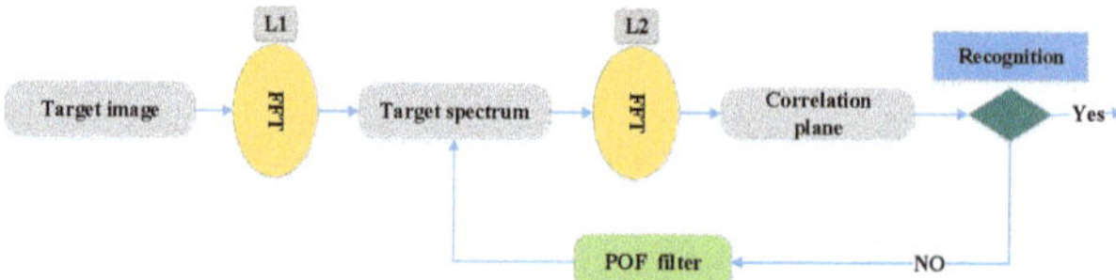

Figure 5. Flowchart of the VanderLugt correlator (VLC) technique [4]. FFT, fast Fourier transform; POF, phase-only filter.

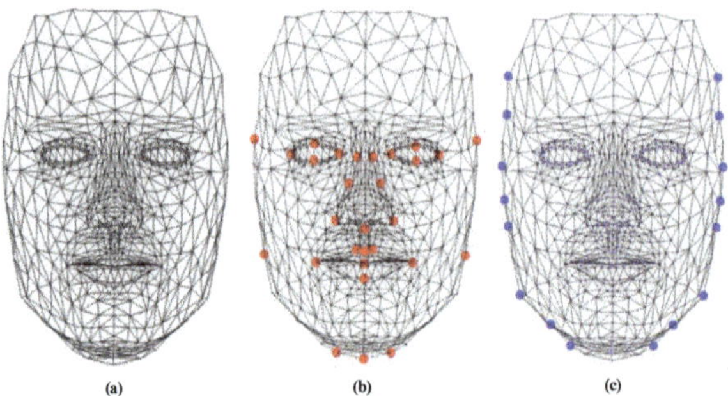

Figure 6. (**a**) Creation of the 3D face of a person, (**b**) results of the detection of 29 landmarks of a face using the active shape model, (**c**) results of the detection of 26 landmarks of a face [3].

3.2. Key-Points-Based Techniques

The key-points-based techniques are used to detect specific geometric features, according to some geometric information of the face surface (e.g., the distance between the eyes, the width of the head). These techniques can be defined by two significant steps, key-point detection and feature extraction [3,30,54,55]. The first step focuses on the performance of the detectors of the key-point features of the face image. The second step focuses on the representation of the information carried with the key-point features of the face image. Although these techniques can solve the missing parts and occlusions, scale invariant feature transform (SIFT), binary robust independent elementary features (BRIEF), and speeded-up robust features (SURF) techniques are widely used to describe the feature of the face image.

- Scale invariant feature transform (SIFT) [56,57]: SIFT is an algorithm used to detect and describe the local features of an image. This algorithm is widely used to link two images by their local descriptors, which contain information to make a match between them. The main idea of the SIFT descriptor is to convert the image into a representation composed of points of interest. These points contain the characteristic information of the face image. SIFT presents invariance to scale and rotation. It is commonly used today and fast, which is essential in real-time applications, but one of its disadvantages is the time of matching of the critical points. The algorithm is realized in four steps: (1) detection of the maximum and minimum points in the space-scale, (2) location of characteristic points, (3) assignment of orientation, and (4) a descriptor of the characteristic point. A framework to detect the key-points based on the SIFT descriptor was proposed by L. Lenc et al. [56], where they use the SIFT technique in combination with a Kepenekci approach for the face recognition.
- Speeded-up robust features (SURF) [29,57]: the SURF technique is inspired by SIFT, but uses wavelets and an approximation of the Hessian determinant to achieve better performance [29]. SURF is a detector and descriptor that claims to achieve the same, or even better, results in terms of repeatability, distinction, and robustness compared with the SIFT descriptor. The main advantage of SURF is the execution time, which is less than that used by the SIFT descriptor.

Besides, the SIFT descriptor is more adapted to describe faces affected by illumination conditions, scaling, translation, and rotation [57]. To detect feature points, SURF seeks to find the maximum of an approximation of the Hessian matrix using integral images to dramatically reduce the processing computational time. Figure 7 shows an example of SURF descriptor for face recognition using AR face datasets [58].

- Binary robust independent elementary features (BRIEF) [30,57]: BRIEF is a binary descriptor that is simple and fast to compute. This descriptor is based on the differences between the pixel intensity that are similar to the family of binary descriptors such as binary robust invariant scalable (BRISK) and fast retina keypoint (FREAK) in terms of evaluation. To reduce noise, the BRIEF descriptor smoothens the image patches. After that, the differences between the pixel intensity are used to represent the descriptor. This descriptor has achieved the best performance and accuracy in pattern recognition.

- Fast retina keypoint (FREAK) [57,59]: the FREAK descriptor proposed by Alahi et al. [59] uses a retinal sampling circular grid. This descriptor uses 43 sampling patterns based on retinal receptive fields that are shown in Figure 8. To extract a binary descriptor, these 43 receptive fields are sampled by decreasing factors as the distance from the thousand potential pairs to a patch's center yields. Each pair is smoothed with Gaussian functions. Finally, the binary descriptors are represented by setting a threshold and considering the sign of differences between pairs.

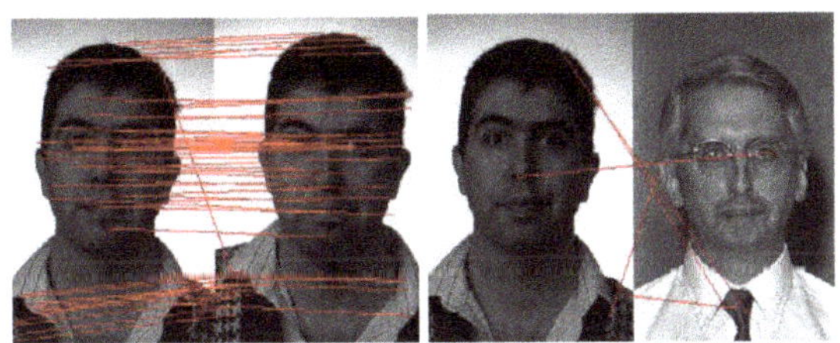

Figure 7. Face recognition based on the speeded-up robust features (SURF) descriptor [58]: recognition using fast library for approximate nearest neighbors (FLANN) distance.

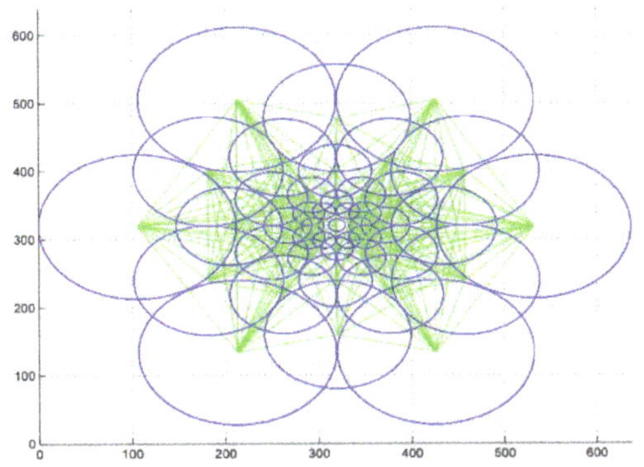

Figure 8. Fast retina keypoint (FREAK) descriptor used 43 sampling patterns [19].

3.3. Summary of Local Approaches

Table 1 summarizes the local approaches that we presented in this section. Various techniques are introduced to locate and to identify the human faces based on some regions of the face, geometric features, and facial expressions. These techniques provide robust recognition under different illumination conditions and facial expressions. Furthermore, they are sensitive to noise, and invariant to translations and rotations.

Table 1. Summary of local approaches. SIFT, scale-invariant feature transform; SURF, scale-invariant feature transform; BRIEF, binary robust independent elementary features; LBP, local binary pattern; HOG, histogram of oriented gradients; LPQ, local phase quantization; PCA, principal component analysis; LDA, linear discriminant analysis; KPCA, kernel PCA; CNN, convolutional neural network; SVM, support vector machine; PLBP, pyramid of LBP; KNN, k-nearest neighbor; MLBP, multiscale LBP; LTP, local ternary pattern.; PHOG, pyramid HOG; VLC, VanderLugt correlator; LFW, Labeled Faces in the Wild; FERET, Face Recognition Technology; PHPID, Pointing Head Pose Image Database; PCE, peak to correlation energy; POF, phase-only filter; PSR, peak-to-sidelobe ratio.

Author/Technique Used		Database	Matching	Limitation	Advantage	Result
Local Appearance-Based Techniques						
Khoi et al. [20]	LBP	TDF CF1999 LFW	MAP	Skewness in face image	Robust feature in fontal face	5% 13.03% 90.95%
Xi et al. [15]	LBPNet	FERET LFW	Cosine similarity	Complexities of CNN	High recognition accuracy	97.80% 94.04%
Khoi et al. [20]	PLBP	TDF CF LFW	MAP	Skewness in face image	Robust feature in fontal face	5.50% 9.70% 91.97%
Laure et al. [40]	LBP and KNN	LFW CMU-PIE	KNN	Illumination conditions	Robust	85.71% 99.26%
Bonnen et al. [42]	MRF and MLBP	AR (Scream) FERET (Wearing sunglasses)	Cosine similarity	Landmark extraction fails or is not ideal	Robust to changes in facial expression	86.10% 95%
Ren et al. [43]	Relaxed LTP	CMU-PIE Yale B	Chisquare distance	Noise level	Superior performance compared with LBP, LTP	95.75% 98.71%
Hussain et al. [60]	LPQ	FERET/ LFW	Cosine similarity	Lot of discriminative information	Robust to illumination variations	99.20% 75.30%
Karaaba et al. [44]	HOG and MMD	FERET LFW	MMD/MLPD	Low recognition accuracy	Aligning difficulties	68.59% 23.49%
Arigbabu et al. [46]	PHOG and SVM	LFW	SVM	Complexity and time of computation	Head pose variation	88.50%
Leonard et al. [50]	VLC correlator	PHPID	ASPOF	The low number of the reference image used	Robustness to noise	92%
Napoléon et al. [38]	LBP and VLC	YaleB YaleB Extended	POF	Illumination	Rotation + Translation	98.40% 95.80%
Heflin et al. [54]	correlation filter	LFW/PHPID	PSR	Some pre-processing steps	More effort on the eye localization stage	39.48%

Table 1. *Cont.*

Author/Technique Used		Database	Matching	Limitation	Advantage	Result
Local Appearance-Based Techniques						
Zhu et al. [55]	PCA–FCF	CMU-PIE FRGC2.0	Correlation filter	Use only linear method	Occlusion-insensitive	96.60% 91.92%
Seo et al. [27]	LARK + PCA	LFW	Cosine similarity	Face detection	Reducing computational complexity	78.90%
Ghorbel et al. [61]	VLC + DoG	FERET	PCE	Low recognition rate	Robustness	81.51%
Ghorbel et al. [61]	uLBP + DoG	FERET	chi-square distance	Robustness	Processing time	93.39%
Ouernari et al. [18]	VLC	PHPID	PCE	Power	Processing time	77%
Key-Points-Based Techniques						
Lenc et al. [56]	SIFT	FERET AR LFW	a posterior probability	Still far to be perfect	Sufficiently robust on lower quality real data	97.30% 95.80% 98.04%
Du et al. [29]	SURF	LFW	FLANN distance	Processing time	Robustness and distinctiveness	95.60%
Vinay et al. [23]	SURF + SIFT	LFW Face94	FLANN distance	Processing time	Robust in unconstrained scenarios	78.86% 96.67%
Calonder et al. [30]	BRIEF	–	KNN	Low recognition rate	Low processing time	48%

4. Holistic Approach

Holistic or subspace approaches are supposed to process the whole face, that is, they do not require extracting face regions or features points (eyes, mouth, noses, and so on). The main function of these approaches is to represent the face image by a matrix of pixels, and this matrix is often converted into feature vectors to facilitate their treatment. After that, these feature vectors are implemented in low dimensional space. However, holistic or subspace techniques are sensitive to variations (facial expressions, illumination, and poses), and these advantages make these approaches widely used. Moreover, these approaches can be divided into categories, including linear and non-linear techniques, based on the method used to represent the subspace.

4.1. Linear Techniques

The most popular linear techniques used for face recognition systems are Eigenfaces (principal component analysis; PCA) technique, Fisherfaces (linear discriminative analysis; LDA) technique, and independent component analysis (ICA).

- Eigenface [34] and principal component analysis (PCA) [27,62]: Eigenfaces is one of the popular methods of holistic approaches used to extract features points of the face image. This approach is based on the principal component analysis (PCA) technique. The principal components created by the PCA technique are used as Eigenfaces or face templates. The PCA technique transforms a number of possibly correlated variables into a small number of incorrect variables called "principal components". The purpose of PCA is to reduce the large dimensionality of the data space (observed variables) to the smaller intrinsic dimensionality of feature space (independent variables), which are needed to describe the data economically. Figure 9 shows how the face can be represented by a small number of features. PCA calculates the Eigenvectors of the covariance matrix, and projects the original data onto a lower dimensional feature space, which are defined by Eigenvectors with large Eigenvalues. PCA has been used in face representation and recognition, where the Eigenvectors calculated are referred to as Eigenfaces (as shown in Figure 10).

An image may also be considering the vector of dimension $M \times N$, so that a typical image of size 4×4 becomes a vector of dimension 16. Let the training set of images be $\{X_1, X_2, X_3 \dots X_N\}$. The average face of the set is defined by the following:

$$X = \frac{1}{N} \sum_{i=1}^{N} X_i.$$ (9)

Calculate the estimate covariance matrix to represent the scatter degree of all feature vectors related to the average vector. The covariance matrix Q is defined by the following:

$$Q = \frac{1}{N} \sum_{i=1}^{N} \left(X - X_i\right)\left(X - X_i\right)^{\mathrm{T}}.$$ (10)

The Eigenvectors and corresponding Eigen-values are computed using

$$CV = \lambda V, \quad (V \epsilon R_n, \ V \neq 0),$$ (11)

where V is the set of eigenvectors matrix Q associated with its eigenvalue λ. Project all the training images of i_{th} person to the corresponding Eigen-subspace:

$$y_k^i = w^T \ (x_i), \quad (i = 1, 2, 3 \dots N),$$ (12)

where the y_k^i are the projections of x and are called the principal components, also known as eigenfaces. The face images are represented as a linear combination of these vectors' "principal components". In order to extract facial features, PCA and LDA are two different feature extraction algorithms that are used. Wavelet fusion and neural networks are applied to classify facial features. The ORL database is used for evaluation. Figure 10 shows the first five Eigenfaces constructed from the ORL database [63].

- Fisherface and linear discriminative analysis (LDA) [64,65]: The Fisherface method is based on the same principle of similarity as the Eigenfaces method. The objective of this method is to reduce the high dimensional image space based on the linear discriminant analysis (LDA) technique instead of the PCA technique. The LDA technique is commonly used for dimensionality reduction and face recognition [66]. PCA is an unsupervised technique, while LDA is a supervised learning technique and uses the data information. For all samples of all classes, the within-class scatter matrix S_W and the between-class scatter matrix S_B are defined as follows:

$$S_B = \sum_{I=1}^{C} M_i(x_i - \mu)(x_i - \mu)^T, \tag{13}$$

$$S_w = \sum_{I=1}^{C} \sum_{x_k \epsilon X_i} M_i(x_k - \mu)(x_k - \mu)^T, \tag{14}$$

where μ is the mean vector of samples belonging to class i, X_i represents the set of samples belonging to class i with x_k being the number image of that class, c is the number of distinct classes, and M_i is the number of training samples in class i. S_B describes the scatter of features around the overall mean for all face classes and S_w describes the scatter of features around the mean of each face class. The goal is to maximize the ratio $det|S_B|/det|S_w|$, in other words, minimizing S_w while maximiz ing S_B. Figure 11 shows the first five Eigenfaces and Fisherfaces obtained from the ORL database [63].

- Independent component analysis (ICA) [35]: The ICA technique is used for the calculation of the basic vectors of a given space. The goal of this technique is to perform a linear transformation in order to reduce the statistical dependence between the different basic vectors, which allows the analysis of independent components. It is determined that they are not orthogonal to each other. In addition, the acquisition of images from different sources is sought in uncorrelated variables, which makes it possible to obtain greater efficiency, because ICA acquires images within statistically independent variables.

- Improvements of the PCA, LDA, and ICA techniques: To improve the linear subspace techniques, many types of research are developed. Z. Cui et al. [67] proposed a new spatial face region descriptor (SFRD) method to extract the face region, and to deal with noise variation. This method is described as follows: divide each face image in many spatial regions, and extract token-frequency (TF) features from each region by sum-pooling the reconstruction coefficients over the patches within each region. Finally, extract the SFRD for face images by applying a variant of the PCA technique called "whitened principal component analysis (WPCA)" to reduce the feature dimension and remove the noise in the leading eigenvectors. Besides, the authors in [68] proposed a variant of the LDA called probabilistic linear discriminant analysis (PLDA) to seek directions in space that have maximum discriminability, and are hence most suitable for both face recognition and frontal face recognition under varying pose.

- Gabor filters: Gabor filters are spatial sinusoids located by a Gaussian window that allows for extracting the features from images by selecting their frequency, orientation, and scale. To enhance the performance under unconstrained environments for face recognition, Gabor filters are transformed according to the shape and pose to extract the feature vectors of face image combined with the PCA in the work of [69]. The PCA is applied to the Gabor features to remove the redundancies and to get the best face images description. Finally, the cosine metric is used to evaluate the similarity.

- Frequency domain analysis [70,71]: Finally, the analysis techniques in the frequency domain offer a representation of the human face as a function of low-frequency components that present high energy. The discrete Fourier transform (DFT), discrete cosine transform (DCT), or discrete wavelet transform (DWT) techniques are independent of the data, and thus do not require training.
- Discrete wavelet transform (DWT): Another linear technique used for face recognition. In the work of [70], the authors used a two-dimensional discrete wavelet transform (2D-DWT) method for face recognition using a new patch strategy. A non-uniform patch strategy for the top-level's low-frequency sub-band is proposed by using an integral projection technique for two top-level high-frequency sub-bands of 2D-DWT based on the average image of all training samples. This patch strategy is better for retaining the integrity of local information, and is more suitable to reflect the structure feature of the face image. When constructing the patching strategy using the testing and training samples, the decision is performed using the neighbor classifier. Many databases are used to evaluate this method, including Labeled Faces in Wild (LFW), Extended Yale B, Face Recognition Technology (FERET), and AR.
- Discrete cosine transform (DCT) [71] can be used for global and local face recognition systems. DCT is a transformation that represents a finite sequence of data as the sum of a series of cosine functions oscillating at different frequencies. This technique is widely used in face recognition systems [71], from audio and image compression to spectral methods for the numerical resolution of differential equations. The required steps to implement the DCT technique are presented as follows.

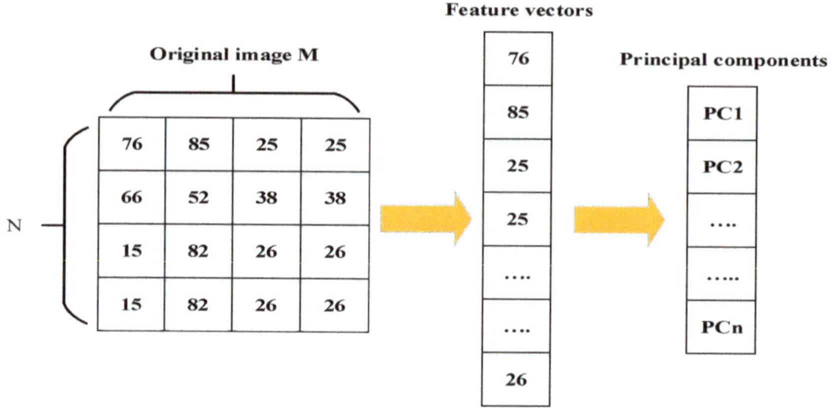

Figure 9. Example of dimensional reduction when applying principal component analysis (PCA) [62].

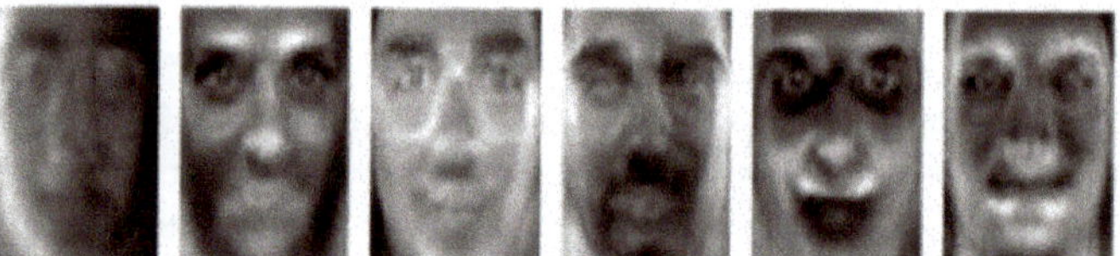

Figure 10. The first five Eigenfaces built from the ORL database [63].

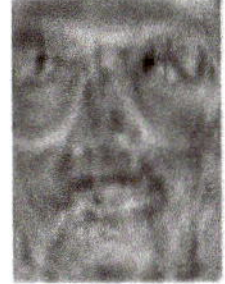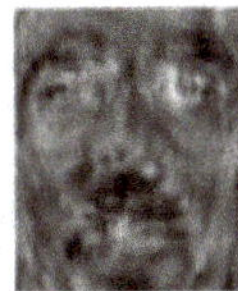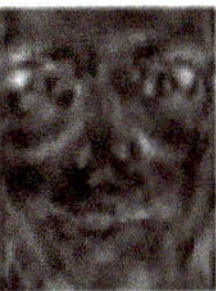

Figure 11. The first five Fisherfaces obtained from the ORL database [63].

Owing to their limitations in managing the linearity in face recognition, the subspace or holistic techniques are not appropriate to represent the exact details of geometric varieties of the face images. Linear techniques offer a faithful description of face images when the data structures are linear. However, when the face images data structures are non-linear, many types of research use a function named "kernel" to construct a large space where the problem becomes linear. The required steps to implement the DCT technique are presented as Algorithm 1.

Algorithm 1. DCT Algorithm

---

1. *The input image is N by M;*
2. *$f(i,j)$ is the intensity of the pixel in row i and column j;*
3. *$F(u,v)$ is the DCT coefficient in row u and column v of the DCT matrix:*

$$F(u,v) = \frac{2C(u)C(v)}{N} \sum_{i=1}^{N} \sum_{j=1}^{N} f(i,j)\cos\left(\frac{(2i-1)(u-1)\pi}{2N}\right)\cos\left(\frac{(2j-1)(v-1)\pi}{2N}\right)$$

$$= \frac{2C(u)C(v)}{N} \sum_{i=0}^{N-1} \sum_{j=0}^{N-1} f(i,j)\cos\left(\frac{(2i+1)u\pi}{2N}\right)\cos\left(\frac{(2j+1)v\pi}{2N}\right)$$

$$where\ 0 \le i, v \le N - 1\ and\ C(n) = \begin{cases} \frac{1}{\sqrt{2}} & (n = 0) \\ 1 & (n \neq 0) \end{cases}$$

4. *For most images, much of the signal energy lies at low frequencies; these appear in the upper left corner of the DCT.*
5. *Compression is achieved since the lower right values represent higher frequencies, and are often small - small enough to be neglected with little visible distortion.*
6. *The DCT input is an 8 by 8 array of integers. This array contains each pixel's grayscale level;*
7. *8-bit pixels have levels from 0 to 255.*

4.2. Nonlinear Techniques

- Kernel PCA (KPCA) [28]: is an improved method of PCA, which uses kernel method techniques. KPCA computes the Eigenfaces or the Eigenvectors of the kernel matrix, while PCA computes the covariance matrix. In addition, KPCA is a representation of the PCA technique on the high-dimensional feature space mapped by the associated kernel function. Three significant steps of the KPCA algorithm are used to calculates the function of the kernel matrix K of distribution consisting of n data points $x_i \in R^d$, after which the data points are mapped into a high-dimensional feature space F, as shown in Algorithm 2.

  The performance of the KPCA technique depends on the choice of the kernel matrix K. The Gaussian or polynomial kernel are linear typically-used kernels. KPCA has been successfully used for novelty detection [72] or for speech recognition [62].

- Kernel linear discriminant analysis (KDA) [73]: the KLDA technique is a kernel extension of the linear LDA technique, in the same kernel extension of PCA. Arashloo et al. [73] proposed a nonlinear binary class-specific kernel discriminant analysis classifier (CS-KDA) based on the spectral regression kernel discriminant analysis. Other nonlinear techniques have also been used in the context of facial recognition:

- Gabor-KLDA [74].
- Evolutionary weighted principal component analysis (EWPCA) [75].
- Kernelized maximum average margin criterion (KMAMC), SVM, and kernel Fisher discriminant analysis (KFD) [76].
- Wavelet transform (WT), radon transform (RT), and cellular neural networks (CNN) [77].
- Joint transform correlator-based two-layer neural network [78].
- Kernel Fisher discriminant analysis (KFD) and KPCA [79].
- Locally linear embedding (LLE) and LDA [80].
- Nonlinear locality preserving with deep networks [81].
- Nonlinear DCT and kernel discriminative common vector (KDCV) [82].

Algorithm 2. Kernel PCA Algorithm

- *Step 1: Determine the dot product of the matrix K using kernel function:* $K_{ij} = k(x_i, x_j)$.
- *Step 2: Calculate the Eigenvectors from the resultant matrix K and normalize with the function:* $\gamma k(\alpha k \alpha k) = 1$.
- *Step 3: Calculate the test point projection on to Eigenvectors Vk using kernel function:*
  $kPCk(x) = (Vk\varphi(x)) = \sum_{i}^{m} \alpha k_i \, k(x_i, x)$

4.3. Summary of Holistic Approaches

Table 2 summarizes the different subspace techniques discussed in this section, which are introduced to reduce the dimensionality and the complexity of the detection or recognition steps. Linear and non-linear techniques offer robust recognition under different lighting conditions and facial expressions. Although these techniques (linear and non-linear) allow a better reduction in dimensionality and improve the recognition rate, they are not invariant to translations and rotations compared with local techniques.

Table 2 Subspace approaches. ICA, independent component analysis; DWT, discrete wavelet transform; FFT, fast Fourier transform; DCT, discrete cosine transform.

Author/Techniques Used		Databases	Matching	Limitation	Advantage	Result
	Linear Techniques					
Seo et al. [27]	LARK and PCA	LFW	L2 distance	Detection accuracy	Reducing computational complexity	85.10%
Annalakshmi et al. [35]	ICA and LDA	LFW	Bayesian Classifier	Sensitivity	Good accuracy	88%
Annalakshmi et al. [35]	PCA and LDA	LFW	Bayesian Classifier	Sensitivity	Specificity	59%
Hussain et al. [36]	LQP and Gabor	FERET LFW	Cosine similarity	Lot of discriminative information	Robust to illumination variations	99.2% 75.3%
Gowda et al. [17]	LPQ and LDA	MEPCO	SVM	Computation time	Good accuracy	99.13%
Z. Cui et al. [67]	BoW	AR ORL FERET	ASM	Occlusions	Robust	99.43% 99.50% 82.30%
Khan et al. [83]	PSO and DWT	CK MMI JAFFE	Euclidienne distance	Noise	Robust to illumination	98.60% 95.50% 98.80%
Huang et al. [70]	2D-DWT	FERET LFW	KNN	Pose	Frontal or near-frontal facial images	90.63% 97.10%
Perlibakas and Vytautas [69]	PCA and Gabor filter	FERET	Cosine metric	Precision	Pose	87.77%
Hafez et al. [84]	Gabor filter and LDA	ORL C. YaleB	2DNCC	Pose	Good recognition performance	98.33% 99.33%
Sufyanu et al. [71]	DCT	ORL Yale	NCC	High memory	Controlled and uncontrolled databases	93.40%
Shanbhag et al. [85]	DWT and BPSO	– –	– –	Rotation	Significant reduction in the number of features	88.44%
Ghorbel et al. [61]	Eigenfaces and DoG filter	FERET	Chi-square distance	Processing time	Reduce the representation	84.26%
Zhang et al. [12]	PCA and FFT	YALE	SVM	Complexity	Discrimination	93.42%
Zhang et al. [12]	PCA	YALE	SVM	Recognition rate	Reduce the dimensionality	84.21%

Table 2. *Cont.*

Author/Techniques Used		Databases	Matching	Limitation	Advantage	Result
				Nonlinear Techniques		
Fan et al. [86]	RKPCA	MNIST ORL	RBF kernel	Complexity	Robust to sparse noises	–
Vinay et al. [87]	ORB and KPCA	ORL	FLANN Matching	Processing time	Robust	87.30%
Vinay et al. [87]	SURF and KPCA	ORL	FLANN Matching	Processing time	Reduce the dimensionality	80.34%
Vinay et al. [87]	SIFT and KPCA	ORL	FLANN Matching	Low recognition rate	Complexity	69.20%
Lu et al. [88]	KPCA and GDA	UMIST face	SVM	High error rate	Excellent performance	48%
Yang et al. [89]	PCA and MSR	HELEN face	ESR	Complexity	Utilizes color, gradient, and regional information	98.00%
Yang et al. [89]	LDA and MSR	FRGC	ESR	Low performances	Utilizes color, gradient, and regional information	90.75%
Ouanan et al. [90]	FDDL	AR	CNN	Occlusion	Orientations, expressions	98.00%
Vankayalapati and Kyamakya [77]	CNN	ORL	– –	Poses	High recognition rate	95%
Devi et al. [63]	2FNN	ORL	– –	Complexity	Low error rate	98.5

5. Hybrid Approach

5.1. Technique Presentation

The hybrid approaches are based on local and subspace features in order to use the benefits of both subspace and local techniques, which have the potential to offer better performance for face recognition systems.

- Gabor wavelet and linear discriminant analysis (GW-LDA) [91]: Fathima et al. [91] proposed a hybrid approach combining Gabor wavelet and linear discriminant analysis (HGWLDA) for face recognition. The grayscale face image is approximated and reduced in dimension. The authors have convolved the grayscale face image with a bank of Gabor filters with varying orientations and scales. After that, a subspace technique 2D-LDA is used to maximize the inter-class space and reduce the intra-class space. To classify and recognize the test face image, the k-nearest neighbour (k-NN) classifier is used. The recognition task is done by comparing the test face image feature with each of the training set features. The experimental results show the robustness of this approach in different lighting conditions.
- Over-complete LBP (OCLBP), LDA, and within class covariance normalization (WCCN): Barkan et al. [92] proposed a new representation of face image based over-complete LBP (OCLBP). This representation is a multi-scale modified version of the LBP technique. The LDA technique is performed to reduce the high dimensionality representations. Finally, the within class covariance normalization (WCCN) is the metric learning technique used for face recognition.
- Advanced correlation filters and Walsh LBP (WLBP): Juefei et al. [93] implemented a single-sample periocular-based alignment-robust face recognition technique based on high-dimensional Walsh LBP (WLBP). This technique utilizes only one sample per subject class and generates new face images under a wide range of 3D rotations using the 3D generic elastic model, which is both accurate and computationally inexpensive. The LFW database is used for evaluation, and the proposed method outperformed the state-of-the-art algorithms under four evaluation protocols with a high accuracy of 89.69%.
- Multi-sub-region-based correlation filter bank (MS-CFB): Yan et al. [94] propose an effective feature extraction technique for robust face recognition, named multi-sub-region-based correlation filter bank (MS-CFB). MS-CFB extracts the local features independently for each face sub-region. After that, the different face sub-regions are concatenated to give optimal overall correlation outputs. This technique reduces the complexity, achieves higher recognition rates, and provides a better feature representation for recognition compared with several state-of-the-art techniques on various public face databases.
- SIFT features, Fisher vectors, and PCA: Simonyan et al. [64] have developed a novel method for face recognition based on the SIFT descriptor and Fisher vectors. The authors propose a discriminative dimensionality reduction owing to the high dimensionality of the Fisher vectors. After that, these vectors are projected into a low dimensional subspace with a linear projection. The objective of this methodology is to describe the image based on dense SIFT features and Fisher vectors encoding to achieve high performance on the challenging LFW dataset in both restricted and unrestricted settings.
- CNNs and stacked auto-encoder (SAE) techniques: Ding et al. [95] proposed multimodal deep face representation (MM-DFR) framework based on convolutional neural networks (CNNs) technique from the original holistic face image, rendered frontal face by 3D face model (stand for holistic facial features and local facial features, respectively), and uniformly sampled image patches. The proposed MM-DFR framework has two steps: a CNNs technique is used to extract the features and a three-layer stacked auto-encoder (SAE) technique is employed to compress the high-dimensional deep feature into a compact face signature. The LFW database is used to

evaluate the identification performance of MM-DFR. The flowchart of the proposed MM-DFR framework is shown in Figure 12.

- PCA and ANFIS: Sharma et al. [96] propose an efficient pose-invariant face recognition system based on PCA technique and ANFIS classifier. The PCA technique is employed to extract the features of an image, and the ANFIS classifier is developed for identification under a variety of pose conditions. The performance of the proposed system based on PCA–ANFIS is better than ICA–ANFIS and LDA–ANFIS for the face recognition task. The ORL database is used for evaluation.

- DCT and PCA: Ojala et al. [97] develop a fast face recognition system based on DCT and PCA techniques. Genetic algorithm (GA) technique is used to extract facial features, which allows to remove irrelevant features and reduces the number of features. In addition, the DCT–PCA technique is used to extract the features and reduce the dimensionality. The minimum Euclidian distance (ED) as a measurement is used for the decision. Various face databases are used to demonstrate the effectiveness of this system.

- PCA, SIFT, and iterative closest point (ICP): Mian et al. [98] present a multimodal (2D and 3D) face recognition system based on hybrid matching to achieve efficiency and robustness to facial expressions. The Hotelling transform is performed to automatically correct the pose of a 3D face using its texture. After that, in order to form a rejection classifier, a novel 3D spherical face representation (SFR) in conjunction with the SIFT descriptor is used, which provide efficient recognition in the case of large galleries by eliminating a large number of candidates' faces. A modified iterative closest point (ICP) algorithm is used for the decision. This system is less sensitive and robust to facial expressions, which achieved a 98.6% verification rate and 96.1% identification rate on the complete FRGC v2 database.

- PCA, local Gabor binary pattern histogram sequence (LGBPHS), and GABOR wavelets: Cho et al. [99] proposed a computationally efficient hybrid face recognition system that employs both holistic and local features. The PCA technique is used to reduce the dimensionality. After that, the local Gabor binary pattern histogram sequence (LGBPHS) technique is employed to realize the recognition stage, which proposed to reduce the complexity caused by the Gabor filters. The experimental results show a better recognition rate compared with the PCA and Gabor wavelet techniques under illumination variations. The Extended Yale Face Database B is used to demonstrate the effectiveness of this system.

- PCA and Fisher linear discriminant (FLD) [100,101]: Sing et al. [101] propose a novel hybrid technique for face representation and recognition, which exploits both local and subspace features. In order to extract the local features, the whole image is divided into a sub-regions, while the global features are extracted directly from the whole image. After that, PCA and Fisher linear discriminant (FLD) techniques are introduced on the fused feature vector to reduce the dimensionality. The CMU-PIE, FERET, and AR face databases are used for the evaluation.

- SPCA–KNN [102]: Kamencay et al. [102] develop a new face recognition method based on SIFT features, as well as PCA and KNN techniques. The Hessian–Laplace detector along with SPCA descriptor is performed to extract the local features. SPCA is introduced to identify the human face. KNN classifier is introduced to identify the closest human faces from the trained features. The results of the experiment have a recognition rate of 92% for the unsegmented ESSEX database and 96% for the segmented database (700 training images).

- Convolution operations, LSTM recurrent units, and ELM classifier [103]: Sun et al. [103] propose a hybrid deep structure called CNN–LSTM–ELM in order to achieve sequential human activity recognition (HAR). Their proposed CNN–LSTM–ELM structure is evaluated using the OPPORTUNITY dataset, which contains 46,495 training samples and 9894 testing samples, and each sample is a sequence. The model training and testing runs on a GPU with 1536 cores, 1050 MHz clock speed, and 8 GB RAM. The flowchart of the proposed CNN–LSTM–ELM structure is shown in Figure 13 [103].

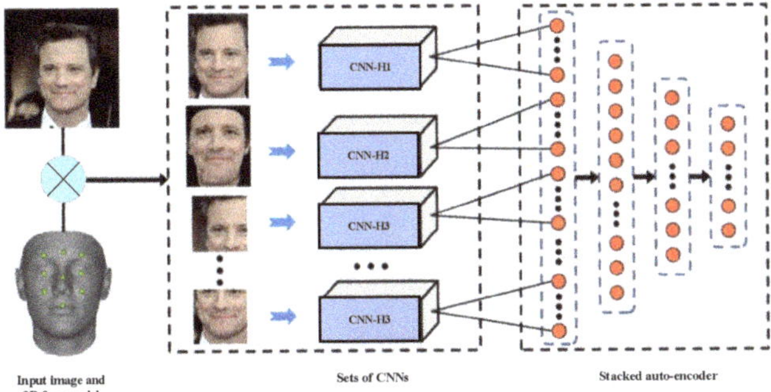

Figure 12. Flowchart of the proposed multimodal deep face representation (MM-DFR) technique [95]. CNN, convolutional neural network.

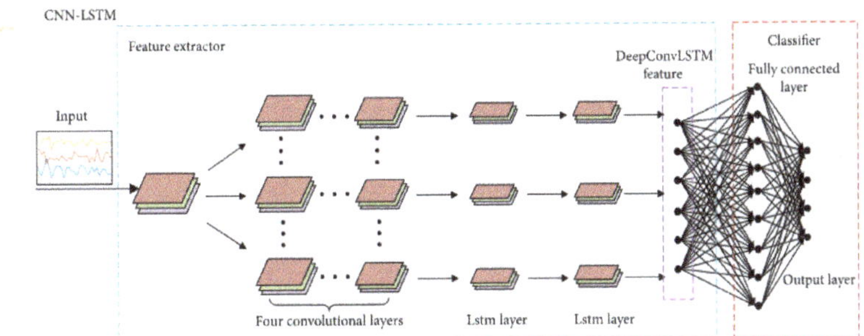

Figure 13. The proposed CNN–LSTM–ELM [103].

5.2. Summary of Hybrid Approaches

Table 3 summarizes the hybrid approaches that we presented in this section. Various techniques are introduced to improve the performance and the accuracy of recognition systems. The combination between the local approaches and the subspace approach provides robust recognition and reduction of dimensionality under different illumination conditions and facial expressions. Furthermore, these technologies are presented to be sensitive to noise, and invariant to translations and rotations.

Table 3. Hybrid approaches. GW, Gabor wavelet; OCLBP, over-complete LBP; WCCN, within class covariance normalization; WLBP, Walsh LPB; ICP, iterative closest point; LGBPHS, local Gabor binary pattern histogram sequence; FLD, Fisher linear discriminant; SAE, stacked auto-encoder.

Author/Technique Used		Database	Matching	Limitation	Advantage	Result
Fathima et al. [91]	GW-LDA	AT&T FACES94 MITINDIA	k-NN	High processing time	Illumination invariant and reduce the dimensionality	88% 94.02% 88.12%
Barkan et al., [92]	OCLBP, LDA, and WCCN	LFW	WCCN	–	Reduce the dimensionality	87.85%
Juefei et al. [93]	ACF and WLBP	LFW		Complexity	Pose conditions	89.69%
Simonyan et al. [64]	Fisher + SIFT	LFW	Mahalanobis matrix	Single feature type	Robust	87.47%
Sharma et al. [96]	PCA–ANFIS ICA–ANFIS LDA–ANFIS	ORL	ANFIS ANFIS ANFIS	Sensitivity-specificity	Pose conditions	96.66% 71.30% 68%
Ojala et al. [97]	DCT–PCA	ORL UMIST YALE	Euclidian distance	Complexity	Reduce the dimensionality	92.62% 99.40% 95.50%
Mian et al. [98]	Hotelling transform, SIFT, and ICP	FRGC	ICP	Processing time	Facial expressions	99.74%
Cho et al. [99]	PCA–LGBPHS PCA–GABOR Wavelets	Extended Yale Face	Bhattacharyya distance	Illumination condition	Complexity	95%
Sing et al. [101]	PCA–FLD	CMU FERET AR	SVM	Robustness	Pose, illumination, and expression	71.98% 94.73% 68.65%
Kamencay et al. [102]	SPCA-KNN	ESSEX	KNN	Processing time	Expression variation	96.80%
Sun et al. [103]	CNN–LSTM–ELM	OPPORTUNITY	LSTM/ELM	High processing time	Automatically learn feature representations	90.60%
Ding et al. [95]	CNNs and SAE	LFW	– –	Complexity	High recognition rate	99%

6. Assessment of Face Recognition Approaches

In the last step of recognition, the face extracted from the background during the face detection step is compared with known faces stored in a specific database. To make the decision, several techniques of comparison are used. This section describes the most common techniques used to make the decision and comparison.

6.1. Measures of Similarity or Distances

- Peak-to-correlation energy (PCE) or peak-to-sidelobe ratio (PSR) [18]: The PCE was introduced in (8).
- Euclidean distance [54]: The Euclidean distance is one of the most basic measures used to compute the direct distance between two points in a plane. If we have two points P1 and P2, with the coordinates $(x1, y1)$ and $(x2, y2)$, respectively, the calculation of the Euclidean distance between them would be as follows:

$$d_E(P1, P2) = \sqrt{(x2 - x1)^2 + (y2 - y1)^2}. \tag{15}$$

In general, the Euclidean distance between two points $P = (1, p2, \ldots, pn)$ and $Q = (q1, q2, \ldots, qn)$ in the n-dimensional space would be defined by the following:

$$d_E(P, Q) = \sqrt{\sum_i^n (pi - qi)^2}. \tag{16}$$

- Bhattacharyya distance [104,105]: The Bhattacharyya distance is a statistical measure that quantifies the similarity between two discrete or continuous probability distributions. This distance is particularly known for its low processing time and its low sensitivity to noise. For the probability distributions p and q defined on the same domain, the distance of Bhattacharyya is defined as follows:

$$D_B(p, q) = -ln(BC(p, q)), \tag{17}$$

$$BC(p, q) = \sum_{x \in X} \sqrt{p(x)q(x)}\ (a\ );\ BC(p, q) = \int \sqrt{p(x)q(x)}dx\ (b), \tag{18}$$

where BC is the Bhattacharyya coefficient, defined as Equation (18a) for discrete probability distributions and as Equation (18b) for continuous probability distributions. In both cases, $0 \leq BC \leq 1$ and $0 \leq DB \leq \infty$. In its simplest formulation, the Bhattacharyya distance between two classes that follow a normal distribution can be calculated from a mean (μ) and the variance (σ^2):

$$DB(p, q) = \frac{1}{4}ln\left(\frac{1}{4}\left(\frac{\sigma_p^2}{\sigma_q^2} + \frac{\sigma_q^2}{\sigma_p^2} + 2\right)\right) + \frac{1}{4}\left(\frac{\left(\mu_p - \mu_q\right)}{\sigma_q^2 + \sigma_p^2}\right). \tag{19}$$

- Chi-squared distance [106]: The Chi-squared $\left(X^2\right)$ distance was weighted by the value of the samples, which allows knowing the same relevance for sample differences with few occurrences as those with multiple occurrences. To compare two histograms $S_1 = (u_1, \ldots \ldots \ldots u_m)$ and $S_2 = (w_1, \ldots \ldots \ldots w_m)$, the Chi-squared $\left(X^2\right)$ distance can be defined as follows:

$$\left(X^2\right) = D(S_1, S_2) = \frac{1}{2}\sum_{i=1}^{m} \frac{(u_i - w_i)^2}{u_i + w_i}. \tag{20}$$

6.2. Classifiers

There are many face classification techniques in the literature that allow to select, from a few examples, the group or class to which the objects belong. Some of them are based on statistics, such as the Bayesian classifier and correlation [18], and so on, and others based on the regions that generate the different classes in the decision space, such as K-means [9], CNN [103], artificial neural networks (ANNs) [37], support vector machines (SVMs) [26,107], k-nearest neighbors (K-NNs), decision trees (DTs), and so on.

- Support vector machines (SVMs) [13,26]: The feature vectors extracted by any descriptor are classified by linear or nonlinear SVM. The SVM classifier may realize the separation of the classes with an optimal hyperplane. To determine the last, only the closest points of the total learning set should be used; these points are called support vectors (Figure 14).

  There is an infinite number of hyperplanes capable of perfectly separating two classes, which implies to select a hyperplane that maximizes the minimal distance between the learning examples and the learning hyperplane (i.e., the distance between the support vectors and the hyperplane). This distance is called "margin". The SVM classifier is used to calculate the optimal hyperplane that categorizes a set of labels training data in the correct class. The optimal hyperplane is solved as follows:

  $$D = \left\{ (x_i, y_i) \big| x_i \in R^n , \; y_i \in \{-1, 1\}, \; i = 1 \ldots \ldots l \right\}. \tag{21}$$

  Given that x_i are the training features vectors and y_i are the corresponding set of l (1 or -1) labels. An SVM tries to find a hyperplane to distinguish the samples with the smallest errors. The classification function is obtained by calculating the distance between the input vector and the hyperplane.

  $$wx_i - b = C_f, \tag{22}$$

  where w and b are the parameters of the model. Shen et al. [108] proposed the Gabor filter to extract the face features and applied the SVM for classification. The proposed FaceNet method achieves a good record accuracy of 99.63% and 95.12% using the LFW YouTube Faces DB datasets, respectively.
- k-nearest neighbor (k-NN) [17,91]: k-NN is an indolent algorithm because, in training, it saves little information, and thus does not build models of difference, for example, decision trees.
- K-means [9,109]: It is called K-means because it represents each of the groups by the average (or weighted average) of its points, called the centroid. In the K-means algorithm, it is necessary to specify a priori the number of clusters k that one wishes to form in order to start the process.
- Deep learning (DL): An automatic learning technique that uses neural network architectures. The term "deep" refers to the number of hidden layers in the neural network. While conventional neural networks have one layer, deep neural networks (DNN) contain several layers, as presented in Figure 15.

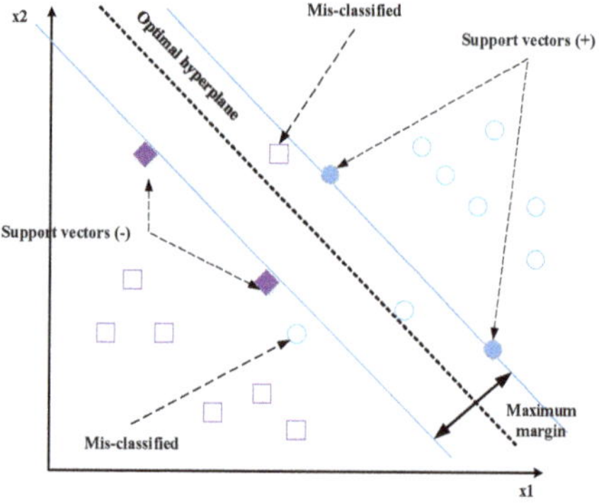

Figure 14. Optimal hyperplane, support vectors, and maximum margin.

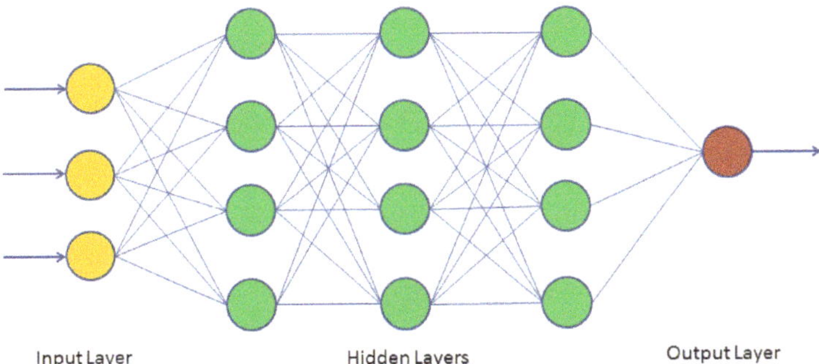

Figure 15. Artificial neural network.

Various variants of neural networks have been developed in the last years, such as convolutional neural networks (CNN) [14,110] and recurrent neural networks (RNN) [111], which very effective for image detection and recognition tasks. CNNs are a very successful deep model and are used today in many applications [112]. From a structural point of view, CNNs are made up of three different types of layers: convolution layers, pooling layers, and fully-connected layers.

1. *Convolutional layer*: sometimes called the feature extractor layer because features of the image are extracted within this layer. Convolution preserves the spatial relationship between pixels by learning image features using small squares of the input image. The input image is convoluted by employing a set of learnable neurons. This produces a feature map or activation map in the output image, after which the feature maps are fed as input data to the next convolutional layer. The convolutional layer also contains rectified linear unit (ReLU) activation to convert all negative value to zero. This makes it very computationally efficient, as few neurons are activated each time.

2. *Pooling layer:* used to reduce dimensions, with the aim of reducing processing times by retaining the most important information after convolution. This layer basically reduces the number of parameters and computation in the network, controlling over fitting by progressively reducing the spatial size of the network. There are two operations in this layer: average pooling and maximum pooling:

    -   Average-pooling takes all the elements of the sub-matrix, calculates their average, and stores the value in the output matrix.
    -   Max-pooling searches for the highest value found in the sub-matrix and saves it in the output matrix.

3. *Fully-connected layer*: in this layer, the neurons have a complete connection to all the activations from the previous layers. It connects neurons in one layer to neurons in another layer. It is used to classify images between different categories by training.

Wen et al. [113] introduce a new supervision signal, called center loss, for the face recognition task in order to improve the discriminative power of the deeply learned features. Specifically, the proposed center loss function is trainable and easy to optimize in the CNNs. Several important face recognition benchmarks are used for evaluation including LFW, YTF, and MegaFace Challenge. Passalis and Tefas [114] propose a supervised codebook learning method for the bag-of-features representation able to learn face retrieval-oriented codebooks. This allows using significantly smaller codebooks enhancing both the retrieval time and storage requirements. Liu et al. [115] and Amato et al. [116] propose a deep face recognition technique under open-set protocol based on the CNN technique. A face dataset composed of 39,037 faces images belonging to 42 different identities is used to perform the experiments.

Taigman et al. [117] present a system (DeepFace) able to outperform existing systems with only very minimal adaptation. It is trained on a large dataset of faces acquired from a population vastly different than the one used to construct the evaluation benchmarks. This technique achieves an accuracy of 97.35% on the LFW. Ma et al. [118] introduce a robust local binary pattern (LBP) guiding pooling (G-RLBP) mechanism to improve the recognition rates of the CNN models, which can successfully lower the noise impact. Koo et al. [119] propose a multimodal human recognition method that uses both the face and body and is based on a deep CNN. Cho et al. [120] propose a nighttime face detection method based on CNN technique for visible-light images. Koshy and Mahmood [121] develop deep architectures for face liveness detection that uses a combination of texture analysis and a CNN technique to classify the captured image as real or fake. Elmahmudi and Ugail [122] present the performance of machine learning for face recognition using partial faces and other manipulations of the face such as rotation and zooming, which we use as training and recognition cues. The experimental results on the tasks of face verification and face identification show that the model obtained by the proposed DNN training framework achieves 97.3% accuracy on the LFW database with low training complexity. Seibold et al. [123] proposed a morphing attack detection method based on DNNs. A fully automatic face image morphing pipeline with exchangeable components was used to generate morphing attacks, train neural networks based on these data, and analyze their accuracy. Yim et al. [124] propose a new deep architecture based on a novel type of multitask learning, which can achieve superior performance in rotating to a target-pose face image from an arbitrary pose and illumination image while preserving identity. Nguyen et al. [111] propose a new approach for detecting presentation attack face images to enhance the security level of a face recognition system. The objective of this study was the use of a very deep stacked CNN–RNN network to learn the discrimination features from a sequence of face images. Finally, Bajrami et al. [125] present experiment results with LDA and DNN for face recognition, while their efficiency and performance are tested on the LFW dataset. The experimental results show that the DNN method achieves better recognition accuracy, and the recognition time is much faster than that of the LDA method in large-scale datasets.

6.3. Databases Used

The most commonly used databases for face recognition systems under different conditions are Pointing Head Pose Image Database (PHPID) [126], Labeled Faces in Wild (LFW) [127], FERET [15,16], ORL, and Yale. The last are used for face recognition systems under different conditions, which provide information for supervised and unsupervised learning. Supervised learning is based on two training modules: image unrestricted training setting and image restricted training setting. For the first model, only "same" or "not same" binary labels are used in the training splits. For the second model, the identities of the person in each pair are provided in the training splits.

- LFW (Labeled Faces in the Wild) database was created in October 2007. It contains 13,333 images of 5749 subjects, with 1680 subjects with at least two images and the rest with a single image. These face images were taken on the Internet, pre-processed, and localized by the Viola–Jones detector with a resolution of 250 × 250 pixels. Most of them are in color, although there are also some in grayscale and presented in JPG format and organized by folders.
- FERET (Face Recognition Technology) database was created in 15 sessions in a semi-controlled environment between August 1993 and July 1996. It contains 1564 sets of images, with a total of 14,126 images. The duplicate series belong to subjects already present in the series of individual images, which were generally captured one day apart. Some images taken from the same subject vary overtime for a few years and can be used to treat facial changes that appear over time. The images have a depth of 24 bits, RGB, so they are color images, with a resolution of 512 × 768 pixels.
- AR face database was created by Aleix Martínez and Robert Benavente in the computer vision center (CVC) of the Autonomous University of Barcelona in June 1998. It contains more than 4000 images of 126 subjects, including 70 men and 56 women. They were taken at the CVC under

a controlled environment. The images were taken frontally to the subjects, with different facial expressions and three different lighting conditions, as well as several accessories: scarves, glasses, or sunglasses. Two imaging sessions were performed with the same subjects, 14 days apart. These images are a resolution of 576×768 pixels and a depth of 24 bits, under the RGB RAW format.

- ORL Database of Faces was performed between April 1992 and April 1994 at the AT & T laboratory in Cambridge. It consists of a total of 10 images per subject, out of a total of 40 images. For some subjects, the images were taken at different times, with varying illumination and facial expressions: eyes open/closed, smiling/without a smile, as well as with or without glasses. The images were taken under a black homogeneous background, in a vertical position and frontally to the subject, with some small rotation. These are images with a resolution of 92×112 pixels in grayscale.
- Extended Yale Face B database contains 16,128 images of 640×480 grayscale of 28 individuals under 9 poses and 64 different lighting conditions. It also includes a set of images made with the face of individuals only.
- Pointing Head Pose Image Database (PHPID) is one of the most widely used for face recognition. It contains 2790 monocular face images of 15 persons with tilt angles from $-90°$ to $+90°$ and variations of pan. Every person has two series of 93 different poses (93 images). The face images were taken under different skin color and with or without glasses.

6.4. Comparison between Holistic, Local, and Hybrid Techniques

In this section, we present some advantages and disadvantages of holistic, local, and hybrid approaches to identifying faces during the last 20 years. DL approaches can be considered as a statistical approach (holistic method), because the training procedure scheme usually searches for statistical structures in the input patterns. Table 4 presents a brief summary of the three approaches.

Table 4. General performance of face recognition approaches.

Approaches		Databases Used	Advantages	Disadvantages	Performances	Challenges Handled
Local	Local Appearance	TDF, CF1999, LFW, FERET, CMU-PIE, AR, Yale B, PHPID, YaleB Extended, FRGC2.0, Face94.	• Easy to implement, allowing an analysis of images in a difficult environment in real-time [38]. • Invariant to size, orientation, and lighting [47,48].	• Lack discrimination ability. • It is difficult to automatic detect feature in this approach.	• High performance in terms of processing time and recognition rate [15,38].	• Pose variations [42], various lighting conditions[60], facial expressions [38], and low resolution.
	Key-Points		• Does not require prior knowledge of the images [56]. • Different illumination conditions, scaling, aging effects, facial expressions, face occlusions, and noisy images [57].	• More affected by orientation changes or the expression of the face [23].	• High processing time [29]. • Low recognition rate [30].	• Different illumination conditions, facial expressions, aging effects, scaling, face occlusions and noisy images [56].
Holistic	Linear	LFW, FERET, MEPCO, AR, ORL, CK, MMI, JAFFE, C. Yale B, Yale, MNIST, ORL, UMIST face, HELEN face, FRGC.	• When frontal views of faces are used, these techniques provide good performance [35,70]. • Recognition is effective and simple. • Dimensionality reduction, represent global information [17,27,67,70].	• Sensitive to the rotation and the translation of the face images. • Can only classify a face that is "known" to the database. • Low speed in the face recognition caused by a long feature vector [36].	• Processed with larger size features. • High processing time [17]. • High performance in terms of recognition rate [67].	• Different illumination conditions [36,83], scaling, facial expressions.
	Non-Linear		• Dimensionality reduction [86–88]. • They are because of supervised classification problems. • Automatically detect feature in this approach (CNN and RNN) [63,77,90].	• The recognition performance depends on the chosen kernel [88]. • More difficult to implement than the local technique. • Recognition rate unsatisfying [87,88].	• Complexity [88]. • Computationally expensive and require a high degree of correlation between the test and training images (SVM, CNN) [88,90].	• Different illumination [36,83], poses [70], conditions, scaling, facial expressions.
Hybrid		AT&T, FACES94, MITINDIA, LFW, ORL, UMIST, YALE, FRGC Extended Yale, CMU FERET, AR, ESSEX.	• Provides faster systems and efficient recognition [95].	• More difficult to implement. • Complex and computational cost [93,95,97].	• High recognition rate [95]. • High computational complexity [97].	• Pose, illumination conditions, and facial expressions [101,102].

7. Discussion about Future Directions and Conclusions

7.1. Discussion

In the past decade, the face recognition system has become one of the most important biometric authentication methods. Many techniques are used to develop many face recognition systems based on facial information. Generally, the existing techniques can be classified into three approaches, depending on the type of desired features.

- Local approaches: use features in which the face described partially. For example, some system could consist of extracting local features such as the eyes, mouth, and nose. The features' values are calculated from the lines or points that can be represented on the face image for the recognition step.
- Holistic approaches: use features that globally describe the complete face as a model, including the background (although it is desirable to occupy the smallest possible surface).
- Hybrid approaches: combine local and holistic approaches.

In particular, recognition methods performed on static images produce good results under different lighting and expression conditions. However, in most cases, only the face images are processed at the same size and scale. Many methods require numerous training images, which limits their use for real-time systems, where the response time is an important aspect.

The main purpose of techniques such as HOG, LBP, Gabor filters, BRIEF, SURF, and SIFT is to discover distinctive features, which can be divided into two parts: (1) local appearance-based techniques, which are used to extract local features when the face image is divided into small regions (including HOG, LBP, Gabor filters, and correlation filters); and (2) key-points-based techniques, which are used to detect the points of interest in the face image, after which features' extraction is localized based on these points, including BRIEF, SURF, and SIFT. In the context of face recognition, local techniques only treat certain facial features, which make them very sensitive to facial expressions and occlusions [4,14,37,50–53]. The relative robustness is the main advantage of these feature-based local techniques. Additionally, they take into account the peculiarity of the face as a natural form to recognize a reduced number of parameters. Another advantage is that they have a high compaction capacity and a high comparison speed. The main disadvantages of these methods are the difficulty of automating the detection of facial features and the fact that the person responsible for the implementation of these systems must make an arbitrary decision on really important points.

Unlike the local approaches, holistic approaches are other methods used for face recognition, which treat the whole face image and do not require extracting face regions or features points (eyes, mouth, noses, and so on). The main function of these approaches is to represent the face image with a matrix of pixels. This matrix is often converted into feature vectors to facilitate their treatment. After that, the feature vectors are applied in a low-dimensional space. In fact, subspace techniques are sensitive to different variations (facial expressions, illumination, and different poses), which make them easy to implement. Many subspace techniques are implemented to represent faces such as Eigenface, Eigenfisher, PCA, and LDA, which can be divided into two categories: linear and non-linear techniques. The main advantage of holistic approaches is that they do not destroy image information by focusing only on regions or points of interest. However, this property represents a disadvantage because it assumes that all the pixels of the image have the same importance. As a result, these techniques are not only computationally expensive, but also require a high degree of correlation between the test and the training images. In addition, these approaches generally ignore local details, which means they are rarely used to identify faces.

Hybrid approaches are based on local and global features to exploit the benefits of both techniques. These approaches combine the two approaches described above into a single system to improve the performance and accuracy of recognition. The choice of the required method to be used must take into account the application in which it was applied. For example, in the face recognition systems that use very small images, methods based on local features are a bad choice. Another consideration in the algorithm selection process is the number of training examples needed. Finally, we can remember that the tendency is to develop hybrid methods that combine the advantages of local and holistic approaches, but these methods are very complex and require more processing time.

A notable limitation that we found in all the publications reviewed is methodological: despite that the 2D facial recognition has reached a significant level of maturity and a high success rate, it is not surprising that it continues to be one of the most active research areas in computer vision. Considering the results published to date, in the opinion of these authors, three particularly promising techniques for further development of this area stand out: (i) the development of 3D face recognition methods; (ii) the use of multimodal fusion methods of complementary data types, in particular those based on visible and infrared images; and (iii) the use of DL methods.

1. Three-dimensional face recognition: In 2D image-based techniques, some features are lost owing to the 3D structure of the face. Lighting and pose variations are two major unresolved problems of 2D face recognition. Recently, 3D facial recognition for facial recognition has been widely studied by the scientific community to overcome unresolved problems in 2D facial recognition and to achieve significantly higher accuracy by measuring geometry of rigid features on the face. For this reason, several recent systems based on 3D data have been developed [3,93,95,128,129].

2. Multimodal facial recognition: sensors have been developed in recent years with a proven ability to acquire not only two-dimensional texture information, but also facial shape, that is, three-dimensional information. For this reason, some recent studies have merged the two types of 2D and 3D information to take advantage of each of them and obtain a hybrid system that improves the recognition as the only modality [98].

3. Deep learning (DL): a very broad concept, which means that it has no exact definition, but studies [14,110–113,121,130,131] agree that DL includes a set of algorithms that attempt to model high level abstractions, by modeling multiple processing layers. This field of research began in the 1980s and is a branch of automatic learning where algorithms are used in the formation of deep neural networks (DNN) to achieve greater accuracy than other classical techniques. In recent progress, a point has been reached where DL performs better than people in some tasks, for example, to recognize objects in images.

Finally, researchers have gone further by using multimodal and DL facial recognition systems.

7.2. Conclusions

Face recognition system is a popular study task in the field of image processing and computer vision, owing to its potentially enormous application as well as its theoretical value. This system is widely deployed in many real-world applications such as security, surveillance, homeland security, access control, image search, human-machine, and entertainment. However, these applications pose different challenges such as lighting conditions and facial expressions. This paper highlights the recent research on the 2D or 3D face recognition system, focusing mainly on approaches based on local, holistic (subspace), and hybrid features. A comparative study between these approaches in terms of processing time, complexity, discrimination, and robustness was carried out. We can conclude that local feature techniques are the best choice concerning discrimination, rotation, translation, complexity, and accuracy. We hope that this survey paper will further encourage researchers in this field to participate and pay more attention to the use of local techniques for face recognition systems.

Author Contributions: Y.K. highlights the recent research on the 2D or 3D face recognition system, focusing mainly on approaches based on local, holistic, and hybrid features. M.J., A.A.F. and M.A. supervised the research and helped in the revision processes. All authors have read and agreed to the published version of the manuscript.

Funding: The paper is co-financed by L@bISEN of ISEN Yncrea Ouest Brest, France, Dept Ai-DE, Team Vision-AD and by FSM University of Monastir, Tunisia with collaboration of the Ministry of Higher Education and Scientific Research of Tunisia. The context of the paper is the PhD project of Yassin Kortli.

Conflicts of Interest: The authors declare no conflict of interest.

References

1. Liao, S.; Jain, A.K.; Li, S.Z. Partial face recognition: Alignment-free approach. *IEEE Trans. Pattern Anal. Mach. Intell.* **2012**, *35*, 1193–1205. [CrossRef] [PubMed]
2. Jridi, M.; Napoléon, T.; Alfalou, A. One lens optical correlation: Application to face recognition. *Appl. Opt.* **2018**, *57*, 2087–2095. [CrossRef] [PubMed]
3. Napoléon, T.; Alfalou, A. Pose invariant face recognition: 3D model from single photo. *Opt. Lasers Eng.* **2017**, *89*, 150–161. [CrossRef]
4. Ouerhani, Y.; Jridi, M.; Alfalou, A. Fast face recognition approach using a graphical processing unit "GPU". In Proceedings of the 2010 IEEE International Conference on Imaging Systems and Techniques, Thessaloniki, Greece, 1–2 July 2010; IEEE: Piscataway, NJ, USA, 2010; pp. 80–84.
5. Yang, W.; Wang, S.; Hu, J.; Zheng, G.; Valli, C. A fingerprint and finger-vein based cancelable multi-biometric system. *Pattern Recognit.* **2018**, *78*, 242–251. [CrossRef]
6. Patel, N.P.; Kale, A. Optimize Approach to Voice Recognition Using IoT. In Proceedings of the 2018 International Conference on Advances in Communication and Computing Technology (ICACCT), Sangamner, India, 8–9 February 2018; IEEE: Piscataway, NJ, USA, 2018; pp. 251–256.
7. Wang, Q.; Alfalou, A.; Brosseau, C. New perspectives in face correlation research: A tutorial. *Adv. Opt. Photonics* **2017**, *9*, 1–78. [CrossRef]
8. Alfalou, A.; Brosseau, C.; Kaddah, W. Optimization of decision making for face recognition based on nonlinear correlation plane. *Opt. Commun.* **2015**, *343*, 22–27. [CrossRef]
9. Zhao, C.; Li, X.; Cang, Y. Bisecting k-means clustering based face recognition using block-based bag of words model. *Opt. Int. J. Light Electron Opt.* **2015**, *126*, 1761–1766. [CrossRef]
10. HajiRassouliha, A.; Gamage, T.P.B.; Parker, M.D.; Nash, M.P.; Taberner, A.J.; Nielsen, P.M. FPGA implementation of 2D cross-correlation for real-time 3D tracking of deformable surfaces. In Proceedings of the 2013 28th International Conference on Image and Vision Computing New Zealand (IVCNZ 2013), Wellington, New Zealand, 27–29 November 2013; IEEE: Piscataway, NJ, USA, 2013; pp. 352–357.
11. Kortli, Y.; Jridi, M.; Al Falou, A.; Atri, M. A comparative study of CFs, LBP, HOG, SIFT, SURF, and BRIEF techniques for face recognition. In *Pattern Recognition and Tracking XXIX*; International Society for Optics and Photonics; SPIE: Bellingham, WA, USA, 2018; Volume 10649, p. 106490M.
12. Dehai, Z.; Da, D.; Jin, L.; Qing, L. A pca-based face recognition method by applying fast fourier transform in pre-processing. In *3rd International Conference on Multimedia Technology (ICMT-13)*; Atlantis Press: Paris, France, 2013.
13. Ouerhani, Y.; Alfalou, A.; Brosseau, C. Road mark recognition using HOG-SVM and correlation. In *Optics and Photonics for Information Processing XI*; International Society for Optics and Photonics; SPIE: Bellingham, WA, USA, 2017; Volume 10395, p. 103950Q.
14. Liu, W.; Wang, Z.; Liu, X.; Zeng, N.; Liu, Y.; Alsaadi, F.E. A survey of deep neural network architectures and their applications. *Neurocomputing* **2017**, *234*, 11–26. [CrossRef]
15. Xi, M.; Chen, L.; Polajnar, D.; Tong, W. Local binary pattern network: A deep learning approach for face recognition. In Proceedings of the 2016 IEEE International Conference on Image Processing (ICIP), Phoenix, AZ, USA, 25–28 September 2016; IEEE: Piscataway, NJ, USA, 2016; pp. 3224–3228.
16. Ojala, T.; Pietikäinen, M.; Harwood, D. A comparative study of texture measures with classification based on featured distributions. *Pattern Recognit.* **1996**, *29*, 51–59. [CrossRef]
17. Gowda, H.D.S.; Kumar, G.H.; Imran, M. Multimodal Biometric Recognition System Based on Nonparametric Classifiers. *Data Anal. Learn.* **2018**, *43*, 269–278.
18. Ouerhani, Y.; Jridi, M.; Alfalou, A.; Brosseau, C. Optimized pre-processing input plane GPU implementation of an optical face recognition technique using a segmented phase only composite filter. *Opt. Commun.* **2013**, *289*, 33–44. [CrossRef]
19. Mousa Pasandi, M.E. Face, Age and Gender Recognition Using Local Descriptors. Ph.D. Thesis, Université d'Ottawa/University of Ottawa, Ottawa, ON, Canada, 2014.
20. Khoi, P.; Thien, L.H.; Viet, V.H. Face Retrieval Based on Local Binary Pattern and Its Variants: A Comprehensive Study. *Int. J. Adv. Comput. Sci. Appl.* **2016**, *7*, 249–258. [CrossRef]
21. Zeppelzauer, M. Automated detection of elephants in wildlife video. *EURASIP J. Image Video Process.* **2013**, *46*, 2013. [CrossRef] [PubMed]

22. Parmar, D.N.; Mehta, B.B. Face recognition methods & applications. *arXiv* **2014**, arXiv:1403.0485.
23. Vinay, A.; Hebbar, D.; Shekhar, V.S.; Murthy, K.B.; Natarajan, S. Two novel detector-descriptor based approaches for face recognition using sift and surf. *Procedia Comput. Sci.* **2015**, *70*, 185–197.
24. Yang, H.; Wang, X.A. Cascade classifier for face detection. *J. Algorithms Comput. Technol.* **2016**, *10*, 187–197. [CrossRef]
25. Viola, P.; Jones, M. Rapid object detection using a boosted cascade of simple features. In Proceedings of the 2001 IEEE Computer Society Conference on Computer Vision and Pattern Recognition, Kauai, HI, USA, 8–14 December 2001.
26. Rettkowski, J.; Boutros, A.; Göhringer, D. HW/SW Co-Design of the HOG algorithm on a Xilinx Zynq SoC. *J. Parallel Distrib. Comput.* **2017**, *109*, 50–62. [CrossRef]
27. Seo, H.J.; Milanfar, P. Face verification using the lark representation. *IEEE Trans. Inf. Forensics Secur.* **2011**, *6*, 1275–1286. [CrossRef]
28. Shah, J.H.; Sharif, M.; Raza, M.; Azeem, A. A Survey: Linear and Nonlinear PCA Based Face Recognition Techniques. *Int. Arab J. Inf. Technol.* **2013**, *10*, 536–545.
29. Du, G.; Su, F.; Cai, A. Face recognition using SURF features. In *MIPPR 2009: Pattern Recognition and Computer Vision*; International Society for Optics and Photonics; SPIE: Bellingham, WA, USA, 2009; Volume 7496, p. 749628.
30. Calonder, M.; Lepetit, V.; Ozuysal, M.; Trzcinski, T.; Strecha, C.; Fua, P. BRIEF: Computing a local binary descriptor very fast. *IEEE Trans. Pattern Anal. Mach. Intell.* **2011**, *34*, 1281–1298. [CrossRef]
31. Smach, F.; Miteran, J.; Atri, M.; Dubois, J.; Abid, M.; Gauthier, J.P. An FPGA-based accelerator for Fourier Descriptors computing for color object recognition using SVM. *J. Real-Time Image Process.* **2007**, *2*, 249–258. [CrossRef]
32. Kortli, Y.; Jridi, M.; Al Falou, A.; Atri, M. A novel face detection approach using local binary pattern histogram and support vector machine. In Proceedings of the 2018 International Conference on Advanced Systems and Electric Technologies (IC_ASET), Hammamet, Tunisia, 22–25 March 2018; IEEE: Piscataway, NJ, USA, 2018; pp. 28–33.
33. Wang, Q.; Xiong, D.; Alfalou, A.; Brosseau, C. Optical image authentication scheme using dual polarization decoding configuration. *Opt. Lasers Eng.* **2019**, *112*, 151–161. [CrossRef]
34. Turk, M.; Pentland, A. Eigenfaces for recognition. *J. Cogn. Neurosci.* **1991**, *3*, 71–86. [CrossRef] [PubMed]
35. Annalakshmi, M.; Roomi, S.M.M.; Naveedh, A.S. A hybrid technique for gender classification with SLBP and HOG features. *Clust. Comput.* **2019**, *22*, 11–20. [CrossRef]
36. Hussain, S.U.; Napoléon, T.; Jurie, F. *Face Recognition Using Local Quantized Patterns*; HAL: Bengaluru, India, 2012.
37. Alfalou, A.; Brosseau, C. Understanding Correlation Techniques for Face Recognition: From Basics to Applications. In *Face Recognition*; Oravec, M., Ed.; IntechOpen: Rijeka, Croatia, 2010.
38. Napoléon, T.; Alfalou, A. Local binary patterns preprocessing for face identification/verification using the VanderLugt correlator. In *Optical Pattern Recognition XXV*; International Society for Optics and Photonics; SPIE: Bellingham, WA, USA, 2014; Volume 9094, p. 909408.
39. Schroff, F.; Kalenichenko, D.; Philbin, J. Facenet: A unified embedding for face recognition and clustering. In Proceedings of the IEEE conference on computer vision and pattern recognition, Boston, MA, USA, 7–12 June 2015; pp. 815–823.
40. Kambi Beli, I.; Guo, C. Enhancing face identification using local binary patterns and k-nearest neighbors. *J. Imaging* **2017**, *3*, 37. [CrossRef]
41. Benarab, D.; Napoléon, T.; Alfalou, A.; Verney, A.; Hellard, P. Optimized swimmer tracking system by a dynamic fusion of correlation and color histogram techniques. *Opt. Commun.* **2015**, *356*, 256–268. [CrossRef]
42. Bonnen, K.; Klare, B.F.; Jain, A.K. Component-based representation in automated face recognition. *IEEE Trans. Inf. Forensics Secur.* **2012**, *8*, 239–253. [CrossRef]
43. Ren, J.; Jiang, X.; Yuan, J. Relaxed local ternary pattern for face recognition. In Proceedings of the 2013 IEEE International Conference on Image Processing, Melbourne, Australia, 15–18 September 2013; IEEE: Piscataway, NJ, USA, 2013; pp. 3680–3684.

44. Karaaba, M.; Surinta, O.; Schomaker, L.; Wiering, M.A. Robust face recognition by computing distances from multiple histograms of oriented gradients. In Proceedings of the 2015 IEEE Symposium Series on Computational Intelligence, Cape Town, South Africa, 7–10 December 2015; IEEE: Piscataway, NJ, USA, 2015; pp. 203–209.

45. Huang, C.; Huang, J. A fast HOG descriptor using lookup table and integral image. *arXiv* **2017**, arXiv:1703.06256.

46. Arigbabu, O.A.; Ahmad, S.M.S.; Adnan, W.A.W.; Yussof, S.; Mahmood, S. Soft biometrics: Gender recognition from unconstrained face images using local feature descriptor. *arXiv* **2017**, arXiv:1702.02537.

47. Lugh, A.V. Signal detection by complex spatial filtering. *IEEE Trans. Inf. Theory* **1964**, *10*, 139.

48. Weaver, C.S.; Goodman, J.W. A technique for optically convolving two functions. *Appl. Opt.* **1966**, *5*, 1248–1249. [CrossRef] [PubMed]

49. Horner, J.L.; Gianino, P.D. Phase-only matched filtering. *Appl. Opt.* **1984**, *23*, 812–816. [CrossRef] [PubMed]

50. Leonard, I.; Alfalou, A.; Brosseau, C. Face recognition based on composite correlation filters: Analysis of their performances. In *Face Recognition: Methods, Applications and Technology*; Nova Science Pub Inc.: London, UK, 2012.

51. Katz, P.; Aron, M.; Alfalou, A. *A Face-Tracking System to Detect Falls in the Elderly*; SPIE Newsroom; SPIE: Bellingham, WA, USA, 2013.

52. Alfalou, A.; Brosseau, C.; Katz, P.; Alam, M.S. Decision optimization for face recognition based on an alternate correlation plane quantification metric. *Opt. Lett.* **2012**, *37*, 1562–1564. [CrossRef] [PubMed]

53. Elbouz, M.; Bouzidi, F.; Alfalou, A.; Brosseau, C.; Leonard, I.; Benkelfat, B.E. Adapted all-numerical correlator for face recognition applications. In *Optical Pattern Recognition XXIV*; International Society for Optics and Photonics; SPIE: Bellingham, WA, USA, 2013; Volume 8748, p. 874807.

54. Heflin, B.; Scheirer, W.; Boult, T.E. For your eyes only. In Proceedings of the 2012 IEEE Workshop on the Applications of Computer Vision (WACV), Breckenridge, CO, USA, 9–11 January 2012; pp. 193–200.

55. Zhu, X.; Liao, S.; Lei, Z.; Liu, R.; Li, S.Z. Feature correlation filter for face recognition. In *Advances in Biometrics, Proceedings of the International Conference on Biometrics, Seoul, Korea, 27–29 August 2007*; Springer: Berlin/Heidelberg, Germany, 2007; Volume 4642, pp. 77–86.

56. Lenc, L.; Král, P. Automatic face recognition system based on the SIFT features. *Comput. Electr. Eng.* **2015**, *46*, 256–272. [CrossRef]

57. Işık, Ş. A comparative evaluation of well-known feature detectors and descriptors. *Int. J. Appl. Math. Electron. Comput.* **2014**, *3*, 1–6. [CrossRef]

58. Mahier, J.; Hemery, B.; El-Abed, M.; El-Allam, M.; Bouhaddaoui, M.; Rosenberger, C. Computation evabio: A tool for performance evaluation in biometrics. *Int. J. Autom. Identif. Technol.* **2011**, *24*, hal-00984026.

59. Alahi, A.; Ortiz, R.; Vandergheynst, P. Freak: Fast retina keypoint. In Proceedings of the 2012 IEEE Conference on Computer Vision and Pattern Recognition, Providence, RI, USA, 16–21 June 2012; pp. 510–517.

60. Arashloo, S.R.; Kittler, J. Efficient processing of MRFs for unconstrained-pose face recognition. In Proceedings of the 2013 IEEE Sixth International Conference on Biometrics: Theory, Applications and Systems (BTAS), Rlington, VA, USA, 29 September–2 October 2013; IEEE: Piscataway, NJ, USA, 2013; pp. 1–8.

61. Ghorbel, A.; Tajouri, I.; Aydi, W.; Masmoudi, N. A comparative study of GOM, uLBP, VLC and fractional Eigenfaces for face recognition. In Proceedings of the 2016 International Image Processing, Applications and Systems (IPAS), Hammamet, Tunisia, 5–7 November 2016; IEEE: Piscataway, NJ, USA, 2016; pp. 1–5.

62. Lima, A.; Zen, H.; Nankaku, Y.; Miyajima, C.; Tokuda, K.; Kitamura, T. On the use of kernel PCA for feature extraction in speech recognition. *IEICE Trans. Inf. Syst.* **2004**, *87*, 2802–2811.

63. Devi, B.J.; Veeranjaneyulu, N.; Kishore, K.V.K. A novel face recognition system based on combining eigenfaces with fisher faces using wavelets. *Procedia Comput. Sci.* **2010**, *2*, 44–51. [CrossRef]

64. Simonyan, K.; Parkhi, O.M.; Vedaldi, A.; Zisserman, A. Fisher vector faces in the wild. In Proceedings of the BMVC 2013—British Machine Vision Conference, Bristol, UK, 9–13 September 2013.

65. Li, B.; Ma, K.K. Fisherface vs. eigenface in the dual-tree complex wavelet domain. In Proceedings of the 2009 Fifth International Conference on Intelligent Information Hiding and Multimedia Signal Processing, Kyoto, Japan, 12–14 September 2009; IEEE: Piscataway, NJ, USA, 2009; pp. 30–33.

66. Agarwal, R.; Jain, R.; Regunathan, R.; Kumar, C.P. Automatic Attendance System Using Face Recognition Technique. In *Proceedings of the 2nd International Conference on Data Engineering and Communication Technology*; Springer: Singapore, 2019; pp. 525–533.

67. Cui, Z.; Li, W.; Xu, D.; Shan, S.; Chen, X. Fusing robust face region descriptors via multiple metric learning for face recognition in the wild. In Proceedings of the IEEE conference on computer vision and pattern recognition, Portland, OR, USA, 23–28 June 2013; pp. 3554–3561.

68. Prince, S.; Li, P.; Fu, Y.; Mohammed, U.; Elder, J. Probabilistic models for inference about identity. *IEEE Trans. Pattern Anal. Mach. Intell.* **2011**, *34*, 144–157.

69. Perlibakas, V. Face recognition using principal component analysis and log-gabor filters. *arXiv* **2006**, arXiv:cs/0605025.

70. Huang, Z.H.; Li, W.J.; Shang, J.; Wang, J.; Zhang, T. Non-uniform patch based face recognition via 2D-DWT. *Image Vision Comput.* **2015**, *37*, 12–19. [CrossRef]

71. Sufyanu, Z.; Mohamad, F.S.; Yusuf, A.A.; Mamat, M.B. Enhanced Face Recognition Using Discrete Cosine Transform. *Eng. Lett.* **2016**, *24*, 52–61.

72. Hoffmann, H. Kernel PCA for novelty detection. *Pattern Recognit.* **2007**, *40*, 863–874. [CrossRef]

73. Arashloo, S.R.; Kittler, J. Class-specific kernel fusion of multiple descriptors for face verification using multiscale binarised statistical image features. *IEEE Trans. Inf. Forensics Secur.* **2014**, *9*, 2100–2109. [CrossRef]

74. Vinay, A.; Shekhar, V.S.; Murthy, K.B.; Natarajan, S. Performance study of LDA and KFA for gabor based face recognition system. *Procedia Comput. Sci.* **2015**, *57*, 960–969. [CrossRef]

75. Sivasathya, M.; Joans, S.M. Image Feature Extraction using Non Linear Principle Component Analysis. *Procedia Eng.* **2012**, *38*, 911–917. [CrossRef]

76. Zhang, B.; Chen, X.; Shan, S.; Gao, W. Nonlinear face recognition based on maximum average margin criterion. In Proceedings of the 2005 IEEE Computer Society Conference on Computer Vision and Pattern Recognition (CVPR'05), San Diego, CA, USA, 20–25 June 2005; IEEE: Piscataway, NJ, USA, 2005; Volume 1, pp. 554–559.

77. Vankayalapati, H.D.; Kyamakya, K. Nonlinear feature extraction approaches with application to face recognition over large databases. In Proceedings of the 2009 2nd International Workshop on Nonlinear Dynamics and Synchronization, Klagenfurt, Austria, 20–21 July 2009; IEEE: Piscataway, NJ, USA, 2009; pp. 44–48.

78. Javidi, B.; Li, J.; Tang, Q. Optical implementation of neural networks for face recognition by the use of nonlinear joint transform correlators. *Appl. Opt.* **1995**, *34*, 3950–3962. [CrossRef]

79. Yang, J.; Frangi, A.F.; Yang, J.Y. A new kernel Fisher discriminant algorithm with application to face recognition. *Neurocomputing* **2004**, *56*, 415–421. [CrossRef]

80. Pang, Y.; Liu, Z.; Yu, N. A new nonlinear feature extraction method for face recognition. *Neurocomputing* **2006**, *69*, 949–953. [CrossRef]

81. Wang, Y.; Fei, P.; Fan, X.; Li, H. Face recognition using nonlinear locality preserving with deep networks. In Proceedings of the 7th International Conference on Internet Multimedia Computing and Service, Hunan, China, 19–21 August 2015; ACM: New York, NY, USA, 2015; p. 66.

82. Li, S.; Yao, Y.F.; Jing, X.Y.; Chang, H.; Gao, S.Q.; Zhang, D.; Yang, J.Y. Face recognition based on nonlinear DCT discriminant feature extraction using improved kernel DCV. *IEICE Trans. Inf. Syst.* **2009**, *92*, 2527–2530. [CrossRef]

83. Khan, S.A.; Ishtiaq, M.; Nazir, M.; Shaheen, M. Face recognition under varying expressions and illumination using particle swarm optimization. *J. Comput. Sci.* **2018**, *28*, 94–100. [CrossRef]

84. Hafez, S.F.; Selim, M.M.; Zayed, H.H. 2d face recognition system based on selected gabor filters and linear discriminant analysis lda. *arXiv* **2015**, arXiv:1503.03741.

85. Shanbhag, S.S.; Bargi, S.; Manikantan, K.; Ramachandran, S. Face recognition using wavelet transforms-based feature extraction and spatial differentiation-based pre-processing. In Proceedings of the 2014 International Conference on Science Engineering and Management Research (ICSEMR), Chennai, India, 27–29 November 2014; IEEE: Piscataway, NJ, USA, 2014; pp. 1–8.

86. Fan, J.; Chow, T.W. Exactly Robust Kernel Principal Component Analysis. *IEEE Trans. Neural Netw. Learn. Syst.* **2019**. [CrossRef] [PubMed]

87. Vinay, A.; Cholin, A.S.; Bhat, A.D.; Murthy, K.B.; Natarajan, S. An Efficient ORB based Face Recognition framework for Human-Robot Interaction. *Procedia Comput. Sci.* **2018**, *133*, 913–923.

88. Lu, J.; Plataniotis, K.N.; Venetsanopoulos, A.N. Face recognition using kernel direct discriminant analysis algorithms. *IEEE Trans. Neural Netw.* **2003**, *14*, 117–126. [PubMed]

89. Yang, W.J.; Chen, Y.C.; Chung, P.C.; Yang, J.F. Multi-feature shape regression for face alignment. *EURASIP J. Adv. Signal Process.* **2018**, *2018*, 51. [CrossRef]

90. Ouanan, H.; Ouanan, M.; Aksasse, B. Non-linear dictionary representation of deep features for face recognition from a single sample per person. *Procedia Comput. Sci.* **2018**, *127*, 114–122. [CrossRef]

91. Fathima, A.A.; Ajitha, S.; Vaidehi, V.; Hemalatha, M.; Karthigaiveni, R.; Kumar, R. Hybrid approach for face recognition combining Gabor Wavelet and Linear Discriminant Analysis. In Proceedings of the 2015 IEEE International Conference on Computer Graphics, Vision and Information Security (CGVIS), Bhubaneswar, India, 2–3 November 2015; IEEE: Piscataway, NJ, USA, 2015; pp. 220–225.

92. Barkan, O.; Weill, J.; Wolf, L.; Aronowitz, H. Fast high dimensional vector multiplication face recognition. In Proceedings of the IEEE International Conference on Computer Vision, Sydney, Australia, 1–8 December 2013; pp. 1960–1967.

93. Juefei-Xu, F.; Luu, K.; Savvides, M. Spartans: Single-sample periocular-based alignment-robust recognition technique applied to non-frontal scenarios. *IEEE Trans. Image Process.* **2015**, *24*, 4780–4795. [CrossRef]

94. Yan, Y.; Wang, H.; Suter, D. Multi-subregion based correlation filter bank for robust face recognition. *Pattern Recognit.* **2014**, *47*, 3487–3501. [CrossRef]

95. Ding, C.; Tao, D. Robust face recognition via multimodal deep face representation. *IEEE Trans. Multimed.* **2015**, *17*, 2049–2058. [CrossRef]

96. Sharma, R.; Patterh, M.S. A new pose invariant face recognition system using PCA and ANFIS. *Optik* **2015**, *126*, 3483–3487. [CrossRef]

97. Moussa, M.; Hmila, M.; Douik, A. A Novel Face Recognition Approach Based on Genetic Algorithm Optimization. *Stud. Inform. Control* **2018**, *27*, 127–134. [CrossRef]

98. Mian, A.; Bennamoun, M.; Owens, R. An efficient multimodal 2D-3D hybrid approach to automatic face recognition. *IEEE Trans. Pattern Anal. Mach. Intell.* **2007**, *29*, 1927–1943. [CrossRef] [PubMed]

99. Cho, H.; Roberts, R.; Jung, B.; Choi, O.; Moon, S. An efficient hybrid face recognition algorithm using PCA and GABOR wavelets. *Int. J. Adv. Robot. Syst.* **2014**, *11*, 59. [CrossRef]

100. Guru, D.S.; Suraj, M.G.; Manjunath, S. Fusion of covariance matrices of PCA and FLD. *Pattern Recognit. Lett.* **2011**, *32*, 432–440. [CrossRef]

101. Sing, J.K.; Chowdhury, S.; Basu, D.K.; Nasipuri, M. An improved hybrid approach to face recognition by fusing local and global discriminant features. *Int. J. Biom.* **2012**, *4*, 144–164. [CrossRef]

102. Kamencay, P.; Zachariasova, M.; Hudec, R.; Jarina, R.; Benco, M.; Hlubik, J. A novel approach to face recognition using image segmentation based on spca-knn method. *Radioengineering* **2013**, *22*, 92–99.

103. Sun, J.; Fu, Y.; Li, S.; He, J.; Xu, C.; Tan, L. Sequential Human Activity Recognition Based on Deep Convolutional Network and Extreme Learning Machine Using Wearable Sensors. *J. Sens.* **2018**, *2018*, 10. [CrossRef]

104. Soltanpour, S.; Boufama, B.; Wu, Q.J. A survey of local feature methods for 3D face recognition. *Pattern Recognit.* **2017**, *72*, 391–406. [CrossRef]

105. Sharma, G.; ul Hussain, S.; Jurie, F. Local higher-order statistics (LHS) for texture categorization and facial analysis. In *European Conference on Computer Vision*; Springer: Berlin/Heidelberg, Germany, 2012; pp. 1–12.

106. Zhang, J.; Marszałek, M.; Lazebnik, S.; Schmid, C. Local features and kernels for classification of texture and object categories: A comprehensive study. *Int. J. Comput. Vis.* **2007**, *73*, 213–238. [CrossRef]

107. Leonard, I.; Alfalou, A.; Brosseau, C. Spectral optimized asymmetric segmented phase-only correlation filter. *Appl. Opt.* **2012**, *51*, 2638–2650. [CrossRef]

108. Shen, L.; Bai, L.; Ji, Z. A svm face recognition method based on optimized gabor features. In *International Conference on Advances in Visual Information Systems*; Springer: Berlin/Heidelberg, Germany, 2007; pp. 165–174.

109. Pratima, D.; Nimmakanti, N. Pattern Recognition Algorithms for Cluster Identification Problem. *Int. J. Comput. Sci. Inform.* **2012**, *1*, 2231–5292.

110. Zhang, C.; Prasanna, V. Frequency domain acceleration of convolutional neural networks on CPU-FPGA shared memory system. In Proceedings of the 2017 ACM/SIGDA International Symposium on Field-Programmable Gate Arrays, Monterey, CA, USA, 22–24 February 2017; ACM: New York, NY, USA, 2017; pp. 35–44.

111. Nguyen, D.T.; Pham, T.D.; Lee, M.B.; Park, K.R. Visible-Light Camera Sensor-Based Presentation Attack Detection for Face Recognition by Combining Spatial and Temporal Information. *Sensors* **2019**, *19*, 410. [CrossRef] [PubMed]

112. Parkhi, O.M.; Vedaldi, A.; Zisserman, A. Deep face recognition. In Proceedings of the BMVC 2015—British Machine Vision Conference, Swansea, UK, 7–10 September.

113. Wen, Y.; Zhang, K.; Li, Z.; Qiao, Y. A discriminative feature learning approach for deep face recognition. In *European Conference on Computer Vision*; Springer: Berlin/Heidelberg, Germany, 2016; pp. 499–515.

114. Passalis, N.; Tefas, A. Spatial bag of features learning for large scale face image retrieval. In *INNS Conference on Big Data*; Springer: Berlin/Heidelberg, Germany, 2016; pp. 8–17.

115. Liu, W.; Wen, Y.; Yu, Z.; Li, M.; Raj, B.; Song, L. Sphereface: Deep hypersphere embedding for face recognition. In Proceedings of the IEEE Conference on Computer Vision and Pattern Recognition, Honolulu, HI, USA, 21–26 July 2017; pp. 212–220.

116. Amato, G.; Falchi, F.; Gennaro, C.; Massoli, F.V.; Passalis, N.; Tefas, A.; Vairo, C. Face Verification and Recognition for Digital Forensics and Information Security. In Proceedings of the 2019 7th International Symposium on Digital Forensics and Security (ISDFS), Barcelos, Portugal, 10–12 June 2019; IEEE: Piscataway, NJ, USA, 2019; pp. 1–6.

117. Taigman, Y.; Yang, M.; Ranzato, M.A. Wolf, LDeepface: Closing the gap to human-level performance in face verification. In Proceedings of the IEEE conference on computer vision and pattern recognition, Washington, DC, USA, 23–28 June 2014; pp. 1701–1708.

118. Ma, Z.; Ding, Y.; Li, B.; Yuan, X. Deep CNNs with Robust LBP Guiding Pooling for Face Recognition. *Sensors* **2018**, *18*, 3876. [CrossRef] [PubMed]

119. Koo, J.; Cho, S.; Baek, N.; Kim, M.; Park, K. CNN-Based Multimodal Human Recognition in Surveillance Environments. *Sensors* **2018**, *18*, 3040. [CrossRef]

120. Cho, S.; Baek, N.; Kim, M.; Koo, J.; Kim, J.; Park, K. Detection in Nighttime Images Using Visible-Light Camera Sensors with Two-Step Faster Region-Based Convolutional Neural Network. *Sensors* **2018**, *18*, 2995. [CrossRef]

121. Koshy, R.; Mahmood, A. Optimizing Deep CNN Architectures for Face Liveness Detection. *Entropy* **2019**, *21*, 423. [CrossRef]

122. Elmahmudi, A.; Ugail, H. Deep face recognition using imperfect facial data. *Future Gener. Comput. Syst.* **2019**, *99*, 213–225. [CrossRef]

123. Seibold, C.; Samek, W.; Hilsmann, A.; Eisert, P. Accurate and robust neural networks for security related applications exampled by face morphing attacks. *arXiv* **2018**, arXiv:1806.04265.

124. Yim, J.; Jung, H.; Yoo, B.; Choi, C.; Park, D.; Kim, J. Rotating your face using multi-task deep neural network. In Proceedings of the IEEE Conference on Computer Vision and Pattern Recognition, Boston, MA, USA, 7–12 June 2015; pp. 676–684.

125. Bajrami, X.; Gashi, B.; Murturi, I. Face recognition performance using linear discriminant analysis and deep neural networks. *Int. J. Appl. Pattern Recognit.* **2018**, *5*, 240–250. [CrossRef]

126. Gourier, N.; Hall, D.; Crowley, J.L. Estimating Face Orientation from Robust Detection of Salient Facial Structures. Available online: venus.inrialpes.fr/jlc/papers/Pointing04-Gourier.pdf (accessed on 15 December 2019).

127. Gonzalez-Sosa, E.; Fierrez, J.; Vera-Rodriguez, R.; Alonso-Fernandez, F. Facial soft biometrics for recognition in the wild: Recent works, annotation, and COTS evaluation. *IEEE Trans. Inf. Forensics Secur.* **2018**, *13*, 2001–2014. [CrossRef]

128. Boukamcha, H.; Hallek, M.; Smach, F.; Atri, M. Automatic landmark detection and 3D Face data extraction. *J. Comput. Sci.* **2017**, *21*, 340–348. [CrossRef]

129. Ouerhani, Y.; Jridi, M.; Alfalou, A.; Brosseau, C. Graphics processor unit implementation of correlation technique using a segmented phase only composite filter. *Opt. Commun.* **2013**, *289*, 33–44. [CrossRef]

130. Su, C.; Yan, Y.; Chen, S.; Wang, H. An efficient deep neural networks training framework for robust face recognition. In Proceedings of the 2017 IEEE International Conference on Image Processing (ICIP), Beijing, China, 17–20 September 2017; pp. 3800–3804.

131. Coşkun, M.; Uçar, A.; Yildirim, Ö.; Demir, Y. Face recognition based on convolutional neural network. In Proceedings of the 2017 International Conference on Modern Electrical and Energy Systems (MEES), Kremenchuk, Ukraine, 15–17 November 2017; IEEE: Piscataway, NJ, USA, 2017; pp. 376–379.

Article

Enhancing Optical Correlation Decision Performance for Face Recognition by Using a Nonparametric Kernel Smoothing Classification

Matthieu Saumard [1,*], Marwa Elbouz [2], Michaël Aron [1], Ayman Alfalou [2] and Christian Brosseau [3]

[1] Yncrea Ouest, Artificial Intelligence and Emerging data Laboratory, 2 rue de la châtaigneraie, 35510 Cesson-Sevigné, France; michael.aron@isen-ouest.yncrea.com
[2] Yncrea Ouest, Artificial Intelligence and Emerging data Laboratory, 20 rue du Cuirassé Bretagne, 29200 Brest, France; marwa.el-bouz@isen-ouest.yncrea.fr (M.E.); ayman.al-falou@isen-ouest.yncrea.fr (A.A.)
[3] Univ Brest, CNRS, Lab-STICC, 6 avenue Le Gorgeu, 29238 Brest Cedex 3, France; brosseau@univ-brest.fr
* Correspondence: matthieu.saumard@isen-ouest.yncrea.fr

Received: 27 September 2019; Accepted: 19 November 2019; Published: 21 November 2019

Abstract: Optical correlation has a rich history in image recognition applications from a database. In practice, it is simple to implement optically using two lenses or numerically using two Fourier transforms. Even if correlation is a reliable method for image recognition, it may jeopardize decision making according to the location, height, and shape of the correlation peak within the correlation plane. Additionally, correlation is very sensitive to image rotation and scale. To overcome these issues, in this study, we propose a method of nonparametric modelling of the correlation plane. Our method is based on a kernel estimation of the regression function used to classify the individual images in the correlation plane. The basic idea is to improve the decision by taking into consideration the energy shape and distribution in the correlation plane. The method relies on the calculation of the Hausdorff distance between the target correlation plane (of the image to recognize) and the correlation planes obtained from the database (the correlation planes computed from the database images). Our method is tested for a face recognition application using the Pointing Head Pose Image Database (PHPID) database. Overall, the results demonstrate good performances of this method compared to competitive methods in terms of good detection and very low false alarm rates.

Keywords: face verification; optical correlation; Hausdorff distance; image classification

---

1. Introduction

The use of correlation methods [1–4] remains very competitive despite the abundance of purely numerical methods, such as Support Vector Machines and neural networks. Correlation is easy to use in practice because it is based on two Fourier transforms (FTs) and one multiplication in the frequency domain [5].

For comparison, a deep learning-based method has generally good performance but also significant drawbacks due to algorithm complexity, implementation difficulty, time-consuming learning processes, and a high number of computational resources [6]. Most of the developments are devoted to increasing the performance of correlation methods concentrated in the Fourier plane [7–11] by designing innovative correlation filters. On the other hand, there exists a growing scientific community dealing with biometric issues, such as face recognition, fingerprint detection, and early automatic disease detection [12,13]. The primary focus of this paper is to deal with an authentication problem using a database. There are two kinds of issue, i.e., identification and verification. Here, our primary goal is to optimize the solution to verification.

In order to improve the decision performance, our model uses a statistical learning method, i.e., a supervised classification method. The regression function between the binary output (class of person) and the input (correlation plane) is nonparametrically estimated for the learning database by making use of the modified kernel smoothing Nadaraya–Watson algorithm [13,14].

Functional data analysis has recently been developed to statistically analyze curves or objects, see e.g., Reference [15] for a good introduction to this subject. The interested reader may also refer to [16] for an overview of nonparametric estimation with functional data. In [17], the authors defined an extension of the Nadaraya–Watson estimator for objects such as curves by introducing a distance in the kernel between two functional objects. Here, we propose the use of kernel smoothing estimation to cope with the correlation plane, and we choose an appropriate distance, i.e., the Hausdorff distance, to plug in the kernel for estimating the regression function. As a result, it is possible to propose a decision-making protocol which has dual effects for increasing good decision rates and reducing false alarm rates.

The rest of this paper is organized as follows. Section 2 provides the correlation principle. After a short description of the database in Section 3, our overall method is explained in Section 4. Our model is implemented in Section 5. The method's accuracy is checked in Section 6, which provides two series of simulation studies. Section 7 briefly concludes.

2. Modeling Correlation

In essence, a Vander Lugt Correlator (VLC) compared a target image (input plane) with a reference image. The result of this comparison is presented in the form of a correlation plane. More precisely, the spectrum of a target image was obtained with a FT and was multiplied by a correlation filter made from the reference image [1–5]. An inverse FT (FT^{-1}) was then applied to get the output plane containing a noised correlation peak. The measure of the highest peak (i.e., the peak-to-correlation energy (PCE)) characterized the similarities between the reference and the target images. To validate our approach, we used a classical phase-only filter (POF), see Figure 1.

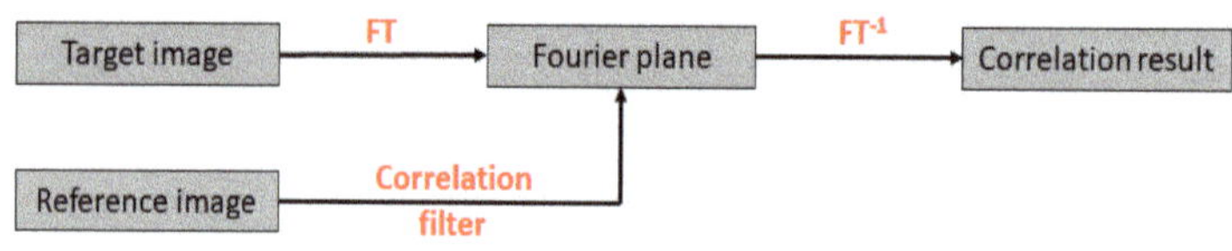

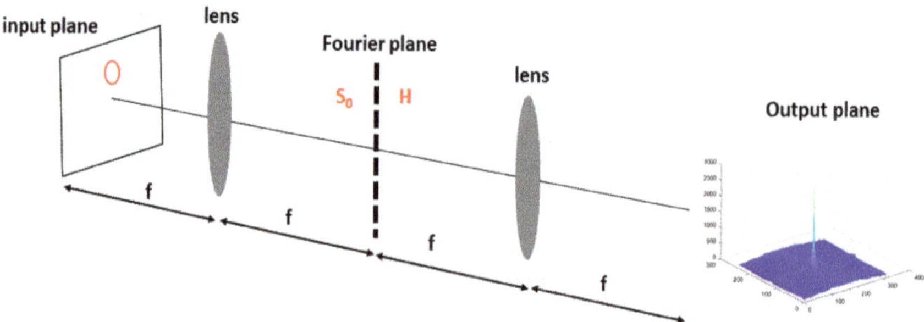

Figure 1. Illustrating the Vander Lugt Correlator (VLC) principle.

3. Dataset

Simulation results were obtained using the Pointing Head Pose Image Database (PHPID) [18]. This dataset includes 1302 face pictures: 14 different persons (93 images per person) with different orientations (from $-90°$ to $+90°$ with respect to the horizontal direction and from $-10°$ and $+10°$ with respect to the vertical direction). The resolution of each image is 314×238 pixels. It is also worth noting that this database includes a variety of persons (Figure 2), e.g., various skin colors, person with glasses or not, etc.

Figure 2. Three selected images from the Pointing Head Pose Image Database (PHPID) dataset.

In this numerical study, different training/testing databases from the PHPID dataset were chosen in order to demonstrate the efficiency of our method. Once the specific dataset was chosen, tests were performed as follows: Firstly, a person from the testing database was chosen as a reference person (person 0). A classical POF filter and an autocorrelation plane were computed for each person in this dataset. The correlation technique, whose principle is shown in Figure 1, was applied to get the output correlation plane using the corresponding POF filter for each person in the database. A classification algorithm based on the Hausdorff distance was then used (Section 4). If the person was recognized, the analysis ended. If the person was not recognized, the procedure was repeated with another person from the training database until the training database was empty. Thus, only two possibilities exist, i.e., either the person is recognized or not.

Several remarks are in order, concerning Figure 3, which is organized in two parts. Part 1 defines the VLC, whose output is a plane of correlation. Part 2 is the decision part. For each face of the database, the VLC is used with the target image, thus resulting in a large collection of correlation planes. Next, the decision-making procedure relies on the Hausdorff distance between the target correlation plane and the correlation planes coming from the database by selecting a specific bandwidth (hereinafter described in Section 4).

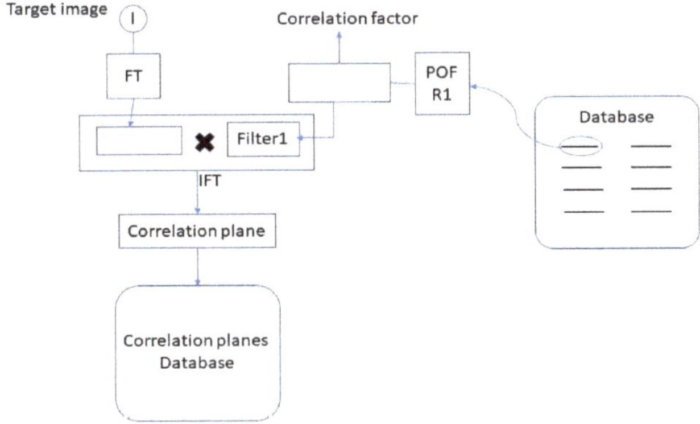

Figure 3. Flowchart illustrating the transition between the database and the correlation planes database.

By adopting a kernel smoothing classification algorithm, we took care of the shape, location, and denoising of the peak of correlation. From this algorithm, we learned which correlation plane was good, and we filtered bad correlation planes. The result of the algorithm was the variable $\hat{Y}$, the value of which is 0 for non-recognition decision, i.e., the person is not in the database and is equal to 1 for recognition, i.e., the person is in the database.

4. Overview of the Method

The method was organized in two parts, see Figures 3–5. The first part (Figure 3) was the construction of a new database containing computed correlation planes. This part was realized by calculating the correlation plane corresponding to a reference person (person 0) and all or a part of the series 1 of the PHPID dataset. The different ways of dividing the series 1 constitute the different training sets, which will be described in Section 6. The testing set is either the whole series 1 or the whole series 2.

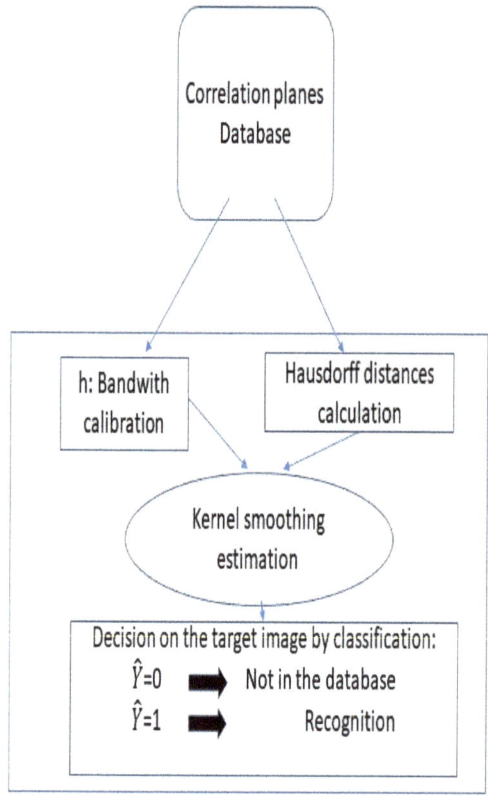

Figure 4. Flowchart illustrating the decision part.

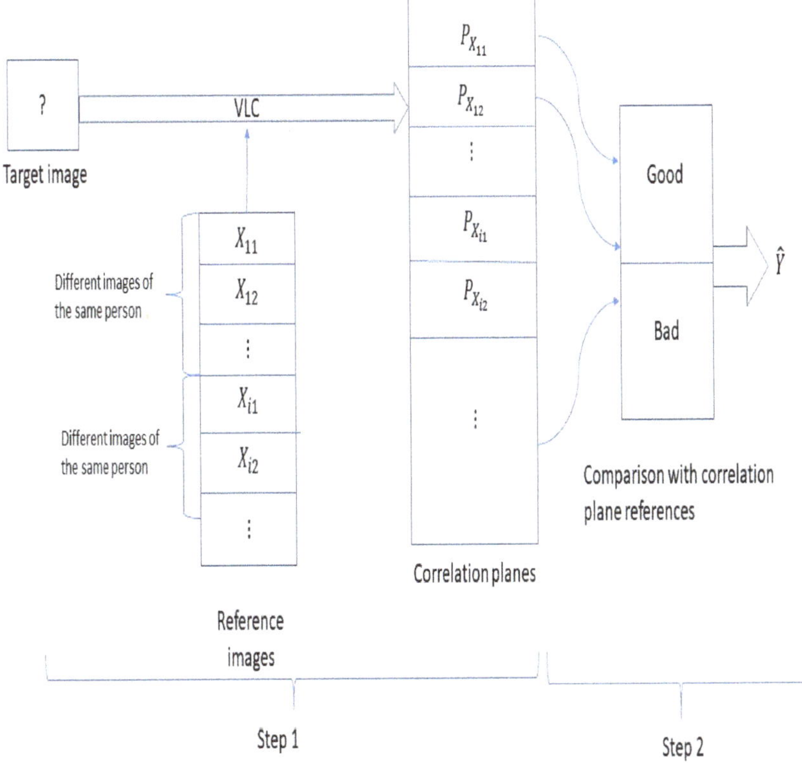

Figure 5. Flowchart illustrating the method.

The second part of our method was the decision part (Figure 4). Considering the correlation planes and the corresponding label indicating if they are the same person or not (in the training set), we made our decision on the testing set by estimating the probability of recognition via a kernel smoothing procedure.

Let us comment on the two-step algorithm represented in Figure 5.

Step 1: Step 1 began with the target image, which is the image to be recognized. The target image was introduced in a VLC correlator to be compared with all reference images of a database. This reference image database contains n different persons (X1,..., Xn) (Xi is the ith person); Xij represents the different variations that the person Xi can have (m is the number of variations considered).

A set of reference images were used with the target image to generate different correlation planes (Px11,...,Pxnm). These correlation planes were then compared in step 2 with pre-computed correlation planes, known as the reference database.

Step 2: The correlation planes (Px11,...,Pxnm) were compared with a correlation plane database realized in step 1. This database was divided in two parts: The first part contained the good correlation plane of references and another part listed the bad correlations, i.e., the false correlation plane of references. We will compare the good and bad correlation planes in Section 5. The construction of these correlation planes of references was made as follows: The correlation planes of various images of person A, the correlation planes of various images of person B, . . ., and the correlations planes of various images of person Z, which constitute the good correlation planes of reference. The reference database of bad correlation planes was constructed as follows: We calculated the correlation planes between various images of person A and various images of person B, etc.

The comparison shown in Figure 5 was then realized with the Hausdorff distance and by making use of the kernel smoothing method, which realizes an estimation of the probability of belonging to the class of a known person (Section 5).

5. Nonparametric Model

The Hausdorff distance is widely used in the context of image recognition, see e.g., [19–21] for reviews. A modified version of the Hausdorff distance has been also applied to matching objects [22]. The Hausdorff distance can be defined as follows: Let E and F be two non-empty subsets of a metric space (M, d). The Hausdorff distance is given as:

$$d_H(E, F) = \max\left\{ \sup_{x \in E} \inf_{y \in F} d(x, y), \ \sup_{x \in F} \inf_{y \in E} d(x, y) \right\}.$$

Here, for the purpose of comparing the target and database correlation planes, the Hausdorff distance between two planes (one with the unknown image and the other one calculated beforehand for the image present in the database) was evaluated. Once the distance was known, a nonparametric classification for decision making was performed.

Next, the decision part illustrated in Figure 4 is described. A kernel smoothing estimate of the regression function was employed. This estimate was used to perform a classification with a given threshold set to 0.5. Here, we made use of the Nadaraya–Watson estimator of the regression function [15] for classification. The principle is described as follows: Assume we have $(Y_1, X_1), \ldots, (Y_n, X_n)$ independent and identically distributed (i.i.d.) random variables coming from (Y, X) where Y is the variable labeled by 1, if the person is detected, and 0 otherwise, X is the corresponding correlation plane, and n denotes the sample size. Let us comment briefly on this i.i.d. sample: This collection of correlation planes was obtained from the learning database with person 0. Considering a new face image from the testing dataset, we computed the correlation plane with person 0. We then performed a nonparametric classification with a kernel estimate of the regression function $E(Y|X)$. Assuming that Y is a Bernoulli random variable, then $P(Y = 1| X = x) = E(Y|X = x)$, where P represents the probability measure. Thus, $P(Y = 1| X = x)$ is the probability of detection, knowing the autocorrelation plane x. $E(Y|X = x)$ is the expected value of Y, knowing the autocorrelation plane x. Now let us define an estimator of this probability as:

$$\hat{Y} = \begin{cases} \dfrac{\sum_{i=1}^{n} Y_i K_h(d(x, X_i))}{\sum_{i=1}^{n} K_h(d(x, X_i))} \\ 0 \ if \ \sum_{i=1}^{n} K_h(d(x, X_i)) = 0 \end{cases} \tag{1}$$

where $\hat{Y}$ is a prediction of Y, knowing the correlation plane x, keeping in mind that in our case it is also an estimate of the probability that Y is equal to 1 at the correlation plane x. Knowing $\hat{Y}$, we can decide if the face image is identical or not. If $\hat{Y}$ is close to 1, there is a high probability that there is a good match between the two persons, and if $\hat{Y}$ is close to zero, the probability that the two persons are not the same is large. In Equation (1), K is a real asymmetrical kernel, h is the bandwidth (calibration parameter), $K_h(.) = \frac{1}{h} K(\frac{.}{h})$, and d is the Hausdorff distance between images.

For the asymmetrical kernel, we used the asymmetrical version of the Epanechnikov kernel, namely $K(x) = \frac{3}{2}(1 - x^2) 1_{[0,1)}(x)$, where $1_{[0,1)}(.)$ stands for the indicator function on the set $[0, 1)$. The use of an asymmetrical kernel is standard in functional data analysis (see Ferraty and Vieu [17]) because the distance is positive for all planes of correlation. Other type of kernels can be used, but our numerical results show that the Epanechnikov kernel performs better than others. From Equation (1), we also see that the value of $\hat{Y}$ ranges from 0 to 1 and represents an estimation of the probability of Y to be of class 1, knowing that the X is at the target point x: $P\left(Y = 1 \mid X = x\right)$. Assume that $Y_{i'}$ is 1 and x is close to $X_{i'}$, then the procedure in Equation (1) will affect a value close to 1 to $\hat{Y}$.

For convenience, a threshold is set to 0.5, i.e., all values of $\hat{Y}$ larger to this threshold are recognized, and vice versa. This threshold value corresponds to an estimation of the Bayes classifiers [23]. The Bayes classifiers maximizes the probability $P\left(Y = 1 \mid X = x\right)$. In our case, it corresponds to set $\hat{Y} = 1$ to the plane of correlation with $\hat{Y} \geq \frac{1}{2}$ and $\hat{Y} = 0$ to $\hat{Y} < \frac{1}{2}$. In the next section, we illustrate this approach by considering two series of faces from the PHPID database.

6. Numerical Results and Discussion

6.1. A Brief Description of the Training Testing Set

Before the study, we needed correlation planes references. The good and bad planes of references were made like it is described in step 2 of Section 4. There are also training and testing sets. In order to clarify these notions, we refer to Figure 5. We call the training set the reference images of step 1 in Figure 5. The testing set represents the set of all the images of the target images.

6.2. Bandwidth Calibration

To achieve a good prediction, we must first find an optimal bandwidth. In other words, the parameter h, which appears in the Equation (1) via the kernel function, must be adjusted in order to eliminate bad behavior of our classification procedure. With the above assumptions in mind, this requires a tradeoff between bias and variance. A value of h close to 0 will give a good estimate of the regression function in the learning database. Otherwise, large values of h will eventually affect the overall error. The optimal bandwidth $\hat{h}$ realized this task: This is a good compromise between a low error term and a good capability of prediction. For this purpose, we used a leave-one-out cross validation procedure to estimate the bandwidth. The training set is made of 126 images: 9 deal with person 0. The optimal bandwidth is calculated as $h = argmin \sum_{i=1}^{n}\left(Y_i - \hat{Y}^{(i)}\right)^2$, where $\hat{Y}^{(i)}$ is the prediction for person i calculated without the i-th observation. We find that the optimal bandwidth $\hat{h}$ is 1.06. Now that the optimal bandwidth $\hat{h}$ is found, we can use the classification algorithm by fixing the value of $\hat{h}$ into Equation (1).

6.3. Simulations for a First Series of Faces from the PHPID Database

With a face image of person 0 chosen beforehand, our goal is to determine if person 0 belongs to a database or not by making use of our classification procedure. The first training set we considered was made of 126 images, where 9 were coming from person 0. We found that when using this training set, a mean square error (MSE) of 4.8% on the whole testing set (541 planes of correlation), corresponding to the first series of the PHPID database, and only one false positive, i.e., a false positive that is an error of prediction when the person must not be recognized, but the person is recognized by the algorithm as person 0, it is a type I error. The false negative is of a type II error, occurring when the result of the algorithm is negative while the true response must be positive. We found that 13 out of 39 images from the person of reference were recognized. If the entire database was used, the MSE was 0.92%, and only 5 images from person 0 were not recognized.

In Figure 6, we plot the MSE calculated with different numbers of images in the training set. The training set was made of $14 \times (2m + 1)$ images, where m is the number of images used from either side of the image centered on the face (horizontal shift) for all 14 persons. $2m + 1$ is the total number of images used for one person: One for the centered face, m images with horizontal shift on the right, and m images with horizontal shift on the left. We observe that when m is in the range 1–11, the MSE decreases linearly, but for larger values of m, the MSE is almost constant at a value of 1%, indicating the good level of performance of our method. Thus, it is not necessary to build a learning process of the algorithm on the whole database. This avoids the so-called problem of overfitting.

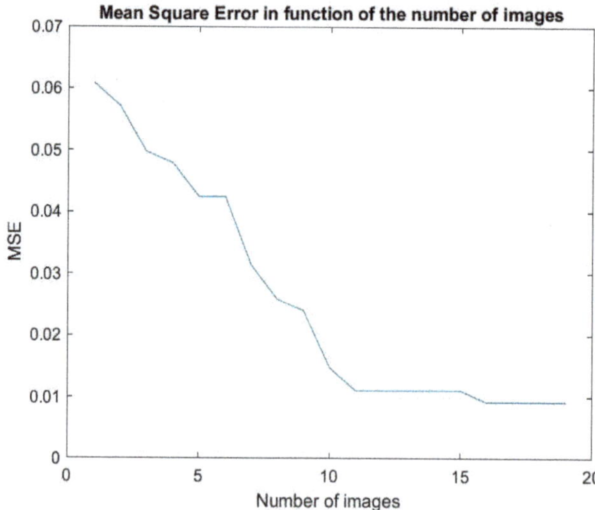

Figure 6. The mean square error (MSE) versus the number of images in the training set.

Figure 7 shows the Receiver Operating Characteristic (ROC) curve to check the ability of this algorithm to classify person 0 and the others correctly. The area under the curve is 0.9996. In order to compare with existing methods, we plotted the ROC curve with the peak-to-correlation energy (PCE) criterion for the same data. We observe that our method leads to much better results. We then have a near perfect classifier which clearly outperforms the standard algorithm using the PCE criterion.

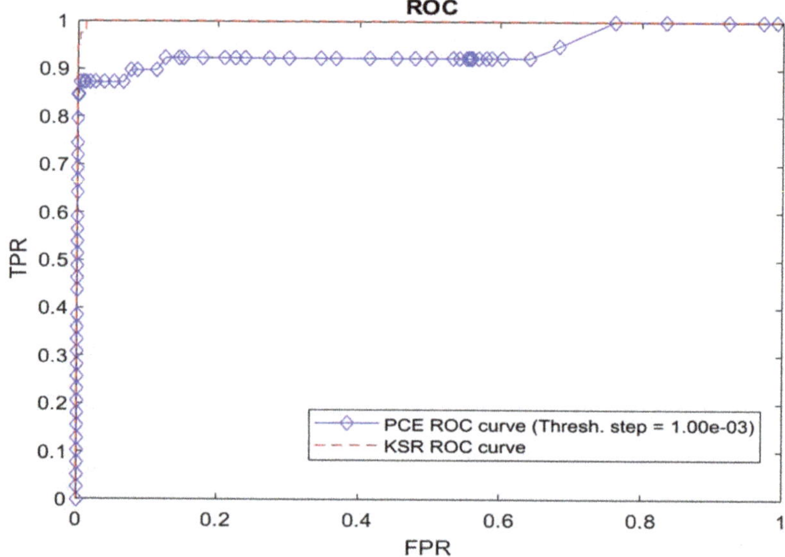

Figure 7. ROC curves on the testing set of our method (KSR) and peak-to-correlation energy (PCE) criterion. Plot of true positive rate (TPR) vs false positive rate (FPR).

6.4. Computation Time

With a training set of 541 images, the running time of the algorithm was 11 s. We now show how to improve the computation time by reducing the image size. One drawback of this method is that the performance of the algorithm can be affected by reducing the size of the correlation plane. The selection of one over two pixels was first considered, then an increase by considering 2 out of 3 was considered, then 3 out of 4 and so on, until 10 out of 11. When selecting 1 out of 2 pixels, it takes almost 0.5 s to perform the algorithm with a MSE of 4.81%. For 4 out of 5 pixels, the MSE is close to 1.5% and the running time is less than 6 s (Figure 8). In Figure 8, the top plot shows the MSE according to the number of pixels used and the bottom plot shows the corresponding running time. As the number of pixels is increased, the MSE decreases due to the loss of information in making the correlation planes. But the computation time increases with the number of pixels used. A good compromise is to use 4 out of 5 pixels. This yields to a running time less than 6 s

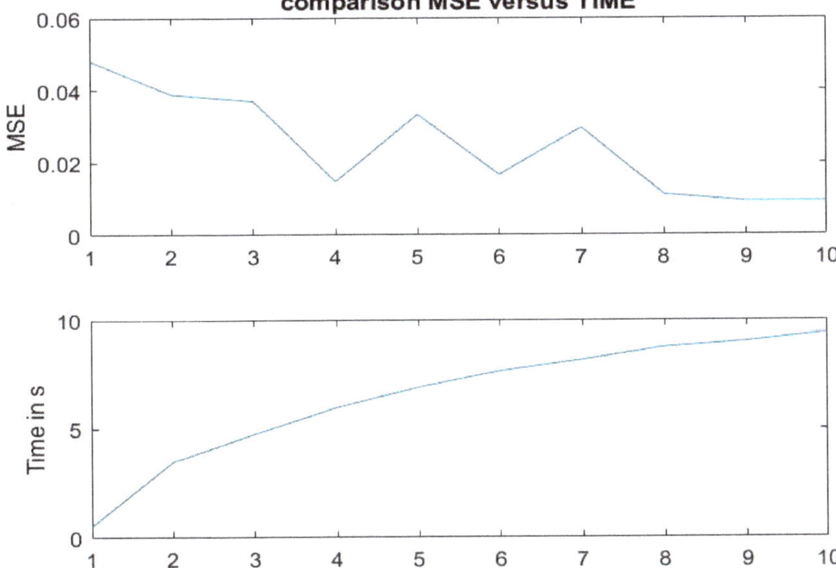

Figure 8. MSE and corresponding running times (in seconds) obtained by reducing the image size in the set of images.

6.5. Simulations for a Second Series of Faces from the PHPID Database

Consider a second series of faces from the PHPID database. These simulations differ from the previous database, since the clothes and haircuts are different and the persons can wear glasses. Here, there are 93 images of person 0 with different poses. For the training set, we use the 541 images coming from the series 1 so that we have 39 images of person 0. We find an error rate of 12.9%. For 93 images, only 12 were not detected. To illustrate the performance of our procedure, Figure 9 shows 4 images and their corresponding correlation planes. The 4 images have been well recognized by our method, where the PCE method gives an error. The first column represents the 4 images and the second column represents the corresponding correlation planes with the image of reference of person 0.

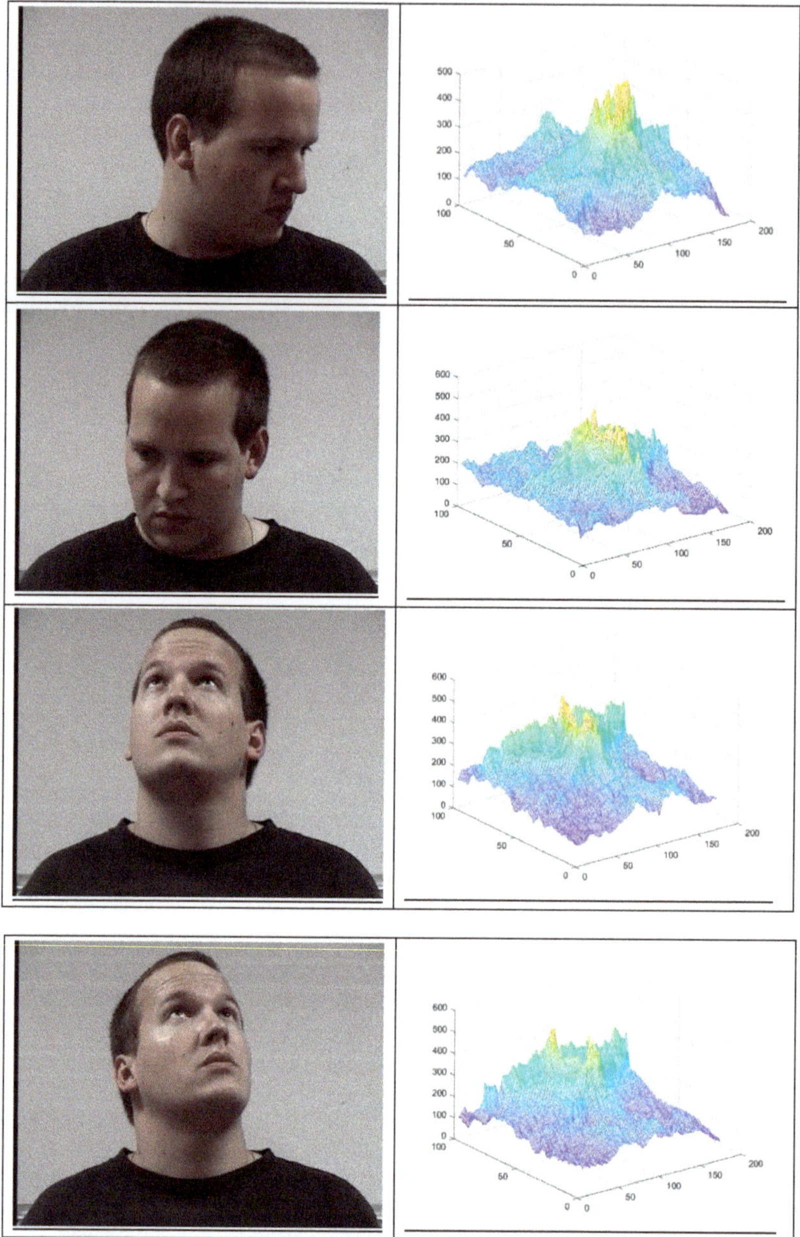

Figure 9. Images and correlation planes well recognized by our method (KSR) and badly recognized by PCE.

In order to compare our results with those obtained using the PCE criterion, we provide a comparison of the ROC curves in Figure 10. We conclude that our method is significantly better than the method using the PCE criterion. To make this plot, we used 93 images of two persons from the second series of faces from the PHPID database as the testing set. The training set comprised 541 images (from which, there were 39 images of person 0) from the first series of faces from the PHPID database.

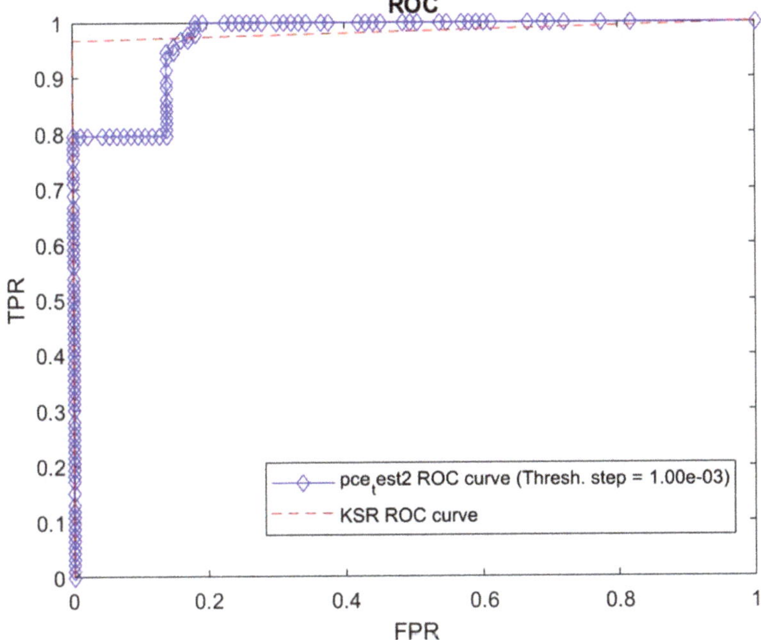

Figure 10. Same as Figure 7 for the second series of faces from the PHPID database.

7. Conclusions

We presented an innovative method specially designed to enhance the decision performance for face recognition applications. This approach is based on a classification algorithm by means of nonparametric estimation of the regression function, which defines the probability of recognition. A two-step procedure was developed, considering first the construction of the correlation planes and then the decision-making based on a kernel smoothing regression algorithm. The results and their discussion show that this easy and fast to implement algorithm performs very well using the PHPID dataset. Our results are useful because we reach a very good level of recognition, namely less than 1% of MSE. This method can be extended in parametrically modelling correlation planes.

Author Contributions: Conceptualization, M.S., M.E. M.A., A.A. and C.B.; methodology, M.S., M.E. and A.A.; software, M.S. and M.A.; validation, A.A. and C.B.; investigation, M.S. and M.E.; resources, M.A.; writing—original draft preparation, M.S., M.E., M.A. and A.A.; writing—review and editing, M.S., M.E. M.A., A.A. and C.B.; supervision, A.A. and C.B.

Funding: This research received no external funding.

Conflicts of Interest: The authors declare no conflict of interest.

References

1. Alfalou, A.; Brosseau, C. Understanding correlation techniques for face recognition: From basics to applications. In *Face Recognition*; Oravec, M., Ed.; In-Tech: Rijeka, Croatia, 2010; pp. 354–380. ISBN 978-953-307-060-5.
2. Elbouz, M.; Alfalou, A.; Brosseau, C. Fuzzy logic and optical correlation-based face recognition method for patient monitoring application in home video surveillance. *Opt. Eng.* **2011**, *50*, 067003. [CrossRef]

3.   Weaver, C.S.; Goodman, J.W. A technique for optically convolving two functions. *Appl. Opt.* **1966**, *5*, 1248. [CrossRef] [PubMed]

4.   Lugt, V.A. Signal detection by complex spatial filtering. *IEEE Trans. Inf. Theory* **1964**, *10*, 139–145. [CrossRef]

5.   Goodman, J.W. *Introduction to Fourier Optics*; McGraw-Hill: New York, NY, USA, 1968.

6.   Goodfellow, I.J.; Bengio, J.; Courville, A. *Deep Learning*; MIT Press: Cambridge, MA, USA, 2016.

7.   Marechale, C.S.; Groce, P. Un filtre de fréquences spatial pour l'amélioration du contraste des images optiques. *C. R. Acad. Sci.* **1953**, *127*, 607.

8.   Horner, J.; Gianino, P. Phase-only matched filtering. *Appl. Opt.* **1984**, *23*, 812. [CrossRef] [PubMed]

9.   Taouche, C.; Batouche, M.C.; Chemachema, M.; Taleb-Ahmed, A.; Berkane, M. New face recognition method based on local binary pattern histogram. In Proceedings of the IEEE International Conference on Sciences and Techniques of Automatic Control and Computer Engineering (STA), Hammamet, Tunisia, 21–23 December 2014; pp. 508–513.

10.   Armitage, J.D.; Lohmann, A.W. Character recognition by incoherent spatial filtering. *Appl. Opt.* **1965**, *4*, 461. [CrossRef]

11.   Dai-Xian, Z.; Zhe, S.; Jing, W. Face recognition method combined with gamma transform and Gabor transform. In Proceedings of the IEEE international Conference on Signal Processing, Communications and Computing (ICSPCC), Ningbo, China, 19–22 September 2015; pp. 1–4.

12.   Marcolin, F.; Vezzetti, E. Novel descriptors for geometrical 3D face analysis. *Multimed. Tools Appl.* **2017**, *76*, 13805–13834. [CrossRef]

13.   Moos, S.; Marcolin, F.; Tornincasa, S.; Vezzetti, E.; Violante, M.G.; Fracastoro, G.; Speranza, D.; Padula, F. Cleft lip pathology diagnosis and foetal landmark extraction via 3D geometrical analysis. *Int. J. Interact. Design Manuf. (IJIDeM)* **2017**, *11*. [CrossRef]

14.   Wassermann, L. *All of Nonparametric Statistics*; Springer: New York, NY, USA, 2006.

15.   Tsybakov, A. *Introduction to Nonparametric Estimation*; Springer: New York, NY, USA, 2009.

16.   Ramsay, J.; Silverman, B.W. *Functional Data Analysis*; Springer: New York, NY, USA, 2005.

17.   Ferraty, F.; Vieu, P. *Nonparametric Functional Data Analysis. Theory and Practice*; Springer: Berlin/Heidelberg, Germany; New York, NY, USA, 2006.

18.   Gourier, N.; Hall, D.; Crowley, J.L. Estimating Face Orientation from Robust Detection of Salient Facial Features. In Proceedings of the Pointing 2004, ICPR, International Workshop on Visual Observation of Deictic Gestures, Cambridge, UK, 23–26 August 2004.

19.   Rucklidge, W. *Efficient Visual Recognition Using the Hausdorff Distance*; Springer: Berlin, Germany, 1996.

20.   Huttenlocher, D.P.; Klanderman, G.A.; Rucklidge, W. Comparing images using the Hausdorff distance. *IEEE Trans. Pattern Anal. Mach. Intell.* **1993**, *15*, 850–863. [CrossRef]

21.   Jesorsky, O.; Kirchberg, K.J.; Frischholz, R.W. Robust Face Detection Using the Hausdorff Distance. In Proceedings of the Third International Conference on Audio- and Video-based Biometric Person Authentication, Halmstad, Sweden, 6–8 June 2001; pp. 90–95.

22.   Dubuisson, M.-P.; Jain, A.K. A Modified Hausdorff Distance for Object Matching. In Proceedings of the 12th International Conference on Pattern Recognition, Jerusalem, Israel, 9–13 October 1994.

23.   Devroye, L.; Gyorfi, L.; Lugosi, G. *A Probabilistic Theory of Pattern Recognition*; Springer: New York, NY, USA, 1996.

 *sensors*

Article

Superpixel-Based Temporally Aligned Representation for Video-Based Person Re-Identification [†]

Changxin Gao [1], Jin Wang [1], Leyuan Liu [2], Jin-Gang Yu [3] and Nong Sang [1,*]

[1]   Key Laboratory of Ministry of Education for Image Processing and Intelligent Control, School of Artificial
    Intelligence and Automation, Huazhong University of Science and Technology, Wuhan 430074, China;
    cgao@hust.edu.cn (C.G.); jinw@hust.edu.cn (J.W.)
[2]   National Engineering Research Center for E-Learning, Central China Normal University, Wuhan 430079,
    China; lyliu@mail.ccnu.edu.cn
[3]   School of Automation Science and Engineering, South China University of Technology, Guangzhou 510640,
    China; jingangyu@scut.edu.cn
*   Correspondence: nsang@hust.edu.cn
†   This paper is an extended version of our conference paper: Gao, C., Wang, J., Liu, L., Yu, J.-G., Sang, N.
    "Temporally aligned pooling representation for video-based person re-identification" Proceedings of 2016
    IEEE International Conference on Image Processing (ICIP) (19 August 2016).

Received: 18 July 2019; Accepted: 3 September 2019; Published: 6 September 2019

Abstract: Most existing person re-identification methods focus on matching still person images across non-overlapping camera views. Despite their excellent performance in some circumstances, these methods still suffer from occlusion and the changes of pose, viewpoint or lighting. Video-based re-id is a natural way to overcome these problems, by exploiting space–time information from videos. One of the most challenging problems in video-based person re-identification is temporal alignment, in addition to spatial alignment. To address the problem, we propose an effective superpixel-based temporally aligned representation for video-based person re-identification, which represents a video sequence only using one walking cycle. Particularly, we first build a candidate set of walking cycles by extracting motion information at superpixel level, which is more robust than that at the pixel level. Then, from the candidate set, we propose an effective criterion to select the walking cycle most matching the intrinsic periodicity property of walking persons. Finally, we propose a temporally aligned pooling scheme to describe the video data in the selected walking cycle. In addition, to characterize the individual still images in the cycle, we propose a superpixel-based representation to improve spatial alignment. Extensive experimental results on three public datasets demonstrate the effectiveness of the proposed method compared with the state-of-the-art approaches.

Keywords: person re-identification; superpixel; temporally aligned pooling; walking cycle

---

1. Introduction

Person re-identification (re-id), as an important technique to automatically match a specific person in a non-overlapping multi-camera network, has been widely used in many applications, such as surveillance [1,2], forensic search [3], and multimedia analysis [4]. It is a challenging problem because of large variations in a person's appearance caused by illumination, pose or viewpoint changes, as well as occlusion. Two fundamental problems in person re-identification intensively studied in the literature [3,5–7] are feature representation [8–11] and metric learning [15–25], and this work is mainly concerned with the former.

Existing works on feature representation mostly rely on still person images across non-overlapping camera views. These works can further be divided into two groups according to the number of images they use, namely, single-shot re-id and multiple-shot re-id. Single-shot re-id methods use one single image to model the person, which are evidently limited in that they do not make full use of the information. Because, in real-world applications (like surveillance), there is usually a sequence of images available for a person in each camera view. As a consequence, the single-shot methods often suffer from some practical challenging factors, like occlusion, and pose, viewpoint or lighting changes. Multiple-shot re-id methods [26–29] can alleviate these issues to some extent by utilizing more images in their feature representations. However, the multiple-shot re-id methods still have two typical limitations: (1) they treat a video sequence as an unordered set of images, where the temporal information is totally lost; (2) they may be computationally costly since appearance features need to be calculated on a large number of frames.

To tackle the limitations mentioned above, some authors recently advocate video-based re-id, which can usually achieve better performance by exploiting the abundant space–time information in feature representation. Compared to still-image-based re-id, where only spatial alignment needs to be considered since the primary challenge in feature representation is to achieve robustness to viewpoint changes, one fundamental but challenging problem in video-based re-id is temporal alignment. Considering its significance, temporal alignment has been studied in very recent literature [30,31], both using Flow Energy Profile (FEP) to align the video sequences temporally. The FEP extracts the motion information the optic flow field. Wang et al. extract fixed-length fragments around the local maxima/minima of FEP [30], and Zhang et al. use the frequency of dominant one in the discrete Fourier transform domain of FEP to extract a more stable walking cycle [31]. However, they still suffer from the following problems: (1) FEP, captured using optic flow, is based on individual pixels, and all the pixels of the lower body are considered. That means FEP is pixel level motion information, which is less accurate or robust due to the heavy noise caused by background clutter or occlusions. (2) These works represent the video using all the motion information [30,31], i.e., around all the local maxima/minima of FEP or all the walking cycles. However, this to some extent introduces redundancy and noise caused by cluttered background and occlusions, which is harmful to robust representations. Thus, we argue that person representation using only one walking cycle may be a more effective strategy to address the problem of redundancy and noise. However, the walking cycle should representative, i.e., with less redundancy and less noise. Although some deep learning-based methods have been proposed recently, they still depend on much more resources of computation, memory, training data, compared to the traditional methods. This makes the deep learning based method can not be used in some resource limited applications.

To this end, we present a superpixel-based Temporally Aligned Representation (STAR) method to address the temporal alignment problem for video-based person re-identification. More precisely, we first extract motion information from the input sequence by tracking the superpixel of the lowest portions of human, to build a candidate set of walking cycles. Second, on the assumption that there is a whole walking cycle in the video of each person, we select the "best" walking cycle to perform temporal alignment across videos. Finally, to extract the representation of the selected walking cycle of a video, we propose a superpixel-based representation for each single image, and a walking cycle based temporally aligned pooling method.

The preliminary version of portions of this paper has been published in [32]. Compared to [32], this manuscript (1) proposes a superpixel based representation for the still images, termed as SPLOMO, and compares it with original LOMO; (2) expands it from 5 pages to more than 17 pages; (3) expands or rewrites all the sections, and adds the "Related work" section; (4) in Experimental Results Section, adds some recent works in the comparison results, and adds the "Evaluation of Components and Parameters", analyzing the results in more detail.

To sum up, the contributions of this paper are as follows:

(1)   We propose a robust temporal alignment method for video-based person re-id, which is featured by the superpixel-based motion information extraction, the effective criterion for candidate walking cycles, and the use of only the best walking cycle to build the appearance representation.

(2)   Superpixel-based support regions and person masks are introduced to still image representation, so as to improve spatial alignment and thereby alleviate the undesired effects of the background.

(3)   The proposed method, performs favorably against the state-of-the-art methods, even deep learning-based ones.

The rest of this paper is organized as follows. Related work is reviewed in Section 2. We introduce the proposed STAR video representation in detail in Section 3. In Section 4, we present extensive experimental results, and we conclude this paper in Section 5.

2. Related Work

Person re-id has attracted much attention in recent years, and we first point the readers to some literature surveys on this topic [3,5–7,33]. Many methods focus on tackling this problem, which can be roughly divided into three categories, i.e., feature representation [8,34,35], metric learning [36–39] and deep learning [40–48]. Since this paper focuses on the video-based re-id task, this section only gives a review of the literature closely related to this work. Particularly, we first compare the difference between multiple-shot re-id and video-based re-id and point out the critical problem in video-based re-id, followed by reviewing the video-based re-id methods according to its main steps.

2.1. Multiple-Shot Re-Id vs. Video-Based Re-Id

In real video surveillance applications, the information for a person is recorded by a video or an image sequence, rather than a single image. Person re-id based on video is an effective way, due to abundant information in videos, such as underlying dynamic information. Some works have validated the superiority of video-based re-id methods [30,31,49–51].

The direct way to use the video information is the multiple-shot re-id methods [26–29,52]. Some state-of-the-art methods select key frames from the video sequence and then process them like the still images [40,52,53]. However, the multiple-shot methods ignore the dynamic information in the videos. Therefore, the video-based re-id methods are recently proposed [30,31,52,54–58]. To avoid confusion, we point out that the difference between the multiple-shot re-id and the video based re-id: the multiple-shot re-id utilizes the video as an unordered image set; while the video based re-id considers the video as an image sequence with the space–time information (i.e., motion information). As mentioned before, temporal information is the main difference between multiple-shot re-id and video-based re-id. However, as the image sequences of different persons are unsynchronized, it is not easy to build a robust representation which contains motion information. Therefore, the critical problem in the video-based re-identification is to synchronize the starting/ending frames of the image sequences of different persons according to the motion information, termed as temporal alignment. For instance, the fragments of complete walking cycles or gait periods can be selected from the video sequences, to build robust spatial–temporal representations.

2.2. Video-Based Re-Id Methods

According to its characteristics, we list three main steps in the video-based re-id methods: temporal alignment, spatial–temporal representation, and metric learning. Many works related to these steps have been proposed recently:

- Temporal alignment. Temporal alignment has been demonstrated to be able to lead to a robust video-based representation in the context of gait recognition [59]. Some recent works try to consider this problem in video-based re-id [30,31]. Inspired by motion energy in gait recognition [59], Wang et al. propose to use Flow Energy Profile (FEP) to describe the motion of the two legs, and employ its local maxima/minima to temporally aligned the video sequences [30]. To improve the robustness of temporal alignment, Liu et al. proposed to use the frequency of dominant one in the discrete Fourier transform domain of FEP [31].

Although the methods [30,31] have demonstrated the effectiveness of FEP based temporal alignment, they still suffer from heavy noise due to cluttered background and occlusions, since they apply optic flow to extract the motion information based on all the pixels of the lower body including the background. Moreover, in real video surveillance applications, the video sequence of a person usually contains several walking cycles, which is redundant.

To address the aforementioned problems, this paper proposes a superpixel based temporal alignment method, by first extracting the superpixels on lowest portions of human in the first frame, and then tracking them to obtain the curves of their horizontal displacements, finally selecting the "best" cycle in the curves. Note that only one cycle is used for person representation, to address the problem of redundancy and cluttered background and occlusions. Our work is partially inspired by two aforementioned video-based re-id approaches [30,31]. However, our proposed method essentially departs from these existing methods in the following three aspects:

(1)   We adopt a superpixel-based strategy to extract the walking cycles of walking person, i.e., by tracking the superpixels of the lowest portions of human (like feet, ankles, or legs near the ankles), instead of relying on the motion information of all the pixels of the lower body as in previous approaches [30,31], which is less accurate or robust due to the heavy noise caused by background clutter or occlusions.

(2)   We propose to use only the best walking cycle for temporal alignment, rather than using all the walking cycles. An effective criterion based on the intrinsic periodicity property of walking persons is proposed to select the best walking cycle from the motion information of all the superpixels. The motion information of a superpixel matches the human walking pattern, to some extent means the superpixel lies on person, and not be occluded.

(3)   Based on our temporal alignment, we introduce a novel temporally aligned pooling method to establish the final feature representation. More specifically, we take superpixels as the region supports to characterize the individual still images. Meanwhile, we then utilize person masks to enhance the robustness further.

- Spatial–temporal representation. The gait based feature is the reliable information for person re-id [59]. However, it suffers from occlusions and cluttered background. Most spatial–temporal representations for video-based re-id are devised by considering the videos as 3D volumes, inspired by the some existing works of action recognition, for instance, 3D-SIFT [60], extended SURF [61], HOG3D [62], local trinary patterns [63], motion boundary histograms (MBH) [64]. Recently, some spatio–temporal representation methods have been proposed for person re-id [30,31,57,65–67]. For instance, Wang et al. use HOG3D [30], Liu et al. aggregate a 3D low-level spatial–temporal descriptors into a single Fisher Vector (STFV3D) [31], and Liu et al. propose Fast Adaptive Spatio-Temporal 3D feature (FAST3D) for video-based re-id [57]. Being built upon 3D volumes, these appearance based spatial–temporal representations also consider the dynamic motion information to some extent. It is worth noting that, combining with temporal alignment, these representations are more robust.

Considering that most 3D representations are extensions to some widely used 2D descriptors, this paper proposes a simple framework to build a 3D representation, by combining single image based representations and temporally aligned pooling. Another reason is that lots of successful single image based representations have been proposed for person re-id, which are person re-id specific, such as LOcal Maximal Occurrence representation (LOMO) [8], Gaussian Of Gaussian (GOG) [35], and so forth. To introduce these features to 3D video data representations, we propose the temporally aligned pooling to integrate all single image based features to form a spatial–temporal representation.

- Metric learning. Learning a reliable metric for video matching is another important factor for the video-based re-id [30,54–56,68,69]. Recently, some works have been proposed for video matching.

For instance, Simonnet et al. introduce Dynamic Time Warping (DTW) distance to metric learning for the video-based re-id [54]; Wang et al. propose Discriminative Video fragments selection and Ranking (DVR) method for video matching [30]. Based on the observation that the inter-class distances with the video-based representation are much smaller than that with single image based representation, You et al. propose a top-push distance learning model (TDL) for the video-based re-id [56].

If we extract a fixed-length representation for a video sequence, the metric learning methods for video-based re-id is same with that for single-image-based re-id. Thus, the single-image-based metric learning methods can be used directly, such as KISSME in [8,31]. In this paper, we obtain the fixed-length representations by temporally aligned pooling and use Cross-view Quadratic Discriminant Analysis (XQDA) [8], a well known single-image-based metric learning method, for video matching.

It is stressed that deep networks based frameworks do not follow the steps mentioned above, but train an end-to-end neural network architecture [28,52,53,58,70–74]. For instance, McLaughlin et al. proposed to represent the appearance of the video sequences by a convolutional neural network (CNN) model, and represent the temporal information by a recurrent layer [70]. Although they achieve good performance, the deep learning based methods are still limited to the resources of computation, memory, and training data.

3. Superpixel-Based Temporally Aligned Representation

This paper focuses on appearance representation for video-based person re-id. In this section, we introduce the proposed superpixel-based temporally aligned representation by (1) extracting motion information based on superpixel tracking (Section 3.1), (2) selecting the "best" walking cycle using an unsupervised method (Section 3.2), and (3) constructing a 3D representation based on superpixel-based representation (Section 3.3) and temporally aligned pooling (Section 3.4). We depict the entire framework in Figure 1.

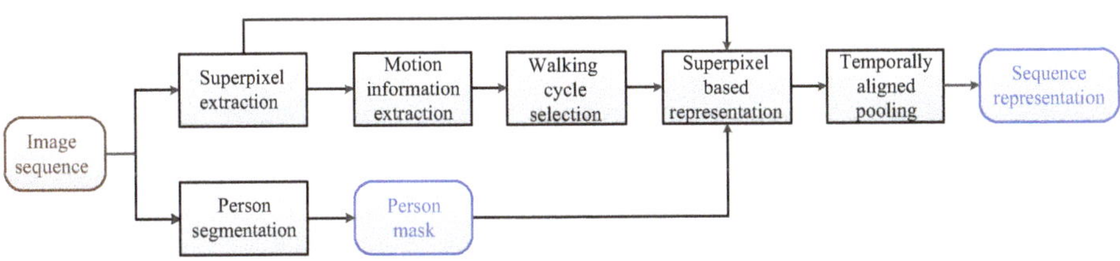

Figure 1. Framework of the proposed superpixel-based temporally aligned representation.

3.1. Motion Information Extraction

We propose a superpixel based motion information extraction, which ought to be more robust than pixel based methods, because (1) it is based on superpixels, dynamic information extraction is robust to noise caused by some individual pixels, while pixels based on methods like FEP suffer from this; (2) to alleviate the effect of occlusions and cluttered background, we utilize the local superpixel to extract motion information, and then select the "best" cycle in the curves of all the superpixels. Note that, although the superpixels may also be on the background, whose curves of motion information are quite easy to be distinguished from that on person, as discussed in Sections 3.2 and 4.3.

Given a video sequence $V = \{I_t\}_{t=1,...,T}$, with T frames, we extract the motion information, as illustrated in Figure 2. In our implementation, we only consider the motion information of the lowest portions of human, because of its amplitude of walking is more significant. Specifically, we first perform superpixel segmentation on the lowest portion of the first frame, using SLIC method [75]. N superpixels $\{S_1^j\}_{j=1,...,N}$, are obtained. Figure 2b shows an example, where the superpixel labeled in red is on the right foot of the person.

Then we track all the N superpixel to extract the motion information. For the jth superpixel S_1^j, we track it throughout the video sequence V, resulting in a set of superpixel $\{ST_t^j\}_{t=1,\ldots,T}$ containing T elements, where $\{ST_t^j\}$ is from the tth frame. Note that, although we use a set to describe the superpixels, it is ordered, and the superpixel in the sets are in the same order as the frames in the corresponding video.

At frame t, we extract N_t superpixels using SLIC, that is $\{S_t^k\}_{k=1,\ldots,N_t}$. For simplicity, we obtain the tracking result ST_t^j, which is the best match in $\{S_t^k\}_{k=1,\ldots,N_t}$ to the initial superpixel S_1^j, with the smallest distance:

$$ST_t^j = S_t^{k^*} = \arg\min_k \left(f(S_1^j) - f(S_t^k)\right)^2, \tag{1}$$

where $f(x)$ denotes the representation of superpixel x. In this paper, we use the color feature, i.e., HSV histogram, to represent the superpixels. Figure 2c shows the superpixel tracking results corresponding to the initial superpixel in Figure 2b. We denote the horizontal positions $\{L_t^j\}_{t=1,\ldots,T,j=1,\ldots,N}$ of the centers of the superpixel set $\{ST_t^j\}_{t=1,\ldots,T,j=1,\ldots,N}$. The final motion information of the region corresponding to the jth superpixel S_1^j can be described as $\{L_t^j\}_{t=1,\ldots,T}$, as shown in Figure 2d. We can see that the superpixels along the entire video sequence is about the right foot. There is a high probability that it is a part of a person with somewhat semantic information, which makes the motion information extraction algorithm very robust.

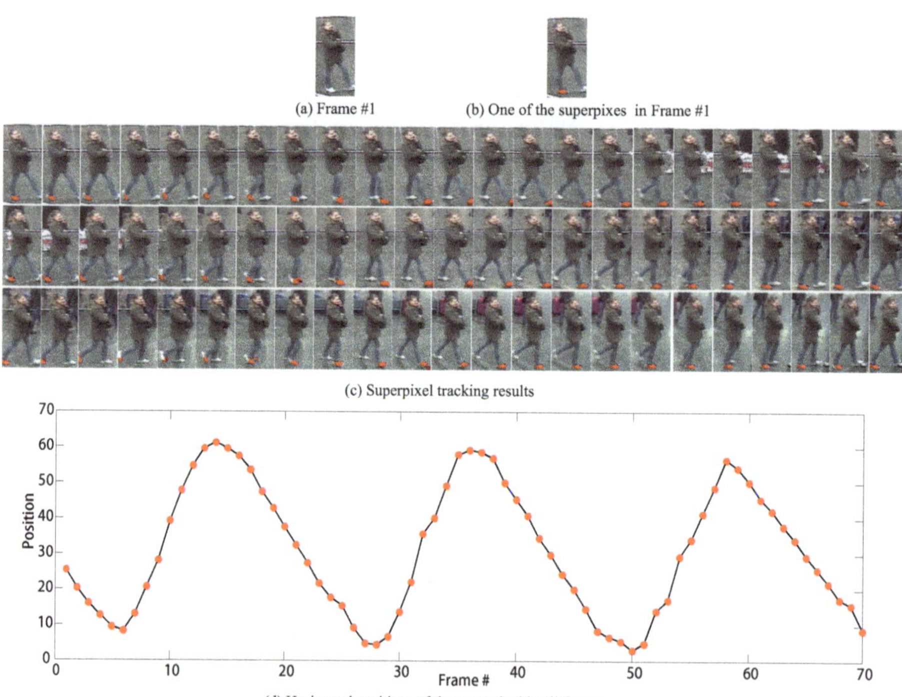

Figure 2. Motion information extraction based on superpixel tracking. (**a**) The image of frame #1; (**b**) One of the superpixels in frame #1, labeled in red color; (**c**) The superpixel tracking results in the images of frame #2 to #70, labeled in red color; (**d**) Horizontal positions (red dots) of the superpixel in all frames. The figure is best viewed in color.

It is worth noting that we track all the superpixel on the lowest portion of an image, without segmenting the lowest portions of human out of the background. The main reasons are two-fold:

(1) foreground segmentation is not dependable since it is difficult due to noise, occlusions, and low-resolution (see the last row in Figure 9); (2) the motion information of the superpixels in the background actually is easy to eliminate, since the motion information of the superpixels from the foreground and from the background in the lowest portion of the frame is quite different. Figure 3 presents the motion information of some superpixels on the background. We can see that their motion information is quite different from that of the superpixels on person as shown in Figure 2, and do not match the intrinsic periodicity property of walking persons. This observation indicates that we can easily select "good" walking cycles from noise and redundancy motion information, as described in Section 3.2.

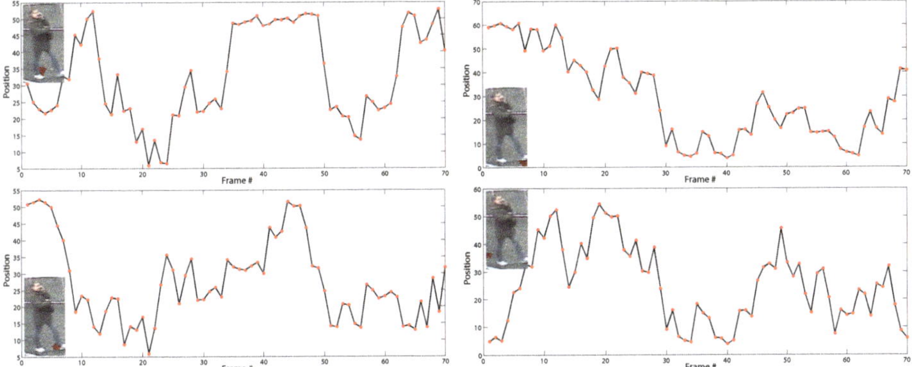

Figure 3. Motion information of four superpixels on the background, from the same sequence in Figure 2. The figure is best viewed in color.

3.2. Walking Cycle Selection

To address the problem of abundance and noise caused by cluttered background and occlusion, we use only one walk cycle of frames to represent a video sequence. Therefore, after obtaining the redundant motion information (i.e., horizontal positions) $\{L_t^j\}_{j=1,...,N, t=1,...,T}$ of the N superpixels, we are trying to select the "best" walking cycle (t_{start}^*, t_{end}^*). The motion information of a superpixel of the lowest portions of human is described as its horizontal displacements with time as in Section 3.1. The fragments of the motion information can be considered as the candidate walking cycles. Then, a natural question is: what is the "best" walking cycle for person representation? The key to this question is to mathematically model the motion information of a walking cycle, for which we adopt the sinusoid function, based on two observations: (1) Many bipedal robots use an intuitive walking method with sinusoidal foot [76]. It is based on the hypothesis that the trajectory of the feet follows sine waves in the x, y and z directions. That is consistent with the biomechanics of gait [77]. (2) The motion information of feet (horizontal positions) annotated on the iLIDS-VID dataset is almost exactly sinusoidal, as shown in Figure 4. Therefore, we model the horizontal displacements of feet as a sinusoid.

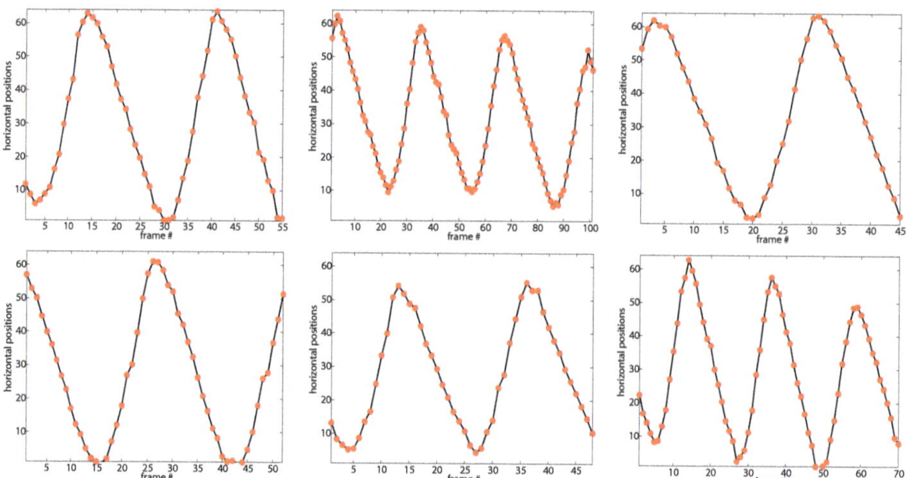

Figure 4. Illustration of motion information of feet with some examples on the iLIDS-VID dataset. We annotate the mean horizontal positions of the right foot of a person's image sequence, and then we show the horizontal positions (red dots) with frame index.

By modeling the horizontal displacements of feet as a sinusoid, we propose an effective criterion to select the "best" walking cycle. We expect the criterion has two characteristics: (1) it should be a complete walking cycle (i.e., a sinusoidal cycle) since it covers the entire dynamic information and variety of poses and shapes, and (2) it should be quite similar to sinusoid, namely, with less noise caused by cluttered background and occlusion. That is, we search the complete candidate walking cycles from the curves of the horizontal displacements according to the prior of walking persons. Then we evaluate how good a candidate walking cycle is, by measuring its fit error to a sinusoid.

Specifically, we first try to find candidate walking cycles from the motion information curves based on extreme points, as shown in Figure 5. Ideally, an extreme point corresponds to the postures when the distance of two legs is maximum. However, in the presence of noise and occlusions in practice, an extreme point with a small distance to the horizontal center line might be a false alarm. To obtain more accurate walking cycles, we process it in two ways: (1) we smooth the curve to extract more accurate extreme points by the least-squares polynomial fitting, and (2) we set upper bound y_up and lower bound y_low to eliminate the false alarms. The second way is based on the observation of the public datasets: the walking person is roughly cropped out in each frame, and approximately at the center of the frame. That means the horizontal center line is the symmetrical axis of two legs in a frame. Thus, we set the upper bound y_up and lower bound y_up with the same distance to the horizontal center line:

$$\begin{cases} y\_up & = c + \lambda \\ y\_low & = c - \lambda \end{cases} \tag{2}$$

where λ is the threshold distance to the horizontal center line, c is the location of the horizontal center line, i.e., $c = W/2$, W is the width of the image. We set the bounds to filter the extreme points on the smooth curve, and alleviate the influence of noise and occlusions.

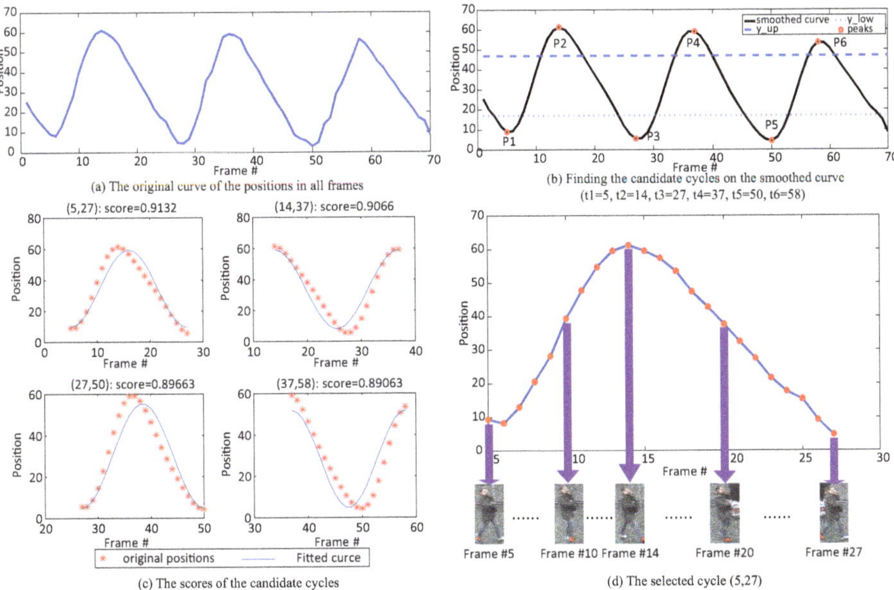

(a) The original curve of the positions in all frames

(b) Finding the candidate cycles on the smoothed curve
(t1=5, t2=14, t3=27, t4=37, t5=50, t6=58)

(c) The scores of the candidate cycles

(d) The selected cycle (5,27)

Figure 5. Walking Cycle Extraction. (**a**) the motion curve of a superpixel; (**b**) the smoothed curve, and the extreme points indicated as the red dot. (**c**) Four candidate cycles, (5,27), (14,37), (27,50), and (37,58), and their scores. (**d**) the final selected walking cycle (5,27).

We denote by $(P_1, P_2, ..., P_K)$ the K extreme points and by t_k the frame number of the k-th extreme point P_k. Then we define a candidate walking cycle (t_{start}, t_{end}) based on a group of three consecutive extreme points (P_k, P_{k+1}, P_{k+2}). If all three of them are bigger than y_up or smaller than y_low, this group will be considered as a candidate cycle $(t_{start} = t_k, t_{end} = t_{k+2})$.

To evaluate how "good" a candidate cycle (t_{start}, t_{end}) of jth superpixel is, we fit its positions $\{L_t^j\}_{t=t_{start},...,t_{end}}$ to sinusoid $\{Q_t^j\}_{t=t_{start},...,t_{end}}$ in a least-squares sense, and calculate the score $R^j(t_{start}, t_{end})$ with its fit error to sinusoid by:

$$R(t_{start}, t_{end}) = log(1 - \frac{\sum\limits_{t=t_{start},...,t_{end}} |L_t^j - Q_t^j|_2^2}{(t_{end} - t_{start} + 1) * W}) \tag{3}$$

where W is the width of the image. The final selected walking cycle (t_{start}^*, t_{end}^*) is the one with the highest score of all the candidate cycles on the motion curves of all superpixels:

$$(t_{start}^*, t_{end}^*) = \underset{(t_{start}, t_{end})}{\operatorname{argmax}} R(t_{start}, t_{end}). \tag{4}$$

We also present an example to visually illustrate our algorithm in Figure 5, corresponding to the scenario in Figure 2. For the jth superpixel $\{S_1^j\}$ shown in Figure 2b, we first smooth its motion information curve (as shown in Figure 5a), the smoothed curve is shown in Figure 5b. Figure 5b also indicates the extreme points with red dot. Next we evaluate these curves, the scores are calculated using their fit error to sinusoid, as shown in Figure 5c. The best cycle is selected with the highest score, as shown in Figure 5d. Note that, it may still suffer from background clutter and occlusions, although the proposed superpixel based method is more robust than pixel based methods. And selecting the best walking cycle can avoid this problem to some extent. Two more examples and more detailed discussion are shown and discussed in Section 4.3.

3.3. Superpixel-Based Representation

As mentioned in Section 1, a temporally aligned representation is proposed for video-based re-identification. In this paper, we refer to local maximal occurrence representation (LOMO) [8] to represent the individual frames. However, to further improve the robustness, we enhance the original LOMO in two aspects: (1) Fixed-sized patches are used in original LOMO. Although it is more robust to noise than aggregating pixel-level information, patches can span multiple distinct image regions, which can degrade the robustness. It is known that superpixels are superior to patches in many tasks because they can be considered as semantic visual primitives by aggregating visually homogeneous pixels. Thus, we propose to use superpixel-based LOMO to describe the still image. (2) The inclusion of background is another factor which may degrade the robustness of representation. To address this problem, person segmentation is employed to extract the masks of persons. And only the superpixels on the masks of the persons are considered for still image representation.

To sum up, we proposed a superpixel-based LOMO (SPLOMO) representation for still images, as shown in Figure 6. Particularly, we first perform superpixel segmentation and person segmentation, using SLIC method [75] and Deep Decompositional Network (DDN) [78] respectively. Note that, the person mask is a binary map, where the semantic information of different parts is not used. Then for each strip, only the superpixels lie in this strip and the person mask are considered to compute the histogram based representation. In our implementation, for a given strip, a superpixel is considered when meeting two conditions: (1) its overlap with the corresponding person mask O is bigger than the threshold $T_O = 0.8$; (2) its center is in the strip. The final feature is obtained by a max operation as in [8].

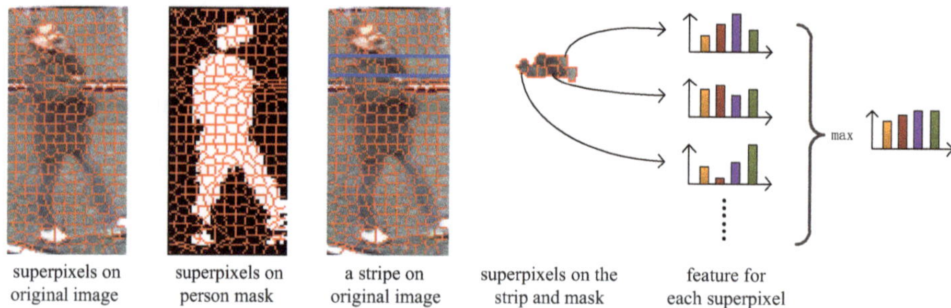

<table>
<tr><td>superpixels on
original image</td><td>superpixels on
person mask</td><td>a stripe on
original image</td><td>superpixels on the
strip and mask</td><td>feature for
each superpixel</td></tr>
</table>

Figure 6. Illustration of the superpixel based representation.

3.4. Temporally Aligned Representation

Given the selected walking cycle (t^*_{start}, t^*_{end}), the next problem for video-based re-identification is how to represent the 3D spatio–temporal video data and learn to match various person videos. This paper focuses on the former one. We propose a representation of a walking cycle, by temporally aligned pooling the descriptors of all the individual frames (or key frames) in the walking cycle. For the still images, we represent them using SPLOMO, introduced in Section 3.3.

Note that the frame numbers of walk cycles in different video sequences are usually different, which is not convenient to learn a metric. An alternative modality is multi-versus-multi (MvsM), in which there is a group of multiple exemplars for each person in the gallery and group of multiple images of each person in the probe set. However, temporal information of a video sequence is missing in the MvsM method, while it is quite important in video-based re-id. To address this problem, we propose to use temporally aligned pooling method to normalize the representations of all the frames, according to the intrinsic periodicity property. That is, we perform temporally aligned pooling according to the sinusoid corresponding to the walking cycle. Specifically, we equally divide the sinusoid into M segments $\{\Phi_m\}_{m=1,...,M}$, and then describe the corresponding phases $\{\Psi_m\}_{m=1,...,M}$ to the M segments in the walking cycle. We describe the phase m as F_m by temporally aligned pooling using the features of the images in Ψ_m. We finally concatenate $\{F_m\}_{m=1,...,M}$ together to form a final representation, termed as superpixel-based temporally aligned representation (STAR). Three pooling manners are proposed for temporal alignment: average pooling, max pooling, and key frame pooling, as shown in Figure 7. It is worth noting that, F_m is the feature of the first frame in the Ψ_m in key frame pooling.

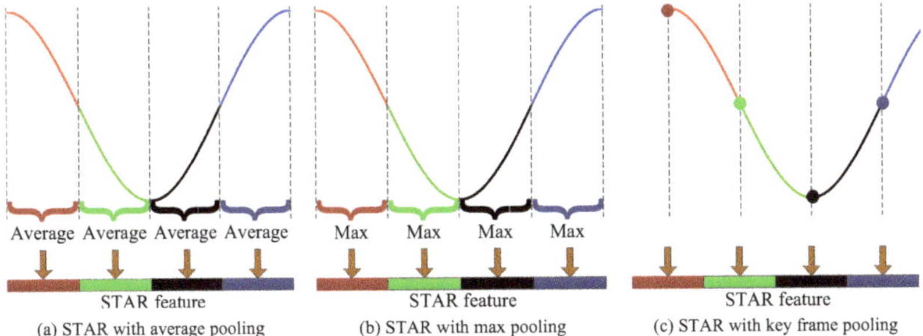

Figure 7. Illumination of temporally aligned pooling representation, with $M = 4$. Three pooling manners for temporal alignment are presented: (**a**) average pooling, (**b**) max pooling, and (**c**) key frame pooling.

4. Experimental Results

In this section, we validate the proposed STAR method (The source code is available on https://github.com/chaunceygao/STAR) and compare it to other state-of-the-art approaches on three public video-based re-id datasets.

4.1. Datasets and Settings

4.1.1. Datasets

In this section, we conduct our experiments on three publicly available video datasets for video-based person re-id: the iLIDS-VID dataset [30], the PRID 2011 dataset [79] and MARS [49], as shown in Figure 8. The iLIDS-VID dataset consists of 600 image sequences for 300 people from two non-overlapping camera views, and each image sequence has variable length consisting of 23 to 192 image frames, with an average number of 73. The dataset is very challenging due to clothing similarities, cluttered background, occlusions, viewpoint variations across camera views (Figure 8a). The PRID 2011 dataset includes 400 images sequences for 200 people from two adjacent camera views. Each image sequence has variable length consisting of 5 to 675 image frames, with an average number of 100. In our experiments, the sequence pairs with more than 21 frames are used to the requirement on the sequence length for extracting walking cycles. The main challenges of the dataset are lighting and viewpoint variations across camera views (Figure 8b). The MARS dataset consists of 1261 identities captured by 2 to 6 cameras. The train and test sets contain 631 and 630 identities respectively. 20,175 tracklets are obtain by DPM detector [80] and GMMCP [81] tracker, among them 3248 are distractors due to false detection or tracking. A large number of tracklets contain 25–50 frames, and most pedestrians have 5–20 tracklets. The MARS dataset is more challenging due to distractors, detected or tracked bounding box, besides the challenges mentioned above (Figure 8c).

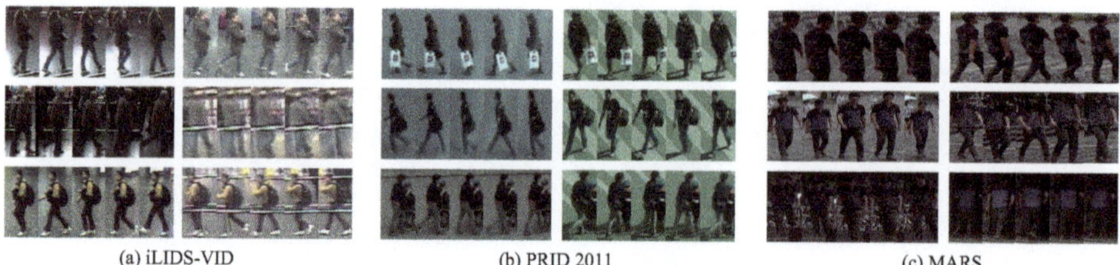

(a) iLIDS-VID   (b) PRID 2011   (c) MARS

Figure 8. Example pairs of the image sequences of the same person in different camera views from two datasets. (**a**,**b**) shows images of one person with two different cameras in each row, in iLIDS-VID and the PRID 2011 dataset, (**c**) shows images of one person with six different cameras in MARS.

4.1.2. Settings

In the proposed STAR method, SPLOMO is used to represent the still images and Cross-view Quadratic Discriminant Analysis (XQDA) [8] is used for metric learning. On iLIDS-VID and PRID 2011, the performance of all the methods is measured by the average Cumulative Matching Characteristics (CMC) curves after 10 trials. On MARS, the performance is evaluated by CMC with a fixed partition as [49], and by Mean Average Precision (mAP). Then, we give the parameters in our implementation. The width and height of the images in both datasets are $W = 64$ and $H = 128$ respectively. We perform superpixel segmentation using SLIC [75], with the maximal number of superpixel is 100, in both motion information extraction and superpixel based representation. The threshold distance to the horizontal center line $\lambda = 17$, according to the datasets. We utilize average pooling manner for temporal pooling and divide a walking cycle into $M = 8$ segments in our STAR method, according to the analysis in Section 4.4.

In SPLOMO, we extract person masks using DDN (The code is available on http://mmlab.ie.cuhk. edu.hk/projects/luoWTiccv2013DDN/index.html) [78]. Figure 9 shows some examples of person segmentation by DDN, the person segmentation result used in this paper indicates the person region (see more details of DDN in [78]).

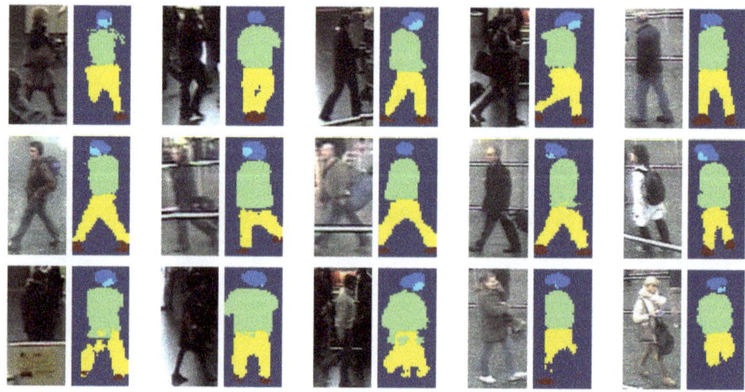

Figure 9. Some examples of person segmentation on the iLIDS-VID dataset.

4.2. Comparison with the State-of-the-Art Methods

In this section, we report the comparison results of STAR with the existing state-of-the-art video-based person re-id approaches on iLIDS-VID, PRID 2011 and MARS datasets. Three groups of the approaches are compared, as shown in Table 1, e.g., (1) traditional methods: GEI + RSVM [59], HOG3D + DVR [30], Color + LFDA [82], STFV3D + KISSME [31], CS-FAST3D + RMLLC [57], SRID [55], TDL [56]; (2) deep network based methods: RNN [70], CNN + XQDA + MQ [49], SPRNN [53], ASTPN [28], DSAN [52]; and (3) TAPR [32] is the preliminary version of our method.

More specifically, GEI+RSVM [59] is a gait based approach, which is not specially designed for person re-id. HOG3D + DVR [30], Color + LFDA [82], STFV3D + KISSME [31], CS-FAST3D + RMLLC [57] focus on appearance based representations for video. SRID [55] formulates the re-id problem as a block sparse recovery problem. TDL [56] mainly focuses on metric learning under the top-push constraint. [49] uses CNN to represent each frame. RNN [70], SPRNN [53], ASTPN [28], and DSAN [52] are end-to-end deep architectures, by incorporating feature leaning and metric learning together.

Table 1. Quantitative comparison of the proposed method and the state of the art methods on iLIDS-VID, PRID 2011 and MARS datasets. Bold and underlined values indicate the best and the second-best performance respectively.

Dataset	iLIDS-VID			PRID 2011			MARS			
Rank R	$R=1$	$R=5$	$R=20$	$R=1$	$R=5$	$R=20$	$R=1$	$R=5$	$R=20$	mAP
GEI + RSVM [59]	2.8	13.1	34.5	-	-	-	-	-	-	-
HOG3D + DVR [30]	23.3	42.4	68.4	28.9	55.3	82.8	-	-	-	-
Color + LFDA [82]	28.0	55.3	88.0	43.0	73.1	90.3	-	-	-	-
STFV3D + KISSME [31]	44.3	71.7	91.7	64.1	87.3	92.0	-	-	-	-
CS-FAST3D + RMLLC [57]	28.4	54.7	78.1	31.2	60.3	88.6	-	-	-	-
SRID [55]	24.9	44.5	66.2	35.1	59.4	79.7	-	-	-	-
TDL [56]	56.3	<u>87.6</u>	98.3	56.7	80.0	93.6	-	-	-	-
RNN [70]	58	84	96	70	90	97	-	-	-	-
CNN + XQDA + MQ [49]	53.0	81.4	95.1	<u>77.3</u>	93.5	<u>99.3</u>	68.3	82.6	89.4	49.3
SPRNN [53]	55.2	86.5	97.0	**79.4**	94.4	<u>99.3</u>	70.6	**90.0**	**97.6**	<u>50.7</u>
ASTPN [28]	<u>62.0</u>	86.0	94.0	77.0	<u>95.0</u>	99.0	44.0	70.0	81.0	-
DSAN [52]	61.9	86.8	<u>98.6</u>	77.0	**96.4**	**99.4**	<u>73.5</u>	85.0	<u>97.5</u>	-
TAPR [32]	55.0	87.5	97.2	68.6	94.6	98.9	-	-	-	-
STAR	**67.5**	**91.7**	**98.8**	69.2	94.9	99.1	**80.0**	<u>89.3</u>	95.1	**70.0**

Table 1 shows that the STAR approach outperforms the other methods in general, especially on the iLIDS-VID and MARS datasets. In particular, on the iLIDS-VID dataset, the proposed method performs significantly better than the other methods, even the deep learning based methods. Specifically, the rank-1 and rank-5 identification rates of our method are 5.5% and 4.1% over the second-best scores, respectively. On the PRID 2011 dataset, although the deep learning based methods perform better, STAR obtains the comparative results and significantly outperforms the traditional methods. On the MARS dataset, although SPRNN [53] obtains the highest scores of rank-5 and rank-20, we achieve the comparable results. Moreover, the rank-1 score of STAR is 6.5% over the second-best score obtained by DSAN [52]. More importantly, the mAP of STAR is 70.0%, about 20% higher over SPRNN [53]. The comparison results on the three datasets demonstrate that the proposed method STAR performs favorably against the state-of-the-art methods, even deep learning-based ones. The comparison results of STAR with [30,31] show the superpixel level motion information is more robust than pixel-level motion information.

We also report the results of TAPR, which is the preliminary version of our method. The main difference of TAPR and STAR is that TAPR describes each frame with LOMO feature, while STAR describes it with a SPLOMO feature. As mentioned above, SPLOMO improves LOMO by introducing the constraint of superpixel and person masks for better spatial alignment. The comparison results of these two methods show that our proposed superpixel based representation improve the performance of LOMO distinctly, especially on the iLIDS-VID dataset. More specifically, the rank-1 identification rate is 67.5% for STAR on the iLIDS-VID dataset, while 55.0% for TAPR. This validates the effectiveness of the SPLOMO representation on spatial alignment.

It is worth noting that we achieve the best performance on both iLIDS-VID dataset and MARS dataset, and a comparable result on the PRID 2011 dataset. We believe this is because the iLIDS-VID and MARS datasets have more occlusions and cluttered background than the PRID 2011 dataset. And the proposed STAR can excavate the "best" walking cycle to reduce their effect and achieve more accurate temporal alignment. Moreover, STAR uses the person masks to avoid the effect of the background, and uses superpixel based representation to achieve better spatial alignment. This means (1) accurate spatial and temporal alignment is quite essential in the video-based re-identification, especially when the scenes are complex; (2) our method has great abilities in video temporal alignment even with complex scenes.

4.3. Examples of Selected Walking Cycles

To alleviate the effect of occlusions and cluttered background, we propose to select the "best" walking cycle (a gait period). To further demonstrate the good performance of the proposed walking cycle selected method, we present two more examples in Figure 10 (we have given an example in Figure 5). All the candidate walking cycles are shown for each example, and the "best" walking cycle with highest score computed by Equation (3) is indicated by the red bounding box. From Figure 10a, we can find that the low portion of human in the first three images of the walking cycle *Frame#4-28* is partially occluded, and the walking cycles *Frame#29-52* and *Frame#40-66* both have cluttered background, while the walking cycle *Frame#14-38*, which is the selected one, has less noise than others. In Figure 10b, the walking cycle *Frame#3-25* suffers from occlusion, and the walking cycle *Frame#3-25* suffers from occlusion and clutter background, while the "best" one *Frame#43-67* has less noise. These examples demonstrate that our walking cycle method can select the ones with fewer occlusions and cluttered background. We believe that building a representation based on the selected walking cycle can alleviate the effect of occlusions and cluttered background.

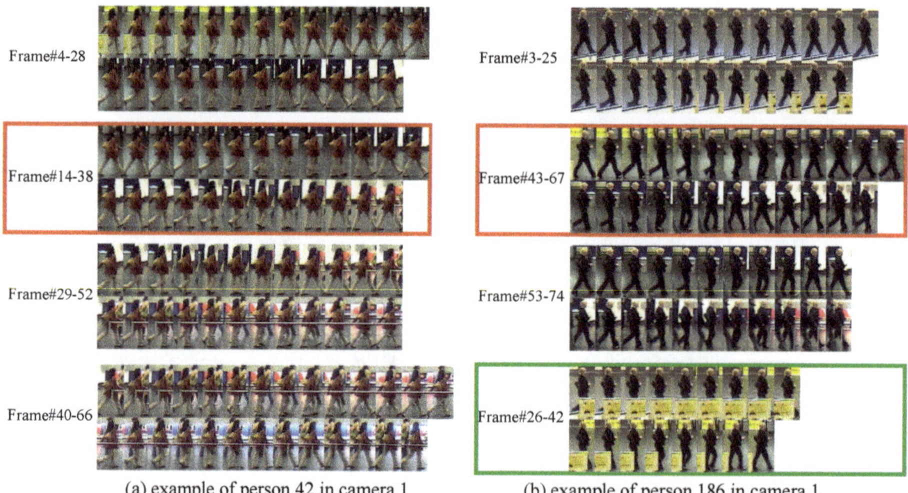

Figure 10. Two examples of selected walking cycles on iLIDS-VID. The candidate walking cycles for each video is presented, and the "best" walking cycle we selected is indicated by the red bounding box. Note that the image sequence in the green bounding box in (**b**) is not a candidate walking cycle. The frame indexes are given on the left side of the corresponding image sequences.

Note that the image sequence *Frame#26-42* with the green bounding box in Figure 10b is not a candidate walking cycle. Since the low portion of a human in this sequence is heavily occluded, it is quite difficult to extract accurate motion information. While we can also observe that all the candidate walking cycles are complete gait periods. This validates the good performance of our motion information extraction method. Combining walking cycle selection and motion information extraction, the proposed method obtains an accurate walking cycle with fewer occlusions and cluttered background, which leads to accurate temporal alignment and robust representation.

4.4. Ablation Studies

4.4.1. Evaluation of Temporally Aligned Pooling Manners

We evaluate the effects of the mentioned three temporally aligned pooling manners of our STAR algorithm on the iLIDS-VID dataset, i.e., average pooling (STAR_avg), max pooling (STAR_max), and key frame pooling (STAR_key), as shown in Table 2. The results show that the performance of STAR with average pooling is slightly better than that with max pooling and much better than that with key frame pooling. STAR_key performs the worst, for which we believe the reason is that it is very difficult to exactly localize the key points, due to discrete frames and noise. To demonstrate the effects of the temporal pooling manner, STAR with no pooling (STAR_no) is also reported in Table 2. We observe that STAR_no performs worse than STAR with a temporal pooling manner, even with key frame pooling. This validates the important role of temporally aligned pooling in video-based representation for re-id, and average pooling or max pooling is a good choice. STAR_avg performs best, thus we employ average pooling manner for quantitative comparison in Section 4.2.

Table 2. Evaluation of three different pooling manners on iLIDS-VID. Bold values indicate the best performance.

Methods	$R = 1$	$R = 5$	$R = 10$	$R = 20$
STAR_avg	**67.5**	**91.7**	**95.9**	**98.8**
STAR_max	65.6	91.3	96.1	98.7
STAR_key	56.2	87.4	94.8	98.3
STAR_no	52.8	83.5	90.8	95.7

Surprisingly, compared with Table 1, we observe that the proposed STAR with key frame pooling can perform comparably with TDL, and even STAR without pooling can outperform GEI + RSVM [59], HOG3D + DVR [30], Color + LFDA [82], STFV3D + KISSME [31], CS-FAST3D + RMLLC [57], and SRID [55]. This demonstrate the outstanding performance of the proposed method.

4.4.2. Influence of Parameters

To devise the temporally aligned representations, we equally divide each walking cycle into N segments according to the corresponding sinusoid curve. Here we evaluate the effects of the number of segments N, as reported in Table 3. The results show that STAR with $N = 8$ obtains the best performance, thus we divide a walking cycle into 8 segments for quantitative comparison in Section 4.2.

Table 3. Evaluation of the number of segments in a walking cycle on iLIDS-VID. "N" in "STAR_N" is the number of segments. Bold values indicate the best performance.

Methods	$R = 1$	$R = 5$	$R = 10$	$R = 20$
STAR_1	52.8	83.5	90.8	95.7
STAR_2	64.7	88.9	93.8	97.9
STAR_4	67.3	90.8	95.5	98.4
STAR_8	**67.5**	**91.7**	**95.9**	**98.8**
STAR_16	66.2	**91.7**	95.7	98.6
STAR 32	65.9	91.5	95.8	98.7

We also evaluate the influence of the number of superpixels in Table 4, which is set as 100 in our previous experiments. Note that we perform superpixel segmentation in both motion information extraction and image representation. The results show that STAR with 100 superpixels obtains the best performance, thus we set it as 100 in our experiments.

Table 4. Evaluation of the number of superpixel on iLIDS-VID. Bold values indicate the best performance.

Superpixel Number	$R = 1$	$R = 5$	$R = 20$
50	65.9	88.0	99.0
75	65.9	89.4	**98.8**
100	**67.5**	**91.7**	**98.8**
125	66.7	90.0	98.7
150	65.8	88.6	98.7

5. Conclusions

We have proposed a novel superpixel-based temporally aligned representation for video-based person re-identification. This representation focuses on both spatial and temporal alignment problems in video-based representations. To achieve temporal alignment, we select a video fragment of a walking cycle and describe the video fragment using temporally aligned pooling. To further improve spatial alignment, a superpixel is introduced to extract motion information and describe a still image. Unlike most previous video-based representations for re-id that use all the frames to build a spatio–temporal feature, we proposed to use only a "best" walking cycle, to reduce redundant information and simultaneously keep the "best" information. The extensive experiments, conducted on iLIDS-VID, PRID 2011 and MARS datasets, demonstrate that our method outperforms the state-of-the-art approaches.

Author Contributions: Conceptualization, C.G. and J.W.; methodology, C.G.; software, C.G.; validation, C.G., L.L. and J.-G.Y.; writing–original draft preparation, C.G.; writing–review and editing, C.G., J.-G.Y. and N.S.; visualization, J.W. and L.L.; supervision, N.S.; project administration, N.S.; funding acquisition, C.G.

Funding: This work was supported by National Natural Science Foundation of China (No. 61876210), and Natural Science Foundation of Hubei Province (No. 2018CFB426).

Conflicts of Interest: The authors declare no conflict of interest.

References

1. Song, B.; Kamal, A.T.; Soto, C.; Ding, C.; Farrell, J.; Roy-Chowdhury, A.K. Tracking and activity recognition through consensus in distributed camera networks. *IEEE Trans. Image Process.* **2010**, *19*, 2564–2579. [CrossRef]
2. Sunderrajan, S.; Manjunath, B. Context-Aware Hypergraph Modeling for Re-identification and Summarization. *IEEE Trans. Multimed.* **2016**, *1*, 51–63. [CrossRef]
3. Vezzani, R.; Baltieri, D.; Cucchiara, R. People re-identification in surveillance and forensics: A survey. *ACM Comput. Surv.* **2013**, *46*, 29. [CrossRef]
4. Li, W.; Wang, X. Locally aligned feature transforms across views. In Proceedings of the IEEE Conference on Computer Vision and Pattern Recognition, Portland, OR, USA, 23–28 Junaury 2013; pp. 3594–3601.
5. Bedagkar-Gala, A.; Shah, S.K. A survey of approaches and trends in person re-identification. *Image Vis. Comput.* **2014**, *32*, 270–286. [CrossRef]
6. Gong, S.; Cristani, M.; Yan, S.; Loy, C.C. *Person Re-Identification*; Springer: Berlin, Germany, 2014.
7. Zheng, L.; Yang, Y.; Hauptmann, A.G. Person Re-identification: Past, Present and Future. *arXiv* **2016**, arXiv:1610.02984.
8. Liao, S.; Hu, Y.; Zhu, X.; Li, S.Z. Person Re-identification by Local Maximal Occurrence Representation and Metric Learning. In Proceedings of the IEEE Conference on Computer Vision and Pattern Recognition, Boston, MA, USA, 7–12 June 2015; pp. 2197–2206.
9. Satta, R. Appearance descriptors for person re-identification: A comprehensive review. *arXiv* **2013**, arXiv:1307.5748.

10. Bazzani, L.; Cristani, M.; Murino, V. Symmetry-driven accumulation of local features for human characterization and re-identification. *Comput. Vis. Image Underst.* **2013**, *117*, 130–144. [CrossRef]

11. Zhao, R.; Ouyang, W.; Wang, X. Person re-identification by salience matching. In Proceedings of the IEEE International Conference on Computer Vision, Portland, OR, USA, 23–28 January 2013; pp. 2528–2535.

12. Ma, B.; Su, Y.; Jurie, F. Covariance descriptor based on bio-inspired features for person re-identification and face verification. *Image Vis. Comput.* **2014**, *32*, 379–390. [CrossRef]

13. An, L.; Chen, X.; Yang, S.; Bhanu, B. Sparse representation matching for person re-identification. *Inf. Sci.* **2016**, *355*, 74–89. [CrossRef]

14. Wei, L.; Zhang, S.; Yao, H.; Gao, W.; Tian, Q. GLAD: Global-Local-Alignment Descriptor for Scalable Person Re-Identification. *IEEE Trans. Multimed.* **2018**, *21*, 986–999. [CrossRef]

15. Yang, L.; Jin, R. *Distance Metric Learning: A Comprehensive Survey*; Michigan State Universiy: East Lansing, MI, USA, 2006.

16. Zheng, W.S.; Gong, S.; Xiang, T. Person re-identification by probabilistic relative distance comparison. In Proceedings of the IEEE Conference on Computer Vision and Pattern Recognition, Colorado Springs, CO, USA, 20–25 June 2011; pp. 649–656.

17. Hirzer, M.; Roth, P.M.; Köstinger, M.; Bischof, H. Relaxed pairwise learned metric for person re-identification. In *European Conference on Computer Vision*; Springer: Berlin/Heidelberg, Germay, 2012; pp. 780–793.

18. An, L.; Kafai, M.; Yang, S.; Bhanu, B. Reference-based person re-identification. In Proceedings of the IEEE Advanced Video and Signal-based Surveillance (AVSS), Krakow, Poland , 27–30 August 2013; pp. 244–249.

19. Ma, L.; Yang, X.; Tao, D. Person re-identification over camera networks using multi-task distance metric learning. *IEEE Trans. Image Process.* **2014**, *23*, 3656–3670.

20. Zhou, T.; Qi, M.; Jiang, J.; Wang, X.; Hao, S.; Jin, Y. Person re-identification based on nonlinear ranking with difference vectors. *Inf. Sci.* **2014**, *279*, 604–614. [CrossRef]

21. Chen, J.; Zhang, Z.; Wang, Y. Relevance Metric Learning for Person Re-Identification by Exploiting Listwise Similarities. *IEEE Trans. Image Process.* **2015**, *24*, 4741–4755. [CrossRef]

22. Wang, Z.; Hu, R.; Liang, C.; Yu, Y.; Jiang, J.; Ye, M.; Chen, J.; Leng, Q. Zero-shot Person Re-identification via Cross-view Consistency. *IEEE Trans. Multimed.* **2016**, *18*, 260–272. [CrossRef]

23. Wang, J.; Sang, N.; Wang, Z.; Gao, C. Similarity Learning with Top-heavy Ranking Loss for Person Re-identification. *IEEE Signal Process. Lett.* **2016**, *23*, 84–88. [CrossRef]

24. Ye, M.; Liang, C.; Yu, Y.; Wang, Z.; Leng, Q.; Xiao, C.; Chen, J.; Hu, R. Person reidentification via ranking aggregation of similarity pulling and dissimilarity pushing. *IEEE Trans. Multimed.* **2016**, *18*, 2553–2566. [CrossRef]

25. Sun, C.; Wang, D.; Lu, H. Person Re-Identification via Distance Metric Learning With Latent Variables. *IEEE Trans. Image Process.* **2017**, *26*, 23–34. [CrossRef]

26. Bazzani, L.; Cristani, M.; Perina, A.; Farenzena, M.; Murino, V. Multiple-shot person re-identification by hpe signature. In Proceedings of the IEEE International Conference on Pattern Recognition, Istanbul, Turkey, 23–26 August 2010; pp. 1413–1416.

27. Farenzena, M.; Bazzani, L.; Perina, A.; Murino, V.; Cristani, M. Person re-identification by symmetry-driven accumulation of local features. In Proceedings of the IEEE Conference on Computer Vision and Pattern Recognition, San Francisco, CA, USA, 13–18 June 2010; pp. 2360–2367.

28. Xu, S.; Cheng, Y.; Gu, K.; Yang, Y.; Chang, S.; Zhou, P. Jointly Attentive Spatial-Temporal Pooling Networks for Video-based Person Re-Identification. In Proceedings of the IEEE International Conference on Computer Vision,Venice, Italy, 22–29 October 2017.

29. Zhou, S.; Wang, J.; Shi, R.; Hou, Q.; Gong, Y.; Zheng, N. Large Margin Learning in Set-to-Set Similarity Comparison for Person Reidentification. *IEEE Trans. Multimed.* **2018**, *20*, 593–604.

30. Wang, T.; Gong, S.; Zhu, X.; Wang, S. Person re-identification by video ranking. In *European Conference on Computer Vision*; Springer: Cham, Switzerland, 2014; pp. 688–703.

31. Zhang, W.; Ma, B.; Liu, K.; Huang, R. Video-based pedestrian re-identification by adaptive spatio–temporal appearance model. *IEEE Trans. Image Process.* **2017**, *26*, 2042–2054. [CrossRef]

32. Gao, C.; Wang, J.; Liu, L.; Yu, J.G.; Sang, N. Temporally aligned pooling representation for video-based person re-identification. In Proceedings of the IEEE International Conference on Image Processing, Phoenix, AZ, USA, 25–28 September 2016; pp. 4284–4288.

33. Wu, D.; Zheng, S.J.; Zhang, X.P.; Yuan, C.A.; Cheng, F.; Zhao, Y.; Lin, Y.J.; Zhao, Z.Q.; Jiang, Y.L.; Huang, D.S. Deep learning-based methods for person re-identification: A comprehensive review. *Neurocomputing* **2019**, *337*, 354–371. [CrossRef]

34. Liu, C.; Zhao, Z. Person re-identification by local feature based on super pixel. In *International Conference on Multimedia Modeling*; Springer: Berlin/Heidelberg, Germay, 2013, pp. 196–205.

35. Matsukawa, T.; Okabe, T.; Suzuki, E.; Sato, Y. Hierarchical gaussian descriptor for person re-identification. In Proceedings of the IEEE Conference on Computer Vision and Pattern Recognition, Las Vegas, NV, USA, 26 June–1 July 2016; pp. 1363–1372.

36. Wang, J.; Wang, Z.; Liang, C.; Gao, C.; Sang, N. Equidistance constrained metric learning for person re-identification. *Pattern Recognit.* **2018**, *74*, 38–51. [CrossRef]

37. Liao, S.; Li, S.Z. Efficient psd constrained asymmetric metric learning for person re-identification. In Proceedings of the IEEE International Conference on Computer Vision, Boston, MA, USA, 7–13 December 2015; pp. 3685–3693.

38. Wang, Z.; Hu, R.; Chen, C.; Yu, Y.; Jiang, J.; Liang, C.; Satoh, S. Person reidentification via discrepancy matrix and matrix metric. *IEEE Trans. Cybern.* **2017**, *48*, 3006–3020. [CrossRef]

39. Li, W.; Wu, Y.; Li, J. Re-identification by neighborhood structure metric learning. *Pattern Recognit.* **2017**, *61*, 327–338. [CrossRef]

40. Liu, Y.; Yan, J.; Ouyang, W. Quality aware network for set to set recognition. In Proceedings of the IEEE Conference on Computer Vision and Pattern Recognition, Honolulu, HI, USA, 21–26 July 2017; pp. 5790–5799.

41. Adaimi, G.; Kreiss, S.; Alahi, A. Rethinking Person Re-Identification with Confidence. *arXiv* **2019**, arXiv:1906.04692.

42. Wang, J.; Wang, Z.; Gao, C.; Sang, N.; Huang, R. Deeplist: Learning deep features with adaptive listwise constraint for person reidentification. *IEEE Trans. Circuits Syst. Video Technol.* **2016**, *27*, 513–524. [CrossRef]

43. Liu, J.; Ni, B.; Yan, Y.; Zhou, P.; Cheng, S.; Hu, J. Pose transferrable person re-identification. In Proceedings of the IEEE Conference on Computer Vision and Pattern Recognition, Salt Lake City, UT, USA, 18–22 Junuary 2018; pp. 4099–4108.

44. Song, G.; Leng, B.; Liu, Y.; Hetang, C.; Cai, S. Region-based quality estimation network for large-scale person re-identification. In Proceedings of the Thirty-Second Aaai Conference on Artificial Intelligence, New Orleans, LA, USA, 2–7 February 2018.

45. Li, M.; Zhu, X.; Gong, S. Unsupervised person re-identification by deep learning tracklet association. In *European Conference on Computer Vision*; Springer: Berlin/Heidelberg, Germay, 2018; pp. 737–753.

46. Sun, Y.; Xu, Q.; Li, Y.; Zhang, C.; Li, Y.; Wang, S.; Sun, J. Perceive Where to Focus: Learning Visibility-aware Part-level Features for Partial Person Re-identification. In Proceedings of the IEEE Conference on Computer Vision and Pattern Recognition, Long Beach, CA, USA, 15–21 January 2019; pp. 393–402.

47. Zhong, Z.; Zheng, L.; Luo, Z.; Li, S.; Yang, Y. Invariance matters: Exemplar memory for domain adaptive person re-identification. In Proceedings of the IEEE Conference on Computer Vision and Pattern Recognition, Long Beach, CA, USA, 15–21 January 2019; pp. 598–607.

48. Zhang, Z.; Lan, C.; Zeng, W.; Chen, Z. Densely Semantically Aligned Person Re-Identification. In Proceedings of the IEEE Conference on Computer Vision and Pattern Recognition, Long Beach, CA, USA, 15–21 January 2019; pp. 667–676.

49. Zheng, L.; Bie, Z.; Sun, Y.; Wang, J.; Su, C.; Wang, S.; Tian, Q. Mars: A video benchmark for large-scale person re-identification. In *European Conference on Computer Vision*; Springer: Berlin/Heidelberg, Germay, 2016; pp. 868–884.

50. Zhang, W.; He, X.; Lu, W.; Qiao, H.; Li, Y. Feature Aggregation With Reinforcement Learning for Video-Based Person Re-Identification. *IEEE Trans. Neural Netw. Learn. Syst.* **2019**, 1–6. [CrossRef]

51. Wu, Y.; Lin, Y.; Dong, X.; Yan, Y.; Ouyang, W.; Yang, Y. Exploit the unknown gradually: One-shot video-based person re-identification by stepwise learning. In Proceedings of the IEEE Conference on Computer Vision and Pattern Recognition, Salt Lake City, UT, USA, 18–22 Junuary 2018; pp. 5177–5186.

52. Wu, L.; Wang, Y.; Gao, J.; Li, X. Where-and-when to look: Deep siamese attention networks for video-based person re-identification. *IEEE Trans. Multimed.* **2018**, *21*, 1412–1424. [CrossRef]

53. Zhou, Z.; Huang, Y.; Wang, W.; Wang, L.; Tan, T. See the forest for the trees: Joint spatial and temporal recurrent neural networks for video-based person re-identification. In Proceedings of the IEEE Conference on Computer Vision and Pattern Recognition, Honolulu, HI, USA 21–26 July 2017; pp. 6776–6785.

54. Simonnet, D.; Lewandowski, M.; Velastin, S.A.; Orwell, J.; Turkbeyler, E. Re-identification of pedestrians in crowds using dynamic time warping. In *European Conference on Computer Vision*; Springer: Berlin/Heidelberg, Germay, 2012; pp. 423–432.

55. Karanam, S.; Li, Y.; Radke, R.J. Sparse re-id: Block sparsity for person re-identification. In Proceedings of the IEEE Conference on Computer Vision and Pattern Recognition Workshops, Boston, MA, USA, 7–12 June 2015; pp. 33–40.

56. You, J.; Wu, A.; Li, X.; Zheng, W.S. Top-push Video-based Person Re-identification. In Proceedings of the IEEE Conference on Computer Vision and Pattern Recognition, Las Vegas, NV, USA, 26 June–1 July 2016.

57. Liu, Z.; Chen, J.; Wang, Y. A fast adaptive spatio–temporal 3D feature for video-based person re-identification. In Proceedings of the IEEE International Conference on Image Processing, Phoenix, AZ, USA, 25–28 September 2016; pp. 4294–4298.

58. Ye, M.; Li, J.; Ma, A.J.; Zheng, L.; Yuen, P.C. Dynamic graph co-matching for unsupervised video-based person re-identification. *IEEE Trans. Image Process.* **2019**, *28*, 2976–2990. [CrossRef]

59. Martín-Félez, R.; Xiang, T. Gait recognition by ranking. In Proceedings of the European Conference on Computer Vision, Florence, Italy, 7–13 October 2012; pp. 328–341.

60. Scovanner, P.; Ali, S.; Shah, M. A 3-dimensional sift descriptor and its application to action recognition. In Proceedings of the ACM International Conference on Multimedia, Bavaria, Germany, 24–29 September 2007; pp. 357–360.

61. Willems, G.; Tuytelaars, T.; Van Gool, L. An efficient dense and scale-invariant spatio–temporal interest point detector. In Proceedings of the European Conference on Computer Vision, Marseille, France, 12–18 October 2008; pp. 650–663.

62. Klaser, A.; Marszałek, M.; Schmid, C. A spatio–temporal descriptor based on 3d-gradients. In Proceedings of the British Machine Vision Conference, Leeds, UK, 1–4 September 2008; p. 275.

63. Yeffet, L.; Wolf, L. Local trinary patterns for human action recognition. In Proceedings of the IEEE International Conference on Computer Vision, Kyoto, Japan, 29 September–2 October 2009; pp. 492–497.

64. Dalal, N.; Triggs, B.; Schmid, C. Human detection using oriented histograms of flow and appearance. In Proceedings of the European Conference on Computer Vision, Graz, Austria, 7–13 May 2006, pp. 428–441.

65. Bak, S.; Charpiat, G.; Corvee, E.; Bremond, F.; Thonnat, M. Learning to match appearances by correlations in a covariance metric space. In Proceedings of the European Conference on Computer Vision, Florence, Italy, 7–13 October 2012; pp. 806–820.

66. Bedagkar-Gala, A.; Shah, S.K. Multiple person re-identification using part based spatio–temporal color appearance model. In Proceedings of the IEEE International Conference on Computer Vision Workshops, Barcelona, Spain, 6–13 November 2011; pp. 1721–1728.

67. Gheissari, N.; Sebastian, T.B.; Hartley, R. Person reidentification using spatiotemporal appearance. In Proceedings of the IEEE Conference on Computer Vision and Pattern Recognition, New York, NY, USA, 17–22 June 2006; Volume 2, pp. 1528–1535.

68. Ma, X.; Zhu, X.; Gong, S.; Xie, X.; Hu, J.; Lam, K.; Zhong, Y. Person Re-Identification by Unsupervised Video Matching. *arXiv* **2016**, arXiv:abs/1611.08512.

69. Zhu, X.; Jing, X.Y.; Wu, F.; Feng, H. Video-Based Person Re-Identification by Simultaneously Learning Intra-Video and Inter-Video Distance Metrics. *IEEE Trans. Image Process.* **2016**, *27*, 3552–3559. [CrossRef]

70. McLaughlin, N.; Martinez del Rincon, J.; Miller, P. Recurrent Convolutional Network for Video-based Person Re-Identification. In Proceedings of the IEEE Conference on Computer Vision and Pattern Recognition, Las Vegas, NV, USA, 26 June–1 July 2016.

71. Hou, R.; Ma, B.; Chang, H.; Gu, X.; Shan, S.; Chen, X. VRSTC: Occlusion-Free Video Person Re-Identification. In Proceedings of the IEEE Conference on Computer Vision and Pattern Recognition, Long Beach, CA, USA, 15–21 January 2019; pp. 7183–7192.

72. Hou, R.; Ma, B.; Chang, H.; Gu, X.; Shan, S.; Chen, X. Interaction-And-Aggregation Network for Person Re-Identification. In Proceedings of the IEEE Conference on Computer Vision and Pattern Recognition, Long Beach, CA, USA, 15–21 January 2019; pp. 9317–9326.

73. Zhang, R.; Li, J.; Sun, H.; Ge, Y.; Luo, P.; Wang, X.; Lin, L. SCAN: Self-and-Collaborative Attention Network for Video Person Re-identification. *IEEE Trans. Image Process.* **2019**, *28*, 4870–4882 . [CrossRef]

74. Zhao, Y.; Shen, X.; Jin, Z.; Lu, H.; Hua, X.S. Attribute-driven feature disentangling and temporal aggregation for video person re-identification. In Proceedings of the IEEE Conference on Computer Vision and Pattern Recognition, Long Beach, CA, USA, 15–21 January 2019; pp. 4913–4922.

75. Achanta, R.; Shaji, A.; Smith, K.; Lucchi, A.; Fua, P.; Süsstrunk, S. SLIC superpixels compared to state-of-the-art superpixel methods. *IEEE Trans. Pattern Anal. Mach. Intell.* **2012**, *34*, 2274–2281. [CrossRef]

76. Han, J. Bipedal Walking for a Full-Sized Humanoid Robot Utilizing Sinusoidal Feet Trajectories and Its Energy Consumption. Ph.D. Thesis, Virginia Polytechnic Institute and State University, Blacksburg, VA, USA, 2012.

77. Cross, R. Standing, walking, running, and jumping on a force plate. *Am. J. Phys.* **1999**, *67*, 304–309. [CrossRef]

78. Luo, P.; Wang, X.; Tang, X. Pedestrian parsing via deep decompositional network. In Proceedings of the IEEE International Conference on Computer Vision, Sydney, Australia, 1–8 December 2013; pp. 2648–2655.

79. Hirzer, M.; Beleznai, C.; Roth, P.M.; Bischof, H. Person re-identification by descriptive and discriminative classification. In *Scandinavian Conference on Image Analysis*; Springer: Berlin/Heidelberg, Germany, 2011; pp. 91–102.

80. Felzenszwalb, P.F.; Girshick, R.B.; McAllester, D.; Ramanan, D. Object detection with discriminatively trained part-based models. *IEEE Trans. Pattern Anal. Mach. Intell.* **2010**, *32*, 1627–1645. [CrossRef]

81. Dehghan, A.; Modiri Assari, S.; Shah, M. Gmmcp tracker: Globally optimal generalized maximum multi clique problem for multiple object tracking. In Proceedings of the IEEE Conference on Computer Vision and Pattern Recognition, Boston, MA, USA, 7–12 June 2015; pp. 4091–4099.

82. Pedagadi, S.; Orwell, J.; Velastin, S.; Boghossian, B. Local fisher discriminant analysis for pedestrian re-identification. In Proceedings of the IEEE Conference on Computer Vision and Pattern Recognition, Portland, OR, USA, 23–28 June 2013; pp. 3318–3325.

Article

ECG Signal as Robust and Reliable Biometric Marker: Datasets and Algorithms Comparison

Mariusz Pelc [1,2,*], **Yuriy Khoma** [3] **and Volodymyr Khoma** [1,3]

[1] Faculty of Electrical Engineering, Automatic Control and Informatics, Opole University of Technology, ul. Proszkowska 76, 45-758 Opole, Poland; v.khoma@po.opole.pl

[2] School of Computing and Mathematical Sciences, University of Greenwich, Park Row, London SE10 9LS, UK

[3] Department of Information Measurement Technologies, Lviv Polytechnic National University, 79013 Lviv, Ukraine; yurii.v.khoma@lpnu.ua

* Correspondence: m.pelc@gre.ac.uk; Tel.: +48-77-449-8699

Received: 14 March 2019; Accepted: 14 May 2019; Published: 22 May 2019

Abstract: In this paper, the possibility of using the ECG signal as an unequivocal biometric marker for authentication and identification purposes has been presented. Furthermore, since the ECG signal was acquired from 4 sources using different measurement equipment, electrodes positioning and number of patients as well as the duration of the ECG record acquisition, we have additionally provided an estimation of the extent of information available in the ECG record. To provide a more objective assessment of the credibility of the identification method, some selected machine learning algorithms were used in two combinations: with and without compression. The results that we have obtained confirm that the ECG signal can be acclaimed as a valid biometric marker that is very robust to hardware variations, noise and artifacts presence, that is stable over time and that is scalable across quite a solid (~100) number of users. Our experiments indicate that the most promising algorithms for ECG identification are LDA, KNN and MLP algorithms. Moreover, our results show that PCA compression, used as part of data preprocessing, does not only bring any noticeable benefits but in some cases might even reduce accuracy.

Keywords: human identification; biomarker; ECG; machine learning; Physionet; Lviv Biometric Dataset

---

1. Introduction

Biometrics is a technology that is widely used as means of access control in different application domains, from smartphones and automobiles, to healthcare, e-commerce, security and the military. The main idea behind it is to perform the operation of matching a given value of a biomarker with a reference value that represents an individual [1,2].

Typically, biomarkers are some physiological and/or behavioural attributes that are unique for each human being. The most recent biometric systems rely on various biomarkers, but the most commonly used ones are based on fingerprint scanning, face and/or voice recognition, iris scanning, hand geometry, finger vein, etc.

The main requirements that a biomarker should fulfill are the following [1]:

- Universal (present for all individuals)
- Stability over time
- Easy to measure/acquire
- Low sensitivity to other physiological factors (e.g., stress, fatigue)
- Unique for each person

There are a few additional requirements that are considered a big plus:

- Fraud resistance (difficult to fake)
- Continuous nature (always available to measure)
- Liveness indication (present only from live humans)

As was shown in [3–6], the electrocardiogram signal (ECG) is a very promising biometric marker. Historically, ECG was mostly used for medical (diagnostic) purposes, but recent progress in the fields of consumer electronics and information technologies has already enabled it applications in biometric systems [7–9].

On the other hand, ECG-based identification systems are still not quite widespread in commercial and government services, and many of them are provided as research prototypes or very new commercial products that have just appeared on the market [10,11].

Given this situation, some questions remain open, for example ECG signal reliability and reproducibility over time, its behavior in real-world applications, the potential impact of the measurement process, hardware (sensors) configuration and matching algorithms on the identification performance, etc.

Consequently, the main aim of this paper is to answer some of these questions and provide some estimates and insights on how robust and reliable the ECG biometric markers are. In order to accomplish this task and figure out some common trends and limitations for the ECG-based identification, we carried out some experiments for a broad range of system configurations (various datasets, matching algorithms, records length, lead type, sampling rate, ADC resolution, etc.).

2. Biometric System Architecture

The ECG identification process typically consists of three main stages: data acquisition, data processing (filtering, normalization, and feature extraction) and pattern matching (classification) [4,7,8,12,13].

Data acquisition requires an analogue front-end (a two- or three-electrode measurement circuit based typically on instrumental amplifiers) followed by ADC. The digitized data are then being streamed to MCU/PC.

For the ECG measurement, we decided to choose a third-party e-Health Sensor Platform V2.0 which is based on ATmega328P. The data acquisition was performed using a differential OpAmp schema followed by an 8-bit ADC operating at a 277 Hz sampling rate [14].

The data processing includes filtering (low-pass to remove offsets and respiration, high-pass to remove noise and 50 Hz coupling, movement artifacts), heartbeats segmentation and normalization. To split the ECG waveform into separate heartbeats, it was necessary to detect the R peaks. There are multiple algorithms that have been developed for this specific purpose, but in our case a third-party implementation of the Hamilton algorithm was chosen, being available in the bioscipy library. After segmentation, each heartbeat was normalized to a range of [−1; 1]. Afterwards, the data points on the ends of each heartbeat were omitted, so only the points within the central part (around 60% of the entire heartbeat) were further used in the identification process.

Another important transformation at the data processing stage is the outlier correction. It is expected that the ECG signal is of a regular nature and that the beats tend to be similar to one another. However, for some of the beats, some strong deviations were observed. There might be different kinds of reasons for this to happen, e.g., muscle noise, respiration, non-stable contact impedance, electrodes displacement, etc. In order to detect and correct those corrupted segments (outliers), a special algorithm, proposed in [15], was applied.

The data processing is followed by the classification stage. The classification model should recognize some user-specific patterns in the processed ECG signal and perform a matching with one of the corresponding classes (users). This is the last step of the identification process.

The entire ECG-based identification process is presented in Figure 1. The ECG signal waveforms appearing at different transformation stages are presented in Figures 2 and 3.

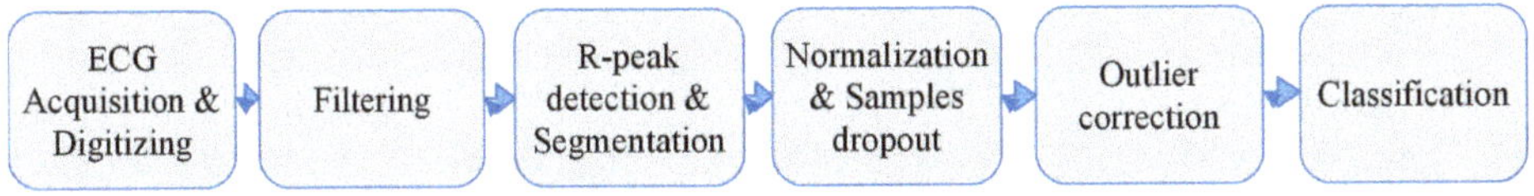

Figure 1. ECG-based identification process.

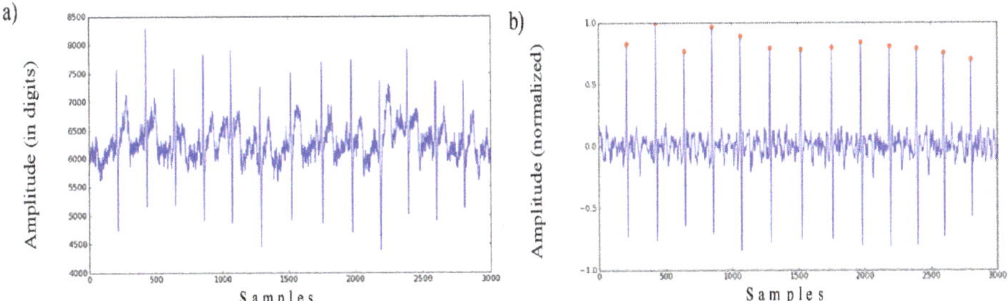

Figure 2. (**a**) Raw ECG signal and (**b**) ECG signal after filtering and normalization with the detected R peaks.

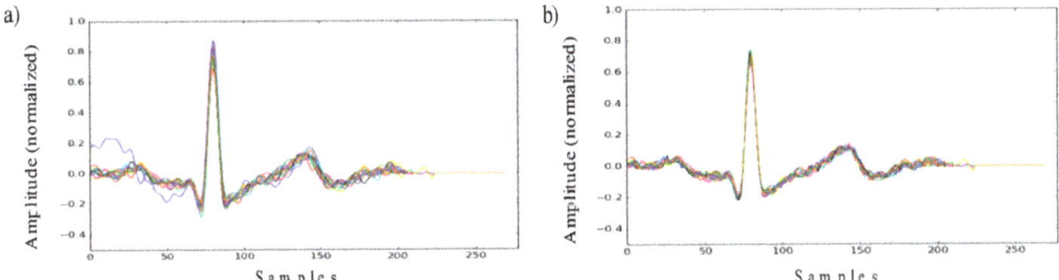

Figure 3. ECG segments (heart beats) aligned to the R peak (**a**) before and (**b**) after the outlier correction.

3. Experimental Methodology

The whole idea of the article is to check whether the ECG signal is sustainable as a biometrics marker. To perform this task, the following methodology is proposed:

- First, different datasets from different sources (both self-collected and publicly available on the Internet) should be included into the study. It is expected that these datasets should have different origins and internal structures. The main parameters that should be taken into consideration are the number of users, total number of records, mean, minimum and maximum number of records per user, records length, etc. There are two constraints related to the dataset selection. The first one is that the dataset should contain records coming only from healthy people with a normal rhythm [16]. The second constraint assumes that the data should be recorded from the same scheme of electrode placement on the patient's body, called the lead in cardiology. The reason for this is that the ECG waveform strongly varies when measured from different parts of the body.
- Second, the records classification should be performed through the use of different algorithms. In our experiments, we expect to use just one single heartbeat (the waveform between the onset of neighboring p-waves) for human identification. Thus, no sequential algorithm will be analysed here, only simple supervised machine learning techniques that map a multidimensional input vector (samples of an ECG heartbeat) onto a specified output vector (number of users). A comparison of various algorithms is required for two main reasons: first, to get some basic intuition on how some complex non-linear algorithms will behave, when compared to simpler

linear algorithms while processing this kind of data; and second, to ensure that there is no bias in the different datasets. This means that the algorithms should demonstrate a similar behaviour on different datasets.

- Third, one of the most important stages while designing machine learning based experiments is to select the most appropriate metrics. In our case, since all datasets are relatively balanced, we decided to choose an identification accuracy (error rate) to estimate the algorithms' performance.
- Finally, it is also proposed to have two-alternative data pre-processing algorithms in place. The first one is described in the section above. It means filtering, normalization and outlier correction. Another one proposes the use of a dimensionality reduction on the top. This trick is commonly used in machine learning and might help to improve the overall classification performance. In our case, it was decided to use PCA, as this is one of the simplest, most commonly used and efficient compression algorithms.

4. Datasets Description

Taking into consideration the requirements outlined above (see Section 3), the following four datasets were chosen for the current study, namely: Lviv Biometric Dataset (self-collected), Physionet ECG-ID dataset, Physionet QT dataset (only some records with a normal sinus rhythm) and Physionet MIT-BIH Normal Sinus Rhythm. A short description of each of these datasets, as well as a comparison of the main parameters, is presented below (see Table 1).

Table 1. Basic parameters of the ECG datasets.

Parameter	LBDS	ECG-ID	QT	Normal Sinus Rhythm
Lead	modified I-lead (from the fingers of the right and left hand)	I-lead	I-lead	I-lead
Number of users	53	90	22	18
Total number of records	545	310	22	18
Records per user	from 3 to 15	from 1 to 22	1	1
Sampling rate	277 Hz	500 Hz	250 Hz	125 Hz
Average record time	~10 seconds	20 seconds	15 minutes	~10:20 hours (from 8:00 to 13:50 hours)

4.1. Lviv Biometric Dataset (LBDS)

This is a self-collected dataset available in [17]. All of the records were acquired using the eHealth Arduino extension board [14]. More details on the measurement procedure can be found in [8].

4.2. Physionet ECG-ID

This dataset was created for human identification purposes, as a part of the MSc thesis [18]. The records were acquired from 44 men and 46 women, between 13 to 75 years old [19]. For some users, there are only a few records available, which means that they were recorded for one day. For other users, there are over 20 records, collected periodically over a 6 months period. The Physionet ECG-ID database is available in [18].

4.3. Physionet QT-Database

This dataset was designed for the evaluation of the ECG heartbeat segmentation algorithms. It has annotations for each record, with boundaries of each heartbeat. This dataset includes not only records of healthy people, but also records of patients with cardiological disorders. Because of this, all annotations were manually reviewed in order to select records with a normal ECG rhythm. The following records were selected: sel103, sel117, sel123, sel16265, sel16272, sel16273, sel16420, sel16483, sel16539, sel16773, sel16786, sel17152, sel17453, sel301, sel302, sel306, sel307, sel310, sele0111, sele0124, sele0133, and sele0210. The Physionet QT database is available in [20,21].

4.4. Physionet MIT-BIH Normal Sinus Rhythm

This database includes ECG records obtained by the Arrhythmia Laboratory at Boston's Beth Israel Hospital. The records originate from healthy people with no significant arrhythmias between 20 to 50 years old. The Physionet MIT-BIH Normal Sinus Rhythm database is available at [22].

5. Results and Discussion

Both the datasets selection and ECG signal pre-processing stages have already been described in the previous sections. The final stage is down to the user identification. It can be considered as a classification task, because the identification algorithm must match each ECG record to one of the existing users (classes). In general, the classification is done using machine learning techniques. The machine learning approach requires the selection of an appropriate algorithm that is powerful enough to model complex internal data relations and dataset splitting for a correct estimation of the classifier performance in real-world applications [23].

Machine learning algorithms have different natures, are based on different ideas and mathematical frameworks and are typically used in different applications. These factors should be taken into consideration when selecting the most suitable algorithm for ECG identification. The following seven algorithms have been chosen as the most promising: Logistic Regression, Support Vector Machine (SVM), Linear Discriminant Analysis (LDA), Naive Bayes, K Nearest Neighbor (KNN), Neural Networks or Multilayer Perceptron (MLP), Extreme Gradient Boosting (xGboost), Random Forest [24,25].

For multi-layer perceptron, the following configurations were used: 1 hidden layer with 50 neurons; 2 hidden layers with 50 and 30 neurons in each layer; and 3 hidden layers with 70, 50, and 30 neurons in each layer, respectively. The Rectified Linear Unit was selected as the activation function for the hidden layers and softmax as the activation function for the output layer. Training algorithm—RMSprop, number of training epochs—1000, learning rate—0.0001, batch size—100, loss function—categorical cross-entropy. For the other algorithms, we used the default configuration recommended by the sklearn framework (for example, in the case of PCA, the number of components was set to 30).

Dataset splitting requires their division into two subsets: a training and a test set. The samples from the training set are used to fit a classification model, while the samples from the test set are used to provide an unbiased evaluation of the model performance. The test set must be carefully prepared, as it should realistically represent the real-world data that the classification model would operate on.

As ECG-ID and LBDS have multiple records per user, we have split the test and the training set based on the records level. Some records will be randomly selected as the training set, while the remaining ones are included in the test set. Experiments will be conducted for the training and test set ratios of 0.7 and 0.3, respectively. To achieve a more realistic identification performance, the dataset split was done 5 times, after which the mean values for each subset was calculated.

For the MIT-BIH Normal Sinus rhythm and QT database, just one record per user is available. However, these records are of quite a substantial length. The idea is to use the time split for the training and the test set. In this case, the training and test ratios were also assumed to be 0.7 and 0.3.

Furthermore, as mentioned in Section 3, the classification models have been trained for two different scenarios: with and without PCA compression. The only exception here is neural networks, because they are complex non-linear models, which can learn efficient data compression in the first hidden layer. Thus, in this case it makes no sense to use PCA here. All results of our experiments are gathered in Table 2.

As one can see in Table 2, all of the algorithms seem to behave similarly across all of the datasets. Simple algorithms, like KNN and linear models (logistic regression, LDA, SVM), proved to work surprisingly well. Some other simple algorithms, like Naive Bayes, gradient boosting and random forest, performed relatively poorly. Neural networks also seem to guarantee a very high accuracy, which was pretty much expected, in view of their complex non-linear nature and modeling capacity.

The PCA compression might slightly improve the accuracy for some datasets, while decreasing it for others. Consequently, it seems that there is no need to include PCA in the data preprocessing pipeline.

Table 2. ECG identification results.

	Physionet ECG-ID	LBDS	Physionet QT	MIT-BIH Normal Sinus Rhythm
Logistic Regression	0.8286	0.9417	0.8809	0.7492
SVM classifier	0.8817	0.9599	0.9174	0.7707
LDA classifier	0.9328	0.9831	0.9659	0.9017
KNN classifier	0.8903	0.9746	0.9686	0.7967
Naive Bayes	0.7003	0.9587	0.9034	0.6607
Random Forest	0.8362	0.9546	0.9278	0.8192
xgboost classifier	0.7352	0.9126	0.9191	0.8591
MLP (1 hidden layer)	0.8933	0.9711	0.9162	0.8925
MLP (2 hidden layer)	0.8976	0.9464	0.9478	0.8744
MLP (3 hidden layer)	0.8406	0.92373	0.9294	0.8808
PCA + Logistic Regression	0.8286	0.9383	0.8465	0.7335
PCA + SVM classifier	0.8865	0.9593	0.8832	0.7472
PCA + LDA classifier	0.9536	0.9833	0.9481	0.8798
PCA + KNN classifier	0.8913	0.9758	0.9675	0.7957
PCA + Naive Bayes	0.6211	0.9511	0.8915	0.6681
PCA + Random Forest	0.7782	0.9199	0.8947	0.7418
PCA + xgboost classifier	0.6723	0.8911	0.9460	0.7305

The best accuracy was achieved by LDA and MLP for all four datasets. KNN shows high results for all datasets, except for the MIT-BIH Normal Sinus Rhythm. Given that this database is much larger compared to the other ones, it is not clear whether KNN would scale well enough for a larger number of users and records. MLP and xGboost were the most time-consuming algorithms to train whilst logistic regression and LDA were among the fastest algorithms.

Another important observation, based on the results from Table 2, is that the hardware parameters (e.g., measurement instrumentation, lead type, and sampling rate) do not affect the identification results significantly. The lowest accuracy was achieved for the ECG-ID database (potentially because of the highly skewed classes and larger number of users) and the MIT-BIH Normal Sinus Rhythm (potentially difficult to scale on a much bigger number of samples).

6. Conclusions

The results we have obtained prove that the ECG signal is a valid biometric marker that is very robust to hardware variations, noise and artifacts presence, that is stable over time, and that is scalable over quite a solid number of users (>90). It is also hard to steal or mimic, is easy to measure, etc.

The biometric system allows for the achievement of a high operational speed, as just one heartbeat (average duration of less than 1 second) is enough to guarantee very good classification results (~90%). On the other hand, the outlier correction requires at least five heartbeats, which means that in a real-world application the overall response time will take at least 5 seconds.

The most promising algorithms for ECG identification are linear discriminant analysis (LDA), k-nearest neighbor (KNN), and neural networks (MLP). Another important conclusion clearly confirmed by our experiments is that PCA compression is not worth using at the data preprocessing stage, as in some cases it might reduce accuracy.

The following ideas might be interesting as potential future research topics: the estimation of system scalability for bigger datasets (e.g., mixed from different sources, and augmented using generative models), optimizing training hyperparameters for artificial neural networks, and performing a sequential analysis of neighboring heartbeats on the classification stage.

Author Contributions: M.P. has reviewed and edited this paper and he is the main corresponding author; Y.K. is the principle investigator who designed and performed the hardware/software implementation and experiments; V.K. presented the initial concept and has wrote article draft preparation.

Funding: This research received no external funding.

Acknowledgments: We thank Dmytro Sabodashko, PhD student at Lviv Polytechnic National University for assistance with collecting data for Lviv Biometric Data Set that was used in current research.

Conflicts of Interest: The authors declare no conflict of interest.

Abbreviations

The following abbreviations are used in this manuscript:

ECG	Electrocardiogram
ADC	Analog to Digital Converter
MCU/PC	Microcontroller Unit/Personal Computer
PCA	Principle Component Analysis
LBDS	Lviv Biometric Dataset
SVM	Support Vector Machine
LDA	Linear Discriminant Analysis
KNN	K Nearest Neighbor
MLP	Multilayer Perceptron

References

1. Jain, A.; Flynn, P.; Ross, A.A. *Handbook of Biometrics*; Springer: New York, NY, USA, 2008; ISBN 978-0-387-71041-9.
2. Kindt, E.J. *Privacy and Data Protection Issues of Biometric Applications: A Comparative Legal Analysis*; Springer: Dordrecht, The Netherlands, 2013; ISBN 978-94-007-7522-0.
3. Fratini, A.; Sansone, M.; Bifulco, P.; Cesarel, M. Individual identification via electrocardiogram analysis. *Biomed. Eng. Online* **2015**, *14*, 1–23. [CrossRef] [PubMed]
4. Kaur, G.; Singh, D.; Kaur, S. Electrocardiogram (ECG) as a Biometric Characteristic: A Review. *Int. J. Emerging Res. Manage. Technol.* **2015**, *4*, 202–206.
5. Lee, W.; Kim, S.; Kim, D. Individual Biometric Identification Using Multi-Cycle Electrocardiographic Waveform Patterns. *Sensors* **2018**, *18*, 1005. [CrossRef]
6. Pal, A.; Singh, Y.N. ECG Biometric Recognition. In *Mathematics and Computing, Proceedings of the 4th International Conference Communications in Computer and Information Science (ICMC 2018), Varanasi, India, 9–11 January 2018*; Ghosh, D., Giri, D., Mohapatra, R.N., Savas, E., Sakurai, K., Singh, L.P., Eds.; Springer: Singapore, 2018; Volume 834, pp. 61–73. [CrossRef]
7. Matos, A.C.; Lourenc, A.; Nascimento, J. Embedded system for individual recognition based on ECG Biometrics. In Proceedings of the Conference on Electronics, Telecommunications and Computers (CETC), Lisbon, Portugal, 5–6 December 2013; pp. 265–272. [CrossRef]
8. Wieclaw, L.; Khoma, Y.; Falat, P.; Sabodashko, D.; Herasymenko, V. Biometric Identification from Raw ECG Signal Using Deep Learning Techniques. In Proceedings of the 9th IEEE International Conference on Intelligent Data Acquisition and Advanced Computing Systems: Technology and Applications, Bucharest, Romania, 21–23 September 2017; pp. 129–133.
9. Cheng, Y.; Ye, Y.; Hou, M.; He, W.; Li, Y.; Deng, X. A Fast and Robust Non-Sparse Signal Recovery Algorithm for Wearable ECG Telemonitoring Using ADMM-Based Block Sparse Bayesian Learning. *Sensors* **2018**, *18*, 2021. [CrossRef] [PubMed]
10. Mawi Band: Stress and Heart Health Monitor Verification. Available online: https://mawi.band/ (accessed on 9 December 2018).
11. SoftServe Biolock. Smart Identity Verification. Available online: https://demo.softserveinc.com/biolock/ (accessed on 9 December 2018).

12. Bassiouni, M.; Khalefa, W.; El-Dahshan, E.S.A.; Salem, A.B.M. A study on the Intelligent Techniques of the ECG-based Biometric Systems. In Proceedings of the International Conference on Communications and Computers (CC 2015) and the International Conference on Circuits, Systems and Signal Processing (CSSP 2015), Agios Nikolaos, Crete, Greece, 17–19 October 2015.

13. Pinto, J.R.; Cardoso, J.S.; Lourenço, A.; Carreiras, C. Towards a Continuous Biometric System Based on ECG Signals Acquired on the Steering Wheel. *Sensors* **2017**, *17*, 2228. [CrossRef] [PubMed]

14. e-Health Sensor Platform V2.0 for Arduino and Raspberry Pi. Available online: https://www.cooking-hacks.com/documentation/tutorials/ehealth-biometric-sensor-platform-arduino-raspberry-pi-medical (accessed on 9 December 2018).

15. Khoma, V.; Pelc, M.; Khoma, Y.; Sabodashko, D. Outlier Correction in ECG-Based Human Identification. In *Biomedical Engineering and Neuroscience, Proceedings of the 3rd International Scientific Conference on Brain-Computer Interfaces (BCI 2018), Opole, Poland, 13–14 March 2018, Advances in Intelligent Systems and Computing*; Hunek, W., Paszkiel, S., Eds.; Springer: Cham, Switzerland, 2018; Volume 720, pp. 11–22.

16. Gertsch, M. *The ECG: A Two-Step Approach to Diagnosis*, 1st ed.; Springer: Berlin/Heidelberg, Germany, 2014; pp. 19–21. [CrossRef]

17. Lviv Biometric Data Set. Available online: https://github.com/YuriyKhoma/Lviv-Biometric-Data-Set/ (accessed on 9 December 2018).

18. The ECG-ID Database. Available online: https://physionet.org/physiobank/database/ecgiddb/ (accessed on 9 December 2018).

19. The Physionet License Terms. Available online: https://physionet.org/faq.shtml#license/ (accessed on 14 May 2019).

20. The QT Database. Available online: https://physionet.org/physiobank/database/qtdb/ (accessed on 9 December 2018).

21. Laguna, P.; Mark, R.G.; Goldberger, A.L.; Moody, G.B. A Database for Evaluation of Algorithms for Measurement of QT and Other Waveform Intervals in the ECG. *Comput. Cardiol.* **1997**, *24*, 673–676.

22. The MIT-BIH Normal Sinus Rhythm Database. Available online: https://physionet.org/physiobank/database/nsrdb/ (accessed on 9 December 2018).

23. Goldberger, A.L.; Amaral, L.A.N.; Glass, L.; Hausdorff, J.M.; Ivanov, P.C.; Mark, R.G.; Mietus, J.E.; Moody, G.B.; Peng, C.-K.; Stanley, H.E. PhysioBank, PhysioToolkit, and PhysioNet: Components of a New Research Resource for Complex Physiologic Signals. *Circulation* **2000**, *101*, e215–e220. [CrossRef] [PubMed]

24. Bishop, C.M. *Pattern Recognition and Machine Learning*; Jordan, M., Kleinberg, J., Scholkopf, B., Eds.; Springer: Singapore, 2006.

25. Pedregosa, F.; Varoquaux, G.; Gramfort, A.; Michel, V.; Thirion, B.; Grisel, O.; Blondel, M.; Prettenhofer, P.; Weiss, R.; Dubourg, V.; et al. Scikit-learn: Machine Learning in Python. *J. Mach. Learn. Res.* **2011**, *12*, 2825–2830.

 *sensors*

Review

Sensor-Based Technology for Social Information Processing in Autism: A Review

Andrea E. Kowallik [1,2,3] **and Stefan R. Schweinberger** [1,2,3,4,5,*]

1   Early Support and Counselling Center Jena, Herbert Feuchte Stiftungsverbund, 07743 Jena, Germany; andrea.kowallik@uni-jena.de
2   Social Potential in Autism Research Unit, Friedrich Schiller University, 07743 Jena, Germany
3   Department of General Psychology and Cognitive Neuroscience, Friedrich Schiller University Jena, Am Steiger 3/Haus 1, 07743 Jena, Germany
4   Michael Stifel Center Jena for Data-Driven and Simulation Science, Friedrich Schiller University, 07743 Jena, Germany
5   Swiss Center for Affective Science, University of Geneva, 1202 Geneva, Switzerland
*   Correspondence: stefan.schweinberger@uni-jena.de; Tel.: +49-(0)-3641-945181; Fax: +49-(0)-3641-945182

Received: 11 October 2019; Accepted: 30 October 2019; Published: 4 November 2019

Abstract: The prevalence of autism spectrum disorders (ASD) has increased strongly over the past decades, and so has the demand for adequate behavioral assessment and support for persons affected by ASD. Here we provide a review on original research that used sensor technology for an objective assessment of social behavior, either with the aim to assist the assessment of autism or with the aim to use this technology for intervention and support of people with autism. Considering rapid technological progress, we focus (1) on studies published within the last 10 years (2009–2019), (2) on contact- and irritation-free sensor technology that does not constrain natural movement and interaction, and (3) on sensory input from the face, the voice, or body movements. We conclude that sensor technology has already demonstrated its great potential for improving both behavioral assessment and interventions in autism spectrum disorders. We also discuss selected examples for recent theoretical questions related to the understanding of psychological changes and potentials in autism. In addition to its applied potential, we argue that sensor technology—when implemented by appropriate interdisciplinary teams—may even contribute to such theoretical issues in understanding autism.

Keywords: automatic recognition; face; voice; body motion; autism spectrum disorder (ASD); assessment; intervention

---

1. Introduction

Throughout the last decades, the number of people diagnosed with an Autism Spectrum Disorder (ASD) increased dramatically [1,2] and so did the need for high-quality diagnostic protocols and therapies. With the ongoing progress in computer sciences and hardware, a lot of creative ideas emerged on how to use sensor data to identify and observe autistic markers, support diagnostic procedures and enhance specific therapies to improve individuals' outcomes.

ASD is a behaviorally defined group of neurodevelopmental disorders that are specified by impaired reciprocal social communication and restricted, repetitive patterns of behavior or activities (DSM-5), [3]. The symptoms are usually apparent from early childhood and tend to persist throughout life [4]. Common social impairments include a lack of social attention as evident in abnormal eye gaze or eye contact [5] and social reciprocity such as in reduced sharing of emotions in facial [6] or vocal behavior [7]. Further, only a minority of the affected people report having mutual friendships [8]. Related to the restricted and repetitive behaviors, stereotyped motor movements and speech are the stand-out features in many people with ASD [9]. Other symptoms are insisting on sameness and routines [10], special interests and hyper- or hyporeactivity to sensory input from various modalities [11]. The exact profile and severity of symptoms in people with ASD as well as their personal strengths and coping capabilities vary to a great degree, and so does their need for support.

Reasons for the increased prevalence over the last decades include a more formalized diagnostic approach and heightened awareness. The current 'gold standard' for a diagnosis of ASD consists of an assessment of current behavior, a biographical anamnesis, and a parental report, all collected and evaluated by a trained multi-professional team [12]. Although the screening and diagnostic methods for ASD improved throughout the last years, many affected people, especially women [13] and high functioning people, still receive a late diagnosis. Since early interventions have been shown to be most effective for improving adaptive behavior, as well as IQ and language skills [14], there is continued demand for methods promoting early assessment in order to avoid follow-up problems. In this context, progress in automatic and sensor-assisted identification of ASD-specific behavioral patterns could make an important contribution to an earlier and less biased diagnosis.

Even beyond assessment, advances in digital technology are highly relevant for autism, and in more than just one way. First, there is some evidence that many autistic people show behavioral tendencies to interact with technology and to potentially prefer such interactions to interactions with humans. It is thought that autistic traits are related to systemizing, the drive to analyze how systems work, as well as to predict, control and construct systems [15]. In this context, high information technology (IT) employment rates are often used as a proxy for higher rates of strong systemizers in a population. Intriguingly, recent research from the Netherlands reported that the prevalence of childhood autism diagnoses, but not of two control neurodevelopmental diagnoses (i.e., ADHD and dyspraxia), was substantially higher in Eindhoven, a classical (IT) region, when compared to two control regions (Utrecht and Haarlem) that had been selected for high demographic and socio-economic similarity in criteria other than the proportion of IT-related jobs [16]. Second, and qualifying any simplistic interpretation of this correlative (but not necessarily causal) relationship, there is evidence that technology can potentially provide powerful social support for children with autism. For instance, children with ASD often perform better with a social robot than a human partner (e.g., in terms of enhanced levels of social behavior towards robots), tend to perceive interactions with robots as positive [17], and subsequently show reduced levels of repetitive or stereotyped actions. For a recent review, see Pennisi et al. [18].

Scientific interest in the utilization of sensor technology to gain an understanding of people with ASD has increased considerably in the recent past. Some fields of research focus on different neurobiological assessments and try to identify autism-specific signals or 'biomarkers' to better understand the neurobiological underpinnings of the disorder. Good overviews covering methods including electroencephalography (EEG), magnetoencephalography (MEG), and functional magnetic resonance imaging (fMRI) are provided by Billeci et al. [19] and by Marco et al. [20]. Other research focused on an autonomic activity such as heart rate variability (HRV) or skin conductance responses (SCR). These can be studied with Wearable devices, typically in the context of emotional monitoring in ASD as seen in a review by Taj-Eldin, et al. [21]. Applications in VR environments [22–24] have also been reviewed as promising methods to train and practice social skills.

The aim of this review is to provide an overview of the current state of research using sensor-based technology in the context of ASD. We focus on sensor technology that is applicable without constraining natural movement, and on sensory input from the face, the voice, or body movements. Accordingly, this review does not consider evidence from wearable technology, VR, or psychophysiological and neurophysiological recordings. Note also that while we provide details of our procedure for identifying relevant original findings to enhance reproducibility, this paper represents a thematic review in which we pre-selected for contents as described below, and in which we occasionally discuss additional relevant findings that were not formally identified by this literature search. For instance, we may refer to some key findings regarding psychological theories of human social or emotional communication where relevant, even when the findings were not obtained with individuals with autism.

2. Literature Search

We performed parallel literature searches on Web of Science, PsychInfo, PubMed, and IEEE Xplore. Considering the rapid advance of technology throughout the last decades, we focussed on the last 10 years (2009–2019) to give an overview of recent developments in this field. Searches were performed on 29 May 2019. Our search terms were (autis* OR ASD OR Asperger) AND (sensor OR sensors) AND (fac* OR voice* OR body motion* OR person* OR regoc* OR identi* OR emotio* OR diagnos*). Furthermore, the considered language was restricted to English. Overall, the search resulted in 386 articles (Web of Science $N = 185$, IEEE Explore $N = 109$, PubMed $N = 52$, PsychInfo $N = 40$). Note that we only included publications with original data from an ASD or high-risk group, such that new algorithms on preexisting data sets were not included. We also included the reference lists of identified articles with respect to additional screening for relevant publications. After removing duplicates, this selection process resulted in a total of 36 articles. These are discussed below, and those articles that focused on assessment or intervention are additionally summarized in Tables 1 and 2, respectively.

3. Supporting Assessment: Identification of ASD Related Features

3.1. Facial Information

3.1.1. Facial Movements

For facial movements, researchers tended to focus on emotional facial information, following the idea that impaired social communication in ASD can be framed as deficits in emotional communication regarding both perception and expression [25], and the finding that people with ASD may show reduced or idiosyncratic emotional expressions [6,26]. Against this context, it may be useful to keep in mind that ASD-specific impairments in face perception are not restricted to emotional expressions, but also affect other aspects such as facial identity [27]. Similarly, expression in social communication may be affected in subtle ways that go beyond emotional expressions [28]. The original articles on sensor-based assessment in terms of identification of autism spectrum disorders (ASD)-related features, regarding facial movements and other forms, represented in this review are listed in Table 1.

Samad et al. [29] compared facial expressions from an ASD and a typically developing (TD) group, with 8 participants each. Computing facial curvatures from 3D point cloud data, they found equally intense but more asymmetrical facial expressions in the ASD group. Another study [30] used a single webcam mounted on a TV screen to record toddlers' spontaneous facial expressions when confronted with emotional cartoons. Comparing descriptive data from small groups with five children each, they reported that the lower face seemed to be more important in distinguishing between the ASD and TD groups. Leo et al. [31] presented a new processing pipeline on 2D video data that aimed at assessing facial expressions in ASD children specifically. They estimated the production skills for individual children based on verbal instructions to express these emotions, separately for individual face parts and emotions. The performance of 17 boys with ASD was variable, and some boys exhibited differential production scores for different categories of emotions. Note that this finding is in broad agreement with theories that emphasize category-specific mechanisms, and with componential approaches to emotion [32,33]. Samad et al. [34] used a storytelling avatar, also with the aim of eliciting spontaneous emotions to differentiate between ASD and TD groups with 10 participants each. Comparing 3D data on the level of facial action units (AUs) they found overall lower AU activations in the ASD group and lower correlations between the AUs. The deviant activation in AUs 6, 12, and 15 (cheek raiser, lip corner puller, and lip corner depressor, respectively), were found to be promising markers for ASD. As a more serious screening approach, the App 'Autism & Beyond' [35] recorded toddlers' facial responses to certain stimuli with the front camera and classified them into positive, neutral, or negative. Using a large database of 1756 children, the associated facial expressions, along with eye gaze and parental reports, with Autism Spectrum risk status. Comparisons between high- and low-risk groups revealed that high risk for ASD was associated with higher frequencies of neutral facial expressions, and with lower frequencies of positive expressions.

3.1.2. Eye Gaze

One major domain in which autistic people are frequently described to behave unusually relates to oculomotor behavior, including low levels of eye contact during communication, and low levels of directional signaling via eye gaze. In fact, a current target article challenges the common belief that autistic people lack social interest in others, and suggests that this interpretation could be erroneously elicited by unusual behavior such as low levels of eye contact [36]. While this may be a useful context to keep in mind when reading this section, there is agreement that the assessment of eye gaze can help to identify behavioral patterns that are relevant for autism [37].

Chawarska and Shic [38] compared toddlers with ASD and TD of different age groups (2 and 4 years old) with an eye-tracking system while they watched neutral faces. Although all groups spent equal time looking at the screen, a restricted scanning pattern in the ASD group was found, with relative neglect of the mouth area. Additionally, the older ASD group spent less time looking at inner facial features in general, even compared to the younger ASD group. According to the authors, this indicates a different scanning pattern of children with ASD that emerges throughout early childhood and suggests less looking at the mouth as the best early predictor for ASD. Liu et al. [39] used a machine-learning framework on eye-tracking data of 29 ASD vs. 29 TD children (mean age 7.90 years) looking at images of faces embedded in a learning task with repeated presentations. Their proposed framework was able to identify ASD children both from age- (mean age 7.86 years) and IQ-matched (mean age 5.74 years) control groups with high classification accuracy (up to 88.5%). Król and Król [40] used eye-tracking to study the effect of including temporal information into spatial eye-tracking of face stimuli, thus creating scan paths for 21 ASD and 23 TD individuals (mean age around 16 years). They found a difference in face-scanning not only in spatial properties but also in temporal aspects of eye gaze, even within short exposures (about 2 s) to a facial image. Classification of group membership based on a machine-learning algorithm on spatial and temporal data led to better accuracy (55.5%) than classification based on spatial data alone (53.9%), although overall accuracy was rather low. Note that discrepancies between classification accuracies in this study and the study by Liu et al. [39] need to be seen in the context of specific conditions of each study, and could potentially be attributed to the facial stimuli and trial numbers used, the experimental task instructions, methodological differences in data analysis, or differences in the samples tested.

In a more general study on the visual scanning of natural scene images, Wang et al. [41] compared an ASD and a TD group (N = 20 and 19, respectively) at different levels of perception. The ASD group was found to fixate more towards the center of an image (pixel level), less on objects in general (object level) and less on certain objects (e.g., faces or objects indicated by social gaze), but more on manipulatable objects (semantic level).

Table 1. Original Articles on Sensor-Based Assessment in Terms of Identification of autism spectrum disorders (ASD)-related features.

Domain of Behavior	Reference	Solution Name	Sensor/Parameters	Sample (Mean Age)	Setting and Stimuli	Results
Facial Movements	Samad et al., 2016 [29]	Sony EVID70 Color Camera, 3dMD	2D, 3D Imaging	ASD: 8 (13 y), TD: 8 (16 y)	Emotion Recognition Task: 12 3D Faces	ASD: Intense, Asymmetrical Facial Expressions with Lack of Differential Facial Muscle Actions
Facial Movements	Del Coco et al., 2017 [30]	Webcam	2D Imaging	ASD: 5 (5 y), TD: 5 (Age-, Gender-Matched)	Watching 9 Emotion Eliciting Videos Taken from Famous Cartoons	ASD Descriptively Exhibited Less Facial Behavioral Complexity; Lower Face Seemed More Significant Than Upper Face to Distinguish Between TD-ASD
Facial Movements	Leo et al., 2018 [31]	Off-The-Shelf Camera	2D Imaging	ASD: 17 (9 y)	Emotion Production Task: 4 Expressions (Happiness, Sadness, Fear, Anger)	Descriptive Scores for Upper and Lower Face Parts
Facial Movements	Egger et al., 2018 [35]	iPhone	2D Imaging (Front Camera)	High Risk: 555 (3 y), Low Risk: 1199 (3 y)	Watching Video Clips	More Neutral and Less Positive Emotional Reactions in High-Risk Group
Facial Movements	Samad et al., 2019 [34]	PrimeSense	3D Imaging	ASD: 10 (14 y), TD: 10 (13 y)	Watching Story Content Narrated by The Animated Avatar	ASD: Overall Lower FAU Activation, Higher Activations of FAU 15, Limited Activation of FAUs In Response to Encountering Negative Emotional States; Concurrent Activations of Several FAU Pairs Are Found to Be Absent
Eye Gaze	Chawarska and Shic, 2009 [38]	iView X™ RED	Eye-Tracking	ASD: 44, TD: 30	Watching Color Images of Affectively Neutral Female Faces	ASD: Attended to Visual Scenes Containing Faces to A Similar Extent; ASD Look Less at Inner Facial Features, Older ASD Spent Less Time with Inner Features Than Younger, ASD Spent Less Time Looking at the Mouth
Eye Gaze	Liu et al., 2016 [39]	Tobii T60	Eye-Tracking	ASD: 29 (8 y), TD-Age: 29 (8 y), TD-Ability: 29 (6 y)	Face Memorization and Recognition Task	Classification Accuracy Predicting ASD: 88.51% (p < 001)
Eye Gaze	Król and Król, 2019 [40]	SMI RED250	Eye-Tracking	ASD: 21 (16 y), TD: 23 (16 y)	Face Perception Tasks with 60 Color Photographs of Faces (FACES Database)	Prediction ASD vs. TD Group Membership: Based on Only "Spatial" Information (M = 53.9%) Was Significantly Smaller Than That of the "Spatial + Temporal" Model (M = 55.5%)

Table 1. *Cont.*

Domain of Behavior	Reference	Solution Name	Sensor/Parameters	Sample (Mean Age)	Setting and Stimuli	Results
Eye Gaze	Wang et al., 2015 [41]	Tobii X300	Eye-Tracking	ASD: 20 (31 y), TD:19 (32 y)	Free Viewing Task with 700 Natural Scene Images (OSIE Dataset)	ASD: Higher Saliency Weights for Low Level Properties, Lower Weights for Object and Semantic-Based Properties
Voice	Min and Tewfik, 2010 [42]	Microphone	Audio, Accelerometer	ASD: 4	No Context Given	22/24 Vocal Stimming Events Detected by Classifier
Voice	Min and Fetzner, 2018 [43]	Microphone	Audio, 2D Imaging	ASD: 4	No Context Given	Trained Dictionaries Detect Vocal Stimming with Sensitivity: 73–93%
Voice	Marchi et al., 2015 [44]	Zoom H1 Handy Recorder (Hebrew, English), Zoom H4 With RØDE NTG-2 Microphone (Swedish)	Audio	ASD: 7 (Hebrew), 11 (Swedish), 9 (English); TD:10 (Hebrew), 9 (Swedish), 9 (English)	Repeating Sentences From 9 Emotional Stories	ASD: Generally Perform Poorer, In English And Swedish Angry Was Poorly Performed, In Hebrew Afraid Was Poorly Performed
Voice	Ringeval et al., 2010 [45]	Logitech USB Desktop Microphone	Audio	AD: 12 (10 y), PDD-NOS: 10 (10 y), SLI: 13 (10 y), TD: 73 (10 y)	Reading 26 Sentences with Certain Prosodic Dependencies (Descending, Falling, Floating, Rising)	AD: Intonation for Falling Floating and Especially Rising Was Worse Compared to TD
Body Movement	Gonçalves et al., 2014 [46]	Microsoft Kinect	Color Depth Sensors, IR Emitter, Microphone	ASD: 5 (9 y)	Playing Session with Robot	Good Detection of Hand Flapping with DTW, But Susceptible to Noise
Body Movement	Jazouli et al., 2019 [47]	Microsoft Kinect V1	Color Sensor, IR Depth Sensors, IR Emitter, Microphone	ASD: 5 (5–10 y), TD: 5 (Training Data)	No Context Given	94% Overall Recognition Rate For Stereotyped Gesture Recognition ($P Algorithm)
Body Movement	Rynkiewicz et al., 2016 [47]	Microsoft Kinect	Color Sensor, IR Depth Sensors, IR Emitter	ASD: 33 (5–10 y)	ADOS-2-Tasks (Cartoon Task, Demonstration Task)	ASD: Females Present Better Non-Verbal Skills (Gestures), Although Communication Skills Were Lower
Body Movement	Anzulewicz et al., 2016 [48]	iPad	Touch Sensor; Accelerometer	ASD: 37 (4 y); TD: 45 (5 y)	Playing 2 Serious Games (Sharing, Coloring)	Differences in Pressure Going into the Device as Well as Differences in Gesture Kinematics and Form
Multimodal	Samad et al., 2017 [49]	Sony EVI-D70, Mirametrix S2	2D Imaging, Eye Tracker	ASD: 8 (13 y); TD: 8 (16 y)	Emotion Recognition and Manipulation Task of 3D Faces	ASD: Have Uncontrolled Manifestation of FAU 12, Spontaneous Facial Responses Are Not Synchronized with Their Visual Engagement with Facial Expressions, Poor

Table 1. *Cont.*

Domain of Behavior	Reference	Solution Name	Sensor/Parameters	Sample (Mean Age)	Setting and Stimuli	Results
Multimodal	Jaiswal et al., 2017 [50]	Microsoft Kinect V2	Color Sensor, IR Depth Sensors, IR Emitter, Microphone	ASD: 22; ADHD: 4, ASD + ADHD: 11, TD: 18	Read and Listen to A Set Of 12 Short Stories (From 'Strange Stories' Task), Accompanied By 2–3 Questions	Classifier Sensitivity for Control vs. Clinical Condition: 96.4%; ASD+ADHD vs. ASD: 93,9%
Social Behavior	Westeyn et al., 2012 [51]	BlueSense, Video Cameras	Touch Sensors, Motion Sensor, Microphone, 2D Imaging	High-Risk: 1, TD: 10 Adults; 12 Children	Playing with Smart Toys with Scripted Play Prompts	Retrieval Score of 59% for a Single Child Using Models Constructed from Adult Play Data
Social Behavior	Anzalone et al., 2014 [52]	Microsoft Kinect; Nao	Color Sensor, IR Depth Sensors, IR Emitter, Microphone	ASD: 16v (9 y); TD: 16 (8 y)	Nao Prompts JA by Gazing/By Gazing and Pointing/By Gazing, Pointing and Vocalizing at Pictures	ASD: Trunk Position Showed Less Stability in 4D Compared to TD Controls, Gazing Exploration Showed Less Accuracy
Social Behavior	Campbell et al., 2019 [53]	iPad	2D Imaging Sensor (Front Camera)	ASD: 22 (2 y); TD: 82 (2 y)	Reaction to Name-Calling While Watching Videos	ASD: Classifying by Atypical Orientation: Sensitivity: 96%, Specificity: 38%
Social Behavior	Petric et al., 2014 [54]	Nao Robot	Cameras, Microphones, Ultrasound Range Sensors, Tactile Sensors, Force Sensitive Resistors, Accelerometers	ASD: 3 (5–8 y), TD: 1 (6 y)	ADOS-Tasks (Name Calling, JA, Play Request, Imitation)	Descriptively High Correspondence of Human Rater and Algorithm

Note: Age in sample description typically refers to mean age of participants per group, with the exception of a few studies which report either age ranges, individual age of single cases, or did not specify exact age.

Overall, facial data suggest that facial movements, either in response to emotional stimuli or as an imitation of seen facial expressions, comprise relevant features that may be markers for autism. Recordings of spontaneous emotional responses could be of particular benefit when assessing small children or nonverbal people, although the importance of age-appropriate task instructions should be considered. Across several studies, eye gaze data reveal different scanning patterns for people with ASD, particularly when viewing faces. However, we believe that further systematic research into fixation patterns and scan paths for more complex natural scenes could enrich our current understanding with additional insights. Similarly, the observation of reduced eye contact/mutual gaze in ASD, reviewed in Jaiswal et al. [50], points at a technologically challenging but theoretically relevant field of future investigations using sensor technology.

3.2. Voices

A scientific metaphor that has become somewhat popular describes the voice as an "auditory face" [55], emphasizing the fact that voices, just like faces, provide rich information about a person's emotions, but also about identity, gender, socioeconomic or regional background, or age (for review, see [56]). In the context of autism, and like for faces, researchers focused on deficits in vocal emotional communication and pointed out that these deficits tend to affect multiple modalities including voices, faces, and body movement [57]. At the same time, deficits in vocal communication may also affect other aspects such as vocal identity perception [58,59], and vocal expression of autistic people in communication could be affected beyond emotional expressions.

One path for detecting auditory markers for autism in voices is linked to the symptom of repetitive behaviors or 'vocal stereotypies'. One study conducted a subspace analysis from acoustic data of autistic children and reported a good detection of vocalized non-word sounds [42]. Subsequently, Min and Fetzner [43] also used subspace learning for vocal stereotypies and trained dictionaries to differentiate between vocal stimming (a nonverbal vocalization often observed in autism) and other noises. Using a small sample of 4 children with ASD who lacked verbal communication (age not reported), the authors could e.g., detect vocal stimming and predict perceived frustration reasonably well, although the study was regarded as preliminary. For verbal children, other potential vocal markers could include prosody. Marchi et al. [44] created an evaluation database in three languages with ASD and TD children's emotionally toned voices which were analyzed using the COMPARE feature set. Groups comprised between 7 and 11 individuals per language, and children were between 5 and 11 years old. Comparisons between groups and emotions showed relatively poor classification performance ASD children's' voices, particularly for 'Anger' for both the Swedish and English dataset, and for 'Afraid' for Hebrew-speaking children with ASD. Although using a pre-existing dataset, a study on automatic voice perception showed that an algorithm successfully classified up to 61.1% of voice samples of children actually diagnosed with and without autism [60]. Ringeval et al. [45] assessed verbal prosody in ASD children, children with pervasive developmental disorder (PDD), children with specific language impairment (SLI), and TD children (aged around 9–10 years, with 10–13 children per clinical group). Specifically, these authors recorded performance during an imitation task for sentences with different intonations (e.g., rising, falling). The rising intonation condition was reported to best discriminate between groups, and the authors interpreted their findings as indicating a pronounced pragmatic impairment to prosodic intonation in ASD.

When compared with facial data, the use of sensors for automatic analysis of markers for autism in voices clearly is still at an early and preliminary stage. The studies reviewed above that use original data typically rely on small data sets from few people. Although initial findings suggest that the systematic search for vocal markers of autistic traits could be highly promising, more research is clearly warranted at this stage. When considering that similar impairments in facial and vocal emotional processing could be present in ASD [57], multisensory assessment of emotional behavior in future studies could be particularly promising.

3.3. Body Movement

The identification of ASD-associated movement patterns has been the subject of intense research, primarily focusing data from accelerometer sensors [42,61]. As the display of stereotypical body movement patterns is a core symptom of ASD, it has been a target feature in computer vision. Gonçalves et al. [46] used a simple gesture recognition algorithm on 3D visual data automatically detecting hand-flapping movements. Validation of these data with 2D video data suggested that automatic 'hand flapping' detection delivers valuable information for monitoring autistic children, as in the case of special needs schools. Jazouli et al. [47] also used the same sensor but based their analysis on a $P Point-Cloud Recogniser to automatically detect body rocking, hand flapping, fingers flapping, hand on the face and hands behind back with an overall mean accuracy of 94%.

Rynkiewicz et al. [47] studied the role of non-verbal communication in the setting of an ADOS-2 assessment for children aged 5–10. They used a 3D sensor for automatic gesture analysis of the upper body, while the boys (N = 17) and girls (N = 16) with high-functioning ASD performed two assessment-related tasks. Females had a higher gesture index than boys although they had less verbal communication skills and a more impaired ability to read mental states from faces. The authors suggested that the vivid use of gestures in girls (generally less common in our current understanding of the autistic phenotype) may contribute to possible under-diagnosis of autism in females. This might further lead to the more general use of automatic gesture analysis that is currently performed by professional human raters in the course of these assessments. A study by Anzulewicz et al. [48] used the touch and inertial sensors of a tablet to assess the specific movement patterns of children with autism (N = 37) and an age- and gender-matched TD group (N = 45) while playing simple games. Apart from interesting findings including the use of greater force, larger and more distal gestures and faster screen taps in the ASD group, they also tested different machine learning algorithms to classify between groups, with promising results.

In summary, the use of 3D data has been preferred by researchers investigating body movements. While it seems to be possible to detect certain stereotypical movements, there is still a lack of large-scale studies. However, even coarser movement indices (including, for instance, gesture indices or general movement patterns) may also provide meaningful information for the identification and differentiation of autistic behavioral markers.

3.4. Multimodal Information

Multimodal approaches to automatic analysis of behavior in ASD are still infrequent. Samad et al. [49] used facial motion, eye-gaze and hand movement analysis of adolescents with (N = 8) and without ASD (N = 8) on tasks involving 3D expressive faces. They found a reduced synchronization of facial expression and visual engagement with the stimuli in the ASD group, as well as poorer correlations in eye-gaze and hand movements. Jaiswal et al. [50] designed an automatic approach based on 3D data of facial features and head movements that detected ASD (vs. TD and vs. comorbid ASD and ADHD, with N between 11 and 22 per group) in adults with high accuracy in classification performance. Their recorded data consisted of participants reading, listening to and answering to questions taken from the 'Strange Stories' task [62] often used in ASD-assessments.

Overall, the integration of multimodal information, unfortunately, is not yet common in this field, despite the fact that there are reasons to expect that multimodal assessments will provide ample additional information (for instance, on synchronization or complementarity of signals from different channels). Thus, it can be expected that future use of multimodal assessments has the potential to substantially improve both the identification of markers for autistic traits and classification results.

3.5. Cognition and Social Behavior

A substantial body of research has suggested that autism is characterized by changes in cognitive information processing. For instance, autistic people may have difficulties in cognitively simulating the observed actions of communication partners – a process that has been related to the so-called human mirror neuron system, and that has been inferred via neurophysiological recordings [63]. Via experiments with social behavioral assessments, autism has also been related to a deficit in forming a theory of mind about other people [64]. Researchers in this field have long focused on inferences about "core" cognitive deficits from observational data, even though results can be very inconsistent across situations or studies [65,66]. We also welcome an increasing awareness of the danger that researchers wrongly infer putative core deficits by misinterpreting observational data [36]. Of course, inferences about core cognitive processes most typically need to be made from observational information. It is this context in which we foresee that sensor data, beyond their immediate value for an individual study, might eventually contribute to resolving theoretical controversies. One of these is the degree to which we should frame cognitive changes in autism in terms of general "core" deficits, or rather in terms of domain-specific deficits that emerge in a concrete perceptual and interactional context.

The automatic analysis of playing behavior could be a promising screening tool to identify stages of development, developmental delays and specific forms of deficits. In a preliminary study with only a single completely analyzed data set of a TD child [51], play data from smart toys with embedded acceleration sensors, as well as simultaneous video and audio recordings, were automatically characterized as certain forms of play behavior (exploratory, relational, functional), roughly indicating a certain stage of development. While this approach seems interesting, current limitations include the verbal prompting nature of their setting, which only allows children with good verbal skills to be assessed.

Joint attention is a basic communicational skill of sharing attention between communicational agents towards an object and has long been considered to be a precursor of a theory of mind [67] which may be reduced in people with ASD. In a study investigating a robot-assisted joint attention task [52] in children with and without ASD (N = 16 each), they captured participants' orientation using 3D sensors over a timespan. In addition to findings of significantly less joint attention in the ASD group's interaction with the robot, the ASD group also showed decreased micro-stability in the trunk area that was tentatively interpreted as a consequence of increased cognitive cost. In another study [53], attentional and orienting responses to name calls were studied in toddlers, aged 16–31 months, with ASD (N = 22) and TD (N = 82). Achieving a high intra-class-correlation with human raters, automated coding offered a reliable method to detect the differential social behavior of toddlers with ASD, who responded to name calls less often and with longer latency. Petric et al. [54] designed and further developed a robot-assisted subset of the ADOS assessment. They included certain social tasks (name-calling, joint attention) to assess information on eye gaze, gestures, and vocal utterances. The agreement of the robot's classifier and clinicians' judgments were evaluated as promising, but the results should be regarded with caution given the very small sample of children (ASD N = 3, TD N = 1).

Overall, while researchers have begun to assess joint attention in ASD, sensor-based assessment of other functional domains of cognition and social behavior still awaits systematic research. In particular, research on a few putative "core" areas of social cognition, including on observational tasks that probe theory of mind in ASD, is currently lacking.

4. Supporting Interventions

While accurate early diagnosis and an assessment of specific impairments are crucial, they also are prerequisites that inform environmental adjustment, intervention, and training approaches which ultimately can be valuable for the individual person with ASD following diagnosis. Technically-assisted training often has the benefit of being readily available, problem-specific, cost-effective, and widely accepted by affected children. Additionally, smart responses of training systems that give reliable, immediate feedback and appraisal can be highly beneficial for fast learning results. Note that although we identified many publications on interventions aiming at autism as a target condition, many of these reported conceptual or technological contributions and few of them presented original data from people with ASD that qualified them for inclusion in this review (cf. Section 2). The original articles presented in this review, regarding sensor-based supporting interventions for ASD, are listed in Table 2.

4.1. Emotion Expression and Recognition

Emotion expression, especially from the face, is considered highly relevant in autism research. A game called FaceMaze [68], in combination with automatic online expression recognition of the user was specifically developed to improve facial expression production. In this game, children played a maze game while posing 'happy' or 'angry' facial expressions to overcome obstacles in the game. In a pre-post-rating with naive human raters, quality ratings for both trained expressions (happy and angry expression) in ASD children ($N = 17$, aged 6–18) increased in the post-test while ratings for an untrained emotion (surprise) did not change. Another smaller study created a robot-child-interaction and tested it with three children with ASD that were to imitate a robot's facial expression [69]. The robot correctly recognized the children's' imitated expressions through an embedded camera in half of the cases. In these cases, it was able to give immediate positive feedback. It also correctly did not respond in about one-third of the trials where there was no imitation by the participants. Piana, et al. [70] designed a serious game with online 3D data acquisition that trained children with ASD ($N = 10$) in several sessions to recognize and express emotional body-movements. Both emotion expression (mean accuracy gain = 21%) and recognition (mean accuracy gain = 28%) increased throughout the sessions. Interestingly, performance in the T.E.C. (Test of Emotion Comprehension, assessing emotional understanding more generally), increased as well (mean gain = 14%).

4.2. Social Skills

Robins et al. [71] created an interactive robot (KASPAR) with force sensitive resistor sensors. They later planned to use KASPAR for robot-assisted play to teach touch, joint attention and body awareness [72], although conclusive data on experimental results from interactions between individuals with autism and KASPAR may still be in the pipeline. Learning social skills also presupposes attention to potential social cues and social engagement. Costa et al. [73] reported preliminary research on using the LEGO mindstorm robots with adolescents with ASD ($N = 2$), in an attempt to increase openness and induce communication since the participants actively had to provide verbal commands or instructional acts. They reported that the two participants behaved differently, one being indifferent, and one being increasingly interested in the interaction. Wong and Zhong [74] used a robotic platform (polar bear) to teach children with ASD ($N = 8$) social skills. They found, that within five sessions an increase in turn-taking, joint attention and eye contact was observable, resulting in overall 90% achievement of individually defined goals.

Greeting is a basic element of communication. In a greeting game with 3D body movement as well as voice acquisition [75], a participant would play an avatar with his own face, learning to greet (vocalization, eye contact and waving) and get immediate appraisal upon success. A single case study suggested that this intervention can be effective at teaching greeting behavior. As a more complex pilot intervention, Mower et al. [76] created the embodied conversational agent 'Rachel' that acted as an emotional coach in guiding children through emotional problem-solving tasks. Of their two participants with ASD, audio and video data were acquired for post hoc analysis and tentatively suggested that the interface could elicit interactive behavior.

Overall, there is some evidence that sensor technology can improve social skills in people with autism, and the use of sophisticated robotic platforms can be regarded as particularly promising. As limitations, it needs to be noted that all studies that met the criteria to be included in this review only tested very few participants, and that there typically were no real-world follow-up tests reported. As a result, a systematic quantitative assessment of treatment effects and effect sizes, as well as a comparison with more conventional interventions (e.g., social competence training) will require substantial cross-disciplinary research. Moreover, most studies were driven by a combination of theoretically interesting and technically advanced approaches, and from the perspective of typical development. Designing more user-centered and irritation-free approaches could promote both usability and motivation for people with autism to engage in technology-driven interventions.

Table 2. Original Articles on Sensor-Based Supporting Interventions in ASD.

Domain of Behavior	Reference	Solution Name	Sensor/Parameters	Sample (Mean Age)	Setting and Stimuli	Results
Emotion	Gordon et al., 2014 [68]	Webcam	2D Imaging Sensor	ASD: 30 (11 y), TD: 23 (11 y)	Playing Computer Game (FaceMaze)	ASD: Increase in Happy and Angry Expression Performance
Emotion	Leo et al., 2015 [69]	Camera, Robokind™ R25 Robot	2D Imaging Sensor	ASD: 3	Imitate Expression from Robot (Happiness, Sadness, Anger, And Fear)	31/60 Interactions Recognized; 19/60 No Imitation
Emotion	Piana et al., 2019 [70]	Microsoft Kinect V2	Color Sensor, IR depth Sensors, IR Emitter	ASD: 10 (10 y)	10 Sessions Body Emotion Expression and Recognition Task	Increased Accuracy in Expression and Recognition After Training Sessions In The Trained Group, Transfer Effect On Facial Expression Recognition
Social Skills	Robins et al., 2010 [71]	KASPAR, Video Cameras	Touch Sensor, 2D Imaging	ASD: 3	Unconstrained Interaction with The Robot	Interaction Evaluation Through Sensor Activation, Differential Interaction and Maintenance of Interaction
Social Skills	Costa et al., 2009 [73]	LEGO Mindstorms TM, Video Cameras	Touch Sensor, Sound Sensor	ASD: 2 (17, 19 y)	4- 5 Sessions of Feedbacked Interaction with Robot	Interaction Evaluation Through Sensor Activation, Differential Interaction and Maintenance of Interaction
Social Skills	Wong and Zhong, 2016 [74]	CuDDler Robot, Video Camera	Microphone, Contact Microphone, Tactile and Posture Sensors	ASD: 8 (5 y)	5 Sessions of ABA (Didactic Teaching Followed by Role Modeling by Either A Robot (RT) Or Human (CT)	Robot Training Significantly Facilitated Verbal and Gestural Communicative Skills, Increased Eye Contact Duration
Social Skills	Uzuegbunam et al., 2015 [75]	Microsoft Kinect	Color Sensor, IR depth Sensors, IR Emitter, Microphone	ASD: 3 (7–12 y)	Greeting Game with Participants Face, Reacting to Participant, Getting Appraisal	All 3 Showed Increased Social Greeting Behavior Throughout and After Intervention
Social Skills	Mower et al., 2011 [76]	HDR-SR12 High Definition Handycam Camcorders	Microphones, 2D Video Sensors	ASD: 2 (6, 12 y)	4 Sessions with Embodied Conversational Agent RACHEL Going Through Emotional Scenarios	Tool for Eliciting Interactive Behavior

Note: Age in sample description, where specified in a study, either refer to mean age of participants per group, to age ranges, or to individual ages in small samples of single cases.

5. Monitoring

Monitoring a child's emotional state or behavioral changes can be crucial for the outcomes of a learning environment. As discussed above, emotional expressions from people with ASD may differ in several respects from those of TD people. As a result, there is a higher risk that caregivers or interaction partners overlook or misinterpret the emotional state of people with autism.

Del Coco et al. [77] created a humanoid and tablet-assisted therapy setup that was trained to monitor behavioral change in children with ASD via a video processing module. Besides creating a visual output of behavioral cues, they computed a score for affective engagement (happiness related features) from visual cues such as facial AUs, head pose and gaze that provides the practitioner with a behavioral trend along with the treatment. Dawood et al. [78] used facial expressions, eye gaze and head movements to identify five discrete emotional states of young adults with ASD in learning situations (e.g., anxiety, engagement, uncertainty). Their resulting model yielded a high validity in identifying emotional states of participants with high-functional ASD. At the same time, a lower validity was found for TD participants, suggesting differential facial expressions of certain emotional states in ASD. For monitoring social interactions, Winoto et al. [79] created a machine-learning-based social interaction coding of 3D data around a target user. Kolakowska et al. [80] approached automatic progress recognition with different tablet games. Over a 6-month time window, they were able to identify movement patterns in their study group of children with ASD ($N = 40$), that not only related to development in fine motor skills but also other fields like communication and socio-emotional skills. Overall, these initial studies suggest that sensor-based monitoring of emotional and behavioral changes may support caregivers in optimizing learning outcomes.

6. Discussion

The studies discussed above demonstrate substantial research activities towards using sensor based technology in the context of autism overall, with attention to multiple aspects including diagnosis/classification and intervention. At the same time, it appears that much current research is largely driven by fast technological progress in terms of innovative engineering and data analysis methods. It remains a significant challenge to reconcile these developments with the specific testing of psychological or neuroscientific theories regarding functional changes and potentials in autism. Similarly, systematic studies with theory-driven protocols and larger samples are required to evaluate in more detail both the diagnostic and interventional potential of sensor-based technology. For the ultimate goal of evaluating its practical relevance, quantitative assessments of diagnostic sensitivities and specificities, or of treatment effect sizes, will be as important as will be comparative studies with more traditional approaches to diagnosis and intervention.

One of many examples of how sensor technology has the potential to go beyond application, and to contribute to current neurocognitive theories of communication is related to the theory of a tight link between perception and motor action in communication. This link now has been firmly established in speech communication [81], but there are reasons to believe that perception and action are also closely linked in nonverbal emotional and social communication. For instance, listening to laughter normally activates premotor and primary motor cortex [82], and may involuntarily elicit orofacial responses in a perceiver in parallel. In turn, there also is initial evidence that voluntary motor imitation can actually facilitate facial emotion recognition, particularly in people with high levels of autistic traits [83] who are thought to engage less in spontaneous imitation. A consistent theoretical account for such findings is that imitation, and covert sensorimotor simulation of others' actions, may be based in part on the so-called mirror neuron system. This system consists of neurons that fire not only when a person performs an action, but also when s/he observes the same action in another individual. However, the human mirror neuron system is thought be specifically impaired in autism [63], and a subset of promising intervention approaches for autism using neurofeedback [84] are based on this theory. However, it should be noted that the underlying theory remains disputed [85].

Findings such as those by Lewis and Dunn [83] may be taken to suggest that interventions that promote facial imitation of emotions in autistic people should also support their abilities for emotion recognition and bidirectional communication. However, it is technically challenging to objectively quantify the degree of facial imitation, and in fact, a limitation of the study by Lewis and Dunn was that these authors failed to quantify imitation beyond simply asking participants to rate their own degree of imitation. Other studies measured facial imitation more objectively but typically did so by measuring the facial muscle response for selected target action units with electromyography (EMG, e.g., [86,87]). Although this can provide an objective measure of facial imitation, the fact that the method uses recording electrodes attached to facial muscles has many drawbacks. For instance, one concern is that this technology could draw the participants' attention to their own facial behavior, which in turn could influence facial action. We believe that contact- and irritation-free assessment of imitation as provided by modern sensor and real-time facial emotion recognition technologies is the method of choice to promote better understanding not only the role of spontaneous facial imitation in emotion recognition in normal communication, but also to determine the potential role of impaired links between perception and action for communication difficulties in people with autism.

While the research discussed in this review appears to underline a great potential for the use of sensor technology, in particular in the context of autism, it is equally clear that many current tests of assessments or interventions will benefit in validity from a clear conceptual framework of autism spectrum disorders in the developmental perspective. At present, and honoring findings of large individual variability within both people with ASD and TD, results that were obtained with only a few participants (not always well described, and sometimes obtained in the absence of a TD group) or with experimental groups that are not comparable with respect to their basic characteristics (e.g., age, gender, IQ) need to be interpreted with caution in order to avoid biased or overgeneralized interpretation of individual study findings.

Other potential obstacles relate to sophisticated developments (and costs) of some of the systems used, which make them unlikely to become available in greater quantities. Moreover, even readily available systems may get discontinued or run out of support, such as in the case of Microsoft's Kinect in 2017, and this provides great challenges for large-sample research in autism which often takes years to complete. Research aiming at training and modeling behavior of people with ASD also will increasingly need to consider usability, to the extent that the relevant systems are to be used by individuals with ASD, their parents, caregivers, and therapists.

Finally, compared to the typical approach of developing sensor-based technology with neurotypical individuals before applying it to people with autism, a more promising strategy may be one in which technology design originates from a user-centered perspective, with autistic people as users actively involved in the process. Such an approach has been forcefully advocated by Rajendran [88], who argues that this may both enhance our understanding of autism and promote better inclusivity of people with autism in an increasingly digital world. At the same time, such technologies ultimately can be useful for people without autism as well. This is because autism is seen as a unique window into social communication and social learning more generally.

7. Conclusions

Technical advancements and the ongoing developments in sensor technology and data science promise to unlock huge potentials for the diagnosis and understanding of autism, and for supporting affected people with training or intervention programs that can be tailored to their specific needs. At the same time, living up to these potentials calls for a concerted and interdisciplinary effort in which computer scientists, engineers, psychologists, and neuroscientists jointly collaborate in large-scale research projects that can uncover, in a quantitative manner, the efficiency of these approaches. In our view, this will be the route not only for establishing routine contributions to evidence-based diagnosis and interventions in autism [89] but also to ensure that more people with autism can genuinely benefit from tailor-made technology.

Funding: Previous research by SRS on related topics has been funded by a grant from the Bundesministerium für Bildung und Forschung (BMBF), in a project on an irritation-free and emotion-sensitive training system (IRESTRA; Grant Reference: 16SV7210), and another BMBF project on the psychological measurement of anxiety in human-robot interaction (3DimIR, Grant Reference 03ZZ0459B).

Acknowledgments: AEK and SRS would like to thank the Herbert Feuchte Stiftungsverbund for supporting the research in the Social Potentials in Autism Research Unit (www.autismus.uni-jena.de). SRS would like to thank the Swiss Center for Affective Sciences at the University of Geneva, Switzerland, for hosting a sabbatical leave in summer 2019 during which this paper was written.

Conflicts of Interest: The authors declare no conflict of interest. In particular, funding bodies had no role in the planning, collection, or interpretation of evidence reviewed in this paper; in the writing of the manuscript, or in the decision to publish the results.

References

1. Matson, J.L.; Kozlowski, A.M. The increasing prevalence of autism spectrum disorders. *Res. Autism Spectr. Disord.* **2011**, *5*, 418–425. [CrossRef]
2. Weintraub, K. The prevalence puzzle: Autism counts. *Nature* **2011**, *479*, 22–24. [CrossRef] [PubMed]
3. American Psychiatric Association. *Diagnostic and Statistical Manual of Mental Disorders (DSM-5®)*; American Psychiatric Pub.: Washington, DC, USA, 2013.
4. Seltzer, M.M.; Shattuck, P.; Abbeduto, L.; Greenberg, J.S. Trajectory of development in adolescents and adults with autism. *Ment. Retard. Dev. Disabil. Res. Rev.* **2004**, *10*, 234–247. [CrossRef] [PubMed]
5. Dawson, G.; Toth, K.; Abbott, R.; Osterling, J.; Munson, J.; Estes, A.; Liaw, J. Early Social Attention Impairments in Autism: Social Orienting, Joint Attention, and Attention to Distress. *Dev. Psychol.* **2004**, *40*, 271–283. [CrossRef] [PubMed]
6. Brewer, R.; Biotti, F.; Catmur, C.; Press, C.; Happé, F.; Cook, R.; Bird, G. Can neurotypical individuals read autistic facial expressions? Atypical production of emotional facial expressions in autism spectrum disorders. *Autism Res.* **2016**, *9*, 262–271. [CrossRef] [PubMed]
7. Green, H.; Tobin, Y. Prosodic analysis is difficult … but worth it: A study in high functioning autism. *Int. J. Speech Lang. Pathol.* **2009**, *11*, 308–315. [CrossRef]
8. Orsmond, G.I.; Krauss, M.W.; Seltzer, M.M. Peer relationships and social and recreational activities among adolescents and adults with autism. *J. Autism Dev. Disord.* **2004**, *34*, 245–256. [CrossRef]
9. Leekam, S.R.; Prior, M.R.; Uljarevic, M. Restricted and repetitive behaviors in autism spectrum disorders: A review of research in the last decade. *Psychol. Bull.* **2011**, *137*, 562–593. [CrossRef]
10. Szatmari, P.; Georgiades, S.; Bryson, S.; Zwaigenbaum, L.; Roberts, W.; Mahoney, W.; Goldberg, J.; Tuff, L. Investigating the structure of the restricted, repetitive behaviours and interests domain of autism. *J. Child Psychol. Psychiatry* **2006**, *47*, 582–590. [CrossRef]
11. Leekam, S.R.; Nieto, C.; Libby, S.J.; Wing, L.; Gould, J. Describing the sensory abnormalities of children and adults with autism. *J. Autism Dev. Disord.* **2007**, *37*, 894–910. [CrossRef]
12. Falkmer, T.; Anderson, K.; Falkmer, M.; Horlin, C. Diagnostic procedures in autism spectrum disorders: A systematic literature review. *Eur. Child Adolesc. Psychiatry* **2013**, *22*, 329–340. [CrossRef]
13. Green, R.M.; Travers, A.M.; Howe, Y.; McDougle, C.J. Women and Autism Spectrum Disorder: Diagnosis and Implications for Treatment of Adolescents and Adults. *Curr. Psychiatry Rep.* **2019**, *21*, 22. [CrossRef] [PubMed]
14. Oono, I.P.; Honey, E.J.; McConachie, H. Parent-mediated early intervention for young children with autism spectrum disorders (ASD). *Evid.-Based Child Heal. A Cochrane Rev. J.* **2013**, *8*, 2380–2479. [CrossRef]
15. Baron-Cohen, S.; Richler, J.; Bisarya, D.; Gurunathan, N.; Wheelwright, S. The systemizing quotient: An investigation of adults with Asperger syndrome or high-functioning autism, and normal sex differences. *Philos. Trans. R. Soc. B Boil. Sci.* **2003**, *358*, 361–374. [CrossRef] [PubMed]
16. Roelfsema, M.T.; Hoekstra, R.A.; Allison, C.; Wheelwright, S.; Brayne, C.; Matthews, F.E.; Baron-Cohen, S. Are autism spectrum conditions more prevalent in an information-technology region? A school-based study of three regions in the Netherlands. *J. Autism Dev. Disord.* **2012**, *42*, 734–739. [CrossRef]
17. Dautenhahn, K. Socially intelligent robots: Dimensions of human–robot interaction. *Philos. Trans. R. Soc. B Boil. Sci.* **2007**, *362*, 679–704. [CrossRef]

18. Pennisi, P.; Tonacci, A.; Tartarisco, G.; Billeci, L.; Ruta, L.; Gangemi, S.; Pioggia, G. Autism and social robotics: A systematic review. *Autism Res.* **2016**, *9*, 165–183. [CrossRef]

19. Billeci, L.; Sicca, F.; Maharatna, K.; Apicella, F.; Narzisi, A.; Campatelli, G.; Calderoni, S.; Pioggia, G.; Muratori, F. On the Application of Quantitative EEG for Characterizing Autistic Brain: A Systematic Review. *Front. Hum. Neurosci.* **2013**, *7*, 442. [CrossRef]

20. Marco, E.J.; Hinkley, L.B.N.; Hill, S.S.; Nagarajan, S.S.; Hinkley, L.B.N. Sensory processing in autism: A review of neurophysiologic findings. *Pediatr. Res.* **2011**, *69*, 48R–54R. [CrossRef]

21. Taj-Eldin, M.; Ryan, C.; O'Flynn, B.; Galvin, P. A Review of Wearable Solutions for Physiological and Emotional Monitoring for Use by People with Autism Spectrum Disorder and Their Caregivers. *Sensors* **2018**, *18*, 4271. [CrossRef]

22. Parsons, S.; Mitchell, P. The potential of virtual reality in social skills training for people with autistic spectrum disorders. *J. Intellect. Disabil. Res.* **2002**, *46*, 430–443. [CrossRef] [PubMed]

23. Bellani, M.; Fornasari, L.; Chittaro, L.; Brambilla, P. Virtual reality in autism: State of the art. *Epidemiol. Psychiatr. Sci.* **2011**, *20*, 235–238. [CrossRef]

24. Pan, X.; Hamilton, A.F.D.C. Why and how to use virtual reality to study human social interaction: The challenges of exploring a new research landscape. *Br. J. Psychol.* **2018**, *109*, 395–417. [CrossRef] [PubMed]

25. Bölte, S.; Hubl, D.; Feineis-Matthews, S.; Prvulovic, D.; Dierks, T.; Poustka, F. Facial affect recognition training in autism: Can we animate the fusiform gyrus? *Behav. Neurosci.* **2006**, *120*, 211–216. [CrossRef] [PubMed]

26. Kasari, C.; Sigman, M.; Mundy, P.; Yirmiya, N. Affective sharing in the context of joint attention interactions of normal, autistic, and mentally retarded children. *J. Autism Dev. Disord.* **1990**, *20*, 87–100. [CrossRef] [PubMed]

27. Weigelt, S.; Koldewyn, K.; Kanwisher, N. Face identity recognition in autism spectrum disorders: A review of behavioral studies. *Neurosci. Biobehav. Rev.* **2012**, *36*, 1060–1084. [CrossRef]

28. Sheppard, E.; Pillai, D.; Wong, G.T.-L.; Ropar, D.; Mitchell, P. How Easy is it to Read the Minds of People with Autism Spectrum Disorder? *J. Autism Dev. Disord.* **2016**, *46*, 1247–1254. [CrossRef]

29. Samad, M.D.; Bobzien, J.L.; Harrington, J.W.; Iftekharuddin, K.M. [INVITED] Non-intrusive optical imaging of face to probe physiological traits in Autism Spectrum Disorder. *Opt. Laser Technol.* **2016**, *77*, 221–228. [CrossRef]

30. Del Coco, M.; Leo, M.; Carcagni, P.; Spagnolo, P.; Luigi Mazzeo, P.; Bernava, M.; Marino, F.; Pioggia, G.; Distante, C. A Computer Vision based Approach for Understanding Emotional Involvements in Children with Autism Spectrum Disorders. In Proceedings of the IEEE International Conference on Computer Vision, Venice, Italy, 22–29 October 2017; pp. 1401–1407.

31. Leo, M.; Carcagnì, P.; Distante, C.; Spagnolo, P.; Mazzeo, P.; Rosato, A.; Petrocchi, S.; Pellegrino, C.; Levante, A.; De Lumè, F. Computational Assessment of Facial Expression Production in ASD Children. *Sensors* **2018**, *18*, 3993. [CrossRef]

32. Phillips, M.L.; Young, A.W.; Senior, C.; Brammer, M.; Andrew, C.; Calder, A.J.; Bullmore, E.T.; Perrett, D.I.; Rowland, D.; Williams, S.C.R.; et al. A specific neural substrate for perceiving facial expressions of disgust. *Nature* **1997**, *389*, 495–498. [CrossRef]

33. Sander, D.; Grandjean, D.; Scherer, K.R. An Appraisal-Driven Componential Approach to the Emotional Brain. *Emot. Rev.* **2018**, *10*, 219–231. [CrossRef]

34. Samad, M.D.; Diawara, N.; Bobzien, J.L.; Taylor, C.M.; Harrington, J.W.; Iftekharuddin, K.M. A pilot study to identify autism related traits in spontaneous facial actions using computer vision. *Res. Autism Spectr. Disord.* **2019**, *65*, 14–24. [CrossRef]

35. Egger, H.L.; Dawson, G.; Hashemi, J.; Carpenter, K.L.; Sapiro, G. 23.1 Autism and Beyond: Lessons from an Iphone Study of Young Children. *J. Am. Acad. Child Adolesc. Psychiatry* **2018**, *57*, S33–S34. [CrossRef]

36. Jaswal, V.K.; Akhtar, N. Being versus appearing socially uninterested: Challenging assumptions about social motivation in autism. *Behav. Brain Sci.* **2019**, *42*, e82. [CrossRef]

37. Tanaka, J.W.; Sung, A. The "Eye Avoidance" Hypothesis of Autism Face Processing. *J. Autism Dev. Disord.* **2016**, *46*, 1538–1552. [CrossRef]

38. Chawarska, K.; Shic, F. Looking but not seeing: Atypical visual scanning and recognition of faces in 2 and 4-year-old children with autism spectrum disorder. *J. Autism Dev. Disord.* **2009**, *39*, 1663–1672. [CrossRef]

39. Liu, W.; Li, M.; Yi, L. Identifying children with autism spectrum disorder based on their face processing abnormality: A machine learning framework. *Autism Res.* **2016**, *9*, 888–898. [CrossRef]

40. Król, M.; Król, M.E. A Novel Eye Movement Data Transformation Technique that Preserves Temporal Information: A Demonstration in a Face Processing Task. *Sensors* **2019**, *19*, 2377. [CrossRef]

41. Wang, S.; Jiang, M.; Duchesne, X.M.; Laugeson, E.A.; Kennedy, D.P.; Adolphs, R.; Zhao, Q. Atypical Visual Saliency in Autism Spectrum Disorder Quantified through Model-Based Eye Tracking. *Neuron* **2015**, *88*, 604–616. [CrossRef]

42. Min, C.-H.; Tewfik, A.H. Novel pattern detection in children with autism spectrum disorder using iterative subspace identification. In Proceedings of the 2010 IEEE International Conference on Acoustics, Speech and Signal Processing, Dallas, TX, USA, 14–19 March 2010; pp. 2266–2269.

43. Min, C.-H.; Fetzner, J. Vocal Stereotypy Detection: An Initial Step to Understanding Emotions of Children with Autism Spectrum Disorder. In Proceedings of the 2018 40th Annual International Conference of the IEEE Engineering in Medicine and Biology Society (EMBC), Honolulu, HI, USA, 18–21 July 2018; pp. 3306–3309.

44. Marchi, E.; Schuller, B.; Baron-Cohen, S.; Lassalle, A.; O'Reilly, H.; Pigat, D.; Golan, O.; Friedenson, S.; Tal, S.; Bolte, S. Voice Emotion Games: Language and Emotion in the Voice of Children with Autism Spectrum Conditio. In Proceedings of the 3rd International Workshop on Intelligent Digital Games for Empowerment and Inclusion (IDGEI 2015) as part of the 20th ACM International Conference on Intelligent User Interfaces, IUI 2015, Atlanta, GA, USA, 29 March–1 April 2015; p. 9.

45. Ringeval, F.; DeMouy, J.; Szaszak, G.; Chetouani, M.; Robel, L.; Xavier, J.; Cohen, D.; Plaza, M. Automatic Intonation Recognition for the Prosodic Assessment of Language-Impaired Children. *IEEE Trans. Audio Speech Lang. Process.* **2010**, *19*, 1328–1342. [CrossRef]

46. Gonçalves, N.; Costa, S.; Rodrigues, J.; Soares, F. Detection of stereotyped hand flapping movements in Autistic children using the Kinect sensor: A case study. In Proceedings of the 2014 IEEE international conference on autonomous robot systems and competitions (ICARSC), Espinho, Portugal, 14–15 May 2014; pp. 212–216.

47. Jazouli, M.; Majda, A.; Merad, D.; Aalouane, R.; Zarghili, A. Automatic detection of stereotyped movements in autistic children using the Kinect sensor. *Int. J. Biomed. Eng. Technol.* **2019**, *29*, 201–220. [CrossRef]

48. Anzulewicz, A.; Sobota, K.; Delafield-Butt, J.T. Toward the Autism Motor Signature: Gesture patterns during smart tablet gameplay identify children with autism. *Sci. Rep.* **2016**, *6*, 31107. [CrossRef] [PubMed]

49. Samad, M.D.; Diawara, N.; Bobzien, J.L.; Harrington, J.W.; Witherow, M.A.; Iftekharuddin, K.M. A Feasibility Study of Autism Behavioral Markers in Spontaneous Facial, Visual, and Hand Movement Response Data. *IEEE Trans. Neural Syst. Rehabil. Eng.* **2017**, *26*, 353–361. [CrossRef] [PubMed]

50. Jaiswal, S.; Valstar, M.F.; Gillott, A.; Daley, D. Automatic detection of ADHD and ASD from expressive behaviour in RGBD data. In Proceedings of the 2017 12th IEEE International Conference on Automatic Face & Gesture Recognition (FG 2017), Washington, DC, USA, 30 May–3 June 2017; pp. 762–769.

51. Westeyn, T.L.; Abowd, G.D.; Starner, T.E.; Johnson, J.M.; Presti, P.W.; Weaver, K.A. Monitoring children's developmental progress using augmented toys and activity recognition. *Pers. Ubiquitous Comput.* **2012**, *16*, 169–191. [CrossRef]

52. Anzalone, S.M.; Tilmont, E.; Boucenna, S.; Xavier, J.; Jouen, A.-L.; Bodeau, N.; Maharatna, K.; Chetouani, M.; Cohen, D.; Group, M.S. How children with autism spectrum disorder behave and explore the 4-dimensional (spatial 3D+ time) environment during a joint attention induction task with a robot. *Res. Autism Spectr. Disord.* **2014**, *8*, 814–826. [CrossRef]

53. Campbell, K.; Carpenter, K.L.; Hashemi, J.; Espinosa, S.; Marsan, S.; Borg, J.S.; Chang, Z.; Qiu, Q.; Vermeer, S.; Adler, E. Computer vision analysis captures atypical attention in toddlers with autism. *Autism* **2019**, *23*, 619–628. [CrossRef]

54. Petric, F.; Hrvatinić, K.; Babić, A.; Malovan, L.; Miklić, D.; Kovačić, Z.; Cepanec, M.; Stošić, J.; Šimleša, S. Four tasks of a robot-assisted autism spectrum disorder diagnostic protocol: First clinical tests. In Proceedings of the IEEE Global Humanitarian Technology Conference (GHTC 2014), San Jose, CA, USA, 10–13 October 2014; pp. 510–517.

55. Belin, P.; Fecteau, S.; Bedard, C. Thinking the voice: neural correlates of voice perception. *Trends Cogn. Sci.* **2004**, *8*, 129–135. [CrossRef]

56. Schweinberger, S.R.; Kawahara, H.; Simpson, A.P.; Skuk, V.G.; Zäske, R. Speaker perception. *Wiley Interdiscip. Rev. Cogn. Sci.* **2014**, *5*, 15–25. [CrossRef]

57. Philip, R.C.M.; Whalley, H.C.; Stanfield, A.C.; Sprengelmeyer, R.; Santos, I.M.; Young, A.W.; Atkinson, A.P.; Calder, A.J.; Johnstone, E.C.; Lawrie, S.M.; et al. Deficits in facial, body movement and vocal emotional processing in autism spectrum disorders. *Psychol. Med.* **2010**, *40*, 1919–1929. [CrossRef]

58. Schelinski, S.; Roswandowitz, C.; von Kriegstein, K. Voice identity processing in autism spectrum disorder. *Autism Res.* **2017**, *10*, 155–168. [CrossRef]

59. Skuk, V.G.; Palermo, R.; Broemer, L.; Schweinberger, S.R. Autistic traits are linked to individual differences in familiar voice identification. *J. Autism Dev. Disord.* **2019**, *49*, 2747–2767. [CrossRef] [PubMed]

60. Fruhholz, S.; Marchi, E.; Schuller, B. The Effect of Narrow-Band Transmission on Recognition of Paralinguistic Information from Human Vocalizations. *IEEE Access* **2016**, *4*, 6059–6072. [CrossRef]

61. Gilchrist, K.H.; Hegarty-Craver, M.; Christian, R.B.; Grego, S.; Kies, A.C.; Wheeler, A.C. Automated detection of repetitive motor behaviors as an outcome measurement in intellectual and developmental disabilities. *J. Autism Dev. Disord.* **2018**, *48*, 1458–1466. [CrossRef] [PubMed]

62. Happé, F.G.E. An advanced test of theory of mind: Understanding of story characters' thoughts and feelings by able autistic, mentally handicapped, and normal children and adults. *J. Autism Dev. Disord.* **1994**, *24*, 129–154. [CrossRef]

63. Oberman, L.M.; Hubbard, E.M.; Mccleery, J.P.; Altschuler, E.L.; Ramachandran, V.S.; Pineda, J.A. EEG evidence for mirror neuron dysfunction in autism spectrum disorders. *Cogn. Brain Res.* **2005**, *24*, 190–198. [CrossRef]

64. Schneider, D.; Slaughter, V.P.; Bayliss, A.P.; Dux, P.E. A temporally sustained implicit theory of mind deficit in autism spectrum disorders. *Cognition* **2013**, *129*, 410–417. [CrossRef]

65. Low, J.; Apperly, I.A.; Butterfill, S.A.; Rakoczy, H. Cognitive Architecture of Belief Reasoning in Children and Adults: A Primer on the Two-Systems Account. *Child Dev. Perspect.* **2016**, *10*, 184–189. [CrossRef]

66. Kulke, L.; Von Duhn, B.; Schneider, D.; Rakoczy, H. Is Implicit Theory of Mind a Real and Robust Phenomenon? Results from a Systematic Replication Study. *Psychol. Sci.* **2018**, *29*, 888–900. [CrossRef]

67. Premack, D.; Woodruff, G. Does the chimpanzee have a theory of mind? *Behav. Brain Sci.* **1978**, *1*, 515–526. [CrossRef]

68. Gordon, I.; Pierce, M.D.; Bartlett, M.S.; Tanaka, J.W. Training Facial Expression Production in Children on the Autism Spectrum. *J. Autism Dev. Disord.* **2014**, *44*, 2486–2498. [CrossRef]

69. Leo, M.; Del Coco, M.; Carcagni, P.; Distante, C.; Bernava, M.; Pioggia, G.; Palestra, G. Automatic emotion recognition in robot-children interaction for ASD treatment. In Proceedings of the IEEE International Conference on Computer Vision Workshops, Santiago, Chile, 11–18 December 2015; pp. 145–153.

70. Piana, S.; Malagoli, C.; Usai, M.C.; Camurri, A. Effects of Computerized Emotional Training on Children with High Functioning Autism. *IEEE Trans. Affect. Comput.* **2019**, *1*, 1. [CrossRef]

71. Robins, B.; Amirabdollahian, F.; Ji, Z.; Dautenhahn, K. Tactile interaction with a humanoid robot for children with autism: A case study analysis involving user requirements and results of an initial implementation. In Proceedings of the 19th International Symposium in Robot and Human Interactive Communication, Viareggio, Italy, 13–15 September 2010; pp. 704–711.

72. Mengoni, S.E.; Irvine, K.; Thakur, D.; Barton, G.; Dautenhahn, K.; Guldberg, K.; Robins, B.; Wellsted, D.; Sharma, S. Feasibility study of a randomised controlled trial to investigate the effectiveness of using a humanoid robot to improve the social skills of children with autism spectrum disorder (Kaspar RCT): A study protocol. *BMJ Open* **2017**, *7*, e017376. [CrossRef] [PubMed]

73. Costa, S.; Resende, J.; Soares, F.O.; Ferreira, M.J.; Santos, C.P.; Moreira, F. Applications of simple robots to encourage social receptiveness of adolescents with autism. In Proceedings of the 2009 Annual International Conference of the IEEE Engineering in Medicine and Biology Society, Minneapolis, MN, USA, 3–6 September 2009; pp. 5072–5075.

74. Wong, H.; Zhong, Z. Assessment of robot training for social cognitive learning. In Proceedings of the 2016 16th International Conference on Control, Automation and Systems (ICCAS), Gyeongju, South Korea, 16–19 October 2016; pp. 893–898.

75. Uzuegbunam, N.; Wong, W.-H.; Cheung, S.-C.S.; Ruble, L. In MEBook: Kinect-based self-modeling intervention for children with autism. In Proceedings of the 2015 IEEE International Conference on Multimedia and Expo (ICME), Turin, Italy, 29 June–3 July 2015; pp. 1–6.

76. Mower, E.; Black, M.P.; Flores, E.; Williams, M.; Narayanan, S. Rachel: Design of an emotionally targeted interactive agent for children with autism. In Proceedings of the 2011 IEEE International Conference on Multimedia and Expo, Barcelona, Spain, 11–15 July 2011; pp. 1–6.

77. Del Coco, M.; Leo, M.; Carcagnì, P.; Fama, F.; Spadaro, L.; Ruta, L.; Pioggia, G.; Distante, C. Study of mechanisms of social interaction stimulation in autism spectrum disorder by assisted humanoid robot. *IEEE Trans. Cogn. Dev. Syst.* **2017**, *10*, 993–1004. [CrossRef]

78. Dawood, A.; Turner, S.; Perepa, P. Affective Computational Model to Extract Natural Affective States of Students with Asperger Syndrome (AS) in Computer-Based Learning Environment. *IEEE Access* **2018**, *6*, 67026–67034. [CrossRef]

79. Winoto, P.; Chen, C.G.; Tang, T.Y. The development of a Kinect-based online socio-meter for users with social and communication skill impairments: A computational sensing approach. In Proceedings of the 2016 IEEE International Conference on Knowledge Engineering and Applications (ICKEA), Singapore, Singapore, 28–30 September 2016; pp. 139–143.

80. Kołakowska, A.; Landowska, A.; Anzulewicz, A.; Sobota, K. Automatic recognition of therapy progress among children with autism. *Sci. Rep.* **2017**, *7*, 13863. [CrossRef]

81. Pickering, M.J.; Garrod, S. An integrated theory of language production and comprehension. *Behav. Brain Sci.* **2013**, *36*, 329–347. [CrossRef]

82. Warren, J.E.; Sauter, D.A.; Eisner, F.; Wiland, J.; Dresner, M.A.; Wise, R.J.S.; Rosen, S.; Scott, S.K. Positive Emotions Preferentially Engage an Auditory–Motor "Mirror" System. *J. Neurosci.* **2006**, *26*, 13067–13075. [CrossRef]

83. Lewis, M.B.; Dunn, E. Instructions to mimic improve facial emotion recognition in people with sub-clinical autism traits. *Q. J. Exp. Psychol.* **2017**, *70*, 1–14. [CrossRef]

84. Pineda, J.A.; Carrasco, K.; Datko, M.; Pillen, S.; Schalles, M. Neurofeedback training produces normalization in behavioural and electrophysiological measures of high-functioning autism. *Philos. Trans. R. Soc. B Boil. Sci.* **2014**, *369*, 20130183. [CrossRef]

85. Caramazza, A.; Anzellotti, S.; Strnad, L.; Lingnau, A. Embodied Cognition and Mirror Neurons: A Critical Assessment. *Annu. Rev. Neurosci.* **2014**, *37*, 1–15. [CrossRef]

86. Dimberg, U.; Thunberg, M.; Elmehed, K. Unconscious facial reactions to emotional facial expressions. *Psychol. Sci.* **2000**, *11*, 86–89. [CrossRef] [PubMed]

87. Korb, S.; With, S.; Niedenthal, P.; Kaiser, S.; Grandjean, D. The Perception and Mimicry of Facial Movements Predict Judgments of Smile Authenticity. *PLoS ONE* **2014**, *9*, e99194. [CrossRef] [PubMed]

88. Rajendran, G. Virtual environments and autism: A developmental psychopathological approach. *J. Comput. Assist. Learn.* **2013**, *29*, 334–347. [CrossRef]

89. Wong, C.; Odom, S.L.; Hume, K.A.; Cox, A.W.; Fettig, A.; Kucharczyk, S.; Brock, M.E.; Plavnick, J.B.; Fleury, V.P.; Schultz, T.R. Evidence-Based Practices for Children, Youth, and Young Adults with Autism Spectrum Disorder: A Comprehensive Review. *J. Autism Dev. Disord.* **2015**, *45*, 1951–1966. [CrossRef]

MDPI
St. Alban-Anlage 66
4052 Basel
Switzerland
Tel. +41 61 683 77 34
Fax +41 61 302 89 18
www.mdpi.com

Sensors Editorial Office
E-mail: sensors@mdpi.com
www.mdpi.com/journal/sensors